Springer Proceedings in Mathematics & Statistics

Volume 339

Springer Proceedings in Mathematics & Statistics

This book series features volumes composed of selected contributions from workshops and conferences in all areas of current research in mathematics and statistics, including operation research and optimization. In addition to an overall evaluation of the interest, scientific quality, and timeliness of each proposal at the hands of the publisher, individual contributions are all refereed to the high quality standards of leading journals in the field. Thus, this series provides the research community with well-edited, authoritative reports on developments in the most exciting areas of mathematical and statistical research today.

More information about this series at http://www.springer.com/series/10533

Michele La Rocca · Brunero Liseo ·
Luigi Salmaso
Editors

Nonparametric Statistics

4th ISNPS, Salerno, Italy, June 2018

Springer

Editors
Michele La Rocca
Department of Economics and Statistics
University of Salerno
Salerno, Italy

Brunero Liseo
Department of Methods and Models
for Economics, Terrority and Finance
Sapienza University of Rome
Rome, Italy

Luigi Salmaso
Department of Management
and Engineering
University of Padua
Vicenza, Italy

ISSN 2194-1009 ISSN 2194-1017 (electronic)
Springer Proceedings in Mathematics & Statistics
ISBN 978-3-030-57305-8 ISBN 978-3-030-57306-5 (eBook)
https://doi.org/10.1007/978-3-030-57306-5

Mathematics Subject Classification: 62Gxx, 62G05, 62G08, 62G09, 62G10, 62G15, 62G20, 62G35

This Springer imprint is published by the registered company Springer Nature Switzerland AG
The registered company address is: Gewerbestrasse 11, 6330 Cham, Switzerland

Preface

This book provides a selection of papers developed from talks presented at the Fourth Conference of the International Society for Nonparametric Statistics (ISNPS), held in Salerno (Italy) June 11–15, 2018. The papers cover a wide spectrum of subjects within nonparametric and semiparametric statistics, including theory, methodology, applications, and computational aspects. Among the most common and relevant topics in the volume, we mention nonparametric curve estimation, regression smoothing, models for time series and more generally dependent data, varying coefficient models, symmetry testing, robust estimation, rank-based methods for factorial design, nonparametric and permutation solution for several different data, including ordinal data, spatial data, survival data and the joint modeling of both longitudinal and time-to-event data, permutation and resampling techniques, and practical applications of nonparametric statistics.

ISNPS was founded in 2010 "to foster the research and practice of nonparametric statistics, and to promote the dissemination of new developments in the field via conferences, books, and journal publication". ISNPS had a distinguished Advisory Committee that included R. Beran, P. Bickel, R. Carroll, D. Cook, P. Hall. R. Johnson, B. Lindsay, E. Parzen, P. Robinson, M. Rosen-blatt, G. Roussas, T. SubbaRao, and G. Wahba; an Executive Committee that comprised of M. Akritas, A. Delaigle, S. Lahiri and D. Politis; and a Council that included P. Bertail, G. Claeskens, R. Cao, M. Hallin, H. Koul, J.-P. Kreiss, T. Lee, R. Liu, W. Gonzáles Maintega, G. Michailidis, V. Panaretos, S. Paparoditis, J. Racine, J. Romo, and Q. Yao.

The 4th ISNPS conference focused on recent advances and trends in several areas of nonparametric statistics. It included 12 plenary and special invited sessions, 69 invited sessions, 30 contributed sessions, with about 450 participants from all over the world, thus promoting and facilitating the exchange of research ideas and collaboration among scientists and contributing to the further development of the field.

We would like to thank Dr. Veronika Rosteck and Dr. Tatiana Plotnikova of Springer for their support in this project. Finally, we are also extremely grateful to all Referees who reviewed the papers included in this volume, giving a constructive feedback on a tight schedule for timely publication of the proceedings. Their valuable contribution and their efforts significantly improved the quality of this volume.

Co-editors also wish to thank Chiara Brombin for her great commitment and support in coordinating and managing the referring and editorial process.

Salerno, Italy Michele La Rocca
Rome, Italy Brunero Liseo
Vicenza, Italy Luigi Salmaso
Co-Editors of the book
and Co-Chairs of the Fourth ISNPS Conference

Contents

Portfolio Optimisation via Graphical Least Squares Estimation

Saeed Aldahmani, Hongsheng Dai, Qiao-Zhen Zhang, and Marialuisa Restaino

Abstract In this paper, an unbiased estimation method called GLSE (proposed by Aldahmani and Dai [1]) for solving the linear regression problem in high-dimensional data ($n < p$) is applied to portfolio optimisation under the linear regression framework and compared to the ridge method. The unbiasedness of method helps in improving the portfolio performance by increasing its expected return and decreasing the associated risk when $n < p$, thus leading to a maximisation of the Sharpe ratio. The verification of this method is achieved through conducting simulation and data analysis studies and comparing the results with those of ridge regression. It is found that GLSE outperforms ridge in portfolio optimisation when $n < p$.

Keywords Graphical model · Linear regression · Ridge regression

1 Introduction

In the world of finance, investors usually seek to construct a portfolio that maximises expected returns and minimises their risk through diversifying and computing the correct weights of the assets in that portfolio. This weights computation can be

S. Aldahmani (✉)
Department of Statistics, College of Business and Economics, United Arab Emirates University, Al Ain, UAE
e-mail: saldahmani@uaeu.ac.ae

H. Dai
Department of Mathematical Sciences, University of Essex, Colchester CO4 3SQ, UK
e-mail: hdaia@essex.ac.uk

Q.-Z. Zhang
Institute of Statistics, Nankai University, Tianjin, China
e-mail: zhangqz@nankai.edu.cn

M. Restaino
University of Salerno, Fisciano, Italy
e-mail: mlrestaino@unisa.it

M. La Rocca et al. (eds.), *Nonparametric Statistics*, Springer Proceedings in Mathematics & Statistics 339, https://doi.org/10.1007/978-3-030-57306-5_1

achieved by what is technically known as portfolio optimisation, a problem that was addressed by Markowitz [14] through utilising a model known as Markowitz theory. The Markowitz theory for portfolio optimisation stipulates selecting portfolio weights $\boldsymbol{w}$ that minimise the risk (variance) of the portfolio return for a predetermined target return. This idea assumes that the future performance of asset returns' mean $\boldsymbol{\mu}$ and variance are known. However, in practice, these two factors are unknown and should be estimated using a historical dataset. To select an optimal portfolio, investors need to estimate the covariance matrix $\boldsymbol{\Sigma}$ of the returns and take its inverse. This is a typical inverse problem if the number of assets p is too large in relation to the return observations n; i.e. the inverse of the covariance matrix of the returns is singular. Therefore, many regularisation methods have been proposed in the literature to find covariance matrices and their inverses, such as in Bickel and Levina [2], Huang et al. [10], Wong et al. [19]. However, the estimates of these methods are biased, which might give undesirable weights for some higher return assets in portfolio.

Britten-Jones [3] utilised regression in order to find the portfolio weights as follows:

$$\boldsymbol{w} = \frac{\hat{\boldsymbol{\beta}}}{\hat{\boldsymbol{\beta}}' \mathbf{1}_v}, \tag{1}$$

where $\hat{\boldsymbol{\beta}}$ is the ordinary least squares (OLS) estimate of the coefficient parameter $\boldsymbol{\beta}$ for the linear regression model

$$\boldsymbol{y} = \mathbf{x}\boldsymbol{\beta} + \boldsymbol{\epsilon}, \tag{2}$$

where the response $\boldsymbol{y} = \mathbf{1}_v$.

When $n < p$, the popular ordinary least square method (OLS) becomes ineffective, and this has triggered the proposal of many methods to solve this issue, such as Least Absolute Shrinkage and Selection Operator (LASSO) [18], Least Angle Regression (LARS) [7] and ridge regression [9]. However, all these methods suffer from the limitation of giving biased estimates. In addition, LASSO and LARS suffer from the problem of not selecting more than n covariates [20] and giving a sparse portfolio. Another problem with some of these methods is over-shrinking the final regression coefficients [16], which might lead to inaccuracy in portfolio weights. Apart from these methods, some other related approaches could be found in Candes and Tao [4], Meinshausen and Yu [15], DeMiguel et al. [6], Still and Kondor [17], Carrasco and Noumon [5], Fastrich et al. [8] and Lin et al. [12]. These methods, however, still give biased estimates and perhaps produce inaccurate weights for some higher return and less risk assets in the portfolio.

Aldahmani and Dai [1] proposed an unbiased estimation method called GLSE which can provide unbiased estimates for regression coefficients in high-dimensional data ($n < p$). The GLSE method is closely related to the theory of graphical models, where least square estimation in conjunction with undirected Gaussian graphical models is implemented.

GLSE can give unbiased coefficient estimates for all assets, which helps the low-risk and high return assets maintain their correct weights in the portfolio and consequently assists in maximising their expected returns and lower the associated risk. Such an advantage will lead to increasing the Sharpe ratio and the expected rate of returns and decreasing the risk of the portfolio for both in-and-out-of-sample periods. This is particularly important upon comparison with other regularisation methods such as ridge, where the weights of low-risk and high return assets may be sharply reduced due to the method's biasedness, thus causing the portfolio's expected returns to fall down and its risk to rise. Moreover, unlike other regularisation methods which produce sparse portfolios(such as LASSO and LARS), GLSE and ridge share the advantage of generating diversified portfolios across a large number of stocks, as they produce non-sparse portfolios. This diversification of the portfolio leads to lowering the risk [13] due to the fact that when one or more sectors of the economy fail or decline, the rest of the sectors can then mitigate the significant impact of the loss caused by market fluctuations. However, due to the biasedness of ridge regression, the weights of some low-risk and high return assets may be sharply reduced, which may deprive ridge of its ability to reduce the risk through diversifying the assets. This limitation can clearly be overcome by GLSE due to its unbiasedness feature.

In the rest of the paper, graph theory and Matrices are given in Sect. 2. Section 3 presents the main methodology of GLSE and its properties. Section 4 provides the algorithm of graph structure selection. Simulation studies are given in Sect. 5, and a real data analysis is presented in Sect. 6. The study is concluded in Sect. 6.

2 Graph Theory and Matrices

2.1 Graph Theory

An undirected graph G consists of two sets, a set P and a set $\mathcal{E}$. The set P denotes the vertices representing variables and $\mathcal{E}$ is the set of edges (a subset of $P \times P$) connecting the vertices [11]. The elements in the set P are usually natural numbers, i.e. $P = 1, 2, \ldots, p$, representing the labels of random variables. If all the pairs of vertices in P in a graph G are joined by an edge, then the graph is complete. If $A \subseteq P$, the subset A induces a subgraph $G_A = (A, \mathcal{E}_A)$, where $\mathcal{E}_A = \mathcal{E} \cap (A \times A)$. The subset graph G_A is complete if it induces a complete subgraph from G. This subgraph is maximal if it cannot be extended by including one more neighbouring vertex. A complete subset that is maximal is called a clique.

2.1.1 Decomposition of a Graph

A triple (A, B, C) of disjoint subsets of the vertex set P of an undirected graph G is said to form a *decomposition* of G if $P = A \cup B \cup C$ and the following conditions hold [11]:

- B separates A from C;
- B is a complete subset of P.

An undirected graph G is considered as decomposable if it holds one of the following:

- Graph g is complete.
- There is a proper decomposition (A, B, C) into decomposable subgraphs g_{AB} and g_{BC} where B separates A from C.

Consider a sequence of sets $C_1, \ldots, C_q$ that are the subsets of the vertex set P of an undirected graph g such that $C_1 \cup \cdots \cup C_q = P$. If the following holds, then the given sequence is said to be a perfect sequence [11]:

$$S_j = C_j \cap (C_1 \cup C_2 \cup \cdots \cup C_{j-1}) \subseteq C_i,$$

where $j = 2, \ldots, q$ and $i \in \{1, \ldots, j-1\}$. The sets S_j are the separators. These orderings, if they exist, might not be unique.

2.2 *Matrices*

A $p \times p$ matrix $\mathbf{F}$ can be written as $(F_{kj})_{k,j \in P}$. $F \in R^p$ represent a vector. Denote $\mathbf{F}_{AB} = (F_{kj})_{k \in A, j \in B}$, a submatrix of $\mathbf{F}$. Denote $[\mathbf{F}_{AB}]^{\Gamma}$ as a $p \times p$-dimensional matrix obtained by filling up 0s, with

$$\left([\mathbf{F}_{AB}]^{\Gamma}\right)_{jk} = \begin{cases} F_{jk} & \text{if } j \in A,\ k \in B \\ 0 & \text{otherwise.} \end{cases} \tag{3}$$

Similarly, let $\mathbf{x}_A$ is a matrix only having covariates with indices in set A and $ssd_A = \mathbf{x}_A' \mathbf{x}_A$. Then $[(ssd_A)^{-1}]^{\Gamma}$ represents a $p \times p$-dimensional matrix obtained by filling up 0s, with

$$\left([(ssd_A)^{-1}]^{\Gamma}\right)_{jk} = \begin{cases} \left((ssd_A)^{-1}\right)_{jk} & \text{if } j, k \in A \\ 0 & \text{otherwise.} \end{cases} \tag{4}$$

3 The Idea of GLSE

Suppose that the graph G is decomposable and let $\mathcal{C}$ denote the set of cliques and $\mathcal{S}$ denote the set of separators [1]. Then the GLSE is given as follows:

$$\hat{\boldsymbol{\beta}} = \left[\sum_{C \in \mathcal{C}} [(ssd_C)^{-1}]^{\Gamma} - \sum_{S \in \mathcal{S}} [(ssd_S)^{-1}]^{\Gamma} \right] \mathbf{x}' \boldsymbol{y}. \tag{5}$$

For the existence of the GLSE, the following condition must hold

Condition 3.1 *The sample size* $n > \max_{C \in \mathcal{C}}\{|C|\}$.

For unbiasedness of $\hat{\boldsymbol{\beta}}$, based on Aldahmani and Dai [1], the following condition is imposed:

Condition 3.2 *Write the cliques and separators of g in the perfect ordering, as* $C_1, \cdots, C_q$ *and* $S_2, \cdots, S_q$, *such that*

$$\mathbf{x}_{C_1 \setminus S_2} = \mathbf{x}_{S_2} \cdot \mathbf{r}_{S_2, C_1 \setminus S_2} + \boldsymbol{\xi}_1, \quad E(\boldsymbol{\xi}_1) = \mathbf{0},$$
$$\mathbf{x}_{C_k \setminus S_k} = \mathbf{x}_{S_k} \cdot \mathbf{r}_{S_k, C_k \setminus S_k} + \boldsymbol{\xi}_k, \quad E(\boldsymbol{\xi}_k) = \mathbf{0}, \quad k = 2, \cdots, q,$$

where $\mathbf{r}_{S_k, C_k \setminus S_k}$ *are constant matrices with dimensions* $|s_k \times (c_k - s_k)|$;

Under Conditions 3.1 and 3.2, Aldahmani and Dai [1] show that the above estimator is unbiased;

$$\mathbb{E}(\hat{\boldsymbol{\beta}}) = \boldsymbol{\beta}.$$

4 Model Selection

A stepwise selection algorithm has been used by Aldahmani and Dai [1] to find which graph G is the best for the data. The method considers adding/deleting edges one by one to/from the current graph. When an edge under consideration is not in the current graph, it will be added if the addition makes an improvement in terms of the predetermined criteria; otherwise it will not be added. The same applies to the case of edge deletion. According to Aldahmani and Dai [1], the best graph is given by minimising a target function $\mathbb{T}(\boldsymbol{\beta}, g, \lambda_g)$:

$$(\hat{\boldsymbol{\beta}}, \hat{g}, \hat{\lambda}_g) = \arg \min_{\boldsymbol{\beta}, g \in \mathcal{G}, \lambda_g} \mathbb{T}(\boldsymbol{\beta}, g, \lambda_g) \tag{6}$$

$$\mathbb{T}(\boldsymbol{\beta}, g, \lambda_g) = ||\boldsymbol{y} - \mathbf{x}\boldsymbol{\beta}||^2 + \lambda_g |\mathcal{E}_g|, \tag{7}$$

where $\mathcal{G}$ is the set of all possible graphs, λ_g is a penalty term and $|\mathcal{E}_g|$ is the number of edges in graph G. The following pseudocode is the algorithm used by Adlahmani and Dai [1] to find the optimal graph that best fits the data:

Algorithm 1 Pseudocode of the GLSE graph selection

1: Start graph $g = (P, \mathcal{E})$, which can be an empty (or a given decomposable) graph such that $n > \max_{C \in \mathcal{C}} |C|$.
2: Generate all possible graphs, g_i, such that there is only one edge difference between g_i and the current graph g. All such g_i are decomposable and $n > \max_{C \in \mathcal{C}} |C|$.
3: Find the graph g_i^* and the associated $\hat{\boldsymbol{\beta}}$ such that g_i^* minimises the target function $\mathbb{T}(.)$ (given in (7)).
4: Go to step 2 with the selected graph g_i^* and iterate until the best one is found.
5: Output g and $\hat{\boldsymbol{\beta}}$.

It is worth noting that step 2 of Algorithm 1 can be improved significantly via parallel computation.

5 Simulation Study

The aims of this simulation study are to investigate the performance of GLSE in constructing a saturated optimal portfolio compared to ridge. The graph structure for the covariates used in generating the dataset under this simulation study is presented in Fig. 1.

This simulation involves a total of $n = 48$ observations corresponding to $p = 60$ variables derived from multivariate normal distribution, with mean 0.01 and variance covariance matrix Σ, where 36 observations are used for the in-sample period through estimating the portfolio's weight and performance (Sharpe ratios, expected returns and risk), and the remaining observations are used to find the performance of the portfolio for the out-of-sample period. The true weight of the portfolio $\boldsymbol{w}$ is derived based on the true covariance matrix Σ.

Table 1 gives the means of 500 simulated data for the in-and-out-of-sample portfolio's Sharpe ratios, expected returns and risk. It shows that out of the 500 simulated data, the GLSE yields higher means of the portfolio's Sharpe ratio and lower risk than the ridge does for the out-of-sample period. However, for the in-sample period, the ridge gives higher means of the portfolio's expected returns than the GLSE does. It should be noted that the ridge portfolio's risk is very high compared to this under the GLSE. In addition, the in-sample portfolio's Sharpe ratio is negative for the ridge but positive for the GLSE, which is desirable in finance.

The computational burden for the proposed algorithm is not too heavy with modern parallel computing technology. The computational times for one run of the above

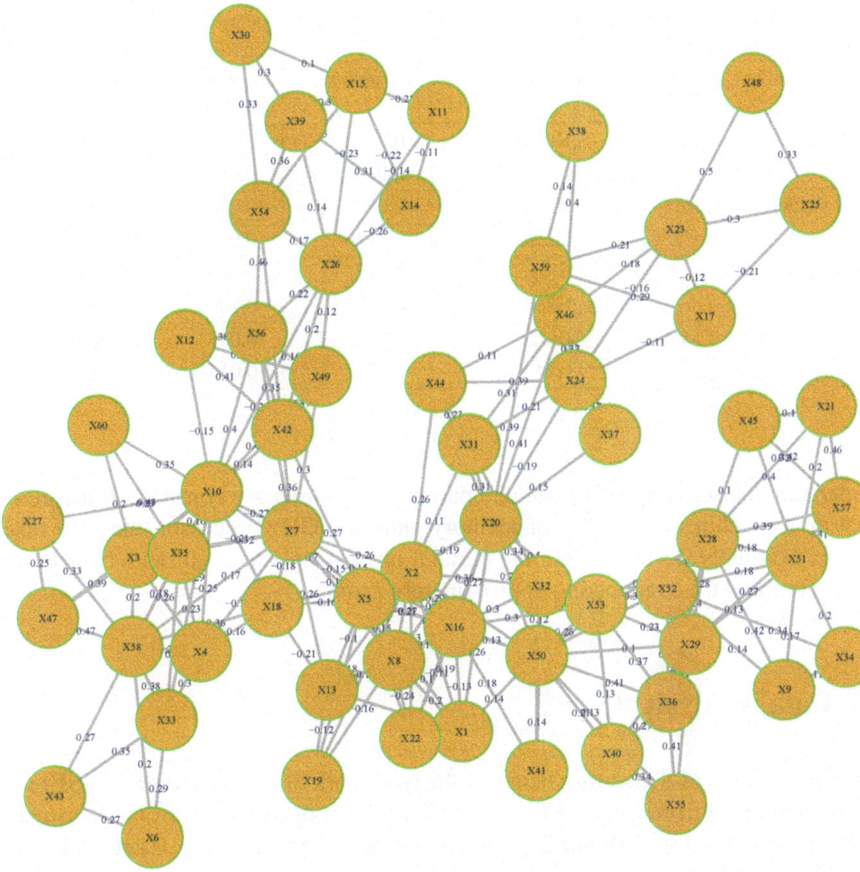

Fig. 1 Graph structure for covariates under the simulation study

Table 1 The in-and-out-of-sample portfolio's Sharpe ratios, expected returns and risk from the simulated data

	Ridge		GLSE	
	In sample	Out of sample	In sample	Out of sample
Sharpe ratio	−0.007	0.005	0.733	0.570
Expected returns	0.149	0.030	0.127	0.107
Portfolio's risk	1.282	1.236	0.526	0.516

Table 2 Portfolio size and in- and out-of sample portfolio's Sharpe ratios, expected returns and risk find by the ridge and GLSE

Portfolio size	Methods	Sharpe ratio	Expected returns	Portfolio's risk
150 stocks (in sample)	Ridge	0.719	0.097	0.134
	GLSE	2.023	0.061	0.030
150 stocks (out of sample)	Ridge	−0.074	−0.010	0.135
	GLSE	0.117	0.015	0.130
200 stocks (in sample)	Ridge	0.792	0.056	0.071
	GLSE	0.963	0.046	0.047
200 stocks (out of sample)	Ridge	0.150	0.013	0.086
	GLSE	0.224	0.015	0.068

generated datasets for both serial and parallel computing are considered. It is noted that on a machine with 8 GB of memory and 3.60 GHz processor, the time taken is approximately 20 min. When the parallel processing was used, with 5 cores, the computational time reduced to approximately 2 min.

6 Data Analysis

Monthly returns of 875 stocks listed on the New York Stock Exchange (NYSE) covering the period from 02/12/2007 to 02/12/2017 are downloaded from Datastream. Out of these stocks, 150 and 200 stocks are selected at random. Then, ridge and GLSE are applied to construct two portfolios for the selected stocks. The in-sample period for the above constructed portfolios is from 02/12/2007 to 01/12/2016. The out-of-sample period, on the other hand, is from 02/12/2016 to 01/12/2017. For ridge, cross validation is used for obtaining the penalty parameter. The in-and-out-of sample average returns, risk and Sharpe ratio are used to evaluate the performance of the obtained portfolios. The results are shown in Table 2 and they reveal that the GLSE method performs better than ridge in term of average returns, risk and the Sharpe ratio of portfolios for both in-and-out-of-sample periods.

7 Conclusion

The unbiased GLSE method was used in this paper to construct a saturated optimal portfolio in high-dimensional data ($n < p$). The results of applying this method were compared to those of ridge and they showed that GLSE outperforms ridge in terms of its ability to reduce the portfolio's risk and increase its expected returns, consequently maximising the Sharpe ratio. While both ridge and GLSE have practical implications in the world of finance in that they both lead to a non-sparse portfolio with diversified assets, the GLSE overcomes ridge's shortcoming where the weights of low-risk and high return assets may be reduced due to its biasedness. Due to its unbiasedness, GLSE thus maintains the higher weights of low-risk and high return assets, which, as a result, minimises the chances of risk increase and income reduction in the portfolio.

References

1. Aldahmani, S., Dai, H.: Unbiased estimation for linear regression when $n<v$. Int. J. Stat. Probab. **4**(3), p61 (2015)
2. Bickel, P.J., Levina, E.: Regularized estimation of large covariance matrices. Ann. Stat. 199–227 (2008)
3. Britten-Jones, M.: The sampling error in estimates of mean-variance efficient portfolio weights. J. Finan. **54**(2), 655–671 (1999)
4. Candes, E., Tao, T.: The dantzig selector: statistical estimation when p is much larger than n. Ann. Stat. **35**(6), 2313–2351 (2007)
5. Carrasco, M., Noumon, N.: Optimal portfolio selection using regularization. Technical report, University of Montreal (2011)
6. DeMiguel, V., Garlappi, L., Nogales, F.J., Uppal, R.: A generalized approach to portfolio optimization: improving performance by constraining portfolio norms. Manage. Sci. **55**(5), 798–812 (2009)
7. Efron, B., Hastie, T., Johnstone, I., Tibshirani, R., et al.: Least angle regression. Ann. Stat. **32**(2), 407–499 (2004)
8. Fastrich, B., Paterlini, S., Winker, P.: Constructing optimal sparse portfolios using regularization methods. Comput. Manage. Sci. **12**(3), 417–434 (2015)
9. Hoerl, A.E., Kennard, R.W.: Ridge regression: Biased estimation for nonorthogonal problems. Technometrics **12**(1), 55–67 (1970)
10. Huang, J.Z., Liu, N., Pourahmadi, M., Liu, L.: Covariance matrix selection and estimation via penalised normal likelihood. Biometrika **93**(1), 85–98 (2006)
11. Lauritzen, S.L.: Graphical Models. Oxford University Press, Oxford (1996)
12. Lin, W., Shi, P., Feng, R., Li, H.: Variable selection in regression with compositional covariates. Biometrika asu 031 (2014)
13. Malkiel, B.G., Xu, Y.: Risk and return revisited. J. Portfolio Manage. **23**(3), 9–14 (1997)
14. Markowitz, H.: Portfolio selection. J. Finan. **7**(1), 77–91 (1952)
15. Meinshausen, N., Yu, B.: Lasso-type recovery of sparse representations for high-dimensional data. Ann. Stat. 246–270 (2009)
16. Radchenko, P., James, G.M., et al.: Improved variable selection with forward-lasso adaptive shrinkage. Ann. Appl. Stat. **5**(1), 427–448 (2011)
17. Still, S., Kondor, I.: Regularizing portfolio optimization. New J. Phys. **12**(7), 075034 (2010)
18. Tibshirani, R.: Regression shrinkage and selection via the lasso: a retrospective. J. R. Stat. Soc. Ser. B (Statistical Methodology) **58**(1), 267–288 (2011)
19. Wong, F., Carter, C.K., Kohn, R.: Efficient estimation of covariance selection models. Biometrika **90**(4), 809–830 (2003)
20. Zou, H., Hastie, T.: Regularization and variable selection via the elastic net. J. R. Stat. Soc. Ser. B (Statistical Methodology) **67**(2), 301–320 (2005)

Change of Measure Applications in Nonparametric Statistics

Mayer Alvo

Abstract Neyman [7] was the first to propose a change in measure in the context of goodness of fit problems. This provided an alternative density to the one for the null hypothesis. Hoeffding introduced a change of measure formula for the ranks of the observed data which led to obtaining locally most powerful rank tests. In this paper, we review these methods and propose a new approach which leads on the one hand to new derivations of existing statistics. On the other hand, we exploit these methods to obtain Bayesian applications for ranking data.

Keywords Ranks · Change of measure · Bayesian methods

Mathematics Subject Classification (2010) 62F07 · 62G86 · 62H11

1 Introduction

In a landmark paper, [7] considered the nonparametric goodness of fit problem and introduced the notion of smooth tests of fit by proposing a parametric family of alternative densities to the null hypothesis. In this article, we describe a number of applications of this change of measure. Hence, we obtain a new derivation of the well-known Friedman statistic as the locally most powerful test in an embedded family of distributions.

2 Smooth Models

Suppose that the probability mass function of a discrete k-dimensional random vector $\boldsymbol{X}$ is given by

M. Alvo (✉)
Department of Mathematics, Statistics University of Ottawa, Ottawa, ON, Canada
e-mail: malvo@uottawa.ca

M. La Rocca et al. (eds.), *Nonparametric Statistics*, Springer Proceedings in Mathematics & Statistics 339, https://doi.org/10.1007/978-3-030-57306-5_2

$$\pi\left(\boldsymbol{x}_j;\boldsymbol{\theta}\right)=\exp\left(\boldsymbol{\theta}'\boldsymbol{x}_j-K(\boldsymbol{\theta})\right)p_j,\ j=1,\ldots,m, \tag{1}$$

where $\boldsymbol{x}_j$ is the jth value of $\boldsymbol{X}$ and $\boldsymbol{p}=\left(p_j\right)'$ denotes the vector of probabilities when $\boldsymbol{\theta}=\boldsymbol{\theta}_0$. Here $K(\boldsymbol{\theta})$ is a normalizing constant for which

$$\sum_j \pi\left(\boldsymbol{x}_j;\boldsymbol{\theta}\right)=1.$$

We see that the model in (1) prescribes a change of measure from the null to the alternative hypothesis. Let $\boldsymbol{T}=[\boldsymbol{x}_i,\ldots,\boldsymbol{x}_m]$ be the $k\times m$ matrix of possible vector values of $\boldsymbol{X}$. Then under the distribution specified by $\boldsymbol{p}$,

$$\boldsymbol{\Sigma}\equiv Cov_{\boldsymbol{p}}(\boldsymbol{X})=E_{\boldsymbol{p}}\left[(\boldsymbol{X}-E[\boldsymbol{X}])(\boldsymbol{X}-E[\boldsymbol{X}])'\right] \tag{2}$$
$$=\boldsymbol{T}(diag(\boldsymbol{p}))\boldsymbol{T}'-(\boldsymbol{T}\boldsymbol{p})(\boldsymbol{T}\boldsymbol{p})', \tag{3}$$

where the expectations are with respect to the model (1). This particular situation arises often when dealing with the nonparametric randomized block design. Define

$$\boldsymbol{\pi}(\boldsymbol{\theta})=\left(\pi(\boldsymbol{x}_1;\boldsymbol{\theta}),\ldots,\pi(\boldsymbol{x}_m;\boldsymbol{\theta})\right)'$$

and suppose that we would like to test

$$H_0:\boldsymbol{\theta}=\boldsymbol{0}\text{ vs }H_1:\boldsymbol{\theta}\neq\boldsymbol{0}.$$

Letting $\boldsymbol{N}$ denote a multinomial random vector with parameters $(n,\boldsymbol{\pi}(\boldsymbol{\theta}))$, we see that the log likelihood as a function of $\boldsymbol{\theta}$ is, apart from a constant, proportional to

$$\sum_{j=1}^m n_j\log\left(\pi\left(\boldsymbol{x}_j;\boldsymbol{\theta}\right)\right)=\sum_{j=1}^m n_j\left(\boldsymbol{\theta}'\boldsymbol{x}_j-K(\boldsymbol{\theta})\right)$$
$$=\boldsymbol{\theta}'\left(\sum_{j=1}^m n_j\boldsymbol{x}_j\right)-nK(\boldsymbol{\theta}).$$

The score vector under the null hypothesis is then given by

$$U(\boldsymbol{\theta};\boldsymbol{X})=\sum_{j=1}^m N_j\left(\frac{1}{\pi_j(\boldsymbol{\theta})}\frac{\partial\pi_j(\boldsymbol{\theta})}{\partial\boldsymbol{\theta}}\right)$$
$$=\boldsymbol{T}(\boldsymbol{N}-n\boldsymbol{p}).$$

Under the null hypothesis,

$$E[U(\boldsymbol{\theta};\boldsymbol{X})]=0,$$

and the score statistic is given by

$$\frac{1}{n}\left[\boldsymbol{T}(\boldsymbol{N}-n\boldsymbol{p})\right]'\boldsymbol{\Sigma}^{-1}\left[\boldsymbol{T}(\boldsymbol{N}-n\boldsymbol{p})\right]=\frac{1}{n}(\boldsymbol{N}-n\boldsymbol{p})'\left(\boldsymbol{T}'\boldsymbol{\Sigma}^{-1}\boldsymbol{T}\right)(\boldsymbol{N}-n\boldsymbol{p})\xrightarrow{\mathcal{L}}\chi_r^2, \tag{4}$$

where $r = rank\left(\boldsymbol{T}'\boldsymbol{\Sigma}^{-1}\boldsymbol{T}\right)$.

In the one-sample ranking problem whereby a group of judges are each asked to rank a set of t objects in accordance with some criterion, let $\mathcal{P} = \left\{\boldsymbol{\nu}_j, j = 1, \ldots, t!\right\}$ be the space of all $t!$ permutations of the integers $1, 2, \ldots, t$ and let the probability mass distribution defined on $\mathcal{P}$ be given by

$$\boldsymbol{p} = (p_1, \ldots, p_{t!}),$$

where $p_j = \Pr\left(\boldsymbol{\nu}_j\right)$. Conceptually, each judge selects a ranking $\boldsymbol{\nu}$ in accordance with the probability mass distribution $\boldsymbol{p}$. In order to test the null hypothesis that each of the rankings are selected with equal probability, that is,

$$H_0 : \boldsymbol{p} = \boldsymbol{p}_0 \text{ vs } H_1 : \boldsymbol{p} \neq \boldsymbol{p}_0, \tag{5}$$

where $\boldsymbol{p}_0 = \frac{1}{t!}\mathbf{1}$, define a k-dimensional vector score function $\boldsymbol{X}(\boldsymbol{\nu})$ on the space $\mathcal{P}$ and following (1), let its smooth probability mass function be given as

$$\pi(\boldsymbol{x}_j;\boldsymbol{\theta}) = \exp\left(\boldsymbol{\theta}'\boldsymbol{x}_j - K(\boldsymbol{\theta})\right)\frac{1}{t!}, \quad j = 1, \ldots, t! \tag{6}$$

where $\boldsymbol{\theta}$ is a t-dimensional vector, $K(\boldsymbol{\theta})$ is a normalizing constant and $\boldsymbol{x}_j$ is a t-dimensional score vector to be specified in (8). Since

$$\sum_{j=1}^{t!}\pi\left(\boldsymbol{x}_j;\boldsymbol{\theta}\right) = 1$$

it can be seen that $K(\mathbf{0}) = 0$ and hence the hypotheses in (5) are equivalent to testing

$$H_0 : \boldsymbol{\theta} = \mathbf{0} \text{ vs } H_1 : \boldsymbol{\theta} \neq \mathbf{0}. \tag{7}$$

It follows that the log likelihood function is proportional to

$$l(\boldsymbol{\theta}) \sim n\left[\boldsymbol{\theta}'\hat{\boldsymbol{\eta}} - K(\boldsymbol{\theta})\right],$$

where

$$\hat{\boldsymbol{\eta}} = \left[\sum_{j=1}^{t!}\boldsymbol{x}_j\hat{p}_{nj}\right], \quad \hat{p}_{nj} = \frac{n_j}{n}$$

and n_j represents the number of observed occurrences of the ranking $\boldsymbol{\nu}_j$. The Rao score statistic evaluated at $\boldsymbol{\theta} = \mathbf{0}$ is

$$U(\boldsymbol{\theta}; \boldsymbol{X}) = n\frac{\partial}{\partial\boldsymbol{\theta}}\left[\boldsymbol{\theta}'\hat{\boldsymbol{\eta}} - K(\mathbf{0})\right]$$
$$= n\left[\hat{\boldsymbol{\eta}} - \frac{\partial}{\partial\boldsymbol{\theta}}K(\mathbf{0})\right],$$

whereas the information matrix is

$$\boldsymbol{I}(\boldsymbol{\theta}) = -n\left[\frac{\partial^2}{\partial\boldsymbol{\theta}^2}K(\mathbf{0})\right].$$

The test then rejects the null hypothesis whenever

$$n^2\left[\hat{\boldsymbol{\eta}} - \frac{\partial}{\partial\boldsymbol{\theta}}K(\mathbf{0})\right]'\boldsymbol{I}^{-1}(\mathbf{0})\left[\hat{\boldsymbol{\eta}} - \frac{\partial}{\partial\boldsymbol{\theta}}K(\mathbf{0})\right] > \chi_f^2(\alpha),$$

where $\chi_f^2(\alpha)$ is the upper $100(1-\alpha)\,\%$ critical value of a chi square distribution with $f = \text{rank}(\boldsymbol{I}(\boldsymbol{\theta}))$ degrees of freedom. We note that the test just obtained is the locally most powerful test of H_0.

Specializing this test statistic to the Spearman score function of adjusted ranks

$$\boldsymbol{x}_j = \left(\nu_j(1) - \frac{t+1}{2}, \ldots, \nu_j(t) - \frac{t+1}{2}\right)', \; j = 1, \ldots, t!, \tag{8}$$

we can show that the Rao score statistic is the well-known Friedman test [5].

$$W = \frac{12n}{t(t+1)}\sum_{i=1}^{t}\left[\bar{R}_i - \frac{t+1}{2}\right]^2, \tag{9}$$

where $\bar{R}_i$ is the average of the ranks assigned to the ith object.

2.1 The Two-Sample Ranking Problem

The approach just described can be used to deal with the two-sample ranking problem assuming again the Spearman score function. Let $\boldsymbol{X}_1, \boldsymbol{X}_2$ be two independent random vectors whose distributions as in the one sample case are expressed for simplicity as

$$\pi\left(\boldsymbol{x}_j; \boldsymbol{\theta}_l\right) = \exp\left\{\boldsymbol{\theta}_l'\boldsymbol{x}_j - K(\boldsymbol{\theta}_l)\right\} p_l(j), \; j = 1, \ldots, t!, l = 1, 2,$$

where $\boldsymbol{\theta}_l = (\theta_{l1}, \ldots, \theta_{lt})'$ represents the vector of parameters for population l. We are interested in testing

$$H_0 : \boldsymbol{\theta}_1 = \boldsymbol{\theta}_2 \text{ vs } H_1 : \boldsymbol{\theta}_1 \neq \boldsymbol{\theta}_2.$$

The probability distribution $\{p_l(j)\}$ represents an unspecified null situation. Define

$$\hat{\boldsymbol{p}}_l = \left(\frac{n_{l1}}{n_l}, \ldots, \frac{n_{lt!}}{n_l}\right)',$$

where n_{ij} represents the number of occurrences of the ranking $\boldsymbol{\nu}_j$ in sample l.

Also, for $l = 1, 2$, set $\sum_j n_{ij} \equiv n_l$, $\boldsymbol{\gamma} = \boldsymbol{\theta}_1 - \boldsymbol{\theta}_2$ and

$$\boldsymbol{\theta}_l = \boldsymbol{m} + b_l \boldsymbol{\gamma},$$

where

$$\boldsymbol{m} = \frac{n_1\boldsymbol{\theta}_1 + n_2\boldsymbol{\theta}_2}{n_1 + n_2}, b_1 = \frac{n_{2,}}{n_1 + n_2}, b_2 = -\frac{n_1}{n_1 + n_2}.$$

Let $\boldsymbol{\Sigma}_l$ be the covariance matrix of $\boldsymbol{X}_l$ under the null hypothesis defined as

$$\boldsymbol{\Sigma}_l = \boldsymbol{\Pi}_l - \boldsymbol{p}_l \boldsymbol{p}_l',$$

where $\boldsymbol{\Pi}_l = diag\,(p_l(1), \ldots, p_l(t!))$ and $\boldsymbol{p}_l = (p_l(1), \ldots, p_l(t!))'$. The logarithm of the likelihood L as a function of $(\boldsymbol{m}, \boldsymbol{\gamma})$ is proportional to

$$\log L(\boldsymbol{m}, \boldsymbol{\gamma}) \sim \sum_{l=1}^{2} \sum_{j=1}^{t!} n_{lj} \left\{(\boldsymbol{m} + b_l\boldsymbol{\gamma})' \boldsymbol{x}_j - K(\boldsymbol{\theta}_l)\right\}.$$

In order to test

$$H_0 : \boldsymbol{\theta}_1 = \boldsymbol{\theta}_2 \text{ vs } H_1 : \boldsymbol{\theta}_1 \neq \boldsymbol{\theta}_2$$

we calculate the Rao score test statistic which is given by

$$n\left(\boldsymbol{T}_S\hat{\boldsymbol{p}}_1 - \boldsymbol{T}_S\hat{\boldsymbol{p}}_2\right)' \hat{\boldsymbol{D}} \left(\boldsymbol{T}_S\hat{\boldsymbol{p}}_1 - \boldsymbol{T}_S\hat{\boldsymbol{p}}_2\right). \tag{10}$$

It can be shown to have asymptotically a χ^2_f whenever $n_l/n \to \lambda_l > 0$ as $n \to \infty$, where $n = n_1 + n_2$. Here $\hat{\boldsymbol{D}}$ is the Moore–Penrose inverse of $\boldsymbol{T}_S\hat{\boldsymbol{\Sigma}}\boldsymbol{T}_S'$ and $\hat{\boldsymbol{\Sigma}}$ is a consistent estimator of $\boldsymbol{\Sigma} = \frac{\boldsymbol{\Sigma}_1}{\lambda_1} + \frac{\boldsymbol{\Sigma}_2}{\lambda_2}$ and f is the rank of $\hat{\boldsymbol{D}}$, as required.

2.2 The Use of Penalized Likelihood

In the previous sections, it was possible to derive test statistics for the one and two-sample ranking problems by means of the change of measure paradigm. This paradigm may be exploited to obtain new results for the ranking problems. Specifically, we consider a negative penalized likelihood function defined to be the negative log likelihood function subject to a constraint on the parameters which is then minimized with respect to the parameter. This approach yields further insight into ranking problems.

For the one-sample ranking problem, let

$$\Lambda(\boldsymbol{\theta}, c) = -\boldsymbol{\theta}'\left[\sum_{j=1}^{t!} n_j \boldsymbol{x}_j\right] + nK(\boldsymbol{\theta}) + \lambda\left(\sum_{i=1}^{t} \theta_i^2 - c\right) \tag{11}$$

represent the penalizing function for some prescribed values of the constant c. We shall assume for simplicity that $\|\boldsymbol{x}_j\| = 1$. When t is large (say $t \geq 10$), the computation of the exact value of the normalizing constant $K(\boldsymbol{\theta})$ involves a summation of $t!$ terms. [6] noted the resemblance of (6) to the continuous von Mises-Fisher density

$$f\left(\boldsymbol{x};\boldsymbol{\theta}\right) = \frac{\|\boldsymbol{\theta}\|^{\frac{t-3}{2}}}{2^{\frac{t-3}{2}} t! I_{\frac{t-3}{2}}(\|\boldsymbol{\theta}\|)\Gamma(\frac{t-1}{2})} \exp\left(\boldsymbol{\theta}'\boldsymbol{x}\right),$$

where $\|\boldsymbol{\theta}\|$ is the norm of $\boldsymbol{\theta}$ and $\boldsymbol{x}$ is on the unit sphere and $I_\upsilon(z)$ is the modified Bessel function of the first kind given by

$$I_\upsilon(z) = \sum_{k=0}^{\infty} \frac{1}{\Gamma(k+1)\Gamma(\upsilon+k+1)}\left(\frac{z}{2}\right)^{2k+\nu}.$$

This seems to suggest the approximation of the constant $K\left(\boldsymbol{\theta}\right)$ by

$$\exp\left(-K(\boldsymbol{\theta})\right) \approx \frac{1}{t!} \cdot \frac{\|\boldsymbol{\theta}\|^{\frac{t-3}{2}}}{2^{\frac{t-3}{2}} I_{\frac{t-3}{2}}(\|\boldsymbol{\theta}\|)\Gamma(\frac{t-1}{2})}.$$

In [1], penalized likelihood was used in ranking situations to obtain further insight into the differences between groups of rankers.

3 Bayesian Models for Ranking Data

The fact that the model in (1) is itself parametric in nature leads one to consider an extension to Bayesian considerations. Let $\boldsymbol{R} = (R(1), \ldots, R(t))'$ be a ranking t items, labeled $1, \ldots, t$ and define the standardized rankings as

$$y = \left(R - \frac{t+1}{2}\mathbf{1}\right) / \sqrt{\frac{t(t^2-1)}{12}},$$

where y is the $t \times 1$ vector with $\|y\| \equiv \sqrt{y'y} = 1$. We consider the following more general ranking model:

$$\pi(y|\kappa, \theta) = C(\kappa, \theta) \exp\left\{\kappa\theta' y\right\},$$

where the parameter θ is a $t \times 1$ vector with $\|\theta\| = 1$, parameter $\kappa \geq 0$, and $C(\kappa, \theta)$ is the normalizing constant. This model has a close connection to the distance-based models considered in [3]. Here, θ is a real-valued vector, representing a consensus view of the relative preference of the items from the individuals. Since both $\|\theta\| = 1$ and $\|y\| = 1$, the term $\theta' y$ can be seen as $\cos\phi$ where ϕ is the angle between the consensus score vector θ and the observation y. The probability of observing a ranking is proportional to the cosine of the angle from the consensus score vector. The parameter κ can be viewed as a concentration parameter. For small κ, the distribution of rankings will appear close to a uniform whereas for larger values of κ, the distribution of rankings will be more concentrated around the consensus score vector. We call this new model an *angle-based ranking model*.

To compute the normalizing constant $C(\kappa, \theta)$, let P_t be the set of all possible permutations of the integers $1, \ldots, t$. Then

$$(C(\kappa, \theta))^{-1} = \sum_{y \in \mathcal{P}} \exp\left\{\kappa\theta^T y\right\}. \tag{12}$$

Notice that the summation is over the $t!$ elements in $\mathcal{P}$. When t is large, say greater than 15, the exact calculation of the normalizing constant is prohibitive. Using the fact that the set of $t!$ permutations lie on a sphere in $(t-1)$-space, our model resembles the continuous von Mises-Fisher distribution, abbreviated as $vMF(x|m, \kappa)$, which is defined on a $(p-1)$ unit sphere with mean direction m and concentration parameter κ:

$$p(x|\kappa, m) = V_p(\kappa) \exp(\kappa m' x),$$

where

$$V_p(\kappa) = \frac{\kappa^{\frac{p}{2}-1}}{(2\pi)^{\frac{p}{2}} I_{\frac{p}{2}-1}(\kappa)},$$

and $I_a(\kappa)$ is the modified Bessel function of the first kind with order a. Consequently, we may approximate the sum in (12) by an integral over the sphere:

$$C(\kappa, \theta) \simeq C_t(\kappa) = \frac{\kappa^{\frac{t-3}{2}}}{2^{\frac{t-3}{2}} t! I_{\frac{t-3}{2}}(\kappa) \Gamma(\frac{t-1}{2})},$$

where $\Gamma(.)$ is the gamma function. In ([9], it is shown that this approximation is very accurate for values of κ ranging from 0.01 to 2 and t ranging from 4 to 11. Moreover, the error drops rapidly as t increases. Note that this approximation allows us to approximate the first and second derivatives of log C which can facilitate our computation in what follows.

3.1 Maximum Likelihood Estimation (MLE) of Our Model

Let $\boldsymbol{Y} = \{\boldsymbol{y}_1, \ldots, \boldsymbol{y}_N\}$ be a random sample of N standardized rankings drawn from $p(\boldsymbol{y}|\kappa, \boldsymbol{\theta})$. The log likelihood of $(\kappa, \boldsymbol{\theta})$ is then given by

$$l(\kappa, \boldsymbol{\theta}) = n \log C_t(\kappa) + \sum_{i=1}^{n} \kappa \boldsymbol{\theta}' \boldsymbol{y}_i. \tag{13}$$

Maximizing (13) subject to $\|\boldsymbol{\theta}\| = 1$ and $\kappa \geq 0$, we find that the maximum likelihood estimator of $\boldsymbol{\theta}$ is given by $\hat{\boldsymbol{\theta}}_{MLE} = \frac{\sum_{i=1}^{N} \boldsymbol{y}_i}{\left\|\sum_{i=1}^{N} \boldsymbol{y}_i\right\|}$, and $\hat{\kappa}$ is the solution of

$$A_t(\kappa) \equiv \frac{-C_t'(\kappa)}{C_t(\kappa)} = \frac{I_{\frac{t-1}{2}}(\kappa)}{I_{\frac{t-3}{2}}(\kappa)} = \frac{\left\|\sum_{i=1}^{N} \boldsymbol{y}_i\right\|}{N} \equiv r. \tag{14}$$

A simple approximation to the solution of (14) following [4] is given by

$$\hat{\kappa}_{MLE} = \frac{r(t-1-r^2)}{1-r^2}.$$

A more precise approximation can be obtained from a few iterations of Newton's method. Using the method suggested by [8], starting from an initial value κ_0, we can recursively update κ by iteration:

$$\kappa_{i+1} = \kappa_i - \frac{A_t(\kappa_i) - r}{1 - A_t(\kappa_i)^2 - \frac{t-2}{\kappa_i} A_t(\kappa_i)}, \quad i = 0, 1, 2, \ldots.$$

3.2 One-Sample Bayesian Method with Conjugate Prior

Taking a Bayesian approach, we consider the following conjugate prior for $(\kappa, \boldsymbol{\theta})$ as

$$p(\kappa, \boldsymbol{\theta}) \propto [C_t(\kappa)]^{\nu_0} \exp\left\{\beta_0 \kappa \boldsymbol{m}_0' \boldsymbol{\theta}\right\}, \tag{15}$$

where $\|\boldsymbol{m}_0\| = 1, \nu_0, \beta_0 \geq 0$. Given $\boldsymbol{y}$, the posterior density of $(\kappa, \boldsymbol{\theta})$ can be expressed by

$$p(\alpha, \boldsymbol{\theta}|\boldsymbol{y}) \propto \exp\left\{\beta\kappa\boldsymbol{m}'\boldsymbol{\theta}\right\} V_t(\beta\kappa) \cdot \frac{[C_t(\kappa)]^{N+\nu_0}}{V_t(\beta\kappa)},$$

where $\boldsymbol{m} = \left(\beta_0\boldsymbol{m}_0 + \sum_{i=1}^N \boldsymbol{y}_i\right)\beta^{-1}$, $\beta = \left\|\beta_0\boldsymbol{m}_0 + \sum_{i=1}^N \boldsymbol{y}_i\right\|$. The posterior density can be factored as

$$p(\kappa, \theta|\boldsymbol{y}) = p(\theta|\kappa, \boldsymbol{y})p(\kappa|\boldsymbol{y}), \tag{16}$$

where $p(\boldsymbol{\theta}|\kappa, \boldsymbol{y}) \sim vMF(\boldsymbol{\theta}|\boldsymbol{m}, \beta\kappa)$ and

$$p(\kappa|\boldsymbol{y}) \propto \frac{[C_t(\kappa)]^{N+\nu_0}}{V_t(\beta\kappa)} = \frac{\kappa^{\frac{t-3}{2}(\upsilon_0+N)} I_{\frac{t-2}{2}}(\beta\kappa)}{\left[I_{\frac{t-3}{2}}(\kappa)\right]^{\nu_0+N} (\beta\kappa)^{\frac{t-2}{2}}}.$$

The normalizing constant for $p(\kappa|\boldsymbol{y})$ is not available in closed form. For reasons explained in [9], we approximate the posterior distribution using the method of variational inference (abbreviated VI from here on). Variational inference provides a deterministic approximation to an intractable posterior distribution through optimization. We first adopt a joint vMF- Gamma distribution as the prior for $(\kappa, \boldsymbol{\theta})$:

$$\begin{aligned} p(\kappa, \boldsymbol{\theta}) &= p(\boldsymbol{\theta}|\kappa)p(\kappa) \\ &= vMF(\boldsymbol{\theta}|\boldsymbol{m}_0, \beta_0\kappa)\, Gamma(\kappa|a_0, b_0), \end{aligned}$$

where $Gamma(\kappa|a_0, b_0)$ is the Gamma density function with shape parameter a_0 and rate parameter b_0 (i.e., mean equal to $\frac{a_0}{b_0}$), and $p(\boldsymbol{\theta}|\kappa) = vMF(\boldsymbol{\theta}|\boldsymbol{m}_0, \beta_0\kappa)$. The choice of $Gamma(\kappa|a_0, b_0)$ for $p(\kappa)$ is motivated by the fact that for large values of κ, $p(\kappa)$ in (15) tends to take the shape of a Gamma density. In fact, for large values of κ, $I_{\frac{t-3}{2}}(\kappa) \simeq \frac{e^\kappa}{\sqrt{2\pi\kappa}}$, and hence $p(\kappa)$ becomes the Gamma density with shape $(\nu_0 - 1)\frac{t-2}{2} + 1$ and rate $\nu_0 - \beta_0$:

$$p(\kappa) \propto \frac{[C_t(\kappa)]^{\nu_0}}{V_t(\beta_0\kappa)} \propto \kappa^{(\nu_0-1)\frac{t-2}{2}} \exp(-(\nu_0 - \beta_0)\kappa).$$

Using the variational inference framework, [9] showed that the optimal posterior distribution of θ conditional on κ is a von Mises-Fisher distribution $vMF(\boldsymbol{\theta}|\boldsymbol{m}, \kappa\beta)$ where

$$\beta = \left\|\beta_0\boldsymbol{m}_0 + \sum_{i=1}^N \boldsymbol{y}_i\right\| \quad \text{and} \quad \boldsymbol{m} = \left(\beta_0\boldsymbol{m}_0 + \sum_{i=1}^N \boldsymbol{y}_i\right)\beta^{-1}.$$

The optimal posterior distribution of κ is a $Gamma(\kappa|a, b)$ with shape a and rate b with

$$a = a_0 + N\left(\frac{t-3}{2}\right) + \beta\bar{\kappa}\left[\frac{\partial}{\partial\beta\kappa}\ln I_{\frac{t-2}{2}}(\beta\bar{\kappa})\right], \tag{17}$$

$$b = b_0 + N\frac{\partial}{\partial\kappa} I_{\frac{t-3}{2}}(\bar{\kappa}) + \beta_0\left[\frac{\partial}{\partial\beta_0\kappa}\ln I_{\frac{t-2}{2}}(\beta_0\bar{\kappa})\right]. \tag{18}$$

Finally, the posterior mode $\bar{\kappa}$ can be obtained from the previous iteration as

$$\bar{\kappa} = \begin{cases} \frac{a-1}{b} & \text{if } a > 1, \\ \frac{a}{b} & \text{otherwise.} \end{cases} \tag{19}$$

3.3 Two-Sample Bayesian Method with Conjugate Prior

Let $\boldsymbol{Y}_i = \left\{\boldsymbol{y}_{i1}, \ldots, \boldsymbol{y}_{iN_i}\right\}$ for $i = 1, 2$, be two independent random samples of standardized rankings each drawn, respectively, from $p(\boldsymbol{y}_i|\kappa_i, \boldsymbol{\theta}_i)$. Taking a Bayesian approach, we assume that conditional on κ, there are independent von Mises conjugate priors, respectively, for $(\boldsymbol{\theta}_1, \boldsymbol{\theta}_2)$ as

$$p(\boldsymbol{\theta}_i|\kappa) \propto [C_t(\kappa)]^{\nu_{i0}} \exp\left\{\beta_{i0}\kappa \boldsymbol{m}_{i0}^T\boldsymbol{\theta}_i\right\},$$

where $\|\boldsymbol{m}_{i0}\| = 1, \nu_{i0}, \beta_{i0} \geq 0$. We shall be interested in computing the Bayes factor when considering two models. Under model 1, denoted M_1, $\boldsymbol{\theta_1} = \boldsymbol{\theta_2}$ whereas under model 2, denoted M_2, equality is not assumed. The Bayes factor comparing the two models is defined to be

$$B_{21} = \frac{\int p(\boldsymbol{y_1}|\kappa, \boldsymbol{\theta_1})p(\boldsymbol{y_2}|\kappa, \boldsymbol{\theta_2})p(\boldsymbol{\theta_1}|\kappa)p(\boldsymbol{\theta_2}|\kappa)d\boldsymbol{\theta_1}d\boldsymbol{\theta_2}d\kappa}{\int p(\boldsymbol{y_1}|\kappa, \boldsymbol{\theta})p(\boldsymbol{y_2}|\kappa, \boldsymbol{\theta})p(\boldsymbol{\theta}|\kappa)d\boldsymbol{\theta}d\kappa}$$
$$= \frac{\int \left[\int p(\boldsymbol{y_1}|\kappa, \boldsymbol{\theta_1})p(\boldsymbol{\theta_1}|\kappa)d\boldsymbol{\theta_1}\right]\left[\int p(\boldsymbol{y_2}|\kappa, \boldsymbol{\theta_2})p(\boldsymbol{\theta_2}|\kappa)d\boldsymbol{\theta_2}\right]d\kappa}{\int p(\boldsymbol{y_1}|\kappa, \boldsymbol{\theta})p(\boldsymbol{y_2}|\kappa, \boldsymbol{\theta})p(\boldsymbol{\theta}|\kappa)d\boldsymbol{\theta}d\kappa}.$$

The Bayes factor enables us to compute the posterior odds of model 2 to model 1. We fist deal with the denominator in B_{21}. Under M_1, we assume a joint von Mises-Fisher prior on θ and a Gamma prior on κ:

$$p(\theta, \kappa) = vMF(\theta|m_0, \beta_0\kappa)\, G(\kappa|a_0, b_0).$$

Hence,

$$\int p(\boldsymbol{y_1}|\kappa, \boldsymbol{\theta})p(\boldsymbol{y_2}|\kappa, \boldsymbol{\theta})p(\boldsymbol{\theta}|\kappa)d\boldsymbol{\theta}d\kappa = \int C_t^N(\kappa)\, exp\left\{\beta\kappa\boldsymbol{\theta}^T\boldsymbol{m}\right\} V_t(\beta_0\kappa)\, G(\kappa|a_0, b_0)\, d\boldsymbol{\theta}d\kappa$$
$$= \int C_t^N(\kappa)\, V_t(\beta_0\kappa)\, V_t^{-1}(\beta\kappa)\, G(\kappa|a_0, b_0)\, d\kappa,$$

where $N = N_1 + N_2$ and

$$\boldsymbol{m} = \left[\beta_0 \boldsymbol{m}_0 + \sum_{i=1}^{2} \sum_{j=1}^{N_i} \boldsymbol{y}_{ij}\right] \beta^{-1}, \beta = \| \boldsymbol{m} \| .$$

Now,

$$\begin{aligned}
\int C_t^N(\kappa)\, V_t(\beta_0\kappa)\, V_t^{-1}(\beta\kappa)\, G(\kappa|a_0, b_0)\, d\kappa &= C\left(\frac{\beta_0}{\beta}\right)^{\frac{t-2}{2}} \int \left[\frac{\kappa^{a_0+N\left(\frac{t-3}{2}\right)-1} e^{-b_0\kappa} I_{\left(\frac{t-2}{2}\right)}(\beta\kappa)}{I_{\left(\frac{t-2}{2}\right)}(\beta_0\kappa)\, I^N_{\left(\frac{t-3}{2}\right)}(\kappa)}\right] d\kappa \\
&\approx C\left(\frac{\beta_0}{\beta}\right)^{\frac{t-2}{2}} \int \kappa^{a-1} e^{-b\kappa} d\kappa,
\end{aligned}$$

where in the last step, we used the method of variational inference as an approximation, with

$$\begin{aligned}
C &= \frac{b_0^{a_0}}{\Gamma(a_0)} \left(2^{N\left(\frac{t-3}{2}\right)} (t!)^N \Gamma^N\left(\frac{t-1}{2}\right)\right)^{-1} \\
a_1 &= a_0 + N\left(\frac{t-3}{2}\right) + \beta\bar{\kappa}\left[\frac{\partial}{\partial \beta\kappa} \ln I_{\frac{t-2}{2}}(\beta\bar{\kappa})\right], \\
b_1 &= b_0 + N\frac{\partial}{\partial\kappa} I_{\frac{t-3}{2}}(\bar{\kappa}) + \beta_0\left[\frac{\partial}{\partial\beta_0\kappa} \ln I_{\frac{t-2}{2}}(\beta_0\bar{\kappa})\right]
\end{aligned}$$

and the posterior mode $\bar{\kappa}$ is

$$\bar{\kappa} = \begin{cases} \frac{a_1-1}{b_1} & \text{if } a_1 > 1, \\ \frac{a_1}{b_1} & \text{otherwise.} \end{cases}$$

It follows that the denominator of B_{21} is

$$C\left(\frac{\beta_0}{\beta}\right)^{\frac{t-2}{2}} \frac{\Gamma(a_1)}{b_1^{a_1}}.$$

For the numerator, we shall assume that conditional on κ, there are independent von Mises conjugate priors, respectively, for $\boldsymbol{\theta}_1, \boldsymbol{\theta}_2$ given by

$$p(\boldsymbol{\theta}_i|\kappa) \propto [C_t(\kappa)] \exp\left\{\beta_0\kappa \boldsymbol{m}_0^T \boldsymbol{\theta}_i\right\}, i = 1, 2$$

where $\|\boldsymbol{m}_0\| = 1$, $\beta_0 \geq 0$. Hence,

$$\int\left[\int p(\mathbf{y_1}|\kappa,\boldsymbol{\theta_1})p(\boldsymbol{\theta_1}|\kappa)d\boldsymbol{\theta_1}\right]\left[\int p(\mathbf{y_2}|\kappa,\boldsymbol{\theta_2})p(\boldsymbol{\theta_2}|\kappa)\mathrm{d}\boldsymbol{\theta_2}\right]\mathrm{d}\kappa$$

$$= \int C_t^N(\kappa)\, exp\left\{\beta_1\kappa\boldsymbol{\theta_1}^T\boldsymbol{m_1}\right\} exp\left\{\beta_2\kappa\boldsymbol{\theta_2}^T\boldsymbol{m_2}\right\} V_t^2(\beta_0\kappa)\, G(\kappa|a_0,b_0)\,\mathrm{d}\boldsymbol{\theta_1}\mathrm{d}\boldsymbol{\theta_2}\mathrm{d}\kappa$$

$$= \int C_t^N(\kappa)\, V_t^{-1}(\beta_1\kappa)\, V_t^{-1}(\beta_2\kappa)\, V_t^2(\beta_0\kappa)\, G(\kappa|a_0,b_0)\,\mathrm{d}\kappa$$

$$= C\left(\frac{\beta_0}{\beta_1}\right)^{\frac{t-2}{2}}\left(\frac{\beta_0}{\beta_2}\right)^{\frac{t-2}{2}}\int\left[\frac{\kappa^{a_0+N\left(\frac{t-3}{2}\right)-1}e^{-b_0\kappa}I_{\left(\frac{t-2}{2}\right)}(\beta_1\kappa)\, I_{\left(\frac{t-2}{2}\right)}(\beta_2\kappa)}{I^2_{\left(\frac{t-2}{2}\right)}(\beta_0\kappa)\, I^N_{\left(\frac{t-3}{2}\right)}(\kappa)}\right]\mathrm{d}\kappa$$

$$= C\left(\frac{\beta_0}{\beta_1}\right)^{\frac{t-2}{2}}\left(\frac{\beta_0}{\beta_2}\right)^{\frac{t-2}{2}}\int\kappa^{a_2-1}e^{-b_2\kappa}\mathrm{d}\kappa$$

where for $i = 1, 2$,

$$\boldsymbol{m_i} = \left[\beta_0\boldsymbol{m_0} + \sum_{j=1}^{N_i}\boldsymbol{y_{ij}}\right]\beta_i^{-1} = \parallel \boldsymbol{m_i} \parallel$$

$$a_2 = a_0 + N\left(\frac{t-3}{2}\right) + \sum_i \beta_i\bar{\kappa}\left[\frac{\partial}{\partial\beta_i\kappa}\ln I_{\frac{t-2}{2}}(\beta_i\bar{\kappa})\right]$$

$$b_2 = b_0 + N\frac{\partial}{\partial\kappa}lnI_{\frac{t-3}{2}}(\bar{\kappa}) + 2\beta_0\left[\frac{\partial}{\partial\beta_0\kappa}\ln I_{\frac{t-2}{2}}(\beta_0\bar{\kappa})\right]$$

and the posterior mode $\bar{\kappa}$ is given recursively:

$$\bar{\kappa} = \begin{cases}\frac{a_2-1}{b_2} & \text{if } a_2 > 1,\\ \frac{a_2}{b_2} & \text{otherwise.}\end{cases}$$

It follows that the numerator of the Bayes factor is

$$C\left(\frac{\beta_0}{\beta_1}\right)^{\frac{t-2}{2}}\left(\frac{\beta_0}{\beta_2}\right)^{\frac{t-2}{2}}\frac{\Gamma(a_1)}{b_1^{a_1}}.$$

The Bayes factor is then given by the ratio

$$B_{21} = \frac{\left(\frac{\beta_0}{\beta}\right)^{\frac{t-2}{2}} \frac{\Gamma(a_1)}{b_1^{a_1}}}{\left(\frac{\beta_0}{\beta_1}\right)^{\frac{t-2}{2}} \left(\frac{\beta_0}{\beta_2}\right)^{\frac{t-2}{2}} \frac{\Gamma(a_2)}{b_2^{a_2}}}$$

$$= \left(\frac{\beta_1\beta_2}{\beta\beta_0}\right)^{\frac{t-2}{2}} \frac{\Gamma(a_1)\, b_2^{a_2}}{\Gamma(a_2)\, b_1^{a_1}}.$$

4 Conclusion

In this article, we have considered a few applications of the change of measure paradigm. In particular, it was possible to obtain a new derivation of the Friedman statistic. As well, extensions to the Bayesian models for ranking data were considered. Further applications as, for example, to the sign and Wilcoxon tests are found in [2].

Acknowledgements Work supported by the Natural Sciences and Engineering Council of Canada, Grant OGP0009068.

References

1. Alvo, M., Xu, H.: The analysis of ranking data using score functions and penalized likelihood. Austrian J. Stat. **46**, 15–32 (2017)
2. Alvo, M., Yu, Philip, L.H.: A Parametric Introduction to Nonparametric Statistics. Spinger, Berlin (2018)
3. Alvo, M., Yu, Philip, L.H.: Statistical Methods for Ranking Data. Springer, Berlin (2014)
4. Banerjee, A., Dhillon, IS., Ghosh, J., Sra, S.: Clustering on the unit hypersphere using von mises-fisher distributions. J. Mach. Learn. Res. **6**, 1345–1382 (2005)
5. Friedman, M.: The use of ranks to avoid the assumption of normality implicit in the analysis of variance. J. Am. Stat. Assoc. **32**, 675–701 (1937)
6. McCullagh, P.: Models on spheres and models for permutations. In: Fligner, M.A., Verducci, J.S. (eds.) Probability Models and Statistical Analyses for Ranking Data, p. 278283. Springer, Berlin
7. Neyman, J.: Smooth test for goodness of fit. Skandinavisk Aktuarietidskrift **20**, 149–199 (1937)
8. Sra, S.: A short note on parameter approximation for von Mises-Fisher distributions: and a fast implementation of is (x). Comput. Stat. **27**(1), 177–190 (2012)
9. Xu, H., Alvo, M., Yu, Philip, L.H.: Angle-based models for ranking data. Comput. Stat. Data Anal. **121**, 113–136 (2018)

Choosing Between Weekly and Monthly Volatility Drivers Within a Double Asymmetric GARCH-MIDAS Model

Alessandra Amendola, Vincenzo Candila, and Giampiero M. Gallo

Abstract Volatility in financial markets has both low- and high-frequency components which determine its dynamic evolution. Previous modelling efforts in the GARCH context (e.g. the Spline-GARCH) were aimed at estimating the low-frequency component as a smooth function of time around which short-term dynamics evolves. Alternatively, recent literature has introduced the possibility of considering data sampled at different frequencies to estimate the influence of macro-variables on volatility. In this paper, we extend a recently developed model, here labelled Double Asymmetric GARCH-MIDAS model, where a market volatility variable (in our context, VIX) is inserted as a daily lagged variable, and monthly variations represent an additional channel through which market volatility can influence individual stocks. We want to convey the idea that such variations (separately) affect the short- and long-run components, possibly having a separate impact according to their sign.

Keywords Volatility · Asymmetry · GARCH-MIDAS · Forecasting

1 Introduction

Volatility modelling has been extensively studied: more than 30 years have gone by since the seminal contributions by [9, 14]. As they have about 25 K citations each (and some pertinent papers do not even mention them), it is clear that GARCH-type models are the standard among academicians and practitioners alike. These models

A. Amendola (✉)
Department of Economics and Statistics, University of Salerno, Via Giovanni Paolo II, Fisciano, Italy
e-mail: alamendola@unisa.it

V. Candila
MEMOTEF Department, Sapienza University of Rome, Rome, Italy
e-mail: vincenzo.candila@uniroma1.it

G. M. Gallo
Italian Court of Audits (Corte dei conti), and NYU in Florence, Florence, Italy
e-mail: giampiero.gallo@nyu.edu

M. La Rocca et al. (eds.), *Nonparametric Statistics*, Springer Proceedings in Mathematics & Statistics 339, https://doi.org/10.1007/978-3-030-57306-5_3

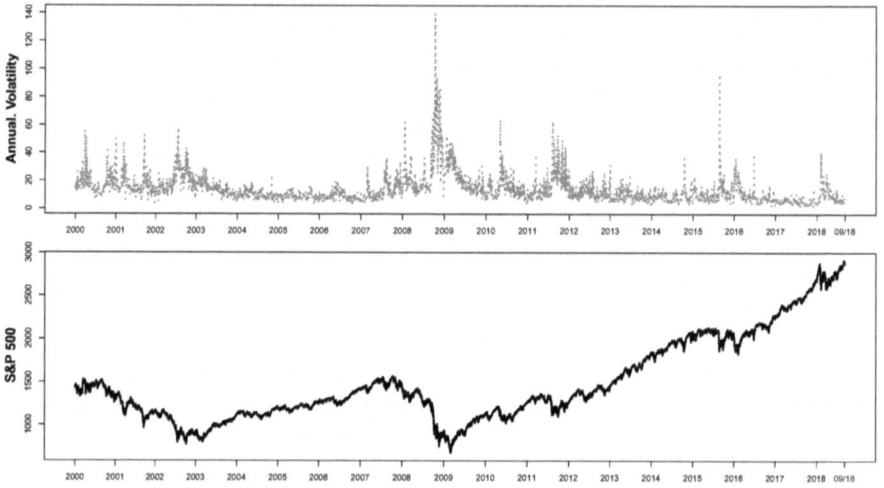

Fig. 1 S&P 500 Index and its realized volatility

build upon stylized facts of persistence in the conditional second moments (volatility clustering), an analysis made easier by the direct measurement of volatility, starting from the availability of ultra-high-frequency data (cf. [7]). Looking directly at the series of the Standard and Poor's (S&P) 500 Index and of its realized volatility, as illustrated in Fig. 1, one encounters two of such stylized facts in need of adequate modelling: the first is that volatility has a slow-moving/state-dependent *average local level* of volatility to be accounted for, and hence its dynamic evolution is driven by two components: a high-frequency and a low-frequency one. Another is that peaks of volatility are recorded in correspondence with streaks of downturns in the index, a sign of well-documented asymmetry in the dynamics.

Various suggestions exist in the literature to model the first of these two stylized facts: in a Markov Switching approach, GARCH parameters are state-dependent ([10, 13, 19], among others). The resulting high-frequency dynamics varies across states and evolves around a constant average level of volatility within states as a low-frequency component. In other contributions, the two components are additive; [12, 15] specify a model in which higher persistence is an identifying condition for the long-run component. The most popular GARCH specification is one in which long- and short-run components combine multiplicatively with the error term. Amado et al. [4] survey the contributions in this field: the common feature is that long run is a term which *smoothly* amplifies or dampens the short-run GARCH dynamics. The long-run term can be a deterministic function of time as in the Spline GARCH [16]; a logistic function of a forcing variable as in the Smooth Transition approach ([1–3], for instance); an exponential function of a one-sided filter of past values of a variable sampled at a lower frequency than the daily series of interest, as in the GARCH-MIDAS of [17]. In this paper, we take a modification of this latter model, called the Double Asymmetric GARCH-MIDAS (DAGM) introduced by [5], where a rate of change at a low frequency is allowed to have differentiated effects according to its sign, determining a local trend around which an asymmetric GARCH that describes the short-run dynamics. A market volatility variable (in our context, we choose the VIX index which is based on implied volatilities from at-the-money option prices) is inserted as a daily lagged variable, and monthly variations represent an additional channel through which market volatility can influence individual stocks.

The issue at stake in this empirically motivated paper is how this information about market-based volatility can help in shaping the MIDAS-GARCH dynamics. The idea we are pursuing is to illustrate

1. how a predetermined daily variable (in lagged levels) adds some significant influence to the short-run component (an asymmetric GARCH in the form of the GJR [18] model—this would be the first asymmetry considered); and, most importantly,
2. how the same variable observed at a lower frequency (in lagged first differences) can determine a useful combination (in the MIDAS sense seen in detail below) for the low-frequency component in the slowly moving level of local average volatility. In particular, it is of interest to explore what frequency (weekly or monthly), works best in this context, and what horizon is informative. In so doing, we maintain that positive changes (an increase in volatility) and negative ones should be treated differently in the model (this is the second asymmetry considered).

The results show that in characterizing the volatility dynamics, our model with monthly data and six months of lagged information works best, together with the contribution of the lagged VIX in the short-run component. Out-of-sample, the model behaves well, with a performance which is dependent on the sub-period considered.

The rest of the paper is organized as follows: Sect. 2 addresses the empirical question, illustrating first how the DAGM works and then we report the results of an application of various GARCH, GARCH-MIDAS and DAGM models on the S&P 500 volatility, both in- and out-of-sample perspectives. Finally, Sect. 3 concludes.

2 Modelling Volatility with the DAGM Model

Let us focus on the GARCH-MIDAS model, here synthetically labelled GM: the paper by [17] defines GARCH dynamics in the presence of mixed frequency variables. The short-run component varies with the same frequency as the dependent variable while the long-run component filters the lower frequency macro-Variable(s) (MV) observations. Recent contributions on (univariate) GARCH-MIDAS model are [6, 8, 11].

The paper by [5] proposes a DAGM where asymmetry in the short run is captured by a GJR-type [18] reaction to the sign of past returns, and positive and negative MV values have different impacts on the long-run.

2.1 The DAGM Framework

The DAGM-X model is defined as

$$r_{i,t} = \sqrt{\tau_t \times g_{i,t}}\varepsilon_{i,t}, \quad \text{with } i = 1, \ldots, N_t, \tag{1}$$

where

- $r_{i,t}$ represents the log-return for day i of the period t;
- N_t is the number of days for period t, with $t = 1, \ldots, T$;
- $\varepsilon_{i,t}|\Phi_{i-1,t} \sim N(0, 1)$, where $\Phi_{i-1,t}$ denotes the information set up to day $i-1$ of period t;
- $g_{i,t}$ follows a unit-mean reverting GARCH(1,1) process (short-run component);
- τ_t provides the slow-moving average level of volatility (long-run component).

The short-run component of the DAGM-X is given by

$$g_{i,t} = (1 - \alpha - \beta - \gamma/2) + \left(\alpha + \gamma \cdot \mathbb{1}_{(r_{i-1,t}<0)}\right) \frac{\left(r_{i-1,t}\right)^2}{\tau_t} + \beta g_{i-1,t} + z V_{i-1,t}, \tag{2}$$

where $\mathbb{1}_{(.)}$ is an indicator function and $V_{i,t}$ is an additional, positive volatility determinant, observed daily, whose importance on $g_{i,t}$ is given by z. In order to assure the positivity of $g_{i,t}$, we impose the constraint $z \geq 0$. In absence of $V_{i,t}$, the DAGM-X model becomes the DAGM specification.

The long-run component of the DAGM-X and DAGM is defined as

$$\tau_t = \exp\left(m + \theta^+ \sum_{k=1}^{K} \delta_k(\omega)^+ X_{t-k} \mathbb{1}_{(X_{t-k}\geq 0)} + \theta^- \sum_{k=1}^{K} \delta_k(\omega)^- X_{t-k} \mathbb{1}_{(X_{t-k}<0)}\right), \tag{3}$$

where

- m plays the role of an intercept;
- θ^+, θ^- represent the asymmetric responses to the one-sided filter;
- $\delta_k(\omega)^+$ and $\delta_k(\omega)^-$ are suitable functions weighing the past K realizations of the additional stationary variable X_t. As in the related literature, we opt for the Beta function, that is

$$\delta_k(\omega) = \frac{(k/K)^{\omega_1-1}(1-k/K)^{\omega_2-1}}{\sum_{j=1}^{K}(j/K)^{\omega_1-1}(1-j/K)^{\omega_2-1}}. \tag{4}$$

Given that we are only interested in the case of larger weights put on the most recent observations, we set $\omega_1 = 1$ and $\omega_2 \geq 1$. Note that the Beta function represented in (4) is readily applicable for both the GM and the DAGM. In this latter case, it is sufficient to replace $\delta_k(\omega)$ with $\delta_k(\omega)^+$ and $\delta_k(\omega)^-$.

Thus, the short-run component includes a term related to negative returns ("bad news" increasing volatility, the well-known *leverage* effect) and potentially a term associated with an additional MV observed with the same frequency of the dependent variable. The long-run component avoids positive and negative compensations within the one-sided filter, separating the positive MV variations from the negative ones.

Typically, MVs can only be observed at low frequency, but here we move out of the classic MV framework where observations are available only at low frequency. Thus we take a variable which is observable daily, but can be sampled at lower frequencies, e.g. weekly or monthly. We take the DAGM to the empirical evaluation of how different frequencies of observations in the MV may change the results both in estimation and forecasting. Besides that, we include the same variable at high frequency in the short-run component ("–X" specifications).

Assuming a conditional normal distribution for the error term $\varepsilon_{i,t}$ allows to apply the standard statistical inference (for details on the asymptotic properties of the GARCH-MIDAS class of models, see [22]) according to the maximization of the following log-likelihood:

$$\ln \mathscr{L} = -\frac{1}{2}\sum_{t=1}^{T}\left[\sum_{i=1}^{N_t}\left[\log(2\pi) + \log(g_{i,t}\tau_t) + \frac{\left(r_{i,t}\right)^2}{g_{i,t}\tau_t}\right]\right]. \tag{5}$$

2.2 *The Role of VIX in the S&P 500 Volatility Dynamics*

The returns of interest are daily log-differences of the S&P 500 Index (also examined on a different sample and context in [5]), annualized on a sample period: 7 January 2000–7 September 2018 (number of daily observations: 4686, collected from Yahoo Finance).

The MV in this paper is VIX (an implied volatility-based index built on the same index, cf. [23]) which in our setup will appear: (i) lagged daily as a predetermined variable in the short-run component g_i of the GARCH-X; (ii) lagged variations—end-of-month or end-of-week (with various choices of K) in the long-run component. All the observations concerning VIX have been collected from the Thomson Reuters Eikon provider. The distance between the estimated volatility, labelled as $\hat{h}_i$, and the chosen volatility proxy, the realized volatility at five minutes, labelled as σ_i and collected from the realized library of the Oxford-Man Institute, are investigated through three loss functions[1]: QLIKE, Root Mean Squared Error (RMSE) and Mean Absolute Errors (MAE), defined as follows:

[1]For ease of notation and because we are only interested in daily estimates, here the suffix t identifying the period at lower frequency has been suppressed.

$$\begin{aligned}
&\text{QLIKE}: E\left(\sigma_i/\hat{h}_i + \log(\hat{h}_i)\right);\\
&\text{RMSE}: \sqrt{E\left(\left(\sigma_i - \hat{h}_i\right)^2\right)};\\
&\text{MAE}: E\left(|\sigma_i - \hat{h}_i|\right).
\end{aligned} \tag{6}$$

2.2.1 Estimation and Diagnostics

The estimation and diagnostics results are shown in Table 1, where we report the coefficients with their standard errors in parenthesis and their significance. GARCH is the standard (1, 1) model; the GJR allows for an asymmetric response to the lagged negative returns; the GARCH-X and GJR-GARCH-X and DAGM-X contain an extra predetermined variable, the lagged daily VIX. The GM and DAGM are built on a one-sided filter for the monthly VIX, while in the DAGM-W we consider the weekly VIX. The last six months of VIX have been used in GM, DAGM, DAGM-X, and DAGM-W, i.e. $K = 6$ and $K_{\text{week}} = 24$. The choice of the adopted MIDAS lags derives from some preliminary estimations aiming at finding the best values according to the Bayesian information criterion (BIC). The number of "*" indicates the significance (10%, 5%, 1%, respectively) of the estimated coefficients heteroscedasticity and autocorrelation consistent ([21], HAC) standard errors in parenthesis). LB_l and LM_l report the p-value of the Ljung-Box and ARCH-LM tests on the squared standardized residuals at lag l, respectively. RMSE is in percentage terms.

A few comments are in order: the GARCH models (first four columns) exhibit customary results, with the possible surprise of the non-significance of the lagged VIX in the X specifications. The GM has non-significant coefficient on the one-sided filter and the wrong sign: as a matter of fact, the information criteria and the QLIKE signal a worse fit of this model, relative to the standard models. When we introduce our DAGM, the signs of the impact coefficients θ^+ and θ^- are the right ones (positive, negative, respectively), and significant. The information criteria and the QLIKE report a marked improvement over the models previously considered, with the *best* model being the DAGM-X model where the significant coefficients on the low-frequency component are, besides the constant, those pertaining to the positive changes (inducing an increase in volatility). Generally, the residual diagnostics show a good fit of the models. In particular, almost all the models fail to reject the null hypotheses of the Ljung-Box and ARCH-LM tests, independently of the order of lags considered. The only model whose p-values are below the significance level of 5% is the DAGM-X, for $l = 12$, for both the Ljung-Box and ARCH-LM tests. Despite this, the conclusion is that the DAGM-X provides the most convincing performance with VIX contributing to a marked improvement over other models. The result can be appraised graphically as in Fig. 2 where we show the close proximity of the fitted values to the realized volatility.

Table 1 DAGM and GARCH estimates

	GARCH	GARCH–X	GJR	GJR–X	GM	DAGM	DAGM–X	DAGM–W
α	0.105***	0.116***	0.001	0.001	0.001	0.001	0.001	0.001
	(0.013)	(0.019)	(0.01)	(0.011)	(0.011)	(0.01)	(0.014)	(0.011)
β	0.884***	0.876***	0.889***	0.878***	0.94***	0.874***	0.852***	0.884***
	(0.015)	(0.018)	(0.015)	(0.02)	(0.013)	(0.015)	(0.018)	(0.015)
γ			0.192***	0.225***	0.11***	0.194***	0.198***	0.19***
			(0.023)	(0.037)	(0.023)	(0.022)	(0.022)	(0.023)
z		0.117		0.165			0.257***	
		(0.091)		(0.104)			(0.039)	
m					5.169***	4.956***	0.686***	5.123***
					(0.315)	(0.192)	(0.121)	(0.205)
θ					−0.004			
					(0.005)			
ω					1.36			
					(1.385)			
θ^+						0.164***	0.101***	0.096***
						(0.042)	(0.027)	(0.028)
ω_2^+						1.372***	1.681***	13.876***
						(0.368)	(0.546)	(0.693)
θ^-						−0.192***	−0.078	−0.431***
						(0.065)	(0.07)	(0.11)
ω_2^-						1.017	1.124	1.455***
						(0.883)	(0.765)	(0.492)
BIC	37586.899	37590.534	37393.477	37394.136	37546.527	37404.797	37367.454	37397.298
QLIKE	−3.867	−3.865	−3.876	−3.873	−3.882	−3.882	−3.882	−3.879
RMSE	0.418	0.433	0.395	0.418	0.402	0.376	0.364	0.382
LB_{12}	0.274	0.388	0.329	0.123	0.506	0.129	0.048	0.186
LB_{24}	0.322	0.361	0.518	0.239	0.416	0.384	0.229	0.384
LB_{36}	0.362	0.383	0.626	0.37	0.278	0.474	0.37	0.482
LM_{12}	0.26	0.381	0.311	0.092	0.526	0.118	0.037	0.17
LM_{24}	0.318	0.349	0.485	0.159	0.41	0.366	0.203	0.345
LM_{36}	0.411	0.391	0.614	0.253	0.366	0.482	0.371	0.479

Notes Annualized scale. Sample period: 7 January 2000–7 September 2018. Number of daily observations: 4686. Ticker: S&P 500. Comparison of the DAGM with other GARCH models. Model definitions and comments in the text. HAC standard errors in parentheses. *, ** and *** denote significance at the 10%, 5% and 1% levels, respectively

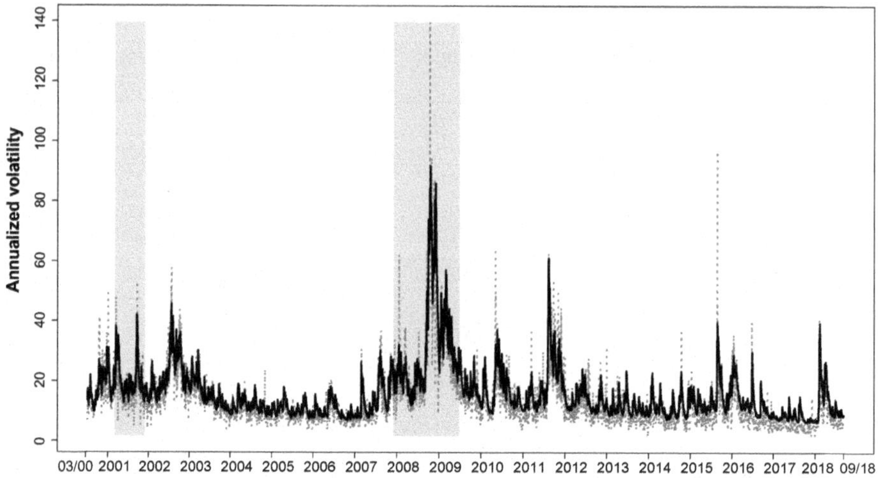

Fig. 2 Realized and DAGM-X volatilities. *Notes* The figure plots the DAGM-X volatility (solid black line) and the S&P 500 realized volatility (dashed grey line). Shaded areas represent NBER recession periods. Annualized scale

2.2.2 Forecasting

Further insights can be had moving to an out-of-sample exercise where we estimate the model over a 10-year period and project one-step ahead for one year and then move forward the estimation and forecasting window. The results are summarized in Fig. 3 where we report the presence in a Model Confidence Set as proposed by [20]. The results (at $\alpha = 10\%$) show that while the DAGM-X has a satisfactory performance, at the same time the standard GARCH or GJR models enter the set.

3 Wrapping Up

The slow-moving feature of conditional volatility can be addressed within a *Double Asymmetric* GARCH-MIDAS framework [5] where the low-frequency variable here is a volatility measure (variations in VIX). The main novelty in this approach is that the same variable can be inserted as a forcing variable (-X in levels) in the short-run component, and we can explore which frequency is the most suitable for the long-run component (in first differences). The fitting capabilities of this approach are comforting, with monthly movements in volatility providing the best in-sample results. In out-of-sample forecasting, though, the model is less satisfactory, in that it gives a performance very similar to a standard GARCH model.

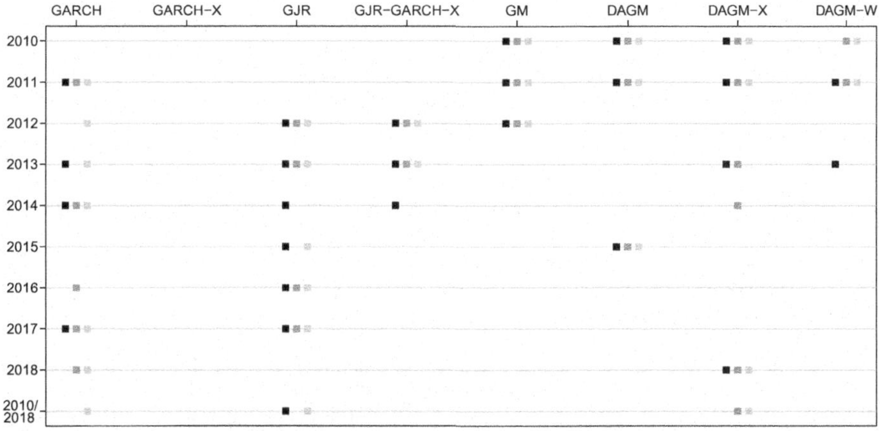

Fig. 3 MCS composition. *Notes* The figure plots the composition of the Model Confidence Set (MCS). For different loss functions, dark (QLIKE), medium-dark (MSE) and light (MAE) shades of grey indicate that a given model is included in the MCS, at a significance level of $\alpha = 0.1$.

References

1. Amado, C., Teräsvirta, T.: Modelling conditional and unconditional heteroskedasticity with smoothly time-varying structure. CREATES Res. Paper **8**, 1–53 (2008)
2. Amado, C., Teräsvirta, T.: Modelling volatility by variance decomposition. J. Econ. **175**(2), 142–153 (2013)
3. Amado, C., Teräsvirta, T.: Modelling changes in the unconditional variance of long stock return series. J. Empir. Finan. **25**, 15–35 (2014)
4. Amado, C., Silvennoinen, A., Teräsvirta, T.: Models with multiplicative decomposition of conditional variances and correlations. CREATES Res. Paper **14**, 1–46 (2018)
5. Amendola, A., Candila, V., Gallo, G.M.: On the asymmetric impact of macro-variables on volatility. Econ. Model. **76**, 135–152 (2019)
6. Amendola, A., Candila, V., Scognamillo, A.: On the influence of US monetary policy on crude oil price volatility. Empir. Econ. **52**(1), 155–178 (2017)
7. Andersen, T.G., Bollerslev, T., Christoffersen, P.F., Diebold, F.X.: Volatility and Correlation Forecasting. In: Elliott, G., Granger, C.W.J., Timmermann, A. (eds.) Handbook of Economic Forecasting, pp. 777–878. North-Holland, Amsterdam (2006)
8. Asgharian, H., Hou, A.J., Javed. F.: The importance of the macroeconomic variables in forecasting stock return variance: a GARCH-MIDAS approach. J. Forecast. **32**(7), 600–612 (2013)
9. Bollerslev, T.: Generalized Autoregressive Conditional Heteroskedasticity. J. Econ. **31**, 307–327 (1986)
10. Brunetti, C., Scotti, C., Mariano, R.S., Tan, A.H.H.: Markov switching GARCH models of currency turmoil in Southeast Asia. Emerg. Mark. Rev. **9**(2), 104–128 (2008)
11. Conrad, C., Loch, K.: Anticipating long-term stock market volatility. J. Appl. Econ. **30**(7), 1090–1114 (2015)
12. Ding, Z., Granger, C.W.J.: Modeling volatility persistence of speculative returns: a new approach. J. Econ. **73**(1), 185–215 (1996)
13. Dueker, M.: Markov switching in GARCH processes and mean-reverting stock-market volatility. J. Bus. Econ. Stat. **15**(1), 26–34 (1997)
14. Engle, R.F.: Autoregressive conditional heteroscedasticity with estimates of the variance of United Kingdom inflation. Econometrica **50**, 987–1007 (1982)

15. Engle, R.F., Lee, G.J.: A permanent and transitory component model of stock return volatility. In: Engle, R.F., White, H. (eds.) Cointegration, Causality, and Forecasting: A Festschrift in Honor of Clive W. J. Granger, pp. 475–497. Oxford University Press, Oxford (1999)
16. Engle, R.F., Rangel, J.G.: The Spline-GARCH model for low frequency volatility and its global macroeconomic causes. Rev. Finan. Stud. **21**, 1187–1222 (2008)
17. Engle, R.F., Ghysels, E., Sohn, B.: Stock market volatility and macroeconomic fundamentals. Rev. Econ. Stat. **95**(3), 776–797 (2013)
18. Glosten, L.R., Jagannathan, R., Runkle, D.E.: On the relation between the expected value and the volatility of the nominal excess return on stocks. J. Finan. **48**(5), 1779–1801 (1993)
19. Hamilton, J., Susmel, R.: Autoregressive conditional heteroskedasticity and changes in regime. J. Econ. **64**(1–2), 307–333 (1994)
20. Hansen, P.R., Lunde, A., Nason, J.M.: The model confidence set. Econometrica **79**(2), 453–497 (2011)
21. Newey, W.K., West, K.D.: A simple, positive semi-definite, heteroskedasticity and autocorrelation consistent covariance matrix. Econometrica **55**(3), 703–708 (1987)
22. Wang, F., Ghysels, E.: Econometric analysis of volatility component models. Econ. Theory. **31**(2), 362–393 (2015)
23. Whaley, R.E.: Understanding the VIX. J. Portfolio Manage. **35**, 98–105 (2009)

Goodness-of-fit Test for the Baseline Hazard Rate

A. Anfriani, C. Butucea, E. Gerardin, T. Jeantheau, and U. Lecleire

Abstract We provide a nonparametric test procedure for the baseline hazard function in the generalized Cox model in presence of fixed given covariates. The test statistic is given by an optimal estimator of the quadratic functional of the same function. Our test procedure attains the rate $n^{-4\alpha/(4\alpha+1)}$ over Besov classes of functions $B_{\alpha}^{2,\infty}(L)$, $\alpha, L > 0$, which is known to be minimax optimal in the context of testing the intensity function of a Poisson processes.

Keywords Baseline hazard rate · Cox model · Goodness-of-fit test · Quadratic functionals · Separation rates

1 Introduction

Let us consider the generalized Cox model defined, for a vector of covariates $Z \in \mathbb{R}^d$,

$$\lambda(t, Z) = h(t) \cdot e^{g(Z)}, \tag{1}$$

A. Anfriani (✉)
Safran Aircraft Engines, 77950 Montereau, France
e-mail: alexandre.anfriani@safrangroup.com

C. Butucea
CREST ENSAE Université Paris Saclay, 5 avenue Henry Le Chatelier, 91120 Palaiseau Cedex, France
e-mail: cristina.butucea@ensae.fr

E. Gerardin
Safran Aircraft Engines, 77950 Montereau, France
e-mail: emilie.gerardin@safrangroup.com

T. Jeantheau
Université Paris-Est Marne-la-Vallée, LAMA(UMR 8050), 77454 Marne-la-Vallée, France
e-mail: thierry.jeantheau@u-pem.fr

U. Lecleire
Safran Aircraft Engines, 77950 Montereau, France
e-mail: uriel.lecleire@safrangroup.com

M. La Rocca et al. (eds.), *Nonparametric Statistics*, Springer Proceedings in Mathematics & Statistics 339, https://doi.org/10.1007/978-3-030-57306-5_4

where $h : [0, \tau] \to [0, +\infty[$ is called the baseline hazard rate function and $g : \mathbb{R}^d \to \mathbb{R}$. The particular case where $g(Z) = \beta^\top Z$ is a linear function which is known as the Cox model.

A lot of attention was devoted to the estimation of both functions h and g. We distinguish methods based on the partial log-likelihood—[1, 8] have obtained nonasymptotic oracle inequalities for g, methods based on maximization of the penalized total likelihood—[3, 9], and kernel methods for estimating h—[4].

Less attention was given to the nonparametric testing of the generalized Cox model. In the case of covariate-free Poisson processes, [6] gave minimax and sharp constants for testing the goodness-of-fit $H_0 : \lambda = \lambda_0$ of the intensity function λ on [0, 1]. In this setup, the intensity λ is supposed to belong to a Sobolev class of functions and the separation from the null hypothesis is measured in $\mathbb{L}_2$ norm, $\|\lambda - \lambda_0\|_2$. Fromont et al. [2] proposed nonasymptotic adaptive tests of homogeneity, i.e., $H_0 : \lambda = I_{[0,1]}$.

In this paper, we want to estimate a quadratic functional of the baseline hazard rate function h and construct a goodness-of-fit test for h based on that functional. More precisely, given h_0 a square integrable function on $[0, \tau]$, that is, h_0 in $\mathbb{L}_2 = \mathbb{L}_2[0, \tau]$, we want to test from our observations that

$$H_0 : \quad h \equiv h_0, \quad \text{against}$$

$$H_1(h_0, \Phi_n) : \quad h \in \mathcal{F} \text{ such that } \int_0^\tau (h - h_0)^2(t)\mathrm{d}t \geq C \cdot \Phi_n, \tag{2}$$

with $C, \Phi_n > 0$ depending on the parameters of nonparametric class of Besov smooth functions $\mathcal{F}$ to be defined. Let us denote the quadratic functional

$$D(h) = \int_0^\tau h^2(t)\mathrm{d}t, \quad \text{for } h \in \mathbb{L}_2.$$

Thus, the separation from the null hypothesis H_0 is measured by $D(h - h_0)$. An estimator of this quadratic functional provides the test statistic for testing $H_0 : h \equiv h_0$ against H_1 in (2). This is now a standard approach in nonparametric testing, as it provides faster rates of testing than the plug-in of an estimator of h, and than the testing in pointwise or sup-norm semi-norms.

We assume that the function g is supposed to be known and proceeds conditionally on the sample of covariates $Z_1, \ldots, Z_n$.

First, we describe an estimator of the quadratic functional $\int_0^\tau h^2(t)\mathrm{d}t$ and study its behavior in Proposition 3. Next, we modify it in order to produce a test statistic and a test whose probabilities of error are controlled in Theorem 1. We note that the behavior of our procedures is of the same order asymptotically as in the covariate-free case. We, therefore, deduce that our procedures are optimal in the minimax sense.

In practice, we can estimate the quadratic functional on $[0, \max_{1\le i\le n} X_i]$. We can show that

$$P(\tau \le \max_{1\le i\le n} X_i) = 1 - (1 - P(T \ge \tau)\cdot P(C \ge \tau))^n$$

tends to 1 as $n \to \infty$, for any fixed $\tau > 0$.

More challenging problems involve to consider unknown function g. However, in the Cox model $g(Z) = \beta^\top Z$ or the generalized Cox model, we can proceed numerically by estimating g using the partial likelihood in presence of an unknown baseline hazard rate. It is interesting but beyond the scope of this paper to study the theoretical impact of plugging the (parametric or nonparametric) estimators of g.

Notation We observe (X_i, Z_i, δ_i), $i = 1, \ldots, n$ over the time interval $\mathcal{T} = [0, \tau]$, for n independent individuals. In our notation, X_i is the censored survival time, that is, $X_i = T_i \wedge C_i$, where T_i is a continuously distributed random time when a failure occurs and C_i is a continuously distributed censoring time. We also observe the failure indicator $\delta_i = I(T_i < C_i)$, which takes the value 1 if a failure occurred and 0 when the censoring occurred. In our setup, the censored survival times are modeled conditionally on the covariates $Z_i = (Z_i^1, \ldots, Z_i^d)$, a d-dimensional vector. We assume that the failure times T_i are independent of the censoring times C_i, conditionally on the covariates Z_i, $i = 1, \ldots, n$.

Our observations allow us to build a marked Poisson process $N_i(t) = I(X_i \le t, \delta_i = 1)$. On our probability space $(\Omega, \mathcal{A}, \mathbb{P})$ we define the filtration

$$\mathcal{F}_t = \sigma\{N_i(s), Z_i, \quad 0 \le s \le t, i = 1, \ldots, n\}. \tag{3}$$

We assume that the counting process $N_i(t)$ satisfies

$$\mathrm{d}N_i(t) = \lambda(t, Z_i)\mathrm{d}t + \mathrm{d}\xi_i(t), \quad 0 \le t \le \tau, i = 1, \ldots, n, \tag{4}$$

where ξ_i is a martingale process on $[0, \tau]$, see [7]. Let us denote it by $\mathrm{d}\Lambda_i(t)$, the random measure $\lambda(t, Z_i)\mathrm{d}t$. We use generic random variables (X, Z, δ) having the same distribution as our sample, and processes $N(t)$, $\Lambda(t)$, and $\xi(t)$ verifying (4).

Here, we consider that the hazard rate satisfies a generalized Cox model (1). Thus,

$$\mathrm{d}\Lambda(t) = \lambda(t, Z)\mathrm{d}t = h(t)e^{g(Z)}\mathrm{d}t.$$

From now on, we shall consider that the design Z_i is fixed and known.

2 Estimation

In this section, the aim is to describe the nonparametric estimation of the functional $D(h)$. In order to do this, it is now established in the literature that a plug-in of the best estimator of h is not the best solution for our problem. Instead, we proceed by making

an $\mathbb{L}_2$ projection on a proper orthogonal basis, express the quadratic functional in terms of the coefficients of the projection and finally estimate it.
Coefficients model Let $\phi : [0, \tau] \to \mathbb{R}$ be square integrable and let us denote it by

$$N[\phi] = \int_0^\tau \phi(t)\mathrm{d}N(t), \quad \xi[\phi] = \int_0^\tau \phi(t)\mathrm{d}\xi(t) \quad \text{and } \Lambda[\phi] = \int_0^\tau \phi(t)\mathrm{d}\Lambda(t)$$

, where $N(t)$ and $\xi(t)$ are the marked point process and the martingale process verifying (4), and $\Lambda(t)$ is given in (1).

Proposition 1 *Let ϕ, ψ be in $\mathbb{L}_2$. The random variables $N_i[\phi]$ for $i = 1, \ldots, n$ are independent and have conditional moments (given Z_i):*

$$E(N_i[\phi]) = \int_0^\tau \phi(t)E[\lambda(t, Z_i)]\mathrm{d}t = e^{g(Z_i)} \int_0^\tau \phi(t)h(t)\mathrm{d}t,$$

$$E[(N_i[\phi])^2] = e^{g(Z_i)} \int_0^\tau \phi^2(t)h(t)\mathrm{d}t + e^{22\,g(Z_i)} \left(\int_0^\tau \phi(t)h(t)\mathrm{d}t\right)^2$$

$$\mathrm{Var}(N_i[\phi]) = e^{g(Z_i)} \int_0^\tau \phi^2(t)h(t)\mathrm{d}t$$

. Moreover,

$$E(N_i[\phi] \cdot N_i[\psi]) = e^{g(Z_i)} \int_0^\tau \phi(t)\psi(t)h(t)\mathrm{d}t + e^{22\,g(Z_i)} \left(\int_0^\tau \phi(t)h(t)dt\right)\left(\int_0^\tau \psi(t)h(t)\mathrm{d}t\right)$$

$$\mathrm{Cov}(N_i[\phi], N_i[\psi]) = \mathrm{Cov}(\xi[\phi], \xi[\psi]) = e^{g(Z_i)} \int_0^\tau \phi(t)\psi(t)h(t)\mathrm{d}t.$$

.

Moreover, it is obvious from the previous result that if ϕ and ψ have disjoint supports then the random variables $N_i[\phi]$ and $N_i[\psi]$ are uncorrelated.

We introduce $\{\phi_j\}_{j=1}^M$ a family of M orthonormal functions of $\mathbb{L}_2[0, \tau]$, which have disjoint supports and the corresponding coefficients of h for $j = \{1, \ldots, M\}$ are given by

$$\theta_j = \int_0^\tau \phi_j(t)h(t)\mathrm{d}t$$

. We project the point process $N_i(t),\ 0 \le t \le \tau$ on these functions to get the random variables

$$N_i[\phi_j] = \Lambda_i[\phi_j] + \xi_i[\phi_j], \quad j \in \{1, \dots, M\}, \tag{5}$$

where $\Lambda_i[\phi_j] := e^{g(Z_i)} \int_0^\tau \phi_j(t)h(t)\mathrm{d}t$. We call (5) the sequence model associated to (4). Random variables $\xi_i[\phi_j]$, for $j \in \{1, \dots, M\}$ are centered, but correlated with the same correlation structure as $N_i[\phi_j],\ j = \{1, \dots, M\}$ as seen in Proposition 1.

The coefficients $\theta_j,\ j \in \{1, \dots, M\}$ are estimated by

$$\hat{\theta}_j = \frac{1}{n}\sum_{i=1}^{n} e^{-g(Z_i)} N_i[\phi_j]. \tag{6}$$

Proposition 2 *The estimator* $\hat{\theta}_j,\ j \in \{1, \dots, M\}$ *defined in (6), is such that*

$$E[\hat{\theta}_j] = \theta_j, \quad Var(\hat{\theta}_j) = \frac{\langle \phi_j^2, h\rangle}{n} \cdot \frac{1}{n}\sum_{i=1}^{n} e^{-g(Z_i)}, \quad \textit{for all} \quad j \in \{1, \dots, M\}$$

and

$$Cov(\hat{\theta}_j, \hat{\theta}_\ell) = \frac{\langle \phi_j \phi_\ell, h\rangle}{n} \cdot \frac{1}{n}\sum_{i=1}^{n} e^{-g(Z_i)}, \quad \textit{for all} \quad j \neq \ell \in \{1, \dots, M\}$$

Note that due to our choice of the functions ϕ_j, j from 1 to M,

$$Cov(\hat{\theta}_j, \hat{\theta}_\ell) = 0, \quad \text{for all} \quad j \neq \ell \text{ in } \{1, \dots, M\}$$

Indeed, $\langle \phi_j \phi_\ell, h\rangle$ is equal to zero if $j \neq \ell$ since ϕ_j and ϕ_ℓ have disjoint support.

Proof Since the $N_i[\phi_j]$ are independent and the $e^{-g(Z_i)}$ fixed, the Proposition 1 gives

$$E[\hat{\theta}_j] = \frac{1}{n}\sum_{i=1}^{n} \langle \phi_j, h\rangle = \theta_j.$$

Then for the variance

$$Var(\hat{\theta}_j) = Var\left(\frac{1}{n}\sum_{i=1}^{n} N_i[\phi_j] \cdot e^{-g(Z_i)}\right) = \frac{1}{n^2}\sum_{i=1}^{n} Var(N_i[\phi_j]) \cdot e^{-2g(Z_i)}$$

$$= \frac{1}{n^2}\sum_{i=1}^{n} \langle \phi_j^2, h\rangle \cdot e^{-g(Z_i)} = \frac{\langle \phi_j^2, h\rangle}{n} \cdot \frac{1}{n}\sum_{i=1}^{n} e^{-g(Z_i)};$$

and, if $j \neq \ell$, for the covariance :

$$Cov(\hat{\theta}_j, \hat{\theta}_\ell) = \frac{1}{n^2}\sum_{\substack{i \neq k \\ i,k=1}}^{n} e^{-g(Z_i)-g(Z_k)} Cov(N_i[\phi_j], N_k[\phi_\ell])$$

$$= \frac{1}{n^2}\sum_{i=1}^{n} e^{-2g(Z_i)} Cov(N_i[\phi_j], N_k[\phi_\ell]) = \frac{\langle \phi_j \phi_\ell, h\rangle}{n} \cdot \frac{1}{n}\sum_{i=1}^{n} e^{-g(Z_i)}.$$

Construction of the estimator We approximate h by $h_M(t) = \sum_{j=1}^{M} \theta_j \phi_j(t)$, with $t \in [0, \tau]$ and $D(h) := \int_0^\tau h^2$ by $\sum_{j=1}^{M} \theta_j^2$. We propose to estimate the quadratic functional $D(h)$ by the U-statistic of order 2:

$$\hat{D}_n = \sum_{j=1}^{M} \frac{1}{n(n-1)} \sum_{\substack{i \neq k \\ i,k=1}}^{n} N_i[\phi_j] e^{-g(Z_i)} \cdot N_k[\phi_j] e^{-g(Z_k)}, \tag{7}$$

where M grows larger with $n \to \infty$.

Proposition 3 *The estimator $\hat{D}_n$ in (7) of the quadratical functional $D(h)$ is such that*

$$E[\hat{D}_n] = \sum_{j=1}^{M} \theta_j^2,$$

and has variance

$$Var[\hat{D}_n] = \frac{2}{n^2(n-1)^2} \sum_{\substack{i \neq k \\ i,k=1}}^{n} e^{-g(Z_i)} e^{-g(Z_k)} \cdot \sum_{j=1}^{M} \langle \phi_j^2, h\rangle^2 + \frac{4}{n^2} \sum_{j=1}^{M} \theta_j^2 \langle \phi_j^2, h\rangle \cdot \sum_{i=1}^{n} e^{-g(Z_i)}.$$

Proof First, let us consider the expected value of the estimator

$$E[\hat{D}_n] = \sum_{j=1}^{M} \frac{1}{n(n-1)} \sum_{\substack{i \neq k \\ i,k=1}}^{n} E\left[N_i[\phi_j] e^{-g(Z_i)} \cdot N_k[\phi_j] e^{-g(Z_k)}\right].$$

The random variables $N_i[\phi_j]$ and $N_k[\phi_j]$ are independent, for $i \neq k$. Thus,

$$E[\hat{D}_n] = \sum_{j=1}^{M} \frac{1}{n(n-1)} \sum_{\substack{i \neq k \\ i,k=1}}^{n} E\left[N_i[\phi_j] e^{-g(Z_i)}\right] \cdot E\left[N_k[\phi_j] e^{-g(Z_k)}\right] = \sum_{j=1}^{M} \theta_j^2.$$

For the variance, let us decompose the sum of indices in two parts starting with the centered expression of $\hat{D}_n$

$$\begin{aligned}
\hat{D}_n - E[\hat{D}_n] &= \sum_{j=1}^{M} \frac{1}{n(n-1)} \sum_{\substack{i \neq k \\ i,k=1}}^{n} (N_i[\phi_j] e^{-g(Z_i)} \cdot N_k[\phi_j] e^{-g(Z_k)} - \theta_j^2) \\
&= \sum_{j=1}^{M} \frac{1}{n(n-1)} \sum_{\substack{i \neq k \\ i,k=1}}^{n} (N_i[\phi_j] e^{-g(Z_i)} - \theta_j)(N_k[\phi_j] e^{-g(Z_k)} - \theta_j) \\
&\quad + \sum_{j=1}^{M} \frac{2}{n} \sum_{i=1}^{n} (N_i[\phi_j] e^{-g(Z_i)} - \theta_j) \cdot \theta_j =: T_1 + T_2, \text{ say.}
\end{aligned}$$

We note that T_1 and T_2 are uncorrelated. Indeed, for any $i \neq k$ or $i \neq \ell$, we have

$$E[(N_i[\phi_j] \cdot e^{-g(Z_i)} - \theta_j)(N_k[\phi_j] \cdot e^{-g(Z_i)} - \theta_j)(N_\ell[\phi_j] e^{-g(Z_i)} - \theta_\ell)] = 0,$$

since at least one term of these three centered terms is independent from the other two. Hence, we can work on the variance of T_1 and T_2 separately. For T_2

$$Var(T_2) = \frac{4}{n^2} \sum_{i=1}^{n} E\left[\left(\sum_{j=1}^{M} (N_i[\phi_j] e^{-g(Z_i)} - \theta_j)\theta_j\right)^2\right].$$

We use the Proposition 1 to get further on

$$\begin{aligned}
&\frac{4}{n^2} \sum_{i=1}^{n} \left(\sum_{j=1}^{M} E[(N_i[\phi_j] e^{-g(Z_i)} - \theta_j)^2 \theta_j^2] + \sum_{j \neq \ell}^{M} \theta_j \theta_\ell e^{-2g(Z_i)} Cov(N_i[\phi_j], N_i[\phi_\ell]) \right) \\
&= \frac{4}{n^2} \sum_{i=1}^{n} \left(\sum_{j=1}^{M} \theta_j^2 Var(N_i[\phi_j]) e^{-2g(Z_i)} + \sum_{j \neq \ell}^{M} \theta_j \theta_\ell e^{-2g(Z_i)} e^{g(Z_i)} \langle \phi_j \phi_\ell, h \rangle \right) \\
&= \frac{4}{n^2} \sum_{i=1}^{n} e^{-g(Z_i)} \left(\sum_{j=1}^{M} \theta_j^2 \langle \phi_j^2, h \rangle + \sum_{j \neq \ell}^{M} \theta_j \theta_\ell \langle \phi_j \phi_\ell, h \rangle \right) \\
&= \frac{4}{n} \sum_{j=1}^{M} \theta_j^2 \langle \phi_j^2, h \rangle \cdot \frac{1}{n} \sum_{i=1}^{n} e^{-g(Z_i)}. \qquad (8)
\end{aligned}$$

Now, let us bound from above the variance of T_1. Let us denote it by $U_j^i = N_i[\phi_j] \cdot e^{-g(Z_i)}$. Then $Var(T_1) = U/(n^2(n-1)^2)$, with

$$U = E\left[\left(\sum_{\substack{i_1 \neq k_1\\ i_1,k_1=1}}^{n} \sum_{j_1=1}^{M} (U_{j_1}^{i_1} - \theta_{j_1})(U_{j_1}^{k_1} - \theta_{j_1})\right)\left(\sum_{\substack{i_2 \neq k_2\\ i_2,k_2=1}}^{n} \sum_{j_2=1}^{M} (U_{j_2}^{i_2} - \theta_{j_2})(U_{j_2}^{k_2} - \theta_{j_2})\right)\right]$$

$$= \sum_{j_1,j_2=1}^{M} \sum_{\substack{i_1 \neq k_1\\ i_1,k_1=1}}^{n} \sum_{\substack{i_2 \neq k_2\\ i_2,k_2=1}}^{n} E\left[(U_{j_1}^{i_1} - \theta_{j_1})(U_{j_1}^{k_1} - \theta_{j_1})(U_{j_2}^{i_2} - \theta_{j_2})(U_{j_2}^{k_2} - \theta_{j_2})\right]. \tag{9}$$

The terms in the previous sum are null except when (i_1, k_1) is equal to either (i_2, k_2) or to (k_2, i_2), thus

$$Var(T_1) = \frac{2}{n^2(n-1)^2} \sum_{j_1,j_2=1}^{M} \sum_{\substack{i \neq k\\ i,k=1}}^{n} E\left[(U_{j_1}^{i} - \theta_{j_1})(U_{j_2}^{i} - \theta_{j_2})\right] E\left[(U_{j_1}^{k} - \theta_{j_1})(U_{j_2}^{k} - \theta_{j_2})\right]$$

and the previous expected value can be written as

$$e^{-2g(Z_i)-2g(Z_k)} Cov(N_i[\phi_{j_1}], N_i[\phi_{j_2}]) Cov(N_k[\phi_{j_1}], N_k[\phi_{j_2}]).$$

Thus,

$$Var(T_1) = \frac{2}{n^2(n-1)^2} \sum_{j=1}^{M} \sum_{\substack{i \neq k\\ i,k=1}}^{n} e^{-g(Z_i)} e^{-g(Z_k)} \left\langle \phi_j^2, h \right\rangle^2$$

$$= \frac{2}{n^2(n-1)^2} \sum_{\substack{i \neq k\\ i,k=1}}^{n} e^{-g(Z_i)} e^{-g(Z_k)} \cdot \sum_{j=1}^{M} \langle \phi_j^2, h \rangle^2. \tag{10}$$

Finally, putting (10) and (8) together, we get the theorem.

3 Goodness-of-Fit Test

In this section, we focus on the nonparametric test. Recall the test problem $H_0 : h \equiv h_0$ against the alternative in (2).

From now on, we assume that for all n and $i = 1, \ldots, n$ there exist $\underline{C}_1$ and $\overline{C}_1 > 0$ such that $\underline{C}_1 \leq e^{-g(Z_i)} \leq \overline{C}_1$.

We consider that $\mathcal{F} = \mathcal{F}_\alpha^{2,\infty}(L)$ a Besov ellipsoid with α, $L > 0$. Functions h belonging to the Besov ellipsoid can be characterized by their coefficients on a wavelet basis with some properties (smoothness, moments, etc.), see [5]. We use as orthonormal basis, a DB2N wavelet basis with resolution level J such that $2^J \sim n^{2/(4\alpha+1)}$. For n large enough, these functions will have disjoint supports.

The separation between h_0 under the null hypothesis H_0 and h under H_1 is measured by $D(h-h_0) := \int_0^\tau (h-h_0)^2$. Let us denote by $\theta_j^0 = \int_0^\tau \phi_j h_0$, $j \geq 1$, the coefficients of the hazard rate function h_0 under H_0. The functional $D(h-h_0)$ is approximated by

$$D_M(h-h_0) := \sum_{j=1}^{M} (\theta_j - \theta_j^0)^2.$$

The test statistic is the U-statistic of order 2 that is an estimator of $D_M(h-h_0)$

$$\hat{D}_n^0 = \sum_{j=1}^{M} \frac{1}{n(n-1)} \sum_{\substack{i \neq k \\ i,k=1}}^{n} (N_i[\phi_j] e^{-g(Z_i)} - \theta_j^0) \cdot (N_k[\phi_j] e^{-g(Z_k)} - \theta_j^0), \tag{11}$$

for some $M = M_n$ of the same order as 2^J. The test procedure is

$$\Delta_n = I(\hat{D}_n^0 \geq r_n), \quad \text{for some } r_n > 0.$$

The test statistic $\hat{D}_n^0$ in (11) has the moments $E(\hat{D}_n^0) = \sum_{j=1}^M (\theta_j - \theta_j^0)^2$, and

$$Var[\hat{D}_n^0] = \frac{2}{n^2(n-1)^2} \sum_{\substack{i \neq k \\ i,k=1}}^{n} e^{-g(Z_i)-g(Z_k)} \sum_{j=1}^{M} \langle \phi_j^2, h \rangle^2 + \frac{4}{n} \sum_{j=1}^{M} (\theta_j - \theta_j^0)^2 \langle \phi_j^2, h \rangle \cdot \frac{1}{n} \sum_{i=1}^{n} e^{-g(Z_i)}.$$

Indeed, this is an easy consequence of Proposition 3. Moreover, using the assumption that $\underline{C}_1 \leq e^{-g(Z)} \leq \overline{C}_1$ for all i from 1 to n, we bound by

$$Var[\hat{D}_n^0] \leq \frac{2}{n(n-1)} \overline{C}_1^2 \sum_{j=1}^{M} \langle \phi_j^2, h \rangle^2 + \frac{4}{n} \sum_{j=1}^{M} (\theta_j - \theta_j^0)^2 \langle \phi_j^2, h \rangle \cdot \overline{C}_1.$$

Since $\{\phi_j\}_{j=1}^M$ is an orthonormal basis, we obtain the following bound of the projection $\langle \phi_j^2, h \rangle < \|h\|_\infty$ and we finally get

$$Var[\hat{D}_n^0] \leq \frac{4\|h\|_\infty \overline{C}_1}{n} \sum_{j=1}^{n} (\theta_j - \theta_j^0)^2 + \frac{2\|h\|_\infty^2 \overline{C}_1^2 \cdot M}{n(n-1)}. \tag{12}$$

The following theorem gives upper bounds for the testing risk.

Theorem 1 *The testing procedure Δ_n based on $\hat{D}_n^0$ in (11) with $M = \lfloor c \cdot n^{\frac{2}{4\alpha+1}} \rfloor$, $\Phi_{n,\alpha} = n^{\frac{-4\alpha}{4\alpha+1}}$, $r_n = r \cdot n^{\frac{-4\alpha}{4\alpha+1}}$, $C_2 \geq 2(L + L_0)$ large enough and convenient choices of $r > 0$ and $c > 0$ is such that there exists $\gamma \in (0, 1)$ giving*

$$P_0(\Delta_n = 1) + \sup_{h \in H_1(h_0, \Phi_{n,\alpha})} P_h(\Delta_n = 0) < \gamma.$$

Proof For the type I error probability, we use the Chebyshev inequality

$$\begin{aligned} P_0(\Delta_n = 1) =& P_0(\hat{D}_n^0 \geq r_n) \leq \frac{Var(\hat{D}_n^0)}{r_n^2} \\ \leq& \left(\frac{4\|h\|_\infty \overline{C}_1}{n} \sum_{j=1}^{n} (\theta_j - \theta_j^0)^2 + \frac{2\|h\|_\infty^2 \overline{C}_1^2 \cdot M}{n(n-1)} \right) \frac{1}{r_n^2}. \end{aligned}$$

Then, using H_0 we have that $\theta_j - \theta_j^0 = 0$ for all $j = 1, \ldots, M$, thus

$$P_0(\Delta_n = 1) \leq \frac{2\|h\|_\infty^2 \overline{C}_1^2 \cdot M}{n(n-1) r_n^2} \leq \frac{2\|h\|_\infty^2 \overline{C}_1^2 \cdot c \cdot n^{\frac{\alpha}{4\alpha+1}}}{r \cdot n^{\frac{-8\alpha}{4\alpha+1}+2}}.$$

For convenient choices of c and r we get

$$P_0(\Delta_n = 1) \leq 2\|h\|_\infty^2 \overline{C}_1^2 \cdot c/r \leq \frac{\gamma}{2}.$$

For the type II error probability, let us first note that

$$D(h - h_0) - E_h[\hat{D}_n^0] = \sum_{j=M+1}^{\infty} (\theta_j - \theta_j^0)^2 \leq \frac{2(L_0 + L)}{M^{2\alpha}}.$$

Then under the alternative hypothesis we use the Chebyshev's inequality and (12):

$$\begin{aligned} P_h(\Delta_n = 0) \leq& P(E_h(\hat{D}_n^0) - \hat{D}_n^0 > C_2 \Phi_{n,\alpha} - 2(L_0 + L)M^{-2\alpha} - r_n) \\ \leq& \frac{Var(\hat{D}_n^0)}{\left(C_2 \cdot \Phi_{n,\alpha} - 2(L_0 + L)M^{-2\alpha} - r_n\right)^2}. \end{aligned}$$

If $\frac{4\|h\|_\infty \overline{C}_1}{n} D_M(h - h_0) \leq \frac{2\|h\|_\infty \overline{C}_1^2}{n(n-1)} M$, then we bound from above

$$P_h(\Delta_n = 0) \leq \frac{4\|h\|_\infty \overline{C}_1^2 M / n(n-1)}{\left(C_2 \cdot \Phi_{n,\alpha} - 2(L_0 + L)M^{-2\alpha} - r_n\right)^2} \leq \frac{\gamma}{2}, sssss$$

for C_2 large enough. If $\frac{4\|h\|_\infty \overline{C}_1}{n} D_M(h-h_0) > \frac{2\|h\|_\infty \overline{C}_1^2}{n(n-1)} M$ this implies that $D_M(h-h_0) > \frac{M\cdot\overline{C}_1}{2(n-1)}$. Let us recall that $E_h(\hat{D}_n^0) = D_M(h-h_0)$ and write

$$P_h(\Delta_n = 0) = P_h(\hat{D}_n^0 < r_n) = P_h\left(\frac{E_h(\hat{D}_n^0) - \hat{D}_n^0}{\sqrt{Var(\hat{D}_n^0)}} > \frac{D_M(h-h_0) - r_n}{\sqrt{Var(\hat{D}_n^0)}}\right)$$

Note that $M/2n \sim c \cdot n^{\frac{-4\alpha+1}{4\alpha+1}} \gg r_n \sim r \cdot n^{\frac{-4\alpha}{4\alpha+1}}$, so $D(h-h_0) \geq 2r_n$ for n large enough. This gives

$$\frac{D_M(h-h_0) - r_n}{\sqrt{Var(\hat{D}_n^0)}} \geq \frac{D_M(h-h_0)/2}{\sqrt{2\cdot 4\|h\|_\infty C_2 D_M(h-h_0)n^{-1}}} \geq \frac{\sqrt{D_M(h-h_0)\cdot n}}{2\sqrt{C_3}} \geq \frac{\sqrt{M\cdot \overline{C}_1}}{2\sqrt{2C_3}}$$

and thus

$$P_h(\Delta_n = 0) \leq P_h\left(Z_n > \frac{\sqrt{M\cdot\overline{C}_1}}{2\sqrt{2C_3}}\right)$$

where we denoted it by Z_n the standardized random variable $-\hat{D}_n^0$. Now let us apply the Chebyshev inequality and that $D_M(h-h_0)$ is uniformly bounded to get

$$P_h(\Delta_n = 0) \leq \frac{Var(Z_n)}{M}\frac{8C_3}{\overline{C}_1} \leq \frac{8C_3}{\overline{C}_1} n^{-\frac{2}{4\alpha+1}} \to 0, \text{ as } n \text{ tends to infinity.}$$

References

1. Bradic, J., Song, R.: Structured estimation for the nonparametric Cox model. Electron. J. Stat. **9**, 492–534 (2015)
2. Fromont, M., Laurent, B., Reynaud-Bouret, P.: Adaptive tests of homogeneity for a Poisson process. Ann. Inst. H. Poincaré Proba. Stat. **47**, 176–213 (2011)
3. Guilloux, A., Lemler, S., Tupin, M.-L.: Adaptive estimation of the baseline hazard function in the Cox model by model selection, with high-dimensional covariates. J. Stat. Planning Inf. **171**, 38–62 (2016)
4. Guilloux, A., Lemler, S., Tupin, M.-L.: Adaptive kernel estimation of the baseline function in the Cox model with high-dimensional covariates. J. Multivar. Anal. **148**, 141–159 (2016)
5. Härdle, W., Kerkyacharian, G., Picard, D., Tsybakov, A.B.: Wavelets, Approximations, and Statistical Applications. Lecture Notes in Statistics, vol. 129, Springer, New York (1998)
6. Ingster, Yu., Kutoyants, YuA: Nonparametric hypothesis testing for intensity of the Poisson process. Math. Meth. Stat. **16**, 217–245 (2007)
7. Karr, A.F.: Point Processes and their Statistical Inference. Marcel Dekker, Inc. (1991)
8. Kong, S., Nan, B.: Non-asymptotic oracle inequalities for the high-dimensional Cox regression via Lasso. Stat. Sinica **24**, 25–42 (2014)
9. Lemler, S.: Oracle inequalities for the Lasso in the high-dimensional Aalen multiplicative intensity model. Ann. Inst. H. Poincaré Proba. Stat. **52**, 981–1008 (2016)

Permutation Tests for Multivariate Stratified Data: Synchronized or Unsynchronized Permutations?

Rosa Arboretti, Eleonora Carrozzo, and Luigi Salmaso

Abstract In the present work, we adopt a method based on permutation tests aimed at facing stratified experiments. The method consists in computing permutation tests separately for each strata and then combining the results. We know that by performing simultaneously permutation tests (synchronized) in different strata, we maintain the underlying dependence structure and we can properly adopt the nonparametric combination of dependent tests procedure. But when strata have different sample sizes, performing the same permutations is not allowed. On the other hand, if units in different strata can be assumed independent we can think to perform permutation tests independently (unsynchronized) for each strata, and then combining the resulting p-values. In this work, we show that when strata are independent we can adopt equivalently both synchronized and unsynchronized permutations.

Keywords Permutation tests · Conditional inference · Multivariate testing · Resampling methods

1 Introduction

The deal with stratified (*pseudo-*) experiments happens quite often in different fields of research. The most typical application examples not only refer to clinical trials, but also industrial problems, social sciences, or demographic studies present a variety of situations in which stratified analysis is required. Literature on stratified experiment is wide and covers different fields (see, e.g., [4–8]). Recently Arboretti et al.

R. Arboretti (✉)
Department of Civil Environmental and Architectural Engineering, University of Padova, Padua, Italy
e-mail: rosa.arboretti@unipd.it

E. Carrozzo · L. Salmaso
Department of Management Engineering, University of Padova, Padua, Italy
e-mail: annaeleonora.carrozzo@unipd.it

L. Salmaso
e-mail: luigi.salmaso@unipd.it

M. La Rocca et al. (eds.), *Nonparametric Statistics*, Springer Proceedings in Mathematics & Statistics 339, https://doi.org/10.1007/978-3-030-57306-5_5

[1] presented a permutation stratified test in a univariate framework based on the nonparametric combination (NPC) methodology [9]. The idea at the basis of this NPC-based procedure is to perform separately although simultaneously different permutation tests, one for each stratum and then suitably combine the results. It is worth noting that a stratified problem may also be of multivariate nature. In this case, in order to properly apply the NPC methodology, we should also take into account the possible dependence among variables.

For the sake of clarity, let us consider a practical example. Suppose, to be interested in evaluating which school among two high schools (say A and B) with different scholarship programs, allows to have more chance to face the entrance exam of a specific University. Suppose also that the entrance exam consists of both written and oral tests. Furthermore, students who attend to entrance exam can choose between two degree courses (say S_1 and S_2).

Schools A and B randomly select a sample of, respectively, n_A and n_B students and simulate the entrance exam. Let us now consider the two following experiments: (1) all students selected from each school perform the tests (written and oral) for both degree courses. (2) half of the students selected from each school performs the test (written and oral) for the degree course S_1 and the other half performs the test (written and oral) for the degree course S_2. Tables 1 and 2 show an example of data structure for the two experiments.

Both experiments (1) and (2) are multivariate because for each student we record the score obtained for written and oral test. The two experiments are stratified because the different degree courses may influence the scores obtained. The difference between two experiments is that in (1) for each school the statistical units in different strata are the same, whereas in (2) we have different units for different strata in each school.

Formalizing $X_{ijs} = \boldsymbol{\mu} + \boldsymbol{\delta}_{js} + \varepsilon_{ijs}$, where X_{ijs} are the multivariate responses, μ is the general mean, δ_{js} is the effect of the j-th treatment in the s-th stratum, and ε_{ijk} are experimental errors, with zero mean from an unknown distribution F_{js}, with $j \in \{A, B\}$, $s = 1, \ldots, S$, where S is the number of strata and $i = 1, \ldots, n_{js}$. We are interested in the following system of hypotheses:

Table 1 Example of data structure for experiment (1)

School	Id student	Written (S1)	Oral (S1)	Written (S2)	Oral (S2)
A	1	87	85	80	78
	2	82	85	80	78
	...	...	...	...	...
	n_A	88	90	85	85
B	1	74	80	80	80
	2	68	74	70	75
	...	...	...	...	...
	n_B	77	85	75	78

Table 2 Example of data structure for experiment (2)

School	Id student	Degree course	Written	Oral
A	1	S_1	87	85
	...	...	...	...
	$\frac{n_A}{2}$	S_1	88	90
	$\frac{n_A}{2}+1$	S_2	80	78
	...	...	...	...
	n_A	S_2	85	85
B	1	S_1	74	80
	...	...	...	...
	$\frac{n_B}{2}$	S_1	77	85
	$\frac{n_B}{2}+1$	S_2	80	80
	...	...	...	...
	n_B	S_2	75	78

$$\begin{cases} H_0^G : \boldsymbol{\delta}_{As} = \boldsymbol{\delta}_{Bs} \text{ for all s} \\ H_1^G : \boldsymbol{\delta}_{As} \underset{(<)}{\overset{(>)}{\neq}} \boldsymbol{\delta}_{Bs} \text{ for at least one s.} \end{cases} \tag{1}$$

In our example $\boldsymbol{\delta}_{js} = (\delta_{js}^{(\text{Written})}, \delta_{js}^{(\text{Oral})})$, $j \in \{A, B\}$, $s \in \{S_1, S_2\}$ denotes the multivariate treatment effect in group j and in stratum s.

If we are in case (1) and we want to solve the problem following the NPC-based procedure, when we perform separately permutation tests we must take into consideration that students in different strata are the same and that for each student correspond two responses. In a permutation framework, this is easily obtainable by permuting the entire rows of Table 1. Let us refer to this type of permutations with the term *synchronized* permutations to emphasize that different tests are simultaneously performed for each stratum and variable. In case (2), when the stratum size in each group is the same it is possible to perform synchronized permutations too. But if different strata in the same group have different sizes this is not allowed. In our example, this may happen if the number of students from the same school who perform the exam for S_1 and S_2 differs. In this paper we wonder if performing independent permutation tests for each stratum, assuming independent strata, affect the results. In this situation, the permutations are performed independently in each stratum but simultaneously for all variables that can be correlated. We refer to this type of permutations with the term *unsynchronized* permutations. In the following sections, we introduce the NPC-based procedure for stratified problems and through a simulation study we investigate the differences between *synchronized* and *unsynchronized* permutations in case of independent strata. Finally, we will show a real-application example.

2 An Algorithm for NPC-based Stratified Tests

In this section, we introduce the NPC-based procedure for multivariate stratified test. For an overview on NPC, its properties, and applications see [2, 3], [9, Chap. 4], [10]. The NPC procedure consists of breaking the problem (1) down into S sub-hypotheses, one for each stratum, i.e,

$$\begin{cases} H_{0(s)} : \boldsymbol{\delta}_{A(s)} = \boldsymbol{\delta}_{B(s)} \\ H_{1(s)} : \boldsymbol{\delta}_{A(s)} \underset{(<)}{\overset{(>)}{\neq}} \boldsymbol{\delta}_{B(s)} \end{cases} \tag{2}$$

Let us denote the K-variate vector of n_{js} observations in group j, in stratum s with $\mathbf{X}_{js} = (\mathbf{X}_{1(js)}, \ldots, \mathbf{X}_{n_{js}(js)})$, $j \in \{A, B\}$, $s = 1, \ldots, S$.

Furthermore, let $\mathbf{X}_s = \mathbf{X}_{As} \biguplus \mathbf{X}_{Bs}$ of size $N_s = n_{As} + n_{Bs}$ denotes the overall sample in stratum s. The steps to achieve the global result are the following:

1 for $s = 1, \ldots, S$

1.1 On $\mathbf{X}_s$ compute a vector of suitable test statistics obtaining

$$\mathbf{T}_s = (T_s^{(1)}, \ldots, T_s^{(K)}).$$

1.2 On $\mathbf{X}_s^*$ obtained after a random rows permutation of $\mathbf{X}_s$, compute the related vector of permuted test statistics.

1.3 Independently repeat step 1.2 a number R of times, with R large enough (i.e., generally $R \gg 1000$). The result estimates the multivariate permutation distribution of the test statistic $\mathbf{T}_s$, denoted by

$$\mathbf{T}_s^* = (T_s^{*(1)}, \ldots, T_s^{*(K)}),$$

where $T_s^{*(k)} = (T_s^{*1(k)}, \ldots, T_s^{*R(k)})$, $k = 1, \ldots, K$.

1.4 Estimate the vector of p-value statistic:

$$\boldsymbol{\lambda}_s = (\lambda_s^{(1)}, \ldots, \lambda_s^{(K)}),$$

where $\lambda_s^{(k)} = \sum_{r=1}^{R} \mathbf{I}(T_s^{*r(k)} \geq T_s^{(k)})/(R+1)$, $k = 1, \ldots, K$.

1.5 Compute the empirical significance level function:

$$\boldsymbol{\lambda}_s^* = (\lambda_s^{*(1)}, \ldots, \lambda_s^{*(K)}),$$

where $\lambda_s^{*(k)} = (\lambda_s^{*1(k)}, \ldots, \lambda_s^{*R(k)})$ with $\lambda_s^{*r(k)} = \sum_{h=1}^{R} \mathbf{I}(T_s^{*h(k)} \geq T_s^{*r(k)})/(R+1)$, $r = 1, \ldots, R$, $k = 1, \ldots, K$.

2 Through a suitable combination function $\Psi(\cdot)$, for each variable combine the p-values statistic related to different strata, obtaining

$$\mathbf{T_\bullet} = \left(T_\cdot^{(1)}, \dots, T_\cdot^{(K)}\right),$$

where for a generic variable k is $T_\cdot^{(k)} = \Psi(\lambda_1^{(k)}, \dots, \lambda_s^{(k)})$ and

$$\mathbf{T_\bullet^*} = \left(T_\cdot^{*(1)}, \dots, T_\cdot^{*(K)}\right),$$

where for a generic variable k is $T_\cdot^{*(k)} = \Psi(\lambda_1^{*(k)}, \dots, \lambda_s^{*(k)})$.

3 Compute the combined p-value statistic as

$$\boldsymbol{\lambda_\bullet} = (\lambda_\cdot^{(1)}, \dots, \lambda_\cdot^{(K)}),$$

where $\lambda_\cdot^{(k)} = \sum_{r=1}^{R} \mathbf{I}(T_\cdot^{*r(k)} \geq T_\cdot^{(k)})/(R+1)$ and the related empirical significance level function.

$$\boldsymbol{\lambda_\bullet^*} = (\lambda_\cdot^{*(1)}, \dots, \lambda_\cdot^{*(K)})$$

where $\lambda_\cdot^{*(k)} = \sum_{h=1}^{R} \mathbf{I}(T_\cdot^{*h(k)} \geq T_\cdot^{*r(k)})/(R+1)$.

4 Combine the p-value statistics related to all variables, obtaining

$$T = \Psi(\lambda_\cdot^{(1)}, \dots, \lambda_\cdot^{(K)})$$

and the related simulated distribution:

$$T^* = \Psi(\lambda_\cdot^{*(1)}, \dots, \lambda_\cdot^{*(K)}).$$

5 Compute the global p-value as $\lambda^{Glob} = \sum_{r=1}^{R} \mathbf{I}(T^{*r} \geq T)/(R+1)$ and reject the null hypothesis (1) if $\lambda^{Glob} \leq \alpha$.

In order to complete the algorithm, let us cite some possible test statistics T and combining function Ψ. At step 1.1, generic test statistics T that can be used are the difference of means

$$T_{DM(s)} = \frac{1}{n_{As}} \sum_{i=1}^{n_{As}} X_{iAs} - \frac{1}{n_{Bs}} \sum_{i=1}^{n_{Bs}} X_{iBs}$$

if X_{js} is continuous or the Anderson–Darling test statistics

$$T_{AD(s)} = \sum_{h=1}^{\upsilon-1} M_{hAs}[M_{h(\bullet s)}(N - M_{h(\bullet s)})]$$

if X_{js} is categorical with υ categories, where $M_{h\bullet s} = M_{hAs} + M_{hBs}$, M_{hAs}, and M_{hBs} are the cumulative frequencies of the category h in stratum s of group A and B, respectively.

For what concern combining function in points 2 and 4, common choices are

- the *Fisher omnibus* combining function defined as $\psi_F = -2 \cdot \sum_i^\nu (\log \lambda_i)$;
- the *Liptak* combining function defined as $\psi_L = \sum_i^\nu \Phi^{-1}(1 - \lambda_i)$;
- the *Tippett* combining function defined as $\psi_T = \max_{i=1}^\nu (1 - \lambda_i)$

where ν represents the number of partial aspects to be combined.

Note that if we cannot assume the independence of the strata, all test statistics have to be computed on the same permutations (*synchronized*), that is, the random permutations obtained at points 1.2–1.3 must be the same for each $s = 1, \ldots, S$ so as to preserve the underlying unknown dependence structure. What we are interested to assess in next section is if, in case of independent strata, performing *unsynchronized* instead of *synchronized* permutations has an effect on the procedure.

3 Synchronized and Unsynchronized Permutations

In this section, we show the results of a simulation study in which NPC-based stratified test is used adopting both synchronized and unsynchronized permutations. In order to assess if the type of permutations affect the analysis it is sufficient a simple case:

- Two treatments A and B;
- $S = 2$ strata;
- $K = 3$ variables correlated ($\rho_{12} = \rho_{13} = \rho_{23} = 0.4$) and uncorrelated;
- $n_{sA} = n_{sB} = 20 \; \forall s = 1, 2$.

As generating data, we considered Normal distribution and Student's t with 3 degrees of freedom. Furthermore, we considered cases in which treatment had the same effect on across all strata and cases with different treatment effect across strata that mimics a situation with a stratum by treatment interaction. Figures 1 and 2 show some examples of interaction plots of generated samples used in the simulation study. Plots in Fig. 1 represent a situation with treatment effect constant across strata. The first plot represents a situation under the null hypothesis (no treatment effect), whereas the second plot is under the alternative. Figure 2 represents a situation with treatment effect interacting with strata.

Results of the simulation study are in the following figures. Figures 3 and 4 show the rejection rates on 1000 simulations of the NPC-based stratified test based on 1000 permutations, when treatment effect is constant across strata, with independent and correlated variables with Normal and Student's t with 3 degrees of freedom distribution, respectively. In this case, the treatment effect δ (on group B) is the same in each stratum. Note that the two power curves of tests based on synchronized and unsynchronized are perfectly overlapped, and under the null hypothesis ($\delta = 0$) the significance α-level is respected.

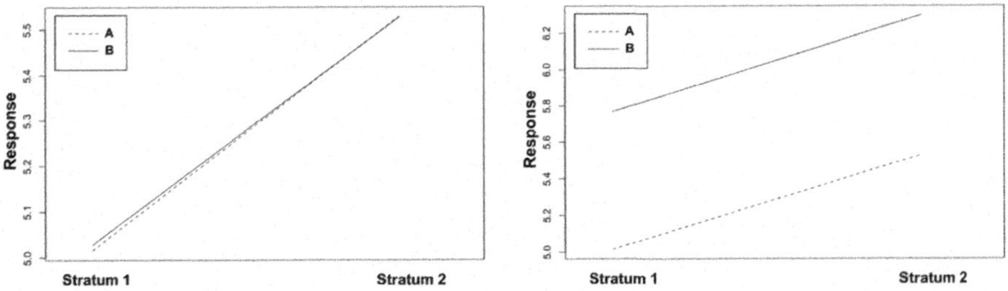

Fig. 1 Interaction plots of some simulated samples with treatment effect constant across strata

Fig. 2 Interaction plots of one simulated sample with treatment effect varying across strata

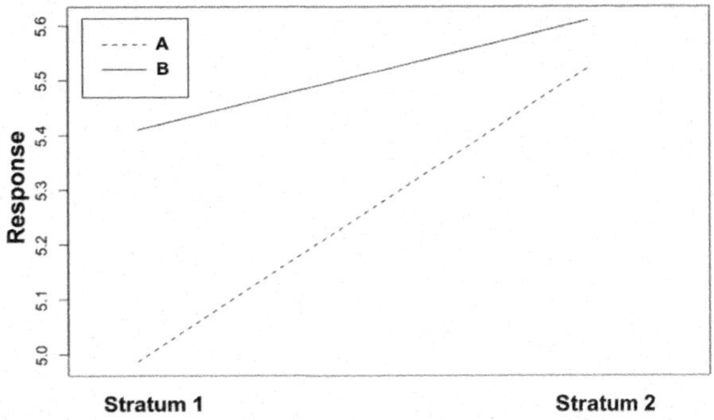

Figures 5 and 6 show the rejection rates on 1000 simulations of the NPC-based stratified test based on 1000 permutations, when treatment effect varies across strata, with variables independent and correlated for Normal and Student's t distribution, respectively. In this case, we show the treatment effect (on group B) in stratum 1 (δ_1) and in stratum 2 (δ_2).

We already performed a simulation with unbalanced stratum size in each group and results are consistent with the balanced case, so that we do not report the corresponding results. From all these results we can conclude that, in case of independent strata, we can use alternatively both *synchronized* and *unsynchronized* permutations strategy. In particular, when we have balanced cases, i.e., strata in the same group have equal size, for computational reasons *synchronized* permutations are preferable. On the other hand, when we have the presence of unbalanced strata, e.g., because of data missed completely at random, we can equivalently perform the NPC-based stratified test using *unsynchronized* permutations.

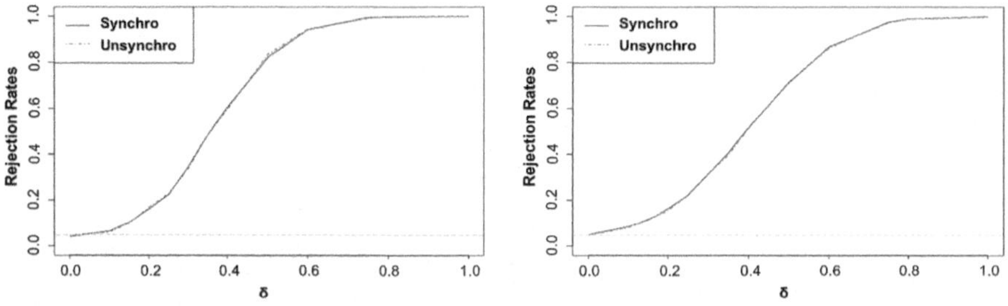

Fig. 3 Rejection rates of NPC-based stratified test when treatment effect is constant across strata, with variables independent (left) and correlated (right), with Normal distributed data

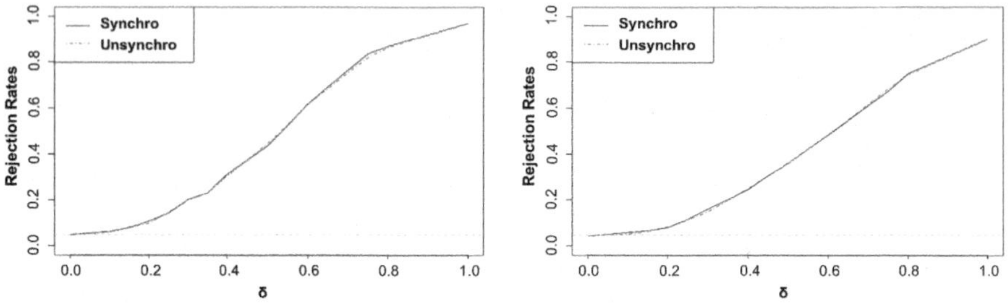

Fig. 4 Rejection rates of NPC-based stratified test when treatment effect is constant across strata, with variables independent (left) and correlated (right), with Student's t distributed data

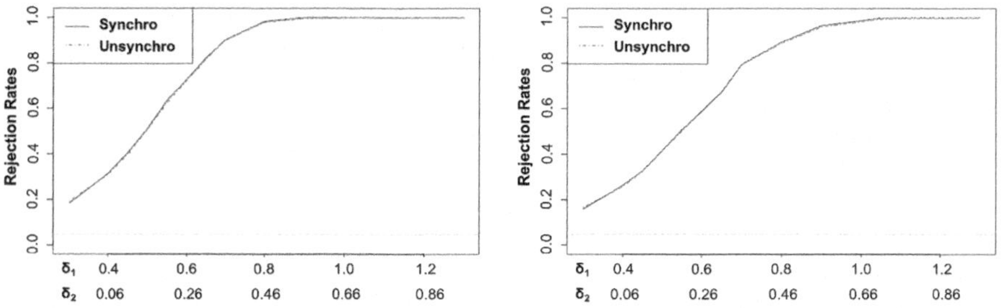

Fig. 5 Rejection rates of NPC-based stratified test when treatment effect varies across strata with variables independent (left) and correlated (right), with Normal distributed data

4 An Example Application

The school of Engineering of the University of Padova (Italy) promoted the analysis of a huge database related to several information on the career of students. The objective was multifold. One of these objectives was to understand if the type of high school from which students come has an impact on the success in the University entrance exam and on the number of university credits reached at the end of the first academic year. In particular, it was of interest to compare schools with a scientific

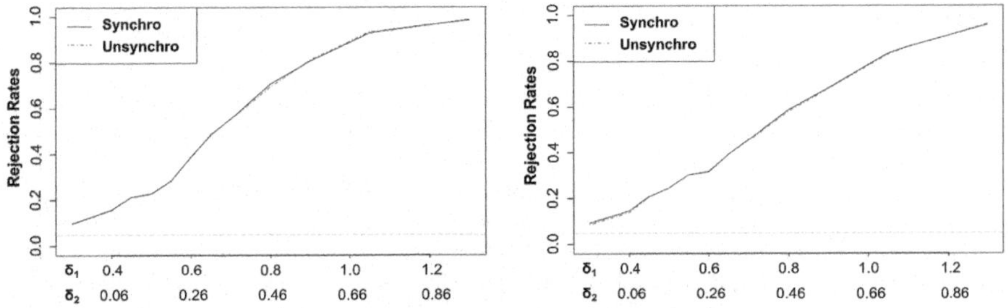

Fig. 6 Rejection rates of NPC-based stratified test when treatment effect varies across strata, with variables independent (left) and correlated (right), with Student's t distributed data

Table 3 Example of strata sizes of the example

School	DC_1	DC_2	DC_3	...	DC_{12}
A	62	106	25	...	40
B	46	211	17	...	107

Table 4 P-values of partial comparisons $H_1 : \delta^k_{As} > \delta^k_{Bs}$, $k \in$ (Score, CFU) and $s \in (DC_1, DC_2, \dots, DC_{12})$

	DC_1	DC_2	DC_3	...	DC_{12}
Score	0.0002	< 0.0001	0.540	...	0.0045
CFU	0.0001	< 0.0001	0.820	...	0.0007

Table 5 Combined p-values of the comparisons $H_1 : \delta^k_{A\cdot} > \delta^k_{B\cdot}$, $k \in$ (Score, CFU)

Score	< 0.0001
CFU	< 0.0001

curriculum (A) with schools with a technical curriculum (B). For this reason, the score at the entrance exam and the number of university credits at the end of the first academic year have been recovered from the database ($K = 2$). Since the School of Engineering has $S = 12$ different degree courses (DC), we considered them as a stratification factor. Table 3 shows an example of the (unbalanced) strata sizes. The NPC-based stratified test has been applied to these data considering a one-sided alternative hypotheses $H_1 : \delta^k_{As} > \delta^k_{Bs}$, $k \in$ (Score, CFU) and $s \in (DC_1, DC_2, \dots, DC_{12})$ obtaining a global p-value $\lambda^{Glob} < 0.0001$. Furthermore, with the NPC procedure, we can investigate all partial aspects as shown in Tables 4 and 5.

What we can see from the analysis is that students with a scientific curriculum look to have better possibilities to face the first year of Engineering at the University of Padova with respect to students with a technical curriculum.

References

1. Arboretti, R., Carrozzo, E., Salmaso, L.: Stratified data: a permutation approach for hypotheses testing. In: Proceeding of the Conference of the Italian Statistical Society, vol. 114, pp. 71–77. Firenze University Press (2017)
2. Arboretti, R., Carrozzo, E., Pesarin, F., Salmaso, L.: Testing for equivalence: an intersection-union permutation solution. Stat. Biopharm. Res. **10**(2), 130–138 (2018)
3. Arboretti, R., Carrozzo, E., Pesarin, F., Salmaso, L.: A multivariate extension of union-intersection permutation solution for two-sample testing. J. Stat. Theory Practice **11**, 436–448 (2017)
4. Boos, D.D., Brownie, C.: A rank-based mixed model approach to multisite clinical trials. Biometrics **48**, 61–72 (1992)
5. Brunner, E., Munzel, U., Puri, M.L.: Rank-score tests in factorial designs with repeated measures. J. Multivar. Anal. **70**, 286–317 (1999)
6. Brunner, E., Puri, M.L., Sun, S.: Nonparametric methods for stratified two-sample designs with application to multiclinic trials. J. Am. Stat. Assoc. **90**(431), 1004–1014 (1995)
7. Gould, A.L.: Multi-center trial analysis revisited. Stat. Med. **17**, 1779–1797 (1998)
8. Mehrotra, D.V., Lu, X., Li, X.: Rank-based analyses of stratified experiments: alternatives to the van Elteren test. Am. Stat. **64**(2), 121–130 (2010)
9. Pesarin, F., Salmaso, L.: Permutation Tests for Complex Data: Theory, Applications and Software. Wiley, Chichester (2010)
10. Pesarin, F., Salmaso, L., Carrozzo, E., Arboretti, R.: Union intersection permutation solution for two-sample equivalence testing. Stat. Comput. **26**(3), 693–701 (2016)

An Extension of the DgLARS Method to High-Dimensional Relative Risk Regression Models

Luigi Augugliaro, Ernst C. Wit, and Angelo M. Mineo

Abstract In recent years, clinical studies, where patients are routinely screened for many genomic features, are becoming more common. The general aim of such studies is to find genomic signatures useful for treatment decisions and the development of new treatments. However, genomic data are typically noisy and high dimensional, not rarely outstripping the number of patients included in the study. For this reason, sparse estimators are usually used in the study of high-dimensional survival data. In this paper, we propose an extension of the differential geometric least angle regression method to high-dimensional relative risk regression models.

Keywords dgLARS · Gene expression data · High-dimensional data · Relative risk regression models · Sparsity · Survival analysis

1 Introduction

In recent years, clinical studies, where patients are routinely screened for many genomic features, are becoming more common. In principle, this holds the promise of being able to find genomic signatures for a particular disease. In particular, cancer survival is thought to be closely linked to the genomic constitution of the tumour. Discovering such signatures will be useful in the diagnosis of the patient, may be used for treatment decisions and, perhaps, even for the development of new treatments. However, genomic data are typically noisy and high dimensional, not rarely outstripping the number of patients included in the study. For this reason, sparse estimators are usually used in the study of high-dimensional survival data.

L. Augugliaro (✉) · A. M. Mineo
SEAS Department Palermo, Viale delle Scienze Edf. 13, Palermo, Italy
e-mail: luigi.augugliaro@unipa.it

A. M. Mineo
e-mail: angelo.mineo@unipa.it

E. C. Wit
Institute of Computational Science, USI, Via Buffi 13, Lugano, Switzerland
e-mail: wite@usi.ch

M. La Rocca et al. (eds.), *Nonparametric Statistics*, Springer Proceedings in Mathematics & Statistics 339, https://doi.org/10.1007/978-3-030-57306-5_6

In the past two decades, sparse inference has been dominated by methods that penalize the likelihood by functions of the parameters that happen to induce solutions with many zero estimates. The Lasso [22], elastic net [24] and the SCAD [6] penalties are only a few examples of such penalties that, depending on a tuning parameter, conveniently shrink estimates to zeros. In [23], the Lasso penalty is applied to the Cox proportional hazards model. Although the Lasso penalty induces sparsity, and it is well known to suffer from a possible inconsistent selection of variables.

In this paper, we will approach sparsity directly from a likelihood point of view. The angle between the covariates and the tangent residual vector within the likelihood manifold provides a direct and scale-invariant way to assess the importance of the individual covariates. The idea is similar to the least angle regression approach proposed by [5]. However, rather than using it as a computational device for obtaining Lasso solutions, we view the method in its own right as in [1]. Moreover, the method extends directly the Cox proportional hazard model. In fact, we will focus on general relative risk regression models.

2 Relative Risk Regression Models

Let T be the (absolutely) continuous random variable associated with the survival time and let $f(t)$ be the corresponding probability density function. The hazard function specifies the instantaneous rate at which failures occur for subjects that are surviving at time t and it is formally defined as $\lambda(t) = f(t)/\{1 - \int_0^t f(s)\mathrm{d}s\}$.

As proposed in [21], we assume that a p-dimensional vector of predictors, possibly time-dependent, say $\boldsymbol{x}(t) = (x_1(t), \ldots, x_p(t))^\top$, can influence the hazard function by the following model:

$$\lambda(t; \boldsymbol{x}) = \lambda_0(t)\psi(\boldsymbol{x}(t); \boldsymbol{\beta}), \tag{1}$$

where $\boldsymbol{\beta} \in B \subseteq \mathbb{R}^p$ is a p-dimensional vector of regression coefficients, $\lambda_0(t)$ is the base line hazard function at time t, which is left unspecified, and $\psi : \mathbb{R}^p \times \mathbb{R}^p \to \mathbb{R}$ is a differentiable function, called the relative risk function, such that $\psi(\boldsymbol{x}(t); \boldsymbol{\beta}) > 0$, for each $\boldsymbol{\beta} \in B$. Model (1) extends the classical Cox regression model [4], and allows us to work with applications in which the exponential form of the relative risk function is not the best choice [12]. Table 1 reports some of the most used relative risk functions (see [10] for more details).

Table 1 Some used relative risk regression functions

	Exponential	Linear	Logit	Excess
$\psi(\boldsymbol{x}(t); \boldsymbol{\beta})$	$\exp\{\boldsymbol{\beta}^\top \boldsymbol{x}(t)\}$	$1 + \boldsymbol{\beta}^\top \boldsymbol{x}(t)$	$\log[1 + \exp\{\boldsymbol{\beta}^\top \boldsymbol{x}(t)\}]$	$\prod_{m=1}^{p}\{1 + x_m(t)\beta_m\}$

Suppose that n observations are available and let t_i be the ith observed failure time. Furthermore, assume that we have k uncensored and untied failure times and let D be the set of indices for which the corresponding failure time is observed; the remaining failure times are right censored. Denote with $R(t)$ the risk set, i.e. the set of indices corresponding to the subjects who have not failed and are still under observation just prior to time t, under the assumption of independent censoring, inference about the $\boldsymbol{\beta}$ can be carried out by the partial likelihood function

$$L_p(\boldsymbol{\beta}) = \prod_{i \in D} \frac{\psi(\boldsymbol{x}_i(t_i); \boldsymbol{\beta})}{\sum_{j \in R(t_i)} \psi(\boldsymbol{x}_j(t_i); \boldsymbol{\beta})}. \tag{2}$$

When the number of predictors exceeds the sample size, a direct maximization of the partial likelihood (2) is not possible. In the next sections, we shall explain how to use the differential geometrical structure of the relative risk regression model to study its sparse structure.

3 DgLARS Method for Relative Risk Regression Models

3.1 Differential Geometrical Structure of a Relative Risk Regression Model

In this section, we study the differential geometrical structure of the relative risk regression model. To do this, we follow the approach proposed in [20], i.e. we relate the partial likelihood (2) with the likelihood function of a logistic regression model for matched case-control studies. The interested reader is also refereed to [16].

Consider an index $i \in D$ and let $\boldsymbol{Y}_i = (Y_{ih})_{h \in R(t_i)}$ be a multinomial random variable with cell probabilities $\boldsymbol{\pi}_i = (\pi_{ih})_{h \in R(t_i)} \in \Pi_i$. Assuming that the random vectors $\boldsymbol{Y}_i$ are independent, the joint probability density function is an element of the set $S = \left\{\prod_{i \in D} \prod_{h \in R(t_i)} \pi_{ih}^{y_{ih}} : (\boldsymbol{\pi}_i)_{i \in D} \in \bigotimes_{i \in D} \Pi_i\right\}$, called the ambient space. We would like to underline that our differential geometric constructions are invariant to the chosen parameterization, which means that S can be equivalently defined by the canonical parameter vector and this will not change the results. In this paper, we prefer to use the mean value parameter vector to specify our differential geometrical description because this will make the relationship with the partial likelihood (2) clearer. If we let

$$E_{\boldsymbol{\beta}}(Y_{ih}) = \pi_{ih}(\boldsymbol{\beta}) = \frac{\psi(\boldsymbol{x}_h(t_i); \boldsymbol{\beta})}{\sum_{j \in R(t_i)} \psi(\boldsymbol{x}_j(t_i); \boldsymbol{\beta})},$$

and we assume that for each $i \in D$, the observed y_{ih} is equal to one if h is equal to i and zero otherwise, it is easy to see that the partial likelihood (2) is formally equivalent to the likelihood function associated with the model space $M = \left\{\prod_{i\in D}\prod_{h\in R(t_i)}\{\pi_{ih}(\boldsymbol{\beta})\}^{y_{ih}} : \boldsymbol{\beta} \in B\right\}$.

From a geometric point of view, the set M can be seen as a differentiable manifold embedded in S, which plays the role of ambient space. To complete the differential geometric framework needed to extend the dgLARS method to the relative risk regression models, we have to introduce the notion of tangent space and equip it with a suitable inner product. This can be done using the approach proposed in [17].

Let $\ell(\boldsymbol{\beta}) = \sum_{i\in D}\sum_{h\in R(t_i)} Y_{ih}\log\pi_{ih}(\boldsymbol{\beta})$ be the log-likelihood function associated to the model space M and $\partial_m\ell(\boldsymbol{\beta}) = \partial\ell(\boldsymbol{\beta})/\partial\beta_m$. The tangent space of M at the model point $\prod_{i\in D}\prod_{h\in R(i)}\{\pi_{ih}(\boldsymbol{\beta})\}^{y_{ih}}$, denoted by $T_{\boldsymbol{\beta}}M$, is defined as the linear vector space spanned by the p elements of the score vectors, formally, $T_{\boldsymbol{\beta}}M = \mathrm{span}\{\partial_1\ell(\boldsymbol{\beta}),\ldots,\partial_p\ell(\boldsymbol{\beta})\}$. In the same way, the tangent space of S at the model point $\prod_{i\in D}\prod_{h\in R(i)}\{\pi_{ih}(\boldsymbol{\beta})\}^{y_{ih}}$, denoted by $T_{\boldsymbol{\beta}}S$, is defined as the linear vector space spanned by the random variables $\partial_{ih}\ell(\boldsymbol{\beta}) = \partial\ell(\boldsymbol{\beta})/\partial\pi_{ih}$. Applying the chain rule, we can see that any tangent vector $v_{\boldsymbol{\beta}} = \sum_{m=1}^p v_m\partial_m\ell(\boldsymbol{\beta})$ belonging to $T_{\boldsymbol{\beta}}M$ can be written as

$$v_{\boldsymbol{\beta}} = \sum_{m=1}^p v_m\partial_m\ell(\boldsymbol{\beta}) = \sum_{i\in D}\sum_{h\in R(i)}\left\{\sum_{m=1}^p v_m\frac{\partial\pi_{ih}(\boldsymbol{\beta})}{\partial\beta_m}\right\}\partial_{ih}\ell(\boldsymbol{\beta}) = \sum_{i\in D}\sum_{h\in R(i)} w_{ih}\partial_{ih}\ell(\boldsymbol{\beta}),$$

which shows that $T_{\boldsymbol{\beta}}M$ is a linear subvector space of $T_{\boldsymbol{\beta}}S$.

Finally, to define the notion of angle between two given tangent vectors belonging to $T_{\boldsymbol{\beta}}M$, say $v_{\boldsymbol{\beta}} = \sum_{m=1}^p v_m\partial_m\ell(\boldsymbol{\beta})$ and $w_{\boldsymbol{\beta}} = \sum_{n=1}^p w_n\partial_n\ell(\boldsymbol{\beta})$, we shall use the information metric [17], in other words, the inner product between $v_{\boldsymbol{\beta}}$ and $w_{\boldsymbol{\beta}}$ is defined as

$$\langle v_{\boldsymbol{\beta}}; w_{\boldsymbol{\beta}}\rangle_{\boldsymbol{\beta}} = E_{\boldsymbol{\beta}}(v_{\boldsymbol{\beta}}w_{\boldsymbol{\beta}}) = \sum_{m,n=1}^p E_{\boldsymbol{\beta}}\{\partial_m\ell(\boldsymbol{\beta})\partial_n\ell(\boldsymbol{\beta})\}v_m w_n = \mathbf{v}^\top I(\boldsymbol{\beta})\mathbf{w}, \quad (3)$$

where $\mathbf{v} = (v_1,\ldots,v_p)^\top$, $\mathbf{w} = (w_1,\ldots,w_p)^\top$ and $I(\boldsymbol{\beta})$ is the Fisher information matrix evaluated at $\boldsymbol{\beta}$. As observed in [11], the matrix $I(\boldsymbol{\beta})$ is not exactly equal to the Fisher information matrix of the relative risk regression model; however, it has the same asymptotic properties for inference. Finally, to complete our differential geometric framework, we need to introduce the notion of tangent residual vector $r_{\boldsymbol{\beta}} = \sum_{i\in D}\sum_{h\in R(i)} r_{ih}(\boldsymbol{\beta})\partial_{ih}\ell(\boldsymbol{\beta})$, where $r_{ih}(\boldsymbol{\beta}) = y_{ih} - \pi_{ih}(\boldsymbol{\beta})$, which is an element of $T_{\boldsymbol{\beta}}S$ and can be used to measure the difference between a model in M and the observed survival data.

As shown in [1], the inner product (3) and the residual vector $r_{\boldsymbol{\beta}}$ can be used to obtain a differential geometric characterization of the classical signed Rao score test statistic for the mth regression coefficient. Formally, denoted by $r_m^u(\boldsymbol{\beta})$ the mth signed Rao score test statistic, we can show that

$$r_m^u(\boldsymbol{\beta}) = I_{mm}^{-1/2}(\boldsymbol{\beta})\partial_m\ell(\boldsymbol{\beta}) = \cos\{\rho_m(\boldsymbol{\beta})\}\|r_{\boldsymbol{\beta}}\|_{\boldsymbol{\beta}}, \quad (4)$$

where $\|r_{\beta}\|_{\beta}^{2} = \sum_{i \in D} \sum_{h,k \in R(i)} E_{\beta}\{\partial_{ih}\ell(\beta)\partial_{ik}\ell(\beta)\} r_{ih}(\beta) r_{ik}(\beta)$ and $I_{mm}(\beta)$ is the mth diagonal element of $I(\beta)$. The quantity $\rho_m(\beta)$ is a generalization of the Euclidean notion of angle between the mth predictor and the tangent residual vector r_{β}, and it is a natural and invariant quantity by means of measuring the strength of the relationship between the mth predictor and the observed data. As we shall show in the next section, characterization (4) establishes the theoretical foundation of the proposed method.

3.2 The Extension of the DgLARS Method

As formalized in [3], dgLARS is a method for constructing a path of solutions, indexed by a positive parameter γ, where the nonzero estimates of each solution can be defined as follows. For any dataset, there exists with probability one a finite decreasing sequence of transition points, denoted by $\{\gamma^{(j)}\}$, such that for any $\gamma \in (\gamma^{(j)}; \gamma^{(j-1)})$ the subvector of nonzero estimates, denoted by $\hat{\beta}_A(\gamma)$, is defined as solution of the following nonlinear equations

$$r_h(\hat{\beta}_A(\gamma)) - s_h\gamma = 0, \quad \forall h \in A, \tag{5}$$

where $A = \{h : \hat{\beta}_h(\gamma) \neq 0\}$ is called active set and $s_h = \text{sign}(r_h(\hat{\beta}_A(\gamma)))$. Furthermore, for any $k \notin A$ we have that $|r_k(\hat{\beta}(\gamma))| < \gamma$. At each transition point we have a change in the active set.

Formally, $\gamma^{(j)}$ is an inclusion transition point if exists a $k \notin A$ such that the following condition is satisfied:

$$|r_k(\hat{\beta}_A(\gamma^{(j)}))| = \gamma^{(j)}. \tag{6}$$

In this case, the active set is updated adding the index k, i.e. the kth predictor is included in the current relative risk model. To gain more insight about the geometrical foundation of the condition (6), let h be an index belonging to A. Then, using equation (5) at the inclusion transition point, we have the identity $|r_h(\hat{\beta}_A(\gamma^{(i)}))| = \gamma^{(i)}$. Combining this identity with the inclusion condition (6) we have that, at $\gamma^{(i)}$, there is a $k \notin A$ such that $|r_k(\hat{\beta}_A(\gamma^{(j)}))| = |r_h(\hat{\beta}_A(\gamma^{(j)}))|$, for any $h \in A$. Finally, using characterization (4), we can conclude that condition (6) is equivalent to

$$\cos\{\rho_k(\hat{\beta}_A(\gamma^{(j)}))\} = \cos\{\rho_h(\hat{\beta}_A(\gamma^{(j)}))\}, \tag{7}$$

for each $h \in A$ and $k \notin A$. Condition (7) is called generalized equiangularity condition [1] because it is a genuine generalization of the equiangularity condition proposed in [5] to define the least angle regression method.

$\gamma^{(j)}$ is an exclusion transition point if exists a $h \in A$ such that the following condition is satisfied:

$$\text{sign}(r_h(\hat{\beta}_A(\gamma^{(j)}))) \neq s_h. \tag{8}$$

In this case, the active set is updated removing the index h and the corresponding predictor is removed from the relative risk regression model. The exclusion condition (8) is inherited from the exclusion condition of the lasso estimator. See Sect. 5 in [5] for more details.

Given the previous definition, the path of solutions defined by the dgLARS method can be constructed in the following way. Since we are working with a class of regression models without intercept term, the starting point of the dgLARS curve is the zero vector this means that, at the starting point, the p predictors are ranked using the signed Rao score test statistics evaluated at zero. Suppose that $h = \arg\max_m |r_m^u(\mathbf{0})|$, then $A = \{h\}$, $\gamma^{(1)} = |r_h^u(\mathbf{0})|$ and the first part of the dgLARS curve is implicitly defined by the nonlinear equation $r_h^u\{\hat{\beta}_h(\gamma)\} - s_h\gamma = 0$. The proposed method traces the first part of the dgLARS curve reducing γ until we find the transition point $\gamma^{(2)}$ corresponding to the inclusion of a new index in the active set, in other words, there exists a predictor, say the kth, satisfying condition (6), then k is included in A and the new part of the dgLARS curve is implicitly defined by the system with nonlinear equations:

$$\begin{cases} r_h^u(\hat{\boldsymbol{\beta}}_A(\gamma)) - s_h\gamma = 0, \\ r_k^u(\hat{\boldsymbol{\beta}}_A(\gamma)) - s_k\gamma = 0, \end{cases}$$

where $\hat{\boldsymbol{\beta}}_A(\gamma) = (\hat{\beta}_h(\gamma), \hat{\beta}_k(\gamma))^\top$. The second part is computed reducing γ and solving the previous system until we find the transition point $\gamma^{(3)}$. At this point, if condition (6) occurs a new index is included in A otherwise condition (8) occurs and an index is removed from A. In the first case, the previous system is updated adding a new nonlinear equation while, in the second case, a nonlinear equation is removed. The curve is traced as previously described until parameter γ is equal to some fixed value that can be zero, if the sample size is large enough, or some positive value, if we are working in a high-dimensional setting. Table 2 reports the pseudocode of the developed algorithm to compute the dgLARS curve for a relative risk regression model.

Table 2 Pseudocode of the dgLARS algorithm for a relative risk regression model

Step	Description
0.	Let $r_m^u(\boldsymbol{\beta})$ be the Rao score statistic associated with the partial likelihood.
1.	Let $\gamma^{(1)} = \max_m |r_m^u(\mathbf{0})|$ and initialize the active set $A = \arg\max_m |r_m^u(\mathbf{0})|$
2.	Repeat the following steps
3.	Trace the segment of the dgLARS curve reducing γ and solving the system
	$r_h^u\{\hat{\boldsymbol{\beta}}_A(\gamma)\} - s_h\gamma = 0, \quad h \in A$
4.	Until γ is equal to the next transition point
5.	If condition (6) is met then include the new index in A
6.	Else (condition (8) is met) remove the index from A
7.	Until γ reaches some small positive value

From a computational point of view, the entire dgLARS curve can be computed using the predictor-corrector algorithm proposed in [1]; for more details about this algorithm, the interested reader is referred to [2, 3, 14].

4 Simulation Studies: Comparison with Other Variable Selection Methods

In this section, we compare the proposed method with three popular variable selection methods: the coordinate descent method [19], named CoxNet; the predictor-corrector method [13], named CoxPath; and the gradient ascent algorithm [9], named CoxPen. These methods are implemented in the R packages `glmnet`, `glmpath` and `penalized`, respectively. Since these methods have only been implemented for Cox regression model, our comparison will focus on this kind of relative risk regression model. In the following of this section, dgLARS method applied to the Cox regression model is referred to as the dgCox model.

The simulation study is based on the following setting. First, we simulated 100 datasets from a Cox regression model where the survival times t_i $(i = 1, \ldots, n)$ follow exponential distributions with parameter $\lambda_i = \exp(\boldsymbol{x}_i^\top \boldsymbol{\beta})$, and $\boldsymbol{x}_i$ is sampled from a p-variate normal distribution $\mathrm{N}(\mathbf{0}, \Sigma)$; the entries of Σ are fixed to $\mathrm{corr}(X_m, X_n) = \rho^{|m-n|}$ with $\rho \in \{0.3, 0.6, 0.9\}$. The censorship is randomly assigned to the survival times with probability $\pi \in \{0.2, 0.4\}$. The number of predictors is equal to 100 and the sample size is equal to 50 and 150. The first value is used to evaluate the behaviour of the methods in a high-dimensional setting. Finally, we set $\beta_m = 0.2$ for $m = 1, \ldots, s$, where $s \in \{5, 10\}$; the remaining parameters are equal to zero.

In order to study the global behaviour of each method, we use the following approach. First, we fitted the models using a sequence of 50 values for the tuning parameter; then, for each fitted model, we computed the false and true positive rate. These quantities are used to compute the ROC curve. A method is declared globally preferable, in the sense that it overcomes the other competitors for any value of the tuning parameter, if its ROC curve is above the others. Table 3 reports some summary measures: for each scenario, we compute the average area under the curve (AUC), the average false positive rate (FPR) and the average true positive rate (TPR). In scenarios where $\rho = 0.3$, CoxNet, CoxPath, and CoxPen exhibit a similar performance, having overlapping curves for both levels of censoring, whereas dgCox method appears to be consistently better with the largest AUC. A similar performance of the methods has been also observed for the other combinations of ρ and π values. In scenarios where the correlation among neighbouring predictors is high ($\rho = 0.9$), the dgCox method is clearly the superior approach for all levels of censoring.

Table 3 Comparison between the considered variable selection models. For each scenario, the variable selection models are evaluated using the average area under the curve (AUC), the average false positive rate (FPR) and the average true positive rate (TPR). First and second part of the table are refereed to the simulation study with sample size equal to 50 and 150, respectively

	s	5						10					
	ρ	0.3		0.6		0.9		0.3		0.6		0.9	
	π	0.2	0.4	0.2	0.4	0.2	0.4	0.2	0.4	0.2	0.4	0.2	0.4
dgCOX	AUC	0.78	0.70	0.82	0.76	0.78	0.74	0.74	0.71	0.80	0.75	0.77	0.77
	FPR	0.19	0.21	0.16	0.19	0.11	0.13	0.18	0.20	0.14	0.17	0.09	0.10
	TPR	0.54	0.48	0.63	0.57	0.53	0.51	0.48	0.48	0.56	0.52	0.52	0.49
CoxNet	AUC	0.72	0.68	0.79	0.71	0.73	0.68	0.71	0.69	0.75	0.70	0.75	0.69
	FPR	0.30	0.27	0.27	0.25	0.21	0.21	0.29	0.26	0.25	0.23	0.18	0.18
	TPR	0.57	0.53	0.63	0.54	0.50	0.46	0.45	0.53	0.56	0.51	0.47	0.45
CoxPath	AUC	0.72	0.68	0.78	0.71	0.73	0.68	0.71	0.69	0.76	0.70	0.74	0.68
	FPR	0.24	0.22	0.23	0.21	0.20	0.19	0.23	0.21	0.21	0.20	0.19	0.18
	TPR	0.57	0.50	0.65	0.54	0.52	0.47	0.54	0.50	0.58	0.52	0.49	0.47
CoxPen	AUC	0.71	0.69	0.76	0.70	0.72	0.66	0.71	0.68	0.74	0.69	0.75	0.69
	FPR	0.12	0.11	0.09	0.09	0.04	0.05	0.10	0.10	0.08	0.08	0.04	0.04
	TPR	0.42	0.36	0.48	0.41	0.41	0.36	0.37	0.34	0.40	0.35	0.36	0.32
dgCOX	AUC	0.90	0.85	0.90	0.89	0.83	0.80	0.90	0.83	0.90	0.87	0.85	0.80
	FPR	0.33	0.22	0.26	0.27	0.14	0.14	0.32	0.22	0.25	0.27	0.12	0.13
	TPR	0.79	0.69	0.77	0.76	0.65	0.61	0.76	0.65	0.75	0.75	0.64	0.58
CoxNet	AUC	0.88	0.83	0.87	0.84	0.76	0.72	0.88	0.81	0.88	0.84	0.78	0.71
	FPR	0.32	0.36	0.34	0.31	0.39	0.32	0.30	0.36	0.35	0.31	0.38	0.30
	TPR	0.88	0.89	0.85	0.86	0.81	0.79	0.85	0.88	0.81	0.88	0.79	0.76
CoxPath	AUC	0.68	0.83	0.87	0.85	0.77	0.73	0.88	0.81	0.88	0.84	0.77	0.72
	FPR	0.30	0.29	0.28	0.28	0.26	0.26	0.28	0.30	0.27	0.28	0.25	0.25
	TPR	0.82	0.75	0.81	0.77	0.67	0.62	0.79	0.73	0.79	0.77	0.67	0.59
CoxPen	AUC	0.87	0.83	0.86	0.84	0.76	0.72	0.88	0.81	0.88	0.83	0.78	0.72
	FPR	0.16	0.17	0.11	0.12	0.05	0.05	0.15	0.17	0.09	0.12	0.04	0.04
	TPR	0.82	0.57	0.62	0.59	0.50	0.48	0.56	0.55	0.56	0.59	0.47	0.42

5 Finding Genetic Signatures in Cancer Survival

In this section, we test the predictive power of the proposed method in two recent studies. In particular, we focus on the identification of genes involved in the regulation of prostate cancer [18] and ovarian cancer [8]. The setup of the two studies was similar. In the patient, cancer was detected and treated. When treatment was complete a follow-up started. In all cases, the expression of several genes was measured in the affected tissue together with the survival times of the patients, which may be censored if the patients were alive when they left the study. Although other socio-economical

Table 4 Description of the studied datasets and summary of the main results

Dataset	Sample size	n. uncensored	p	n. selected genes	p-value
Prostate	61	24	162	24	0.033
Ovarian	103	57	306	43	0.004

variables, such as age, sex, and so on, are available, our analysis only focuses on the impact of the gene expression levels on the patients' survival.

Table 4 contains a brief description of the two datasets used in this section. In each case, the number of predictors is larger than the number of patients. In genomics, it is common to assume that just a moderate number of genes affect the phenotype of interest. To identify such genes in this survival context, we estimate a Cox regression model using the dgLARS method. We randomly select a training sample that contains the 60% of the patients and we save the remaining data to test the models. We calculate the paths of solutions in the two cases and we select the optimal number of components by means of the GIC criterion. For the prostate and ovarian studies, we find gene profiles consisting of, respectively, 24 and 43 genes.

In order to illustrate the prediction performance of the dgLARS method, we classify the test patients into a low-risk group and a high-risk group by splitting the test sample into two subsets of equal size according to the estimated individual predicted excess risk. To test the group separation, we use a non-parametric modification of the Gehan–Wilcoxon test [15]. For the two studies, the difference between the low- and high-risk groups is significant at the traditional 0.05 significance level.

6 Conclusions

In this paper, we have proposed an extension of the differential geometric least angle regression method to relative risk regression models using the relationship existing between the partial likelihood function and a specific generalized linear model. The advantage of this approach is that the estimates are invariant to arbitrary changes in the measurement scales of the predictors. Unlike SCAD or ℓ_1 sparse regression methods, no prior rescaling of the predictors is needed. The proposed method can be used for a large class of survival models, the so-called relative risk models. We have code for the Cox proportional hazards model and the excess relative risk model.

References

1. Augugliaro, L., Mineo, A.M., Wit, E.C.: Differential geometric least angle regression: a differential geometric approach to sparse generalized linear models. J. R. Stat. Soc. Ser. B **75**, 471–498 (2013)
2. Augugliaro, L., Mineo, A.M., Wit, E.C.: dglars: an R package to estimate sparse generalized linear models. J. Stat. Soft. **59**, 1–40 (2014)
3. Augugliaro, L., Mineo, A.M., Wit, E.C.: A differential geometric approach to generalized linear models with grouped predictors. Biometrika **103**, 563–577 (2016)
4. Cox, D.R.: Regression models and life-tables. J. R. Stat. Soc. Ser. B **34**, 187–220 (1972)
5. Efron, B., Hastie, T., Johnstone, I., Tibshirani, R.: Least angle regression. Ann. Stat. **32**, 407–499 (2004)
6. Fan, J., Li, R.: Variable selection via nonconcave penalized likelihood and its oracle properties. J. Am. Stat. Assoc. **96**, 1348–1360 (2001)
7. Fan, Y., Tang, C.Y.: Tuning parameter selection in high dimensional penalized likelihood. J. R. Stat. Soc. Ser. B **75**, 531–552 (2013)
8. Gillet, J.P., Calcagno, A.M., Varma, S., Davison, B., Elstrand, M.B., Ganapathi, R., Kamat, A.A., Sood, A.K., Ambudkar, S.V., Seiden, M.V., others: Multidrug resistance-linked gene signature predicts overall survival of patients with primary ovarian serous carcinoma. Clin. Cancer Res. **18**, 3197–3206 (2012)
9. Goemann, J.J.: L1 penalized estimation in the Cox proportional hazards model. Biom. J. **52**, 70–84 (2010)
10. Kalbfleisch, J.D., Prentice, R.L.: The Statistical Analysis of Failure Time Data. Wiley Series in Probability and Statistics. Wiley, Hoboken, New Jersey (2002)
11. Moolgavkar, S.H., Venzon, D.J.: Confidence regions in curved exponential families: application to matched case-control and survival studies with general relative risk function. Ann. Stat. **15**, 346–359 (1987)
12. Oakes, D.: Survival times: aspects of partial likelihood. Int. Stat. Rev. **49**, 235–252 (1981)
13. Park, M.Y., Hastie, T.: L1-regularization path algorithm for generalized linear models. J. R. Stat. Soc. Ser. B **69**, 659–677 (2007)
14. Pazira, H., Augugliaro, L., Wit, E.C.: Extended differential geometric lars for high-dimensional GLMs with general dispersion parameter. Stat. Comput. **28**, 753–774 (2018)
15. Peto, R., Peto, J.: Asymptotically efficient rank invariant test procedures. J. R. Stat. Soc. Ser. A **13**, 185–207 (1972)
16. Prentice, R.L., Breslow, N.E.: Retrospective studies and failure time models. Biometrika **65**, 153–158 (1978)
17. Rao, C.R.: On the distance between two populations. Sankhyā **9**, 246–248 (1949)
18. Ross, R.W., Galsky, M.D., Scher, H.I., Magidson, J., Wassmann, K., Lee, G.S.M., Katz, L., Subudhi, S.K., Anand, A., Fleisher, M., Kantoff, P.W., others.: A whole-blood RNA transcript-based prognostic model in men with castration-resistant prostate cancer: a prospective study. Lancet Oncol. **13**, 1105–1113 (2012)
19. Simon, N., Friedman, J.H., Hastie, T., Tibshirani, R.: Regularization paths for Cox's proportional hazards model via coordinate descent. J. Stat. Soft. **39**, 1–13 (2011)
20. Thomas, D.C.: Addendum to the paper by Liddell, McDonald, Thomas and Cunliffe. J. R. Stat. Soc. Ser. A **140**, 483–485 (1977)
21. Thomas, D.C: General relative-risk models for survival time and matched case-control analysis. Biometrics **37**, 673–686 (1981)
22. Tibshirani, R.: Regression shrinkage and selection via the lasso. J. R. Stat. Soc. Ser. B **58**, 267–288 (1996)
23. Tibshirani, R.: The lasso method for variable selection in the Cox model. Stat. Med. **16**, 385–395 (1997)
24. Zou, H., Hastie, T.: Regularization and variable selection via the elastic net. J. R. Stat. Soc. Ser. B. **67**, 301–320 (2005)

A Kernel Goodness-of-fit Test for Maximum Likelihood Density Estimates of Normal Mixtures

Dimitrios Bagkavos and Prakash N. Patil

Abstract This article contributes a methodological advance so as to help practitioners decide in selecting between parametric and nonparametric estimates for mixtures of normal distributions. In order to facilitate the decision, a goodness-of-fit test based on the integrated square error difference between the classical kernel density and the maximum likelihood estimates is introduced. Its asymptotic distribution under the null is quantified analytically and a hypothesis test is then developed so as to help practitioners choose between the two estimation options. The article concludes with an example which exhibits the operational characteristics of the procedure.

Keywords Goodness-of-fit · Normal mixtures · Kernel smoothing

1 Introduction

The choice between parametric and nonparametric density estimates is a topic frequently encountered by practitioners. The parametric (maximum likelihood, ML) approach is a natural first choice under strong evidence about the underlying density. However, estimation of normal mixture densities with unknown number of mixture components can become very complicated. Specifically, misidentification of the number of components greatly impairs the performance of the ML estimate and acts incrementally to the usual convergence issues of this technique, e.g., [11]. A robust nonparametric alternative, immune to the above problems is the classical kernel density estimate (kde).

The purpose of this work is to investigate under which circumstances one would prefer to employ the ML or the kde. A goodness-of-fit test is introduced based on the

D. Bagkavos (✉)
Department of Mathematics, University of Ioannina, 45100 Ioannina, Greece
e-mail: dimitrios.bagkavos@gmail.com

P. N. Patil
Department of Mathematics and Statistics, Mississippi State University, Starkville, Mississippi, USA

M. La Rocca et al. (eds.), *Nonparametric Statistics*, Springer Proceedings in Mathematics & Statistics 339, https://doi.org/10.1007/978-3-030-57306-5_7

Integrated Squared Error (ISE) which measures the distance between the true curve and the proposed parametric model. Section 2 introduces the necessary notation and formulates the goodness-of-fit test. Its asymptotic distribution is discussed in Sect. 3 together with the associated criteria for acceptance or rejection of the null. An example is provided in Sect. 4. All proofs are deferred to the last Section.

2 Setup and Notation

Let ϕ denote the standard normal density and $\phi_\sigma(x) = \sigma^{-1}\phi(x\sigma^{-1})$ its scaled version. Let $\boldsymbol{\mu} = (\mu_1, \ldots, \mu_k)$ where for each $\mu_i \in \boldsymbol{\mu}$, $-\infty < \mu_i < +\infty$ and $\boldsymbol{\sigma} = (\sigma_1, \ldots, \sigma_k)$ where each $\sigma_i > 0$. Let also $\boldsymbol{w} = (w_1, w_2, \ldots, w_k)$ be a vector of positive parameters summing to one. The finite positive integer k denotes the number of mixing components. Then,

$$f(x; \boldsymbol{\mu}, \boldsymbol{\sigma}, \boldsymbol{w}) = \sum_{l=1}^{k} w_l \phi_{\sigma_l}(x - \mu_l) \tag{1}$$

is a normal mixture density with location parameter $\boldsymbol{\mu}$, scale parameter $\boldsymbol{\sigma}$, and mixing parameter $\boldsymbol{w}$. The number of mixing components k is estimated prior and separately to estimation of $(\boldsymbol{\mu}, \boldsymbol{\sigma}, \boldsymbol{w})$. Thus it is considered as a fixed constant in the process of ML estimation. Popular estimation methods for k include clustering as in [14] or by multimodality hypothesis testing as in [6] among many others. Regarding $\boldsymbol{\mu}, \boldsymbol{\sigma}, \boldsymbol{w}$, these are considered to belong to the parameter space Ω defined by

$$\Omega = \left\{ \boldsymbol{\mu}, \boldsymbol{\sigma}, \boldsymbol{w} : \sum_{i=1}^{k} w_i = 1, w_i \geq 0, \mu_i \in \mathbb{R}, \sigma_i \geq 0 \text{ for } i = 1, \ldots, k \right\}.$$

The analysis herein assumes that all estimates are based on a random sample $X_1, X_2, \ldots, X_n$ from $f(x; \boldsymbol{\mu}, \boldsymbol{\sigma}, \boldsymbol{w})$. The parametric MLE is denoted by

$$\hat{f}(x; \hat{\boldsymbol{\mu}}, \hat{\boldsymbol{\sigma}}, \hat{\boldsymbol{w}}) = \sum_{l=1}^{k} \hat{w}_l \phi_{\hat{\sigma}_l}(x - \hat{\mu}_l), \tag{2}$$

where $(\hat{\boldsymbol{\mu}}, \hat{\boldsymbol{\sigma}}, \hat{\boldsymbol{w}})$ denote the estimates of $(\boldsymbol{\mu}, \boldsymbol{\sigma}, \boldsymbol{w})$ resulting by maximization of

$$l(\boldsymbol{\mu}, \boldsymbol{\sigma}, \boldsymbol{w}) = \sum_{i=1}^{n} \log \left\{ \sum_{j=1}^{k} w_l \phi_{\sigma_l}(X_i - \mu_l) \right\}, \tag{3}$$

subject to $(\boldsymbol{\mu}, \boldsymbol{\sigma}, \boldsymbol{w}) \in \Omega$. Direct estimation of the density parameters by maximum likelihood is frequently problematic as (3) is not bounded on the parameter space, see [3]. In spite of this, statistical theory guarantees that one local maximizer of the likelihood exists at least for small number of mixtures, e.g., [8] for $k = 2$. Moreover this maximizer is strongly consistent and asymptotically efficient. Several local maximizers can exist for a given sample, and the other major maximum likelihood difficulty is in determining when the correct one has been found. Obviously all these issues, i.e., correct estimation of k, existence and identification of an optimal solution for (3), result in the ML estimation process to perform frequently poorly in practice. A natural alternative is the classical kernel estimate of the underlying density which is given by

$$\hat{f}(x; h) = (nh)^{-1} \sum_{i=1}^{n} K\left\{(x - X_i)h^{-1}\right\}, \tag{4}$$

where h called *bandwidth* controls the amount of smoothing applied to the estimate and K, called *kernel*, is a real function integrating to 1. Attention here is restricted to second-order kernels as from [10], and it is known that using higher order kernels bears little improvement for moderate sample sizes. In estimating $f(x; \boldsymbol{\mu}, \boldsymbol{\sigma}, \boldsymbol{w})$ by $\hat{f}(x; h)$, especially when $K = \phi$, the MISE of the estimate can be quantified explicitly. The purpose of this research is to develop a goodness-of-fit test for

$$H_0 : f(x) = f(x; \boldsymbol{\mu}, \boldsymbol{\sigma}, \boldsymbol{w}) \text{ vs } H_1 : f(x) \neq f(x; \boldsymbol{\mu}, \boldsymbol{\sigma}, \boldsymbol{w}).$$

Its construction is based on the integrated square error of $f(x; \tilde{\boldsymbol{\mu}}, \tilde{\boldsymbol{\sigma}}, \tilde{\boldsymbol{w}})$ given by

$$I_n = \int \left(f(x) - f(x; \tilde{\boldsymbol{\mu}}, \tilde{\boldsymbol{\sigma}}, \tilde{\boldsymbol{w}})\right)^2 \mathrm{d}x,$$

where $\{\boldsymbol{\mu}, \boldsymbol{\sigma}, \boldsymbol{w}\} = \{\tilde{\boldsymbol{\mu}}, \tilde{\boldsymbol{\sigma}}, \tilde{\boldsymbol{w}}\}$ under H_0. Estimation of $f(x)$ by a kernel estimate and $f(x; \tilde{\boldsymbol{\mu}}, \tilde{\boldsymbol{\sigma}}, \tilde{\boldsymbol{w}})$ by $\hat{f}(x; \hat{\boldsymbol{\mu}}, \hat{\boldsymbol{\sigma}}, \hat{\boldsymbol{w}})$ yields the estimate, $\hat{I}_n$ of I_n, defined by

$$\begin{aligned}\hat{I}_n &= \int \left(\hat{f}(x; h) - f(x; \hat{\boldsymbol{\mu}}, \hat{\boldsymbol{\sigma}}, \hat{\boldsymbol{w}})\right)^2 dx \\ &\equiv \int \hat{f}^2(x; \hat{\boldsymbol{\mu}}, \hat{\boldsymbol{\sigma}}, \hat{\boldsymbol{w}})\, \mathrm{d}x - 2\int \hat{f}(x; \hat{\boldsymbol{\mu}}, \hat{\boldsymbol{\sigma}}, \hat{\boldsymbol{w}})\hat{f}(x; h)\, dx + \int \hat{f}^2(x; h)\, \mathrm{d}x. \end{aligned} \tag{5}$$

For $K = \phi$ by Corollary 5.2 in [1],

$$\int \hat{f}^2(x; \hat{\boldsymbol{\mu}}, \hat{\boldsymbol{\sigma}}, \hat{\boldsymbol{w}})\, dx = \sum_{l=1}^{k}\sum_{r=1}^{k} \hat{w}_l \hat{w}_r \phi_{(\hat{\sigma}_l^2 + \hat{\sigma}_r^2)^{\frac{1}{2}}}(\hat{\mu}_l - \hat{\mu}_r). \tag{6}$$

$$\int \hat{f}^2(x; h)\, dx = \int \left\{(nh)^{-1} \sum_{i=1}^{n} \phi(x - X_i)\right\}^2 dx = (nh)^{-2} \sum_{i=1}^{n}\sum_{j=1}^{n} \phi_{\sqrt{2}}(X_i - X_j). \tag{7}$$

Similarly,

$$\int \hat{f}(x; \hat{\boldsymbol{\mu}}, \hat{\boldsymbol{\sigma}}, \hat{\boldsymbol{w}})\hat{f}(x; h)\, dx = (nh)^{-1} \sum_{i=1}^{n} \sum_{l=1}^{k} \phi_{(\hat{\sigma}_l^2+1)^{\frac{1}{2}}}(X_i - \mu_l). \tag{8}$$

Using (6), (7), and (8) back to (5) gives

$$\hat{I}_n = \sum_{l=1}^{k} \sum_{r=1}^{k} \hat{w}_l \hat{w}_r \phi_{(\hat{\sigma}_l^2+\hat{\sigma}_r^2)^{\frac{1}{2}}}(\hat{\mu}_l - \hat{\mu}_r) - 2(nh)^{-1} \sum_{i=1}^{n} \sum_{l=1}^{k} \phi_{(\hat{\sigma}_l^2+1)^{\frac{1}{2}}}(X_i - \hat{\mu}_l) + (nh)^{-2} \sum_{i=1}^{n} \sum_{j=1}^{n} \phi_{\sqrt{2}}(X_i - X_j), \tag{9}$$

which is an equivalent expression for $\hat{I}_n$ that does not require integration.

3 Distribution of $\hat{I}_n$ Under the Null

This section establishes the null distribution of the test statistic I_n. First, the following assumptions are introduced,

1. $h \to 0$ and $nh^2 \to +\infty$ as $n \to +\infty$.
2. The density $f(x; \boldsymbol{\mu}, \boldsymbol{\sigma}, \boldsymbol{w})$ and its parametric estimate $\hat{f}(x; \hat{\boldsymbol{\mu}}, \hat{\boldsymbol{\sigma}}, \hat{\boldsymbol{w}})$ are bounded, and their first two derivatives exist and are bounded and uniformly continuous on the real line.
3. Let $\boldsymbol{s}$ be any of the estimated vectors $\boldsymbol{\mu}, \boldsymbol{\sigma}, \boldsymbol{w}$ and let $\hat{\boldsymbol{s}}$ denote its estimate. Then, there exists a $\boldsymbol{s}^*$ such that $\boldsymbol{s} \to \boldsymbol{s}^*$ almost surely and

$$\boldsymbol{s} - \boldsymbol{s}^* = n^{-1} A(\boldsymbol{s}^*) \sum_{i=1}^{n} D \log f(X_i; \boldsymbol{s}^*) + o_p(n^{-1/2})$$

where $D \log f(X_i; \boldsymbol{s}^*)$ is a vector of the first derivatives of $\log f(X_i; \boldsymbol{s}^*)$ with respect to s_j and evaluated at s_j^* while

$$A(\boldsymbol{s}^*) = \mathbb{E}\left(\frac{\partial^2 \log f(X_i; \boldsymbol{s}^*)}{\partial s_j \partial s_j'}\bigg|_{\boldsymbol{s}=\boldsymbol{s}^*} \right).$$

Theorem 1 *Under assumptions 1–3 and under the null hypothesis,*

$$d(n)\left(\hat{I}_n - c(n)\right) \to \begin{cases} (\sigma_1^2 - \sigma_{30}^2)^{\frac{1}{2}} Z & \text{if } nh^5 \to \infty \\ 2^{1/2}\sigma_2 Z & \text{if } nh^5 \to 0 \\ \left\{\lambda^{\frac{1}{2}}(\sigma_1^2 - \sigma_{30}^2)\lambda^{\frac{4}{5}} + 2\lambda^{-\frac{1}{5}}\sigma_2^2\right\}^{\frac{1}{2}} Z & \text{if } nh^5 \to \lambda \end{cases} \tag{10}$$

with $0 < \lambda < +\infty$,

$$c(n) = \frac{1}{nh}\frac{1}{2\sqrt{\pi}} + \frac{h^4}{4}\left\{\sum_{l=1}^{k}\sum_{r=1}^{k} w_l w_r \phi^{(4)}_{(\sigma_l^2+\sigma_r^2)^{\frac{1}{2}}}(\mu_l - \mu_r)\right\} + o(h^4)$$

$$\sigma_1^2 = \int \{f''(x)\}^2 f(x)\,dx - \left\{\int f''(x) f(x)\,dx\right\}^2 = \mathrm{Var}\{f''(x)\}$$

$$\sigma_2^2 = \frac{1}{2\sqrt{2\pi}}\sum_{l=1}^{k}\sum_{r=1}^{k} w_l w_r \phi_{(\sigma_l^2+\sigma_r^2)^{\frac{1}{2}}}(\mu_l - \mu_r)$$

$$\sigma_{30}^2 = \sigma_{30}^2(\boldsymbol{\mu}, \boldsymbol{\sigma}, \boldsymbol{w}) = \left[\int D' f_0(x, \boldsymbol{\mu}, \boldsymbol{\sigma}, \boldsymbol{w}) f''(x)\,dx\right] A(\boldsymbol{\mu}, \boldsymbol{\sigma}, \boldsymbol{w})^{-1} \times \left[\int D f_0(x, \boldsymbol{\mu}, \boldsymbol{\sigma}, \boldsymbol{w}) f''(x)\,dx\right]$$

and

$$d(n) = \begin{cases} nh^{1/2} & \textit{if } nh^5 \to 0 \\ n^{1/2}h^{-2} & \textit{if } nh^5 \to +\infty \\ n^{9/10} & \textit{if } nh^5 \to \lambda \neq 0. \end{cases}$$

Thus, in testing H_0 against H_1 with significance level α we have

$$\hat{I}_n / \sqrt{\mathbb{V}\mathrm{ar}(\hat{I}_n)} \to N(0, 1),$$

where

$$\mathbb{V}\mathrm{ar}(\hat{I}_n) = \begin{cases} \sigma_1^2 - \sigma_{30}^2 & \text{if } nh^5 \to \infty \\ (2^{1/2}\sigma_2)^2 & \text{if } nh^5 \to 0 \\ \lambda^{\frac{1}{2}}(\sigma_1^2 - \sigma_{30}^2)\lambda^{\frac{4}{5}} + 2\lambda^{-\frac{1}{5}}\sigma_2^2 & \text{if } nh^5 \to \lambda. \end{cases}$$

Consequently, the test suggests rejection of H_0 when

$$\hat{I}_n \left\{\mathbb{V}\mathrm{ar}(\hat{I}_n)\right\}^{-1/2} > z_\alpha,$$

where z_α is the standard normal quantile at level α. Of course rejection of H_0 advises for using a kernel estimate instead of (2) for estimation of the underlying density.

4 An Example

As an illustrative example, the Galaxies data of [14] are used. The data represent velocities in km/sec of 82 galaxies from 6 well-separated conic sections of an unfilled survey of the Corona Borealis region. Multimodality in such surveys is evidence for voids and superclusters in the far universe.

The hypothesis that $k = 6$ is also verified by the multimodality test of [6] and thus it is adopted in the present example as well. Figure 1 contains the ML (solid line) and kernel (dashed red line) estimates after scaling the data by 1000. The null hypothesis of goodness-of-fit of the ML estimate was tested at 5% significance level, using as variance the third component of the variance expression. The test procedure gives

$$\hat{I}_n \left\{ \mathbb{V}\mathrm{ar}(\hat{I}_n) \right\}^{-1/2} = 1.98 > z_{0.95} = 1.64$$

and therefore suggests rejection of the null. This is also supported by Fig. 1 where it is seen that two distinctive patterns around $x = 18$ and $x = 24$ (and one less distinctive at around $x = 28$) are masked by the ML estimate. On the contrary, the fixed bandwidth estimate $\hat{f}(x; h)$ implemented with the Sheather–Jones bandwidth can detect the change in the pattern of the density. It is worth noting that the variable bandwidth estimate $\tilde{f}(x)$ has also been tested with the specific data set and found to perform very similarly to $\hat{f}(x; h)$.

5 Proof of Theorem 1

Write

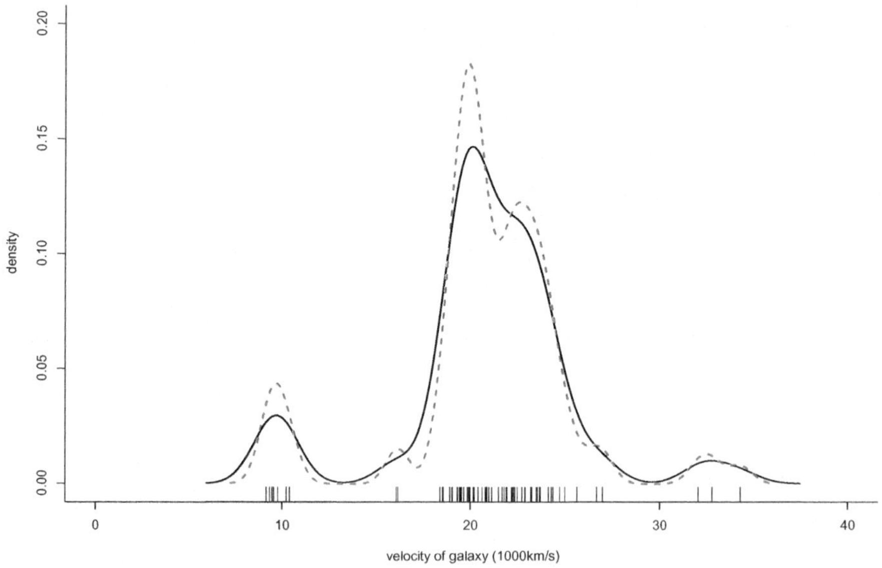

Fig. 1 Variable bandwidth and ML estimates for the Galaxies data

$$
\begin{aligned}
\hat{I}_n &= \int \left\{\hat{f}(x; \hat{\boldsymbol{\mu}}, \hat{\boldsymbol{\sigma}}, \hat{\boldsymbol{w}}) - \hat{f}(x; h)\right\}^2 dx \\
&= \int \left\{\hat{f}(x; \boldsymbol{\mu}, \boldsymbol{\sigma}, \boldsymbol{w}) - f(x; \boldsymbol{\mu}, \boldsymbol{\sigma}, \boldsymbol{w}) + f(x; \boldsymbol{\mu}, \boldsymbol{\sigma}, \boldsymbol{w}) - \hat{f}(x; h)\right\}^2 \mathrm{d}x \\
&= \int \left\{\hat{f}(x; h) - f(x; \boldsymbol{\mu}, \boldsymbol{\sigma}, \boldsymbol{w})\right\}^2 \mathrm{d}x \\
&\quad - 2 \int \left\{\hat{f}(x; h) - f(x; \boldsymbol{\mu}, \boldsymbol{\sigma}, \boldsymbol{w})\right\} \left\{\hat{f}(x; \hat{\boldsymbol{\mu}}, \hat{\boldsymbol{\sigma}}, \hat{\boldsymbol{w}}) - f(x; \boldsymbol{\mu}, \boldsymbol{\sigma}, \boldsymbol{w})\right\} \mathrm{d}x \\
&\quad + \int \left\{\hat{f}(x; \hat{\boldsymbol{\mu}}, \hat{\boldsymbol{\sigma}}, \hat{\boldsymbol{w}}) - f(x; \boldsymbol{\mu}, \boldsymbol{\sigma}, \boldsymbol{w})\right\}^2 \mathrm{d}x \\
&\equiv I_1 - 2I_2 + I_3.
\end{aligned} \tag{11}
$$

Now, under H_0,

$$
\begin{aligned}
I_3 &= \int \left\{\hat{f}(x; \hat{\boldsymbol{\mu}}, \hat{\boldsymbol{\sigma}}, \hat{\boldsymbol{w}}) - f(x; \boldsymbol{\mu}, \boldsymbol{\sigma}, \boldsymbol{w})\right\}^2 \mathrm{d}x \\
&= \int \left\{ \sum_{i=1}^{k} \frac{w_i}{2\sigma_i^3} \left[(x - \mu_i)^2 - (x - \hat{\mu}_i)^2 \right] (1 + o_p(n^{-1})) \right\}^2 \mathrm{d}x = o_p(n^{-1}),
\end{aligned} \tag{12}
$$

since under the null the parameters of the normal converge to the true values. Also,

$$
\begin{aligned}
I_2 &= \int \left\{\hat{f}(x; h) - f(x; \boldsymbol{\mu}, \boldsymbol{\sigma}, \boldsymbol{w})\right\} \left\{\hat{f}(x; \hat{\boldsymbol{\mu}}, \hat{\boldsymbol{\sigma}}, \hat{\boldsymbol{w}}) - f(x; \boldsymbol{\mu}, \boldsymbol{\sigma}, \boldsymbol{w})\right\} \mathrm{d}x \\
&= \int \left\{\hat{f}(x; h) - f(x; \boldsymbol{\mu}, \boldsymbol{\sigma}, \boldsymbol{w})\right\} \left\{\mathbb{E}\hat{f}(x; \hat{\boldsymbol{\mu}}, \hat{\boldsymbol{\sigma}}, \hat{\boldsymbol{w}}) - f(x; \boldsymbol{\mu}, \boldsymbol{\sigma}, \boldsymbol{w})\right\} \mathrm{d}x + o_p(n^{-1}) \\
&\equiv J_2 + o_p(n^{-1}).
\end{aligned} \tag{13}
$$

In (13), we used that under the null

$$
\sup_x |\hat{f}(x; \hat{\boldsymbol{\mu}}, \hat{\boldsymbol{\sigma}}, \hat{\boldsymbol{w}}) - \mathbb{E}\hat{f}(x; \hat{\boldsymbol{\mu}}, \hat{\boldsymbol{\sigma}}, \hat{\boldsymbol{w}})| = o_p(n^{-1}).
$$

Thus, using (12) and (13) back to (11) yields the asymptotically equivalent expression for I_n

$$
\hat{I}_n \equiv I_1 - 2J_2 + o_p(n^{-1}). \tag{14}
$$

Now,

$$I_1 = \int \left\{\hat{f}(x;h) - \mathbb{E}\hat{f}(x;h) + \mathbb{E}\hat{f}(x;h) - f(x;\boldsymbol{\mu},\boldsymbol{\sigma},\mathbf{w})\right\}^2 \mathrm{d}x$$

$$= (nh)^{-2}\sum_{i=1}^{n}\sum_{j=1}^{n} H(X_i,X_j) + \frac{h^4}{4}\mu_2(K)R(f'')$$

$$-2\int \left\{\hat{f}(x;h) - \mathbb{E}\hat{f}(x;h)\right\}\left\{\mathbb{E}\hat{f}(x;h) - f(x;\boldsymbol{\mu},\boldsymbol{\sigma},\mathbf{w})\right\} \mathrm{d}x \quad (15)$$

after using the squared bias expression of $\hat{f}$ from [10], and

$$H(X_i,X_j) = \int \left\{K\left(\frac{x-X_i}{h}\right) - \mathbb{E}K\left(\frac{x-X_i}{h}\right)\right\}\left\{K\left(\frac{x-X_j}{h}\right) - \mathbb{E}K\left(\frac{x-X_j}{h}\right)\right\} \mathrm{d}x.$$

Using the fact that K is a symmetric kernel and separating out the diagonal terms in the double sum in (15) we can write

$$(nh)^{-2}\sum_{i=1}^{n}\sum_{j=1}^{n} H(X_i,X_j) = (nh)^{-2}\sum\sum_{1\le i<j\le n} H(X_i,X_j) + (nh)^{-1}R(K). \quad (16)$$

By (15) and (16),

$$I_{1n} - \frac{h^4}{4}\mu_2(K)R(f'') - \frac{1}{nh}R(K) \equiv I_{1n} - c(n) = (nh)^{-2}\sum_{i=1}^{n}\sum_{j=1}^{n} H(X_i,X_j) \quad (17)$$

$$-2\int \left\{\hat{f}(x;h) - \mathbb{E}\hat{f}(x;h)\right\}\left\{\mathbb{E}\hat{f}(x;h) - f(x;\boldsymbol{\mu},\boldsymbol{\sigma},\mathbf{w})\right\} \mathrm{d}x. \quad (18)$$

Combining (14) and (18) and rearranging yields

$$\hat{I}_n - c(n) = (nh)^{-2}\sum\sum_{i<j} H(X_i,X_j) \quad (19)$$

$$+2\int \left[\left\{\hat{f}(x;h) - \mathbb{E}\hat{f}(x;h)\right\} - \left\{\hat{f}(x;\hat{\boldsymbol{\mu}},\hat{\boldsymbol{\sigma}},\hat{\mathbf{w}}) - f(x;\boldsymbol{\mu},\boldsymbol{\sigma},\mathbf{w})\right\}\right] \times \quad (20)$$

$$\left\{\mathbb{E}\hat{f}(x;h) - f(x;\boldsymbol{\mu},\boldsymbol{\sigma},\mathbf{w})\right\} \mathrm{d}x \quad (21)$$

$$= 2(nh)^{-2}\sum\sum_{i<j} H(X_i,X_j) + 2k_2h^2n^{-1}\sum_{i=1}^{n} Z_i, \quad (22)$$

where Z_i is a term (see [5]) such that

$$h^2n^{-1}\sum_{i=1}^{n} Z_i = O_p(h^2n^{-1/2}).$$

Moreover, under the null and when $nh^5 \to \infty$, this term determines the limiting distribution of the right-hand side of (22). Now, under the null, the fact that

$$\sqrt{n}h^{-2} \int \left\{\hat{f}(x;h) - \mathbb{E}\hat{f}(x;h)\right\} \left\{\mathbb{E}\hat{f}(x;h) - f(x;\boldsymbol{\mu},\boldsymbol{\sigma},\boldsymbol{w})\right\} \mathrm{d}x \to k_2\sigma_1 Z$$

is a standard result. Taking into account that $d(n) = n^{1/2}h^{-2}$ and by applying the Lyapunov Central Limit Theorem yields

$$n^{1/2} \sum_{i=1}^{n} Z_i \to N(0, \sigma_1^2 - \sigma_{30}^2)$$

which proves the first leg of (10). For proving the second leg, we have that under the null and for $nh^5 \to 0$, $d(n) = n\sqrt{h}$. In this case

$$h^2 n^{-1} \sum_{i=1}^{n} Z_i = O_p((nh^5)^{1/2}) = o_p(1).$$

Hence the limit distribution of $d(n)(I_n - c(n))$ has the same distribution as the first term on the right-hand side of (22). By a direct application of Theorem 1 of [7] and taking into account also the proof of Theorem 3.2 in [5], it is straightforward to deduce that

$$n\sqrt{h}\left\{(nh)^{-2}\sum\sum_{1\le i<j\le n} H(X_i, X_j)\right\} \to \sqrt{2}\sigma_2 Z$$

, and thus establish the middle part on the right-hand side of (10). For the remaining part of (10), note that when $nh^5 \to \lambda$, $d(n) = n^{9/10}$ and hence no term on the right-hand side of (22) dominates the other since both are of the same order. Therefore, in this case, the limiting distribution of $d(n)(I_n - c(n))$ is given by the sum of the limit distribution of the two terms since, both terms are uncorrelated to each other.

References

1. Aldershof, B., Marron, J.S., Park, B.U., Wand, M.P.: Facts about the gaussian probability density function. Appl. Anal. **59**, 289–306 (1992)
2. Davies, T.M., Jones, K., Hazelton, M.L.: Symmetric adaptive smoothing regimens for estimation of the spatial relative risk function. Comput. Stat. Data Anal. **101**, 12–28 (2016)
3. Day, N.E.: Estimating the components of a mixture of normal distributions. Biometrika **56**, 463–474 (1969)
4. Eggermont, P.P.B., LaRiccia, V.N.: Maximum Penalized Likelihood Estimation, Volume I: Density Estimation. Springer, New York (2010)

5. Fan, Y.: Testing the goodness of fit of a parametric density function by kernel method **10**, 316–356 (1995)
6. Fisher, N.I., Mammen, E., Marron, J.S.: Testing for multimodality. Comput. Stat. Data Anal. **18**, 499–512 (1994)
7. Hall, P.: Central limit theorem for integrated square error of multivariate nonparametric density estimators. J. Multivar. Anal. **14**, 1–16 (1984)
8. Kiefer, N.M.: Discrete parameter variation: efficient estimation of a switching regression model. Econometrica **46**, 427–434 (1978)
9. Mächler, M.: nor1mix: Normal (1–d) Mixture Models (S3 Classes and Methods). R package version 1.2-2 (2016). https://CRAN.R-project.org/package=nor1mix
10. Marron, J.S., Wand, M.P.: Exact mean integrated squared error. Ann. Stat. **20**, 712–736 (1992)
11. Priebe, C.E., Marchette, D.J.: Alternating kernel and mixture density estimates. Comput. Stat. Data Anal. **35**, 43–65 (2000)
12. Sheather, S.J., Jones, M.C.: A reliable data-based bandwidth selection method for kernel density estimation. J. R. Stat. Soc. B **53**, 683–690, 1991 (2001)
13. Silverman, B.: Density Estimation for Statistics and Data Analysis. Chapman & Hall, New York (1986)
14. Roeder, K.: Density estimation with confidence sets exemplified by superclusters and voids in the galaxies. J. Am. Stat. Assoc. **85**(411), 617–624 (1990)

Robust Estimation of Sparse Signal with Unknown Sparsity Cluster Value

Eduard Belitser, Nurzhan Nurushev, and Paulo Serra

Abstract In the *signal+noise* model, we assume that the signal has a more general sparsity structure in the sense that the majority of signal coordinates are equal to some value which is assumed to be unknown, contrary to the classical sparsity context where one knows the sparsity cluster value (typically, zero by default). We apply an empirical Bayes approach (linked to the penalization method) for inference on the signal, possibly sparse in this more general sense. The resulting method is *robust* in that we do not need to know the sparsity cluster value; in fact, the method extracts as much generalized sparsity as there is in the underlying signal. However, as compared to the case of known sparsity cluster value, the proposed robust method cannot be reduced to thresholding procedure anymore. We propose two new procedures: the empirical Bayes model averaging (EBMA) and empirical Bayes model selection (EBMS) procedures, respectively. The former is procedure realized by an MCMC algorithm based on the partial (mixed) normal–normal conjugacy build in our modeling stage, and the latter is based on a new optimization algorithm of $O(n^2)$-complexity. We perform simulations to demonstrate how the proposed procedures work and accommodate possible systematic error in the sparsity cluster value.

Keywords Empirical Bayes · Sparce signal · Unknow sparsity cluster value

E. Belitser (✉)
Department of Mathematics, VU Amsterdam, Amsterdam, The Netherlands
e-mail: e.n.belitser@vu.nl

N. Nurushev
Korteweg-de Vries Institute for Mathematics, University of Amsterdam Research funded by the Netherlands Organisation for Scientific Research NWO, Amsterdam, The Netherlands
e-mail: n.nurushev@uva.nl

P. Serra
Department of Mathematics and Computer Science, Eindhoven University of Technology, Eindhoven, The Netherlands
e-mail: p.j.serra@tue.nl

M. La Rocca et al. (eds.), *Nonparametric Statistics*, Springer Proceedings in Mathematics & Statistics 339, https://doi.org/10.1007/978-3-030-57306-5_8

1 Introduction

The principle of parsimony, known as Occam's razor, is arguably one of the most fundamental ideas that pervade science, and *sparsity* has become a popular paradigm in statistical analyses, as a particular manifestation of the parsimony principle in the context of modern statistics. Much of this popularity has been driven by the success of frequentist (and Bayesian) methods utilizing the underlying sparsity structure of the unknown parameter of interest.

In general, a *sparse signal* is a high-dimensional parameter which allows a parsimonious representation. In signal processing, this is typically expressed by assuming that it contains only a small number of non-zero elements compared to its dimension. The value zero of the sparsity cluster has the interpretation of being "insignificant" for the corresponding coordinates. Any other value of the sparsity cluster can be handled as well in the analysis (in fact, we can always reduce to zero by subtracting that value) as long as this value is known a priori to the observer.

In the *signal+noise* setting, the best-studied problem is that of signal estimation, and a variety of estimation methods and results are available in the literature: [1, 3–8]. Thresholding strategies are particularly appealing, mainly because thresholding automatically generates sparsity. In addition, the corresponding procedures generally exhibit fast convergence properties. Moreover, thresholding processes the signal in a coordinate-wise fashion, resulting in low complexity algorithms (typically of order n), which are easy to implement in practice.

Many methods have *Bayesian* connections. For example, even some seemingly non-Bayesian estimators can be obtained as certain quantities (like posterior mode for penalized estimators) of the posterior distributions resulting from imposing some specific priors on the parameter; cf. [1, 2, 4, 6, 8, 11]. A common Bayesian way to model sparsity structure is by the two-group priors. Such a prior puts positive mass on vector θ with some exact zero coordinates (zero group) and the remaining coordinates (signal group) are drawn from a chosen distribution. As pointed out by [6] (also by [8]), the prior distributions of non-zero coordinates should not have too light tails; otherwise, one gets sub-optimal convergence rates (or even inconsistency). The important Gaussian case is, for example, excluded, [6, 8] use therefore heavy-tailed priors. On the other hand, in [4], it was shown that normal priors are still usable and lead to strong local results (even for non-iid, non-normal models) if combined with the empirical Bayes approach.

However, all these above-mentioned approaches are based on the essential assumption that the sparsity cluster value is *known* to the observer (which is set to zero by default). In this note, we relax these modeling assumptions by allowing the sparsity cluster value to be an *unknown* constant, obtaining a *robust* formulation of the estimation problem. This situation can occur when, for example, there is a systematic error in the observations and sparsity coordinates get shifted by unknown value (bias of systematic error), leading to what we call *sparsity cluster with unknown cluster value*. It is clear that thresholding procedures are not going to be applicable in this situation, so we need to deal with methodological and computational issues.

We address the first aspect by applying an empirical Bayes approach, which delivers two robust procedures: the empirical Bayes model averaging (EBMA) and empirical Bayes model selection (EBMS) procedures. As to the computational issue, the former procedure is realized by an MCMC algorithm based on the partial (mixed) normal–normal conjugacy in the model, and the latter is based on a new optimization algorithm of $O(n^2)$-complexity (cf. $O(n)$-complexity for typical thresholding procedures). We perform simulations to demonstrate how the proposed procedures work and accommodate possible systematic error in the sparsity cluster value.

2 Setting and Notation

Suppose we observe $X = X^{(\sigma,n)} = (X_1, \dots, X_n)$:

$$X_i = \theta_i + \sigma \xi_i, \quad i \in [n] = \{1, \dots, n\}, \tag{1}$$

where $\theta = (\theta_1, \dots, \theta_n) \in \mathbb{R}^n$ is an unknown high-dimensional parameter of interest, $\xi_i \overset{\text{ind}}{\sim} N(0, 1)$, $\sigma > 0$ is the known noise intensity. The goal is to make inference on the parameter θ based on the data X. We exploit the empirical Bayes approach and make a connection with the penalization method.

Denote the probability measure of X from the model (1) by $\mathrm{P}_\theta = \mathrm{P}_\theta^{(\sigma,n)}$, and by E_θ the corresponding expectation. For notational simplicity, we often skip the dependence on σ and n. Denote by $\mathbb{1}_E = \mathbb{1}\{E\}$ the indicator function of the event E, and by $|S|$ the cardinality of the set S. Let $[n] = \{1, \dots, n\}$ and $[n]_0 = \{0\} \cup [n]$ for $n \in \mathbb{N} = \{1, 2, \dots\}$. For $I \subseteq [n]$, define $I^c = \{i \in [n] : i \notin I\}$. Let $\mathcal{I} = \mathcal{I}_n = \{I : I \subseteq [n]\}$ be the family of all subsets of $[n]$ including the empty set. Throughout, we assume the conventions that $\bar{a}_I = \frac{1}{|I|}\sum_{i \in I} a_i$, $\sum_{i \in \varnothing} a_i = 0$, $\prod_{i \in \varnothing} a_i = 1$, $\sum_a^b a_i = \sum_{a \le i \le b} a_i$, $\sum_i a_i = \sum_{i \in [n]} a_i$, $\sum_I a_I = \sum_{I \in \mathcal{I}} a_I$ for any $a_i, a_I, a, b \in \mathbb{R}$ and $0 \log(c/0) = 0$ (hence $(c/0)^0 = 1$) for any $c > 0$. Let $X_{[1]}^2 \ge \theta_{[2]}^2 \ge \dots \ge X_{[n]}^2$ be the ordered values of $X_1^2, \dots, X_n^2$, introduce also $X_{[0]}^2 = \infty$.

Throughout, $\varphi(x, \mu, \sigma^2)$ will be the density of $\mu + \sigma Z \sim N(\mu, \sigma^2)$ at point x, where $Z \sim N(0, 1)$. By convention, $N(\mu, 0) = \delta_\mu$ denotes a Dirac measure at point μ. Finally, let $\|x\|$ denote the usual norm of $x \in \mathbb{R}^n$.

3 Empirical Bayes Approach

First, we introduce a family of normal priors (similar to priors from [4]). Next, by applying the empirical Bayes approach to the normal likelihood, we derive an empirical Bayes posterior for the case of unknown sparsity cluster value, and use this posterior in further inference on θ.

3.1 Multivariate Normal Prior

To model a possible sparsity cluster of unknown cluster value in the parameter θ, the coordinates of θ can be split into two distinct groups of coordinates of θ: for some $I \in \mathcal{I}$, $\theta_I = (\theta_i, i \in I)$ and $\theta_{I^c} = (\theta_i, i \in I^c)$, so that $\theta = (\theta_I, \theta_{I^c})$. The group $\theta_{I^c} = (\theta_i, i \notin I)$ consists of coordinates that are all assumed to be (almost) equal to some cluster value μ_c, and $\theta_I = (\theta_i, i \in I)$ is the group of coordinates significantly different from μ_c. To model sparsity with unknown sparsity cluster value, we propose a prior on θ given I as follows:

$$\pi_I = \bigotimes_i N\big(\mu_i(I), \tau_i^2(I)\big),$$

where $\mu_i(I) = \mu_i \mathbb{1}\{i \in I\} + \mu_c \mathbb{1}\{i \notin I\}$, $\tau_i^2(I) = \sigma^2 K_n(I) \mathbb{1}\{i \in I\}$, $K_n(I) = (\frac{en}{|I|} - 1)\mathbb{1}\{I \neq \varnothing\}$. The indicators in the above prior ensure the sparsity of the group I^c. The rather specific choice of $K_n(I)$ is made for the sake of concise expressions in later calculations, many other choices are actually possible. By using the normal likelihood $\ell(\theta, X) = (2\pi\sigma^2)^{-n/2} \exp\{-\|X - \theta\|^2/2\sigma^2\}$, the corresponding posterior distribution for θ is readily obtained:

$$\pi_I(\vartheta|X) = \bigotimes_i N\Big(\frac{\tau_i^2(I)X_i + \sigma^2\mu_i(I)}{\tau_i^2(I) + \sigma^2}, \frac{\tau_i^2(I)\sigma^2}{\tau_i^2(I) + \sigma^2}\Big). \tag{2}$$

Next, introduce the prior λ on $\mathcal{I}$. For $\varkappa > 0$, draw a random set from $\mathcal{I}$ with probabilities

$$\lambda_I = c_{\varkappa,n} \exp\big\{ - \varkappa|I| \log(\tfrac{en}{|I|})\big\} = c_{\varkappa,n}(\tfrac{en}{|I|})^{-\varkappa|I|}, \quad I \in \mathcal{I},$$

where $c_{\varkappa,n}$ is the normalizing constant.

Remark 1 A logical choice for λ_I seems to be the uniform prior on $\mathcal{I}$: $\bar{\lambda}_I = \binom{n}{|I|}^{-1}$. However, this prior is not monotone with respect to the cardinality $|I|$, whereas we would like to penalize large cardinalities. As $(\frac{n}{k})^k \leq \binom{n}{k} \leq (\frac{en}{k})^k$ for $k \in [n]_0$, we take the above defined prior λ_I as monotone (in $|I|$) proxy for $\bar{\lambda}_I$, with an extra parameter $\varkappa$ to control the amount of penalization; $\varkappa = 1$ corresponds to the prior $\bar{\lambda}_I$.

Combining the conditional prior π_I with the prior λ_I gives the mixture prior on θ: $\pi = \sum_I \lambda_I \pi_I$. This leads to the marginal distribution of X: $\mathrm{P}_X = \sum_I \lambda_I \mathrm{P}_{X,I}$, with $\mathrm{P}_{X,I} = \bigotimes_i N\big(\mu_i(I), \sigma^2 + \tau_i^2(I)\big)$, and the posterior of θ is

$$\pi(\vartheta|X) = \sum_I \pi_I(\vartheta|X)\pi(I|X), \tag{3}$$

where $\pi_I(\vartheta|X)$ is defined by (2) and the posterior $\pi(I|X)$ for I is

$$\pi(I|X) = \frac{\lambda_I \mathrm{P}_{X,I}}{\mathrm{P}_X} = \frac{\lambda_I \prod_i \varphi\big(X_i, \mu_i(I), \sigma^2 + \tau_i^2(I)\big)}{\sum_{J\in\mathcal{I}} \lambda_J \prod_i \varphi\big(X_i, \mu_i(J), \sigma^2 + \tau_i^2(J)\big)}. \tag{4}$$

3.2 Empirical Bayes Posterior

The parameters $\mu_i(I)$ are yet to be chosen in the prior. We choose $\mu_i(I)$ by using empirical Bayes approach. The marginal likelihood P_X is readily maximized with respect to $\mu_i(I)$: $\hat{\mu}_i(I) = X_i$ for $i \in I$ and $\hat{\mu}_i(I) = \bar{X}_{I^c}$ for $i \in I^c$, where $\bar{X}_{I^c} = \frac{1}{|I^c|}\sum_{i\in I^c} X_i$. We substitute $\hat{\mu} = (\hat{\mu}(I), I \in \mathcal{I})$ instead of $\mu = (\mu(I), I \in \mathcal{I})$ in the expression (3) for $\pi(\vartheta|X)$, obtaining the empirical Bayes posterior (called *empirical Bayes model averaging* (EBMA) posterior)

$$\tilde{\pi}(\vartheta|X) = \sum_I \tilde{\pi}_I(\vartheta|X)\tilde{\pi}(I|X),$$

where the empirical Bayes conditional posterior (recall that $N(\mu, 0) = \delta_\mu$)

$$\tilde{\pi}_I(\vartheta|X) = \prod_{i\in I} N\big(X_i, \tfrac{K_n(I)\sigma^2}{K_n(I)+1}\big) \otimes \prod_{i\in I^c} \delta_{\bar{X}_{I^c}}$$

is obtained from (2) with $\mu_i(I) = \hat{\mu}_i(I) = X_i \mathbb{1}\{i \in I\} + \bar{X}_{I^c}\mathbb{1}\{i \in I^c\}$, and

$$\tilde{\pi}(I|X) = \frac{\lambda_I \mathrm{P}_{X,I}}{\sum_{J\in\mathcal{I}} \lambda_J \mathrm{P}_{X,J}} = \frac{\lambda_I \prod_i \varphi(X_i, \hat{\mu}_i(I), \sigma^2 + \tau_i^2(I))}{\sum_{J\in\mathcal{I}} \lambda_J \prod_i \varphi(X_i, \hat{\mu}_i(J), \sigma^2 + \tau_i^2(J))}$$

is the empirical Bayes posterior for $I \in \mathcal{I}$, obtained from (4) with $\mu_i(I) = \hat{\mu}_i(I)$. Let $\tilde{\mathrm{E}}$ and $\tilde{\mathrm{E}}_I$ be the expectations with respect to the measures $\tilde{\pi}(\vartheta|X)$ and $\tilde{\pi}_I(\vartheta|X)$ respectively. Then $\tilde{\mathrm{E}}_I(\vartheta|X) = \hat{\mu}(I) = (X_i\mathbb{1}\{i \in I\} + \bar{X}_{I^c}\mathbb{1}\{i \in I^c\}, i \in [n])$. Introduce the *EBMA mean* estimator

$$\tilde{\theta}^{EB} = \tilde{\theta}^{EB}(\varkappa, X) = \tilde{\mathrm{E}}(\vartheta|X) = \sum_{I\in\mathcal{I}} \tilde{\mathrm{E}}_I(\vartheta|X)\tilde{\pi}(I|X) = \sum_{I\in\mathcal{I}} \hat{\mu}(I)\tilde{\pi}(I|X). \tag{5}$$

Consider an alternative ("more Bayesian") empirical Bayes posterior. First, derive an empirical Bayes variable selector $\hat{I}$ by maximizing $\tilde{\pi}(I|X)$ over $I \in \mathcal{I}$ (any maximizer will do) as follows:

$$\begin{aligned}\hat{I} &= \operatorname*{argmax}_{I\in\mathcal{I}} \tilde{\pi}(I|X) = \operatorname*{argmax}_{I\in\mathcal{I}} \Big\{-\sum_{i\in I^c} \tfrac{(X_i-\bar{X}_{I^c})^2}{2\sigma^2} - \tfrac{|I|}{2}\log(K_n(I)+1) + \log\lambda_I\Big\} \\ &= \operatorname*{argmin}_{I\in\mathcal{I}} \Big\{\sum_{i\in I^c}(X_i - \bar{X}_{I^c})^2 + (2\varkappa+1)\sigma^2|I|\log\big(\tfrac{en}{|I|}\big)\Big\}. \end{aligned} \tag{6}$$

Now plugging in $\hat{I}$ into $\tilde{\pi}_I(\vartheta|X)$ yields another empirical Bayes posterior (called *empirical Bayes model selection* (EBMS) posterior) and the corresponding *EBMS mean* estimator for θ: with $\hat{\mu}_i(I) = (X_i \mathbb{1}\{i \in I\} + \bar{X}_{I^c}\mathbb{1}\{i \in I^c\}, i \in [n])$,

$$\hat{\pi}(\vartheta|X) = \tilde{\pi}_{\hat{I}}(\vartheta|X), \quad \hat{\theta}^{EB} = \hat{\mathrm{E}}(\vartheta|X) = \hat{\mu}(\hat{I}) = (\hat{\mu}_i(I), i \in [n]), \qquad (7)$$

where $\hat{\mathrm{E}}$ denotes the expectation with respect to the measure $\hat{\pi}(\vartheta|X)$.

4 Known Sparsity Cluster Value: Thresholding Procedures

In the traditional sparsity setting, the sparsity cluster value is assumed to be known and set to be zero without loss of generality. This case is well studied, various estimators are proposed and studied in the literature, see [1, 4–8], and further references therein. Many estimation procedures originate as penalized estimators minimizing the criterion $\mathrm{crit}(X, \theta) = \|X - \theta\|^2 + P(\theta)$, for some appropriately chosen penalties $P(\theta)$, or as (empirical) Bayes estimators according to appropriately chosen priors. An extensive discussion on this can be found in [1].

Notice that whenever the penalty $\mathrm{crit}(X, \theta)$ is of an ℓ_0-type, i.e., $P(\theta) = p(\|\theta\|_0)$ for some function p and $\|\theta\|_0 = \sum_i \mathbb{1}\{\theta_i \neq 0\}$, the resulting penalized estimator is a thresholding estimator $\check{\theta}_i = X_i 1\{|X_i| \geq \check{t}\}$, where $\check{t} = |X_{[\check{k}]}|$ and $\check{k}$ is the minimizer of $\sum_{i=k+1}^n X_{[i]}^2 + p(k)$, $k \in [n]_0$ (recall that $\infty = X_{[0]}^2 > X_{[1]}^2 \geq \ldots \geq X_{[n]}^2$). Thresholding strategies are particularly appealing because thresholding automatically generates sparsity. Besides, thresholding procedures generally exhibit fast convergence properties and process the signal in a coordinate-wise way, which results in low complexity algorithms. There is a vast literature on this topic, see, e.g., [10] and further references therein.

Remark 2 The Bayesian approach can be connected to the penalized estimation by relating the penalties to the corresponding priors on θ. Penalties of ℓ_0-type can be linked to Bayesian procedures involving priors on the number of non-zero entries of θ, see [1].

Within the framework of the present paper, the case of the known sparsity cluster value corresponds to taking $\mu_c = 0$ in the prior π_I. This leads to $\hat{\mu}_i(I) = 0$ for $i \in I^c$ in all the posterior quantities of Sect. 3.2, and the criterion (6) reduces to

$$\check{I} = \operatorname*{argmin}_{I \in \mathcal{I}} \Big\{ \sum_{i \in I^c} X_i^2 + K\sigma^2 |I| \log\big(\tfrac{en}{|I|}\big) \Big\}, \quad K = 2\varkappa + 1,$$

which is reminiscent of the penalization procedure from [5] (cf. also [1]), with the penalty $p(k) = K\sigma^2 k \log(\frac{en}{k})$. Indeed, it can be easily seen that

$$\check{I} = \big\{i \in [n] : |X_i| \geq \check{t} = |X_{[\check{k}]}|\big\}, \quad \check{k} = \arg\min_{k \in [n]_0} \Big\{ \sum_{i=k+1}^{n} X_{[i]}^2 + p(k) \Big\}, \tag{8}$$

and the EBMS procedure yields the corresponding thresholding estimator $\check{\theta}^{HT} = \check{\theta}^{HT}(K, X) = (\check{\theta}_i^{HT}, i \in [n])$, with

$$\check{\theta}_i^{HT} = X_i \mathbb{1}\{i \in \check{I}\} = X_i \mathbb{1}\{|X_i| \geq |X_{[\check{k}]}|\}, \quad i \in [n]. \tag{9}$$

The penalty $p(k)$ corresponds to the *complete variable selection* case in [5]. Recall our rather specific choice of parameter $K_n(I)$ in the prior π_I resulting in this penalty. As we mentioned, other choices of $K_n(I)$ are also possible, which would lead to other penalties. But the main term $\sigma^2 k \log(\frac{en}{k})$ would always be present because of the choice of prior λ_I. The optimality of this kind of penalties (and priors) is discussed in [1, 4, 5]. In [1] it is concluded that essentially only such penalties lead to adaptive penalized estimators over certain *sparsity scales*.

5 EBMA and EBMS Procedures for the Case of Unknown Sparsity Cluster Value

Clearly, the thresholding approach relies very much on the fact that we know the sparsity cluster value, zero by default. Assume now that there is a large (sparsity) cluster of (almost) equal coordinates of θ, but its value is not known. Alternatively, one can think of the so-called *robust inference* in the sense that there may be a systematic error in the "known" sparsity cluster value zero and the true sparsity cluster value may actually deviate from zero. Using a thresholding procedure in such a situation would lead to a big cumulative error, because the sparsity cluster contains most of the coordinates of the high-dimensional signal θ.

Recalling the empirical Bayes approach described in Sect. 3.2 for the case of unknown sparsity cluster value, we immediately see that, unlike (8), the EBMS criterion (6) cannot be reduced to a thresholding procedure. However, the corresponding optimization problem is still feasible from a computational point of view. Indeed, the criterion (6) reduces to

$$\hat{I} = \underset{I \in \mathcal{I}}{\operatorname{argmin}} \Big\{ \sum_{i \in I^c} (X_i - \bar{X}_{I^c})^2 + K\sigma^2 |I| \log\big(\tfrac{en}{|I|}\big) \Big\} = \big\{i \in [n] : X_i \neq X_{[\hat{j}+t]},\ t \in [\hat{k}]\big\},$$

where $K = 2\varkappa + 1$, $X_{[1]} \geq X_{[2]} \geq \ldots \geq X_{[n]}$ are the ordered $X_1, \ldots, X_n$,

$$(\hat{k}, \hat{j}) = \underset{k, j \in [n]_0,\ k+j \leq n}{\operatorname{argmin}} \Big\{ \sum_{i=j+1}^{j+k} (X_{[i]} - \bar{X}_{jk})^2 + K\sigma^2 (n-k) \log\big(\tfrac{en}{n-k}\big) \Big\}, \quad \bar{X}_{jk} = \frac{1}{k} \sum_{i=1+j}^{k+j} X_{[i]}.$$

In this case, the EBMS method yields the robust (the sparsity cluster value is unknown) version of EBMS estimator $\hat{\theta}^{EB} = \hat{\theta}^{EB}(K, X) = (\hat{\theta}_i^{EB}, i \in [n])$ with

$$\hat{\theta}_i^{EB} = X_i \mathbb{1}\{X_i \neq X_{[t+\hat{j}]},\ t \in [\hat{k}]\} + \bar{X}_{\hat{j}\hat{k}} \mathbb{1}\{X_i = X_{[t+\hat{j}]},\ t \in [\hat{k}]\}, \quad i \in [n]. \tag{10}$$

It is not so difficult to see that this procedure has the computational complexity of order n^2, which is of course worse than the procedure (8)–(9), but still computationally feasible. This is demonstrated in the next section.

An alternative is to use the EBMA method. All posterior quantities involved in the construction of the EBMA estimator $\tilde{\theta}^{EB}$ given by (5) are explicit, and the major issue is that the number of terms in (5) is exponential in the dimension so that direct computation is not practically feasible for high dimensions. Therefore, in this case, we have to resort to an *MCMC procedure*.

In the MCMC procedure, each element I in the support of $\tilde{\pi}(I|X)$ is encoded (one-to-one) by a binary state vector $s = (s_1, \dots, s_n) \in \{0, 1\}^n$. The correspondence is that $s_i = 1$ if, and only if, $i \in I$ and $s_i = 0$ if, and only if, $i \notin I$. The proposal s' flips simply one bit chosen uniformly at random from the current state s. This means that we first select j uniformly at random on $\{1, \dots, n\}$, and then set $s'_j = 1 - s_j$ and $s'_i = s_i$, $i \neq j$. The chain moves from s to s' with probability $\alpha = \min\big\{1,\ \tilde{\pi}\big(\{i \in [n] : s'_i = 1\}|X\big)/\tilde{\pi}\big(\{i \in [n] : s_i = 1\}|X\big)\big\}$. The EBMA estimator $\tilde{\theta}^{EB}$ from (5) is the expectation of $\hat{\mu}(I)$ with respect to the posterior $\tilde{\pi}(I|X)$. If $I_1, \dots, I_M$ is a sample drawn from $\tilde{\pi}(I|X)$, or indeed a sample produced by the MCMC procedure from above, then we approximate the EBMA estimator as

$$\tilde{\theta}^{EB} = \tilde{\theta}^{EB}(\varkappa, X) \approx \frac{1}{M} \sum_{i=1}^{M} \hat{\mu}(I_i). \tag{11}$$

6 Comparative Simulation Study

In this section, we present a comparative simulation study for the cases of known and unknown (or shifted) sparsity cluster value.

We generate observations according to the model (1) with $\xi_i \overset{\text{ind}}{\sim} N(0, 1)$, $\sigma = 1$, $n = 500$, where we use signals $\theta = (\theta_1, \dots, \theta_n)$ of the form $\theta = (A_1, \dots, A_p, \delta, \dots, \delta)$. The first p coordinates of θ are "significant," the remaining $n - p$ entries form the sparsity cluster. We consider different "sparsity levels" $p \in \{25, 50, 100\}$ and "signal strengths": $A_i \overset{\text{ind}}{\sim} U[0, 2]$ (signal is undetectable, i.e., comparable to the noise); $A_i \overset{\text{ind}}{\sim} U[2, 4]$ (signal is barely distinct from the noise); $A_i \overset{\text{ind}}{\sim} U[4, 6]$ (signal is well distinct from the noise). Next, we consider two situation: a) known sparsity cluster value $\delta = 0$; b) unknown sparsity cluster value which we set $\delta = -0.5$ in the simulations.

The following estimators are considered: the projection oracle (PO) estimator $\hat{\theta}_i^{PO} = X_i \mathbb{1}\{\theta_i \neq \delta\} + \delta \mathbb{1}\{\theta_i = \delta\}$, $i \in [n]$; the *empirical Bayes mean* (EBMean)

$\hat{\theta}^{EBMean}$ considered in [8] with a standard Laplace prior and realized in the R-package EbayesThresh (see [9]); the classical *universal hard-thresholding* (UHT) (see [7]) $\hat{\theta}_i^{UHT} = X_i \mathbb{1}\{|X_i| > \sqrt{2\log n}\}$, $i \in [n]$; the HT estimator $\check{\theta}^{HT}$ defined by (9), the EBMA estimator $\tilde{\theta}^{EB}$ given by (5), and, finally, the EBMS estimator $\hat{\theta}^{EB}$ defined by (10). Clearly, the PO procedure is not really an estimator as it uses oracle knowledge of the active set $I_*(\theta) = \{i \in [n] : \theta_i \neq \delta\}$ and the sparsity cluster value δ. Clearly, the pseudo-estimator PO cannot be outperformed by any practical estimation procedure. The performance of the pseudo-estimator PO is provided only for reference as a benchmark of the ideal situation.

The estimators EBMean, UHT, and HT are all geared towards the known (zero) sparsity cluster value, whereas our EBMA and EBMS estimators $\tilde{\theta}^{EB}$ and $\hat{\theta}^{EB}$ can also accommodate any unknown sparsity cluster value. To create more competition for our procedures $\tilde{\theta}^{EB}$ and $\hat{\theta}^{EB}$ in the case of unknown sparsity cluster value, we also provide adjusted versions aEBMean, aUHT and aHT of the estimators $\hat{\theta} = \hat{\theta}(X)$, constructed as follows: $\hat{\theta}'(X) = \hat{\theta}(X - \bar{X}1_n) + \bar{X}1_n$, where 1_n is an n-dimensional vector of ones, $\bar{X} = \frac{1}{n}\sum_{i=1}^{n} X_i$ is the empirical mean of the sample X, and $\hat{\theta}$ is the corresponding estimator (respectively, EBMean, UHT, and HT). In the case of unknown non-zero sparsity cluster value, the adjusted versions are clearly biased and only competitive for relatively small p and A_i's. The adjusted versions of the estimators are expected to perform worse (and they do so, as Table 2 shows) for larger values of p and A_i's.

Each of our estimators $\check{\theta}^{HT}(K, X)$, $\hat{\theta}^{EB}(K, X)$ and $\tilde{\theta}^{EB}(\varkappa, X)$ depends on one tuning parameter, K or $\varkappa$. It is possible to choose the parameters K and $\varkappa$ from the data via a *cross-validation* procedure, but this significantly increases the running time for computing $\hat{\theta}^{EB}(K, X)$, and especially $\tilde{\theta}^{EB}(\varkappa, X)$. However, in the simulation results for several various cases, the optimal K did not vary much and appeared to lie mostly in the range $[1.8, 3.2]$. Moreover, the results were good for many choices of K, the performance deteriorates significantly only when K gets close to 1 or becomes too big. This actually agrees with the conclusions (about the penalty constants) from [5]. In the simulations for the EBMS estimators $\check{\theta}^{HT}(K, X)$ and $\hat{\theta}^{EB}(K, X)$, the choice $K = 2.5$ appeared to be fairly good for all considered cases. When computing the EBMA estimator $\tilde{\theta}^{EB}(\varkappa, X)$, we took $\varkappa = 1$ which is a natural choice in the light of Remark 1. The EBMA procedure seemed to be even less sensitive to the choice of parameter $\varkappa$, again many choices are possible as long as $\varkappa$ is not too small (should be larger than 0.7) and not too big. We let the chain burn in for 10000 iterations and then collected 25000 states from the posterior. The final sample of states used to approximate the EBMA estimator was obtained by keeping every 25-th state resulting in $M = 1000$ in (11). This thinning was done to reduce the correlation between the samples from the MCMC procedure.

Tables 1 and 2 contain estimates of the mean square errors $MSE(\hat{\theta}, \theta) = \mathrm{E}_\theta \|\hat{\theta} - \theta\|^2$ for the above-mentioned estimators $\hat{\theta}$ and choices of the signal θ. Tables 1 and 2 concern the cases of the known ($\delta = 0$) and unknown ($\delta = -0.5$) sparsity cluster value, respectively. The $MSE(\hat{\theta}, \theta)$ is evaluated by the average squared error

Table 1 Estimated MSE's for the case (a) of the known sparsity cluster value $\delta = 0$

p	25			50			100		
A_i	U[0,2]	U[2,4]	U[4,6]	U[0,2]	U[2,4]	U[4,6]	U[0,2]	U[2,4]	U[4,6]
PO	25	25	25	50	50	50	99	99	99
EBMean	34	96	91	64	164	172	111	273	319
UHT	39	157	68	75	316	137	141	626	270
HT	37	127	62	72	194	123	138	300	233
EBMA	36	103	79	66	178	162	114	291	254
EBMS	42	132	64	73	204	124	121	313	233

Table 2 Estimated MSE's for the case (b) of an unknown sparsity cluster value $\delta = -0.5$

p	25			50			100		
A_i	U[0,2]	U[2,4]	U[4,6]	U[0,2]	U[2,4]	U[4,6]	U[0,2]	U[2,4]	U[4,6]
PO	25	25	25	50	50	50	99	99	99
EBMean	136	178	176	154	229	240	182	312	348
aEBMean	57	108	118	100	201	254	162	332	441
UHT	162	280	191	191	432	254	245	730	374
aUHT	69	174	96	129	380	285	224	811	916
HT	157	256	206	186	327	268	240	414	360
aHT	68	128	104	127	253	300	222	516	577
EBMA	66	97	79	122	176	161	201	281	251
EBMS	69	107	59	128	170	120	224	275	231

$\widehat{MSE}(\hat{\theta}, \theta) = \frac{1}{l}\sum_{k=1}^{l} \|\hat{\theta}^k - \theta\|^2$ of l estimates $\hat{\theta}^1, \ldots, \hat{\theta}^l$ computed from $l = 100$ data vectors simulated independently from the model (1).

It is not surprising that the shrinkage estimators EBMean and EBMA perform well for weak signals (cases $A_i \sim \mathrm{U}[0, 2]$ and $A_i \sim \mathrm{U}[2, 4]$) in situation a) of known (zero) sparsity cluster value, as one can see from Table 1. Table 2 is for situation b) and is more interesting, it shows a clear advantage of the EBMA and EBMS methods which take into account the unknown sparsity cluster value. Only for the cases with undetectable signal (case $A_i \sim \mathrm{U}[0, 2]$), the adjusted shrinkage estimator aEBMean is still competitive, as this case is very favorable for any shrinkage procedure and a relatively small absolute shift value $|\delta|$.

References

1. Abramovich, F., Grinshtein, V., Pensky, M.: On optimality of Bayesian testimation in the normal means problem. Ann. Stat. **35**, 2261–2286 (2007)

2. Babenko, A., Belitser, E.: Oracle projection convergence rate of posterior. Math. Meth. Stat. **19**, 219–245 (2010)
3. Belitser, E.: On coverage and local radial rates of credible sets. Ann. Stat. **45**, 1124–1151 (2017)
4. Belitser, E., Nurushev, N.: Needles and straw in a haystack: robust empirical Bayes confidence for possibly sparse sequences. Bernoulli 26, 191–225(2020)
5. Birgé, L., Massart, P.: Gaussian model selection. J. Eur. Math. Soc. **3**, 203–268 (2001)
6. Castillo, I., van der Vaart, A.: Needles and straw in a haystack: posterior concentration for possibly sparse sequences. Ann. Stat. **40**, 2069–2101 (2012)
7. Donoho, D.L., Johnstone, I.M.: Minimax risk over ℓ_p-balls for ℓ_q-error. Probab. Theory Rel. Fields **99**, 277–303 (1994)
8. Johnstone, I., Silverman, B.: Needles and straw in haystacks: empirical Bayes estimates of possibly sparse sequences. Ann. Stat. **32**, 1594–1649 (2004)
9. Johnstone, I., Silverman, B.: EbayesThresh: R programs for empirical Bayes thresholding. J. Stat. Softw. 12 (2005)
10. Hastie, T., Tibshirani, R., Wainwright, M.: Statistical Learning with Sparsity: The Lasso and Generalizations. Chapman and Hall, CRC Press (2015)
11. Van der Pas, S.L., Kleijn, B.J.K., van der Vaart, A.W.: The horseshoe estimator: Posterior concentration around nearly black vectors. Electron. J. Stat. **8**, 2585–2618 (2014)
12. Van der Pas, S.L., Szabó, B.T., van der Vaart, A.W.: Uncertainty quantification for the horseshoe (with discussion). Bayesian Anal. **12**, 1221–1274 (2017)

Test for Sign Effect in Intertemporal Choice Experiments: A Nonparametric Solution

Stefano Bonnini and Isabel Maria Parra Oller

Abstract In order to prove the hypothesis of *sign effect* in intertemporal choice experiments, the empirical studies described in the specialized literature apply univariate tests (in most cases parametric t or F tests) even when multivariate inferential procedures are more suitable according to the experimental data, the study design and the goal of the analysis. Furthermore, the used tests do not take into account the possible presence of confounding effects, very common in such kind of experimental studies. In this paper, a multivariate nonparametric method to test for *sign effect in intertemporal choice* is proposed. This method overcomes the mentioned limits of the tests usually applied in previous studies. A case study related to a survey performed at the University of Almeria (Spain) is presented. The methodological solution based on the nonparametric test is described and the results of its application to the data collected in the sample survey performed in Almeria are shown.

Keywords Intertemporal choice · Permutation test · Nonparametric combination · Multivariate test · Confounding factors · Multistrata test

1 Introduction

Intertemporal choice problems concern the study of decision-making processes. Specifically, these problems refer to the case of choices over time. When one wonders whether it is better to save money now in order to consume more in the future or to consume today by giving up a greater future consumption, we are in the presence of an *intertemporal choice*. The decision about how many years to be devoted to the study, i.e., how much time of our life can be focused on (and how much money can

S. Bonnini (✉)
Department of Economics and Management, University of Ferrara, Ferrara, Italy
e-mail: stefano.bonnini@unife.it

I. M. P. Oller
Department of Economics and Business, University of Almeria, Almeria, Spain
e-mail: ipo244@inlumine.ual.es

M. La Rocca et al. (eds.), *Nonparametric Statistics*, Springer Proceedings in Mathematics & Statistics 339, https://doi.org/10.1007/978-3-030-57306-5_9

be invested in) education, is also an *intertemporal choice*: is it better to have a salary (or free time) today or invest money (time) in education and postpone the job market entry, in order to have greater earnings in the future? Another typical example of *intertemporal choice* is related to dietary habits: is it better to eat a good cake or a dish of fried food now or to follow a healthier diet in order to have a better health (and a longer life) in the future?

These problems are typical in Financial Economics, but quite common even in other disciplines such as Neuroscience, Medicine, Marketing, Sport, Economic and Industrial Policy, Fiscal Policy, Monetary Policy, Social and Welfare Policy, and others. There is an extensive scientific literature on *intertemporal choices* in the fields of Psychology and Behavioral Economics. In fact, the decision-making process of individuals, when they make intertemporal choices, is almost always the same, regardless of the specific context.

In this paper, from here on out, we consider the problem according to an economic perspective. In this framework, people tend to prefer immediate gains or rewards and to postpone losses or penalties.

For example, let us assume that the winner of 100€ at the lottery has the possibility of cashing in immediately the gain or postponing it for 1 year. If the winner accepts to postpone the gain only if the amount received after 1 year is greater than 110€, then the discount rate is 10% or equivalently 0.10. The gain will be not postponed if the future amount is less than 110€ and, in the case of future amount exactly equal to 110€, the choice is indifferent. We can also say that 100€ is the discounted value (the current worth) of the future gain of 110€. Definitely, the discount rate is the proportion (percentage) of the increase in value needed to compensate for 1-year delay.

As a consequence, the winner accepts to postpone the gain for $t = 2$ years provided that the amount received will be at least $110 \times (1 + 0.1) = 100 \times (1 + 0.1)^2 =$ 121€. In general, he/she accepts to postpone the gain for t years if he/she will receive at least $(1 + 0.1)^t \times 100$. Hence, the discounted value of a future gain x is given by $(1 + 0.1)^{-t}x$. The winner accepts the reduction of the amount up to the discounted value in order to anticipate the gain today.

A similar reasoning applies to the case of losses (penalties, payment of fines, etc.) because the choice is between a lower amount today or a greater amount in the future. For example, with a discount rate equal to 0.10, the individual prefers to pay the immediate amount 100 (or less), rather than the delayed amount 110 1 year later or $(1 + 0.1)^t \times 100$ after t years. If the amount to be paid today is greater than 100, then it is not worth anticipating the payment.

For an individual, the discount rate of gains could be not equal to the discount rate of losses. Given the future amount x, in the case of awards, the discounted value could be lower than in the case of fines. For example, the winner of a lottery could consider the award of 110€ after 1 year equivalent to 100€ today (discount rate equal to 0.10) but he/she could consider the payment of a 110 euro fine after 1 year equivalent to the payment of 105€ today (discount rate equal to 0.05). Therefore, for a given time horizon, the reduction of the amount for which the anticipation of a payment is accettable, is therefore lower than the reduction of the award for which it is preferable to cash in a gain today. This is what the specialized literature calls the *sign effect* in intertemporal choice.

In order to prove the hypothesis of *sign effect*, the empirical studies reported in the literature apply univariate tests (in most cases parametric t or F tests) even when multivariate inferential procedures are more suitable according to the experimental data, the study design and the goal of the analysis. Furthermore, the used tests do not take into account the possible presence of confounding effects, very common in such kind of experimental studies. In this paper, we propose a multivariate nonparametric method to test for *sign effect in intertemporal choice* that overcomes the mentioned limits of the most common tests. In Sect. 2 the theory of intertemporal choice and the sign effect are formally presented. In order to describe the problem in a clearer and more precise way, in Sect. 3, a case study related to a survey performed at the University of Almeria (Spain) is presented. In Sect. 4, the statistical testing problem is defined. Section 5 is dedicated to the description of the methodological solution based on the nonparametric test. Section 5 includes the results of the application of the nonparametric test to the empirical problem described in Sect. 3. Section 6 concerns the conclusions.

2 Intertemporal Choice and Sign Effect

The basic elements of the *intertemporal choice* are the following:

- It is a problem of allocation between two or more time points.
- There is a tradeoff between earlier pleasure and satisfaction and later wellbeing.
- Some subjective elements can affect the decision.

The first important scientific contribution to explain intertemporal choices is the *Discounting Utility model* (DU model) proposed by Samuelson [1]. Let us imagine the classic "consumption or savings" problem. Consider the case of a subject and her/his decision about how to allocate her/his consumption over time, starting from today ($t = 0$) and considering T different time points in the future ($t = 1, 2, \ldots, T$). In other words, we are interested in the person's comsumption profile over time $(c_0, c_1, \ldots, c_T)$, where c_t is the consumption value at time t, with $t = 0, 1, \ldots, T$. According to the DU model, the utility of a given consumption profile $(c_0, c_1, \ldots, c_T)$ is a linear combination of the utilities of the partial consumptions at different time points and the weights are exponential with respect to time t. Hence, the greater the time horizon represented by t, the lower the weight of the utility of c_t, namely the partial contribution of c_t, to the global utility. Formally, the utility function of a temporal consumption profile is

$$U(c_0, c_1, \ldots, c_T) = \sum_{t=0}^{T} \psi_t u(c_t) \tag{1}$$

where $u(c_t)$ is the partial utility that derives from consuming c_t at time t and ψ_t is the weight of $u(c_t)$ in the overall utility, with $t = 0, 1, \ldots, T$. The weight ψ_t is called *discount factor*. According to the exponential discounting approach, the *discount factor* is given by

$$\psi_t = \delta^t \tag{2}$$

which $\delta \in [0, 1]$. Thus, the weight of the utility of consuming c_t at time t decreases exponentially as t increases.

Let us use the identity function as utility function, i.e., $u(c_t) = c_t$. If, for example, c_t represents the amount of gain (remuneration, reward,...) or loss (penalty, fine,...) at time t and the discount rate is 0.10, the intertemporal choice consists in the decision whether it is better the discounted value $(1 + 0.10)^{-t} c_t$ now or c_t t years later. For a generic discount rate y, the choice is between $(1 + y)^{-t} c_t$ now or c_t at time t. Hence, one way to represent the *discount factor*, in the exponential discounting framework of the Samuelson's DU model, is the following:

$$\psi_t = \frac{1}{(1 + y)^t} \tag{3}$$

where y is the discount rate and $\delta = \frac{1}{1+y}$ is the discount factor corresponding to 1-year delay ($t = 1$).

According to the classic DU model, y is assumed to be constant with respect to the delay t, to the magnitude of the amount and to the sign (gain or loss). Some recent studies have reported empirical evidence against these properties of y. In the current debate emerging from the specialized literature, some new theories that deviate from the hypothesis of constant discount rate are proposed (see [2–13]). These new theories take the name of *intertemporal choice anomalies*. In particular, one of the anomalies, usually called *sign effect* or *gain–loss asymmetry* is that losses are discounted less than gains.

In order to estimate the discount rate and prove the sign effect and other anomalies in intertemporal choice, behavioral experiments are performed. In these studies, a sample of people is asked to make a series of choices concerning amounts of rewards and/or penalties that can be received/paid at different time points. For example, Green et al. [14] performed an experiment where 36 people from three different age groups were asked to choose between a fixed reward, obtainable at time t, and an immediate reward, reduced according to the individual discount rate. The experiment was repeated for two different amounts of the fixed reward (magnitudes) and 8 different time horizons t (delays). To compare the discount rates related to the two magnitudes, several univariate t-paired tests were performed within each age group and for each time horizon. To test for the delay effect, t tests and F tests were performed within each age group and for each amount. To test for the age effect, an ANOVA for each magnitude and time horizon was performed. This statistical approach is not suitable for the following reasons:

- the application of several univariate tests on the effect of age doesn't take into account the multivariate nature of the problem and the dependence of the discount rates of different magnitudes and time horizons
- the application of several univariate tests on the effect of magnitude doesn't take into account the multivariate nature of the problem with respect to time horizon and the confounding effect of age
- the application of several univariate tests on the delay effect doesn't take into account the multivariate nature of the problem with respect to the amount and the confounding role of age
- given the small sample sizes, the use of parametric tests that assume normality is inappropriate.

A similar approach, based on the application of several univariate t-tests, to study magnitude effect and sign effect, ignoring the multivariate nature of the problem and the confounding effects of demographic characteristics such as age, gender and income, is followed by McKerchar et al. [8].

Thus, a suitable method for testing intertemporal choice anomalies, in experiments like those described, should be multivariate and multistrata. Furthermore, to ensure robustness with respect to the family of distributions underlying the data, especially for small samples, a nonparametric approach is preferable.

3 A Sample Survey

In 2016, some *intertemporal choice* experiments were conducted at the University of Almeria. In one of these experiments, a sample of 36 students of the Faculty of Economics was interviewed. These students were asked to take "delay decisions." An example of delay decision is:
"today you won 100€ in the lottery and you can receive this award now or a different amount in three months. What is the minimum amount to delay the receving of the award?".
In another experiment, a different sample of 18 students was asked to take expedite decisions. An example of expedite decision is:
"today you won 100€in the lottery and you can receive this award in three months or a different amount today. What is the minimum amount to expedite the receiving of the award?".
Hence, the set of 54 students involved in these experiments can be classified into two categories according to the factor *decision type*: delay decision and expedite decision.

Each subject filled two 6×4 tables, one for each *scenario* (1. Lottery payout; 2. Payment of a fine). Each table was used in order to communicate the wished delayed/expedited value for six different time horizons t and four different award/fine amounts. The six different time horizons are 3 months, 1 year, 3 years, 5 years,

10 years, and 20 years. The four different amounts are 100€, 2,000€, 25,000€, and 100,000€.

In order to test for the sign effect, two dependent samples defined according to the scenario must be compared. Since each subject, for each scenario, provided 24 values (one for each time–amount combination), the problem is multivariate. In order to take into account the dependence between the 24 different variables, a suitable multivariate testing procedure should be applied. The complexity of the problem is even greater if we consider that the data derive from two different experiments characterized by different decision types. Thus, each decision type identifies a stratum of homogeneous students and a suitable multistrata test should be considered for the problem. The formal definition of the testing problem is presented in the following section.

4 Multivariate Multistrata Test for Sign Effect

As mentioned above, each subject involved in a complex intertemporal choice experiment like the one described in the previous section, filled 48 cells, e.g., two 6×4 tables. Indeed, for each subject, two scenarios, four amounts (in euros) and six time horizons (in years) were considered. Let us denote the subject's answer (in euros) regarding a given decision type and a specific "scenario–amount–time horizon" combination with $x_s^{(d)}(m, t)$, where

- $s \in \{A, F\}$ denotes the *scenario* (A : award; F : fine)
- $m \in \{100, 2\,000, 25\,000, 100\,000\}$ denotes the *amount* in euros of the award or of the fine
- $t \in \{0.25, 1, 3, 5, 10, 20\}$ denotes the *time* horizon in years
- $d \in \{D, E\}$ denotes the *decision type* (D : delay; E : expedite), i.e., the type of experiment, that takes the role of stratification factor.

Consistently with the previous notations, let us denote the subject's discount rate regarding a given decision type and a specific "scenario–amount–time horizon" combination with $y_s^{(d)}(m, t)$. According to (2), in the presence of delay decisions (postponing payments, $d = D$), the discount rate is

$$y_s^{(D)}(m, t) = \left[\frac{x_s^{(D)}(m, t)}{m}\right]^{1/t} - 1 \tag{4}$$

while, the discount rate in case of expedite decisions (anticipating payments, $d = E$), can be computed as follows:

$$y_s^{(E)}(m, t) = \left[\frac{m}{x_s^{(E)}(m, t)}\right]^{1/t} - 1. \tag{5}$$

According to the classic theory based on the Samuelson's DU model, $y_s^{(d)}(m, t)$ is a constant with respect to d, s, m, and t. In the case of sign effect, the discount rate of awards is not equal to the discount rate of fines, *ceteris paribus*. Hence, a test for sign effect is a two-sample test, where the discount rate of awards is compared to the discount rate of fines. The problem is multivariate with respect to amount and time horizon and multistrata with respect to the decision type. Thus, it is a multivariate and multistrata test for repeated measures (or dependent samples). The problem's factor is the *scenario* and the alternative hypothesis is one sided. In the specialized literature, the supporters of the sign effect believe that the direction of the effect depends on the type of decision: in delay decisions, the discount rate of awards is greater than the discount rate of fines; in expedite decisions, the opposite inequality holds. The testing problem is therefore quite complex and can be broken down into a set of sub-problems. Each sub-problem corresponds to a partial test.

Let $y_{s,1}^{(d)}(m, t), y_{s,2}^{(d)}(m, t), \ldots, y_{s,n}^{(d)}(m, t)$ be the computed individual discount rates and assume that the observed value $y_{s,i}^{(d)}(m, t)$, regarding the subject i, is a determination of the random variable $Y_s^{(d)}(m, t)$, with $i = 1, 2, \ldots, n$. The null hypothesis of the test for sign effect is

$$H_0 : \bigcap_m \bigcap_t \bigcap_d \left[Y_A^{(d)}(m, t) =^d Y_F^{(d)}(m, t) \right]. \tag{6}$$

In the null hypothesis, the discount rate of awards and the discount rate of fines follow the same distribution, and this is true for all the amounts, for all the time horizons and for both the decision types.

Under the alternative hypothesis, for at least one combination "amount-time horizon-decision type", there is a sign effect. The sign effect, if present, is opposite for delay and expedite decisions. Formally,

$$H_1 : \bigcup_m \bigcup_t \left\{ \left[Y_A^{(D)}(m, t) >^d Y_F^{(D)}(m, t) \right] \bigcup \left[Y_A^{(E)}(m, t) <^d Y_F^{(E)}(m, t) \right] \right\}, \tag{7}$$

where $>^d$ and $<^d$ denote the classic situations of stochastic dominance (see [15–24]). In short, the problem consists in a multivariate and multistrata stochastic dominance for repeated measures.

5 Nonparametric Solution

According to the specialized literature, the statistical tests usually applied in intertemporal choice problems present some limits. First of all, they are univariate and do not take into account the multivariate nature of the responses. Given that each interviewee must answer several questions, the response variable is obviously multivariate. The main difficulty of multivariate testing problems is to take into account the dependence structure of the marginal responses. Unless the independence between variables, infrequent and not very plausible, is true, the multivariate density function is not equal to the product of the marginal densities. The assumption of normality simplifies the representation of the multivariate distribution, and it implies a linear relationship between variables. But even in this case, pairwise correlation indices must be estimated. When there are not conditions to assume independence or linear dependence and normality, a parametric approach is very difficult if not impossible.

Furthermore, the inferential solutions proposed in the empirical literature on intertemporal choices are not suitable for complex hypotheses such as stochastic dominance and stochastic ordering. Moreover, they do not consider the possible presence of confounding factors like *decision type*. Finally, these tests are not robust with respect to the violation of the assumption about the underlying family of distributions, especially for small sample sizes.

A suitable solution can be found within the family of combined permutation tests [20]. This methodology follows a nonparametric approach because it is based on the nonparametric combination of dependent permutation tests. Let us consider the partial null hypothesis of (6)

$$H^{(d)}_{0,m,t} : Y^{(d)}_A(m,t) =^d Y^{(d)}_F(m,t) \tag{8}$$

and the partial alternative hypotheses of (7)

$$H^{(D)}_{1,m,t} : Y^{(D)}_A(m,t) >^d Y^{(D)}_F(m,t) \tag{9}$$

and

$$H^{(E)}_{1,m,t} : Y^{(E)}_A(m,t) <^d Y^{(E)}_F(m,t). \tag{10}$$

Let $T^{(d)}_{m,t}$ be the test statistic for testing $H^{(d)}_{0,m,t}$ versus $H^{(d)}_{1,m,t}$. Without loss of generality, we can define the partial test statistics in such a way that the null hypothesis is rejected in favor of the alternative when the test statistics take large values. Hence, for delay decisions, a suitable partial test statistic is

$$T^{(D)}_{m,t} = \bar{y}^{(D)}_A(m,t) - \bar{y}^{(D)}_F(m,t) \tag{11}$$

while, for expedite decisions, given that the direction of the alternative hypothesis is the opposite, a suitable partial test statistic is

$$T^{(E)}_{m,t} = \bar{y}^{(E)}_F(m,t) - \bar{y}^{(E)}_A(m,t) \tag{12}$$

where $\bar{y}^{(D)}_A(m,t)$, $\bar{y}^{(D)}_F(m,t)$, $\bar{y}^{(E)}_F(m,t)$ and $\bar{y}^{(E)}_A(m,t)$ are the observed sample means of the subgroup discount rates.

For each partial problem, a permutation test for dependent samples is performed. In order to take into account the dependence between the partial test statistics, the permutations applied for the computation of the null distribution should be the same for all the partial tests. A suitable combining function is then necessary to compute a univariate test statistic for the global multivariate problem and obtain a unique p-value. Let $T_{m,t}^{(d)}(0)$ be the observed value of the partial test statistic, B the number of permutations and $T_{m,t}^{(d)}(b)$ the value of the partial test statistic corresponding to the bth permutation. The significance level function of $T_{m,t}^{(d)}(b)$ is

$$\lambda_{m,t}^{(d)}(b) : P\left[T_{m,t}^{(d)} \geq T_{m,t}^{(d)}(b) | H_{0,m,t}^{(d)}\right], \tag{13}$$

with $b = 0, 1, 2, \ldots, B$ and it is the proportion of values of the partial test statistic greater than or equal to $T_{m,t}^{(d)}(b)$ according to the null permutation distribution.

The combined test statistic for the global problem is obtained through the application of a suitable function $\psi(\cdot)$. The combining function $\psi(\cdot)$ must be non-increasing with respect to each argument $\lambda_{m,t}^{(d)}$. In the case of Tippett combining rule, the combined test statistic is

$$T_{comb} = max_{m,t,d}\left[1 - \lambda_{m,t}^{(d)}\right], \tag{14}$$

and the global p-value is the proportion of values greater than or equal to $T_{comb}(0)$ in the set $\{T_{comb}(0), T_{comb}(1), \ldots, T_{comb}(B)\}$. Instead of considering all the possible permutations of the exact test, for computational convenience, a random sample of B permutations (CMC resampling) can be used to estimate the null distribution of the test statistics.

In the case of significance of the global test, in order to attribute this result to some partial tests, an adjustment of the partial p-values must be done for controlling the Familywise Error Rate.

6 Case Study

The combined permutation test, with Tippett combination, was applied to the data collected in the experiments done at the University of Almeria in 2016. A two-step combination was performed. At the first step, the partial tests were combined with respect to time horizon and amount. At the second step, the two resulting combined tests were combined again to obtain the p-value of the overall test. $B = 1000$ CMC resamplings were considered (Table 1).

The p-value of the global test on sign effect is equal to 0.0001, thus at $\alpha = 0.01$ the null hypothesis that the scenario does not affect the discount rate is rejected in favor of the alternative hypothesis of sign effect (strong significance). For controlling the Familywise Error Rate and avoiding the type first error rate inflation due to the multiplicity of the test, the Bonferroni–Holm method was applied. According to the adjusted p-values of the partial tests, we have a strong significance of the sign effect in the case of delay decisions (adjusted $p = 0.0002$). For expedite decisions, the sign

Table 1 Combined permutation test on sign effect

Hypothesis	p	Adjusted p
Delay decision		
Discount rate of awards greater than discount rate of fines	0.0001	0.0002
Expedite decision		
Discount rate of awards less than discount rate of fines	0.0726	0.0726
Global		
Sign effect	0.0001	

effect presents weak significance because the adjusted $p = 0.0726$ is greater than 0.01 but less than the significance level if $\alpha = 0.10$.

7 Conclusions

The test for sign effect in intertemporal choice experiments needs the application of suitable multivariate testing techniques. In such experiments, a multistrata and multivariate test must be applied. The very frequent practice in the literature of applying univariate (very often parametric) tests is therefore wrong and makes untrustworthy inferential conclusions. In this paper, the application of multiple tests based on the permutation approach and the combination of Tippett is proposed. This solution is suitable for the problem and overcomes the limits of the mentioned inadequate parametric univariate tests.

The proposed multivariate permutation test is suitable for complex hypotheses, takes into account the multivariate nature of the problem and the possible confounding effects due to the presence of stratification factors (e.g., decision type) and does not require the assumption that the underlying distribution is normal or belongs to a known family of probability distributions. Furthermore, this test is consistent, exact and unbiased. It is suitable even in the presence of small sample sizes and when the number of marginal response variables is very large.

The application of this method to the data collected in the survey performed at the University of Almeria in 2016, provides empirical evidence in favor of the hypothesis of sign effect in intertemporal choice: the discount rate of gains seems to be greater than the discount rate of losses for delay decisions, while the discount rate of gains appears less than the discount rate of losses for expedite decisions. The latter property is less evident than the former.

References

1. Samuelson, P.A.: A note on measurement of utility. Rev. Econ. Stud. **4**(2), 155–161 (1935)
2. Andersen, S., Harrison, G.W., Lau, M.I., Rutstrom, E.E.: Discounting behavior and the magnitude effect: evidence from a field experiment in Denmark. Economica **80**, 670–697 (2013)
3. Benhabib, J., Bisin, A., Schotter, A.: Present-bias, quasi-hyperbolic discounting and fixed costs. Game. Econ. Behav. **69**(2), 205–223 (2010)
4. Bocqueho, G., Jacquet, F., Reynaud, A.: Reversal and magnitude effects in long-term time preferences: results from a field experiment. Econ. Lett. **120**, 108–111 (2013)
5. Grace, R.C., Sargisson, R.J., White, K.G.: Evidence for a magnitude effect in temporal discounting with pigeons. J. Exp. Psychol. Anim. B. **38**(1), 102–108 (2012)
6. Green, I., Myerson, J., Oliveira, L., Chang, S.E.: Delay discounting of monetary rewards over a wide range of amounts. J. Exp. Anal. Behav. **100**(3), 269–281 (2013)
7. Hardisty, D.J., Appelt, K.C., Weber, E.U.: Good or bad, we want it now: fixed-cost present bias for gains and losses explains magnitude asymmetries in intertemporal choice. J. Behav. Decis. Mak. **26**, 348–361 (2013)
8. McKerchar, T.L., Pickford, S., Robertson, S.E.: Hyperboloid discounting of delayed outcomes: magnitude effects and the gain-loss asymmetry. Psychol. Rec. **63**(3), 441–451 (2013)
9. Meyer, A.G.: The impacts of elicitation mechanism and reward size on estimated rates of time preference. J. Behav. Exp. Econ. **58**, 132–148 (2015)
10. Mitchell, S.H., Wilson, W.B.: The subjective value of delayed and probabilistic outcomes: outcome size matters for gains but not for losses. Behav. Process. **83**(1), 36–40 (2010)
11. Noor, J.: Intertemporal choice and the magnitude effect. Games Econ. Behav. **72**, 255–270 (2011)
12. Scholten, M., Read, D.: Time and outcome framing in intertemporal tradeoffs. J. Exp. Psychol. Learn. **39**(4), 1192–1212 (2013)
13. Yuki, S., Okanoya, K.: Relatively high motivation for context-evoked reward produces the magnitude effect in rats. Behav. Process. **107**, 22–28 (2014)
14. Green, L., Fry, A.F., Myerson, J.: Discounting of delayed rewards: a life-span comparison. Psychol. Sci. **5**(1), 33–36 (1994)
15. Arboretti, R., Bonnini, S., Corain, L., Salmaso, L.: Dependency and truncated forms of combinations in multivariate combination-based permutation tests and ordered categorical variables. J. Stat. Comput. Simul. **86**(18), 3608–3619 (2016)
16. Bonnini, S.: Multivariate approach for comparative evaluations of customer satisfaction with application to transport services. Commun. Stat.-Simul. C **45**(5), 1554–1568 (2016)
17. Arboretti, R., Bonnini, S., Corain, L., Vidotto, D.: Environmental odor perception: testing regional differences on heterogeneity with application to odor perceptions in the area of Este (Italy). Environmetrics **26**(6), 418–430 (2015)
18. Bonnini, S., Prodi, N., Salmaso, L., Visentin, C.: Permutation approaches for stochastic ordering. Commun. Stat.-Theory Methods **43**(10–12), 2227–2235 (2014)
19. Bonnini, S.: Testing for heterogeneity with categorical data: permutation solution versus bootstrap method. Commun. Stat.-Theory Methods **43**(4), 906–917 (2014)
20. Pesarin, F., Salmaso, L.: Permutation Tests for Complex Data: Theory, Applications and Software. Wiley, Chichester (2010)
21. Arboretti, R., Bonnini, S., Pesarin, F.: A permutation approach for testing heterogeneity in two-sample problems. Stat. Comput. **19**(2), 209–216 (2009)
22. Arboretti, R., Bonnini, S., Salmaso, L.: Employment status and education/employment relationship of PhD graduates from the University of Ferrara. J. Appl. Stat. **36**(12), 1329–1344 (2009)
23. Arboretti, R., Bonnini, S.: Moment-based multivariate permutation tests for ordinal categorical data. J. Nonparametric Stat. **20**(5), 383–393 (2008)
24. Pesarin, F.: Multivariate Permutation Tests with Applications in Biostatistics. Wiley, Chichester (2001)

Nonparametric First-Order Analysis of Spatial and Spatio-Temporal Point Processes

M. I. Borrajo, I. Fuentes-Santos, and W. González-Manteiga

Abstract First-order characteristics are essential functions in point processes representing the distribution of events in the corresponding domain. For decades, the inconsistency of the first-order kernel intensity estimator has been an obstacle to perform inference in the point process context. In this work, we develop different procedures to obtain consistent estimators of the first-order intensity function, and we also propose bootstrap procedures to define effective bandwidth selectors. Moreover, these innovations are used in three testing problems: the goodness-of-fit of an appealing model in the literature of point processes with covariates, the nonparametric comparison of first-order intensity functions and a separability test for spatio-temporal point process. We illustrate the above-mentioned procedures with two wildfire data sets in Galicia (NW Spain) and in Canada.

1 Introduction

The main aim of point processes is to study the geometrical structure of patterns formed by events that are distributed randomly in number and space. Particularly, spatial point processes focus on events located in a planar bounded region $W \subset \mathbb{R}^2$, and spatio-temporal point processes determine the spatial location and time of occurrence of events in a volume, $W \times T \subset \mathbb{R}^2 \times \mathbb{R}^+$, defined by a planar

M. I. Borrajo (✉)
Department of Mathematics and Statistics, Faculty of Science and Technology, Lancaster University, Lancaster LA1 4YF, UK
e-mail: mariaisabel.borrajo@usc.es

I. Fuentes-Santos
Instituto de Investigaciones Marinas - CSIC, Calle de Eduardo Cabello 6, E36280 Vigo, Spain
e-mail: isafusa@gmail.com

W. González-Manteiga
Facultade de Matemáticas, Universidade de Santiago de Compostela, Calle Lope Gómez de Marzoa s/n, E15782 Santiago de Compostela, Spain
e-mail: wenceslao.gonzalez@usc.es

M. La Rocca et al. (eds.), *Nonparametric Statistics*, Springer Proceedings in Mathematics & Statistics 339, https://doi.org/10.1007/978-3-030-57306-5_10

region and a temporal interval. If each event has associated any extra information in the form of a measure, the process is named as marked, but if this extra information exists over the whole observation region, then we are dealing with point processes with covariates.

The analysis of any observed point pattern involves characterizing the spatial distribution of events (first-order characteristics) and interaction between them (second- and higher order characteristics). In this paper, we are focused on the former which has been addressed through parametric models, see Moller and Waagepetersen [19], Bayesian methods, see Illian et al. [18], and nonparametric approaches, see Diggle [11] and Baddeley et al. [2].

Diggle [10] proposed the first kernel intensity estimator, based on the structure of the common kernel density estimator. The main drawback of Diggle's proposal is its lack of consistency, which has almost limited its use to exploratory analysis. Two ideas have been introduced so far to overcome this problem: Cucala [9] introduced the density of event locations and proved the consistency of his estimator, and Guan [17], Baddeley et al. [1] introduced kernel estimators of the first-order intensity based on covariates.

Considering all these approaches, this work addresses important developments in first-order intensity inference: two consistent nonparametric estimators of the first-order intensity, new bandwidth selectors, and different nonparametric tests based on these estimators. This work is organized as follows: in Sect. 2, we use the two strategies referred before to define consistent estimators of the first-order intensity function and we propose bootstrap bandwidth selectors for the two proposals. Section 3 introduces nonparametric tests developed to check for the effect of covariates on the spatial distribution of an observed pattern, compare the intensity of two spatial point processes, and test whether a spatio-temporal point process is separable, and finally in Sect. 4, we illustrate the utility of the techniques introduced above through application to the analysis of wildfire patterns in Galicia (NW Spain) and Canada.

2 First-Order Intensity Estimation

Let X be a spatial point process defined in a bounded region $W \subset \mathbb{R}^2$. Let $X_1, \ldots, X_N$ be a realization of the process with N the random variable counting the number of events. The first-order intensity, from now on referred as intensity, is defined following Diggle [11] as

$$\lambda(x) = \lim_{|dx| \to 0} \frac{E[N(dx)]}{|dx|},$$

where $|dx|$ denotes the area of an infinitesimal region containing the point $x \in \mathbb{R}^2$.

Diggle [10] proposed a kernel intensity estimator for one-dimensional point processes, which has been extended to the plane as

$$\hat{\lambda}_H^D(x) = \frac{\sum_{i=1}^N \mathbf{K}_H(x - X_i)}{p_H(x)}, \quad x \in W \subset \mathbb{R}^2.$$

Here, H is a matrix of bandwidth parameters, $\mathbf{K}_H(x) = |H|^{-1/2}\mathbf{K}\left(H^{-1/2}x\right)$, where $\mathbf{K}$ is a two-dimensional kernel function, and $p_H = \int_W |H|^{-1/2}\mathbf{K}(H^{-1/2}(x - y))dy$ is an edge correction term.

This kernel estimator has been widely used during decades and mostly limited to exploratory analysis due to its lack of consistency. This means that its mean integrated squared error (MISE) does not tend to zero as the expected number of events increases. To better understand this point, let us assume an infill structure or increasing intensity asymptotic framework (see Diggle and Marron [12]), which states that the expected number of events in the observation region W, tends to infinity, and it is equivalent to the asymptotic framework in the classical kernel density estimator. The kernel estimator uses local information around each point to estimate the intensity. If the true intensity is continuous, local smoothing will provide an asymptotically unbiased estimator. However, as the number of events in an infinitesimal region increases, the variance of the estimate does not tend to zero (See details in Fuentes-Santos et al. [14]), then the MISE does not tend to zero either, leading to an inconsistent estimator.

Trying to overcome this lack of consistency, Cucala [9] introduced the concept of "density of events locations" for one-dimensional point processes. He defined such density as $\lambda_0(x) = \lambda(x)/m$, where $m = \int_W \lambda(x)dx$ is the expected number of events lying on W. And, he proposed the following kernel estimator for λ_0:

$$\hat{\lambda}_{0,h}(x) = \frac{1}{N}\sum_{i=1}^{N} K_h(x - X_i)1_{\{N\neq 0\}}, \; x \in \mathbb{R},$$

where $K_h(\cdot) = h^{-1}K(\cdot/h)$, with K being a one-dimensional kernel function and h a scalar bandwidth parameter. Here $1_{\{\}}$ denotes the indicator function. Cucala [9] proved the consistency of its kernel estimator for Poisson point processes under an infill asymptotic framework. In a similar way, we need the Poisson assumption to derive the asymptotic theory for the proposed estimators that will follow.

Following the philosophy of bivariate kernel density estimation, we define a kernel estimator of the density of event locations in two dimensions with a bandwidth matrix:

$$\hat{\lambda}_{0,H}(x) = \frac{\hat{\lambda}_H(x)}{N}1_{\{N\neq 0\}} = (p_H(x)N)^{-1}|H|^{-1/2}\sum_{i=1}^{N}\mathbf{K}\left(H^{-1/2}(x - X_i)\right)1_{\{N\neq 0\}}, \tag{1}$$

where the bandwidth matrix, H, is symmetric and positive-definite and $|H|$ denotes the determinant of H. Fuentes-Santos et al. [14] developed a smooth bootstrap procedure to obtain a consistent estimator of the MISE, which is the basis for the plug-in bandwidth selector proposed in the same work.

Moving on to the framework of point processes with covariates, let $Z : W \subset \mathbb{R}^2 \to \mathbb{R}$ be a spatial continuous covariate that is exactly known in every point of the region of interest W, and $Z_1, \ldots, Z_N$ the realization of the transformed process, i.e, $Z_i = Z(X_i)$. In practice, following the indications of Baddeley et al. [1], this covariate will commonly be known in enough amount of points spread over the region, so the values for the rest of the points can be interpolated and it can be assumed that these values are indeed the real ones.

Following Baddeley et al. [1], let's assume that the intensity can be described from the known covariate through the model:

$$\lambda(u) = \rho(Z(u)), \; u \in W \subset \mathbb{R}^2, \tag{2}$$

where ρ is an unknown function. As Z is known, the only target for intensity estimation is the function ρ. To this purpose, it is considered the transformed one-dimensional point process, $Z(X)$, and established the theoretical relationship between it and the original two-dimensional process X. It has been proved that if X is a Poisson point process in $W \subset \mathbb{R}^2$ with intensity function (2), then $Z(X)$ is a Poisson point process in $\mathbb{R}$ with intensity $\rho g^\star$ and with the same expected number of events, where $g^\star$ is the non-normalized version of the derivative of the spatial cumulative distribution function, see Borrajo et al. [3] for details on this and the extension to multidimensional covariates.

Previously, Guan [17] proposed a closely related kernel estimator that allowed for a multidimensional covariate, $\mathbf{Z} = (Z_1, \ldots, Z_p) : W \subset \mathbb{R}^2 \to \mathbb{R}^p$. This estimator involves measuring the distance between two points by the euclidean distance through their covariates values:

$$\hat{\lambda}_h^G(u) = \frac{\sum_{i=1}^N K_h(||\mathbf{Z}(u) - \mathbf{Z}(X_i)||)}{q_h(u)},$$

with $q_h(u) = \int_W K_h(||\mathbf{Z}(u) - \mathbf{Z}(s)||)ds$ is the edge correction term. Considering the increasing domain asymptotic framework and adding also some suitable assumptions, Guan [17] proved the consistency of the estimator. He also addressed the bandwidth selection problem by a simple, but computationally intense, cross-validation method.

We need to introduce some definitions and additional notation. The spatial cumulative distribution function of Z is defined as

$$G(z) = \frac{1}{|W|} \int_W 1_{\{Z(u) \leq z\}} du,$$

where $|W|$ denotes the area of the region $W \subset \mathbb{R}^2$. Let assume that G has a first derivative g, for which we need Z to be differentiable with non-zero gradient, and let denote by $g^\star(\cdot) = |W| g(\cdot)$ and $G^\star(\cdot) = |W| G(\cdot)$ the unnormalized versions. The results detailed in Borrajo et al. [3] show that $Z(X)$ is indeed a point process with intensity $\rho g^\star$.

To derive our consistent kernel estimator, we follow Cucala [9] and use the relationship between the intensity and the density function. We define the density function for this problem as the relative density of the transformed point process $Z(X)$:

$$f(z) = \frac{\rho(z)g^{\star}(z)}{m}. \tag{3}$$

Our idea is to construct a kernel estimate of f and then plug in it in the expression (3), jointly with an appropriate estimate of m, and to derive an estimator of ρ. This gives the estimator of the intensity λ through Eq. (2).

Following the pre-established notation, we define the following estimator of the relative density f:

$$\hat{f}_h(z) = g^{\star}(z)\frac{1}{N}\sum_{i=1}^{N}\frac{1}{g^{\star}(Z_i)}K_h\left(z - Z_i\right)1_{\{N\neq 0\}}. \tag{4}$$

Now we use (3) to define the final intensity estimator. To this goal, we need to estimate m that recall is the expected number of events. For simplicity, we suggest the sample size N as an estimator and hence derive our kernel intensity estimator from $\hat{f}_h$ as:

$$\hat{\lambda}_h(u) = \hat{\rho}_h(Z(u)) = \frac{N\hat{f}_h(Z(u))}{g^{\star}(Z(u))} = \sum_{i=1}^{N}\frac{1}{g^{\star}(Z_i)}K_h\left(Z(u) - Z_i\right). \tag{5}$$

Remark that, for the particular estimates of the relative density and m, we propose our final intensity estimator shares the same expression as Baddeley et al. [1]'s estimator detailed in the previous section. However, our proposal benefits for being conveniently constructed to guarantee the consistency, to facilitate the theoretical developments and to allow consistent bootstrap methods. This construction also allows intuitive multivariate extensions, including the time dimension, as is discussed in Borrajo et al. [3].

In Borrajo et al. [3], a complete theoretical framework with all the details in terms of mean squared error (MSE) and mean integrated squared error (MISE) is developed, and the expression of an asymptotically optimal bandwidth parameter is also derived.

Bootstrap methodology for bandwidth selection

Nonparametric bootstrap procedures have been widely used in different contexts to perform inference and calibrate the distribution of statistics in tests. The smooth bootstrap procedure for point processes with and without covariates we propose is based on the following works: Cao [7] for kernel density estimation and Cowling et al. [8] for the intensity estimation of a Poisson point process.

Recall $X_1, \ldots, X_n$ is a realization of the spatial point process X, $Z_1, \ldots, Z_n$ the associated realization of the transformed univariate process; let $\hat{f}_b$ be the density estimator in (4) and $\hat{\rho}_b$ the estimator derived from (3) and (4), where b is a pilot bandwidth. Now, conditional on $Z_1, \ldots, Z_n$, let $N^* \sim Poiss(\hat{m})$ with $\hat{m} := \int_{\mathbb{R}} \hat{\rho}_b(z) g^\star(z) dz$, generate n^* a realization of this random variable N^* and then draw $Z_1^*, \ldots, Z_{n^*}^*$ by sampling randomly with replacement n^* times from the distribution with density proportional to $g^\star \hat{\rho}_b$, i.e., $\tilde{f}_b = \frac{\hat{\rho}_b g^\star}{\hat{m}}$.

Using this bootstrap, we have developed a data-driven bandwidth selection procedure for (4); moreover, in Borrajo et al. [3], a specifically designed rule-of-thumb is defined and both selectors are compared with the existing competitors, which to the extent of our knowledge is only the classical Silverman's rule-of-thumb used in Baddeley et al. [1].

3 Testing Problems

Testing first-order intensity model in inhomogeneous Poisson point processes with covariates

In this section, we want to test a null hypothesis $H_0 : \lambda(x) = \rho(Z(x))\, x \in W$, versus a general alternative in which the intensity function is not explained completely through the covariate, for Poisson processes. The idea is to define a test statistic based on a L^2-distance between the classical kernel intensity estimator using only location information and the appealing one using covariate information. To avoid the problem of the lack of consistency, we are using the density of event location and the null hypothesis can be equivalently rewritten as $H_0 : \lambda_0(x) = \rho(Z(x))/m$.

The procedure to construct the statistic isthat we first estimate the relative density with the two-dimensional kernel estimator (1), and then we estimate it using (4). We apply the L^2-distance to obtain a statistic that measures the discrepancy between them:

$$T_1 = \int_W \left(\hat{\lambda}_{0,H}(x) - \hat{\rho}_{0,b}(Z(x)) \right)^2 dx, \tag{6}$$

where $\hat{\rho}_{0,b}(Z(x)) = \frac{\hat{\rho}_b(z)}{N} 1_{\{N \neq 0\}}$ with $\hat{\rho}_b(z) = \hat{f}_b(z) m / g^\star(z)$, with $b \equiv b(m)$ a real bandwidth parameter, see Borrajo et al. [3].

The asymptotic distribution of the statistic (6) under a suitable framework is derived. However, in practice, this asymptotic distribution may not be the best way to calibrate our test since the convergence rate is too slow. Our proposal is to use a bootstrap procedure to perform the calibration, see Borrajo et al. [4] for details.

A complete simulation study including several scenarios and different sample sizes has been carried out in Borrajo et al. [4], showing good values in terms of level and power for this test, that to the extent of our knowledge has yet no competitors.

Nonparametric comparison of first-order intensity functions for Poisson processes

A common question in the analysis of spatial point processes is whether two types of events have the same spatial structure.

Let X_1 and X_2 be spatial patterns of type 1 and type 2 events in a spatial point process X observed in $W \subset \mathbb{R}^2$. We denote, respectively, by $\lambda_1(x)$ and $\lambda_2(x)$ the first-order intensities, and by $\lambda_{01}(x)$, $\lambda_{02}(x)$ their densities of event locations. We can extend the proposal of Duong et al. [13] for multivariate data to the spatial point process framework and use a L^2-distance to test the null hypothesis $\mathcal{H}_0 : \lambda_{01}(x) = \lambda_{02}(x) = \lambda_0(x)$:

$$T_2 = \int_W \left(\hat{\lambda}_{01}(x) - \hat{\lambda}_{02}(x)\right)^2 dx = \hat{\psi}_1 + \hat{\psi}_2 - \left(\hat{\psi}_{12} + \hat{\psi}_{21}\right) \tag{7}$$

where $\hat{\psi}_{ij}$ and $\hat{\psi}_i$ are kernel estimators of$\psi_{ij} = \int_W \lambda_{0i}(x)\, \lambda_{0j}(x)\, dx$ for $i, j = 1, 2$ and $\psi_i = \int_W \lambda_{0i}(x)^2\, dx$.

Fuentes-Santos et al. [15] proved the asymptotic normality of the null distribution of this statistic under some regularity conditions. Again, given that the convergence to the asymptotic distribution is slow, we propose a bootstrap calibration, which good performance was proved through a simulation study in that paper.

This same problem has been extended to the context of point processes with covariates, see Borrajo et al. [5] for details.

Spatio-temporal separability test

Let $\mathbf{S} = \{(X_1, t_1), \ldots, (X_N, t_N)\}$ be a realization of a spatio-temporal point process observed on a bounded domain $W \times T \subset \mathbb{R}^2 \times \mathbb{R}^+$, the spatio-temporal intensity function (STIF) is a natural extension of the first-order intensity function of a spatial point process:

$$\lambda(x, t) = \lim_{|dx \times dt| \to 0} \left\{ \frac{E\left[N(dx, dt)\right]}{|dx \times dt|} \right\}, \tag{8}$$

where $N(dx, dt)$ represents the number of events in the volume $dx \times dt$, dx is an infinitesimal disc containing the location x, and dt is an infinitesimal interval around time t.

One of the first steps in the analysis of any observed pattern is testing whether the STIF is separable, i.e., whether it can be expressed as the product of its spatial and temporal components: $\lambda(x, t) = \lambda_1(x)\lambda_2(t)$. Under separability the ratio between the spatio-temporal and spatial intensities, $r(x, t) = log\,(\lambda(x, t)/\lambda_1(x))$, does not depend on the spatial locations, x, for any $t \in T$. Considering this property, Fuentes-Santos et al. [16] propose using a no-effect test that checks whether the log-ratio function $r(x, t) = \lambda(x, t)/\lambda(x)$ depends on the spatial locations.

To implement the test we first need an estimator of $r(x, t)$. We propose using the log-ratio of the kernel spatio-temporal and spatial intensities with diagonal bandwidth matrices selected by least-squares cross-validation.

Once the log-ratio function has been estimated we have a regression problem where the log-ratio function evaluated at each event, $Y = \{Y_i = \hat{\rho}(X_i, t_i), i = 1, \ldots, n\}$, is a response variable that may depend on the spatial covariate $X = \{X_i = (X_{i1}, X_{i2}), i = 1, \ldots, n\}$ comprising the event locations, and we test for the effect of X on Y. Following Bowman and Azzalini [6], we shall discriminate between two models:

$$\mathcal{H}_0 : E\left[Y_i|X_i\right] = \mu \qquad \text{and} \qquad \mathcal{H}_1 : E\left[Y_i|X_i\right] = m(X_i).$$

We first estimate μ by the empirical mean, $\hat{y} = n^{-1}\sum_{i=1}^{n} Y_i$, and the unknown smooth function, $m(\cdot)$, by kernel regression; then we compute the residual sum of squares for the null, RSS_0, and alternative, RSS_1, models and we define the generalized test:

$$T_3 = \frac{\left(RSS_0 - RSS_1\right) / \left(df_1 - df_0\right)}{RSS_1/df_1}, \tag{9}$$

where df_0, df_1 denote the degrees of freedom for these residuals. Finally, we propose using a permutation test as calibration procedure, see details in Fuentes-Santos et al. [16].

4 Real Data Analysis

Wildfire is the most ubiquitous natural disturbance in the world and represents a problem of considerable social and environmental importance. In this section, we apply the methodology previously presented to two data sets: one consisting of wildfires in Galicia (NW Spain) and the other in Canada. Both regions have a very different background on wildfires. On one hand, Galicia is known to have a low risk of wildfires due to meteorological conditions (it is a very green, rainy region with low to moderate temperatures the whole year), but it has been suffering an extremely high incidence due to arson fires, which have become a major environmental and social problem in the region. On the other hand, Canadian wildfires are known to be studied over decades from different perspectives and meteorological conditions are supposed to be one of the key factors in the incidence.

Galician wildfire data

The first data set comprises the spatial locations and time of occurrence of arson and natural wildfires registered in Galicia during 2006, see Fig. 1. Wildfire data can be obtained through a request to the Wildfire Statistics Department at the Spanish Ministry of Agriculture, Fisheries and Food (https://www.mapa.gob.es/es/desarrollo-rural/estadisticas/Incendios_default.aspx). We have applied kernel intensity estimation and the tests introduced above to characterize the spatial distribution of fires, check whether arson and natural fires have similar behavior and test whether the risk of fire in a given location varies over time.

The kernel intensity estimators in Fig. 2 show that during 2006 the west coast of Galicia registered high incidence of arson fires, where natural fires were more frequent in the east and center of this region. The nonparametric comparison of

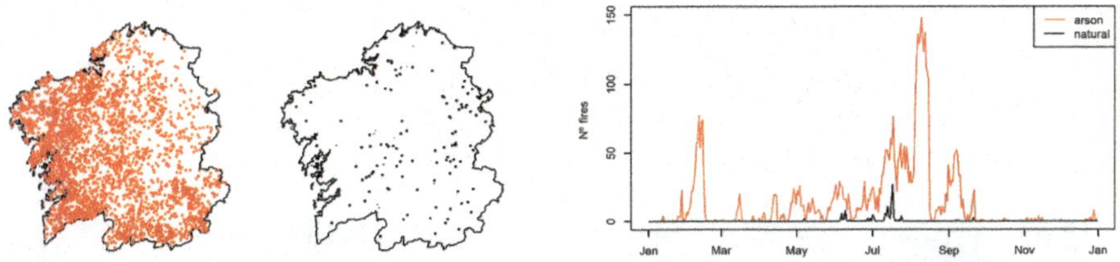

Fig. 1 Spatial pattern of arson (*left*) and natural (*center*) wildfires, and temporal pattern (*right*) of arson (*red*) and natural (*blue*) wildfires registered in Galicia during 2006

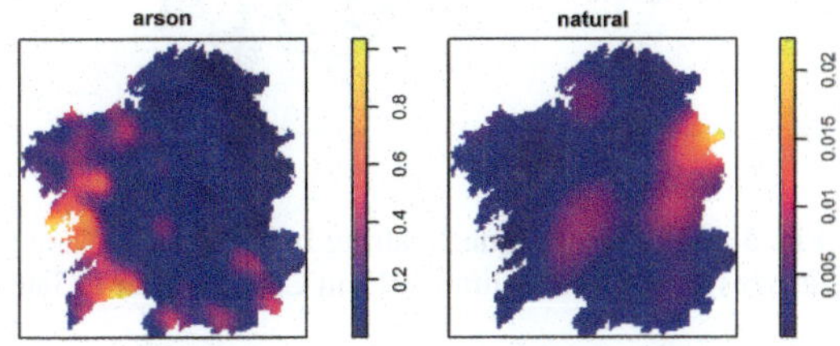

Fig. 2 Kernel intensity estimator of arson (*left*) and natural (*right*) wildfires registered in Galicia during 2006 (different scale)

intensities confirmed that arson and natural wildfires had different intensities. The F-test detected departure from separability in both wildfire patterns. Therefore the spatial distribution of arson and natural wildfires varied over time and support the need for nonseparable models to estimate their spatio-temporal intensity.

Canada wildfire data

Fire activity in Canada mostly relies on meteorological elements such as long periods without rain and high temperatures. We want to study the influence of some of these covariates in the distribution of the process generating the wildfires, particularly on its first-order intensity.

The wildfire data set and also complete meteorological information from the last decades is available at the Canadian Wildland Fire Information System website (http://cwfis.cfs.nrcan.gc.ca/home). We analyze later the influence of meteorological covariates on wildfires during June 2015 (a total number of 1841), see Fig. 3, focusing in this paper our attention on the temperature. It is important to note that for inferential purposes we have removed two regions (Northwest Territories and Nunavut, mostly covered by ice layers) from the whole observation window (Canada) because there are no fires registered on those iced regions.

In Fig. 4, we see the estimations resulting from using the classical kernel intensity estimator, which does not use covariate information, by Diggle [10] and (4). As expected the covariate information seems to be useful in this context because the resulting estimate represents better the pattern. So this might indicate that the temperature has an influence in the distribution of Canadian wildfires.

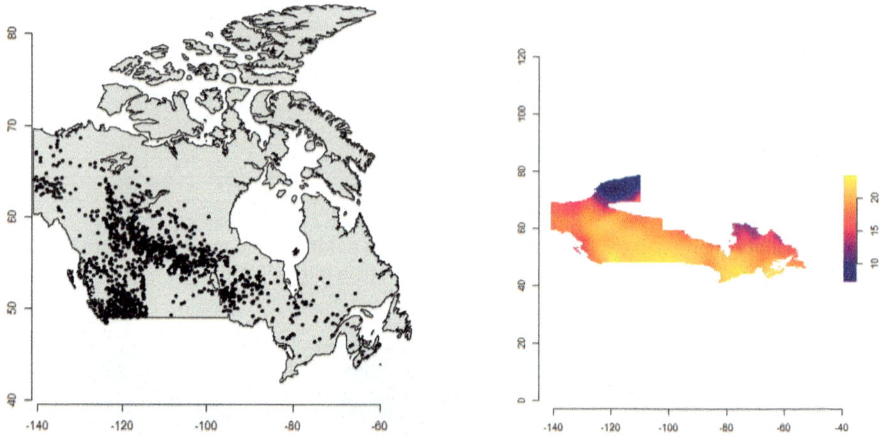

Fig. 3 Wildfires in Canada during June 2015 (*left*) and third quartile of the temperature (in Celsius degrees) registered in June 2015 in Canada, after a Gaussian smoothing with $\sigma = 2$ (*right*)

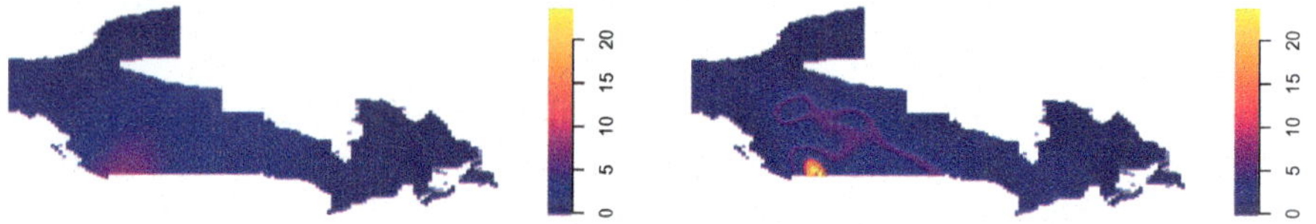

Fig. 4 Estimation without covariate information (*left*), and estimations using temperature as the covariate (*right*)

When we perform the goodness-of-fit test in (6) with this data set, we reject the null hypothesis, so it seems that the temperature is not enough to explain the wildfire distribution, which does not mean that it has o influence. An improvement in this situation is defining indicators using several covariates or applying to those covariates the multidimensional version of the test that is detailed in Borrajo et al. [4].

References

1. Baddeley, A., Chang, Y.M., Song, Y., Turner, R.: Nonparametric estimation of the dependence of a spatial point process on spatial covariates. Stat. Interface **5**, 221–236 (2012)
2. Baddeley, A., Rubak, E., Turner, R.: Spatial Point Patterns: Methodology and Applications with R. CRC Press (2015)
3. Borrajo, M.I., González-Manteiga, W., Martínez-Miranda, M.D.: Bootstrapping kernel intensity estimation for inhomogeneous point processes with spatial covariates. Comput. Stat. Data Anal. **144** (2020)
4. Borrajo, M.I., González-Manteiga, W., Martínez-Miranda, M.D.: Testing first-order intensity model in non-homogeneous poisson point processes with covariates) (2020). (Submitted). arXiv:submit/2316351

5. Borrajo, M.I., González-Manteiga, W., Martínez-Miranda, M.D.: Testing for significant differences between two spatial patterns using covariates. Spat. Stat. (2019)
6. Bowman, A.W., Azzalini, A.: Applied Smoothing Techniques for Data Analysis: The Kernel Approach with S-Plus Illustrations. Oxford Statistical Science Series, vo;. 18 (1997)
7. Cao, R.: Bootstrapping the mean integrated squared error.J. Multivar. Anal. **45**(1), 137–160 (1993)
8. Cowling, A., Hall, P., Phillips, M.J.: Bootstrap confidence regions for the intensity of a poisson point process. J. Am. Stat. Assoc. **91**(436), 1516–1524 (1996)
9. Cucala, L.: Espacements bidimensionnels et données entachés d'erreurs dans l'analyse des procesus ponctuels spatiaux. Ph.D. thesis, Université des Sciences de Toulouse I (2006)
10. Diggle, P.: A kernel method for smoothing point process data. J. R. Stat. Soc. Ser. C **34**(2), 138–147 (1985)
11. Diggle, P.J.: Statistical Analysis of Spatial and Spatio-Temporal Point Patterns. CRC Press (2013)
12. Diggle, P.J., Marron, J.S.: Equivalence of smoothing parameter selectors in density and intensity estimation. J. Am. Stat. Assoc. **83**(403), 793–800 (1988)
13. Duong, T., Goud, B., Schauer, K.: Closed-form density-based framework for automatic detection of cellular morphology changes. Proc. Natl. Acad. Sci. U.S.A. **109**(22), 8382–8387 (2012)
14. Fuentes-Santos, I., González-Manteiga, W., Mateu, J.: Consistent smooth bootstrap kernel intensity estimation for inhomogeneous spatial poisson point processes. Scand. J. Stat. **43**(2), 416–435 (2016)
15. Fuentes-Santos, I., González-Manteiga, W., Mateu, J.: A nonparametric test for the comparison of first-order structures of spatial point processes. Spat. Stat. **22**, 240–260 (2017)
16. Fuentes-Santos, I., González-Manteiga, W., Mateu, J.: A first-order, ratio-based nonparametric separability test for spatiotemporal point processes. Environmetrics **29**(1) (2018)
17. Guan, Y.: On consistent nonparametric intensity estimation for inhomogeneous spatial point processes. J. Am. Stat. Assoc. **103**(483), 1238–1247 (2008)
18. Illian, J., Sørbye, S.H., Rue, H.: A toolbox for fitting complex spatial point process models using integrated nested laplace approximation (INLA). Ann. Appl. Stat. **6**(4), 1499–1530 (2012)
19. Moller, J., Waagepetersen, R.P.: Statistical Inference and Simulation for Spatial Point Processes. CRC Press (2003)

Bayesian Nonparametric Prediction with Multi-sample Data

Federico Camerlenghi, Antonio Lijoi, and Igor Prünster

Abstract In the present paper, we address the problem of prediction within the setting of species sampling models. We consider d populations composed of different species with unknown proportions. Our goal is to predict specific features of additional and unobserved samples from the d populations by adopting a Bayesian nonparametric model. We focus on a broad class of hierarchical priors. These were introduced and investigated in [1], where also an algorithm for drawing predictions is devised, however, without any specific numerical illustration. The aim of this paper is twofold: on the one hand, we provide an illustration with an actual implementation of the algorithm of [1] and, on the other hand, we discuss its relevance with respect to complex prediction problems with species sampling data.

Keywords Bayesian nonparametric · Hierarchical process · Pitman–Yor process · Prediction · Random measure · Species sampling

1 Introduction

A typical problem in statistics relies on forecasting future outcomes of a random experiment, given a set of analogous observations from the past. This is known as the *problem of prediction* and its importance has been emphasized in several contexts (see for example, [6]). In the present paper, we will face this problem within the framework of species sampling models. We will deal with a multiple-populations scenario,

F. Camerlenghi
Department of Economics, Management and Statistics, University of Milano–Bicocca, Piazza dell'Ateneo Nuovo 1, 20126 Milano, Italy
e-mail: federico.camerlenghi@unimib.it

A. Lijoi · I. Prünster (✉)
Department of Decision Sciences and BIDSA, Bocconi University, Via Röntgen 1, 20136 Milano, Italy
e-mail: igor@unibocconi.it

A. Lijoi
e-mail: antonio.lijoi@unibocconi.it

M. La Rocca et al. (eds.), *Nonparametric Statistics*, Springer Proceedings in Mathematics & Statistics 339, https://doi.org/10.1007/978-3-030-57306-5_11

more precisely, we consider d populations of animals composed by different species with unknown proportions and we suppose that the species are shared across the d different populations. Assuming to be provided with a sample for each population, one typically wants to predict the number of new species that will be discovered in future sampling from the populations, the number of new species specific to each population and not shared with the others, the number of shared species across populations, etc.

In order to provide a clear mathematical formulation of the problem, we consider a Polish space $\mathbb{X}$ equipped with its Borel σ-algebra, denoted as $\mathscr{X}$, and a common probability space $(\Omega, \mathscr{A}, \mathbb{P})$, where the data are defined. The jth observation from population i, denoted here as $X_{i,j}$, is an $\mathbb{X}$-valued random element defined on the probability space $(\Omega, \mathscr{A}, \mathbb{P})$. The space $\mathbb{X}$ contains all the possible labels of the species in the d populations, besides the variable $X_{i,j}$ can be interpreted as the species' label of the jth animal from population i, for $i = 1, \dots, d$. In the sequel, we will suppose to be provided with a sample $\boldsymbol{X}_{n_i} := (X_{i,1}, \dots, X_{i,n_i})$ of size n_i for any $i = 1, \dots, d$. The whole sequence of observations is indicated as $\boldsymbol{X} = (\boldsymbol{X}_{n_1}, \dots, \boldsymbol{X}_{n_d})$. We further assume that the $X_{i,j}$s are independent and distributed as $p_i := \sum_{k\geq 1} p_{i,k}\delta_{x_k^*}$, where $p_{i,k}$ denotes the proportion of species k in population i and x_k^* is the corresponding species' label. Since the composition of any population is completely unknown, adopting a Bayesian viewpoint, we need to define a nonparametric prior distribution for the p_is. A good nonparametric prior should take into account the fact that the species' labels are shared across the different populations, but the proportions are not the same. Hence, we are looking for a vector of dependent random probability measures $(\tilde{p}_1, \dots, \tilde{p}_d)$ sharing the same atoms. Among the different Bayesian nonparametric models have been suggested in the literature, one of the most used and known is undoubtedly the hierarchical Dirichlet Process (HDP) defined in [13]. A first generalization of the HDP has been proposed is the hierarchical Pitman–Yor Process (HPY), which allows for much more flexibility in terms of clustering. See [11] for the definition of Pitman–Yor process and [8, 14, 15] for the hierarchical framework. The distribution theory of these processes have been recently studied in [1] within the more general framework of hierarchical transformations of completely random measures (see Sect. 2). Summing up, we consider an ideally infinite sequence of observations $\{(X_{i,j})_{j\geq 1} : i = 1, \dots, d\}$, which are defined on some probability space $(\Omega, \mathscr{A}, \mathbb{P})$ and taking values in the space of species' labels $\mathbb{X}$ equipped with its Borel σ-field $\mathscr{X}$. We further assume that the d sequences of observations are partially exchangeable [5], i.e., by the de Finetti representation theorem they satisfy:

$$\begin{aligned} (X_{1,j_1}, \dots, X_{d,j_d}) \mid (\tilde{p}_1, \dots, \tilde{p}_d) &\overset{\text{iid}}{\sim} \tilde{p}_1 \times \cdots \times \tilde{p}_d \qquad (j_1, \dots, j_d) \in \mathbb{N}^d \\ (\tilde{p}_1, \dots, \tilde{p}_d) &\sim Q_d. \end{aligned} \tag{1}$$

where $(\tilde{p}_1, \dots, \tilde{p}_d)$ is a vector of dependent random probability measures and Q_d is termed the de Finetti measure of the sequence. The specification of Q_d, or equivalently of the dependence structure across $\tilde{p}_1, \dots, \tilde{p}_d$, is a crucial problem in the

Bayesian nonparametric literature. As mentioned before, here we employ hierarchies of priors to define the vector $(\tilde{p}_1, \ldots, \tilde{p}_d)$: such a construction will be better specified in the next section.

The rest of the paper is structured as follows. In Sect. 2, we briefly recall the definition of Completely Random Measures (CRMs) which are employed to define a general class of hierarchical processes $(\tilde{p}_1, \ldots, \tilde{p}_d)$ that can be used in (1). Section 3 is devoted to the problem of prediction, some numerical illustrations are presented in Sect. 3.1 to show the applicability of our results and their performance in simulated scenarios. We conclude the paper with a brief discussion.

2 Hierarchical Processes Based on Completely Random Measures

This section is devoted to the construction of vectors of dependent random probability measures $(\tilde{p}_1, \ldots, \tilde{p}_d)$, which can be used to model the prior opinion in (1). We define a broad class of these vectors, relying on transformations of Completely Random Measures (CRMs). We first recall some basics on CRMs, refer to [4] for a complete account on the subject.

Let $\mathsf{M}_{\mathbb{X}}$ be the space of boundedly finite measures on $(\mathbb{X}, \mathscr{X})$, i.e., $m(A) < +\infty$ for any $m \in \mathsf{M}_{\mathbb{X}}$ and for any bounded set $A \in \mathscr{X}$, equipped with the corresponding Borel σ-algebra $\mathscr{M}_{\mathbb{X}}$. A CRM is a random element $\tilde{\mu}$ defined on some probability space $(\Omega, \mathscr{A}, \mathbb{P})$ and taking values in $(\mathsf{M}_{\mathbb{X}}, \mathscr{M}_{\mathbb{X}})$, such that the random variables $\tilde{\mu}(A_1), \ldots \tilde{\mu}(A_k)$ are independent for any choice of disjoint Borel sets $A_1, \ldots, A_k \in \mathscr{X}$ and for any $k \geq 1$. A nice representation theorem for CRMs has been provided by Kingman [9], who proved that $\tilde{\mu}$ can be written as the sum of three components: (i) a fixed diffuse measure; (ii) an infinite sum of random jumps at fixed locations; (iii) an infinite sum of random jumps at random locations. As most of the current literature, we focus our attention on CRMs of type (iii), therefore represented as $\tilde{\mu} = \sum_{i \geq 1} J_i \, \delta_{Y_i}$. For the sake of simplicity, we further assume that $(J_i)_{i \geq 1}$ and $(Y_i)_{i \geq 1}$ are independent sequences of random elements, leading us to consider the class of homogeneous CRMs. Then, the law of such a $\tilde{\mu}$ may be uniquely characterized through the Laplace functional, which amounts to be

$$\mathbb{E}[e^{-\int_{\mathbb{X}} f(x)\tilde{\mu}(\mathrm{d}x)}] = \exp\left\{-c \int_{\mathbb{X}} \int_0^{\infty} (1 - e^{-sf(x)})\rho(s)\mathrm{d}s \, P(\mathrm{d}x)\right\},$$

where P is a probability on $(\mathbb{X}, \mathscr{X})$, called base measure, c is a positive constant and $\rho : \mathbb{R}^+ \to \mathbb{R}^+$ is a measurable function. In other words, the CRM $\tilde{\mu}$ is a functional of a Poisson process $\{(J_j, Y_i)\}_{i \geq 1}$ on $\mathbb{R}^+ \times \mathbb{X}$ with non-bounded intensity function given by $\rho(s)\mathrm{d}s cP(\mathrm{d}x)$. Noteworthy examples of CRMs are the gamma process, obtained when $\rho(s) = e^{-s}/s$, and the σ-stable process, which corresponds to the choice $\rho(s) = \sigma s^{-1-\sigma}/\Gamma(1 - \sigma)$ for some $\sigma \in (0, 1)$.

Transformations of CRMs can be used to define broad classes of random probability measures, in the sequel, we will focus on two possible transformations leading us to define Normalized Random Measures with Independent Increments and the Pitman–Yor process. Besides, we will further define the corresponding hierarchical structures.

2.1 Hierarchical Normalized CRMs

Let us first focus on random probability measures which are obtained as normalization of a CRM $\tilde{\mu} = \sum_{i\geq 1} J_i \delta_{Y_i}$:

$$\tilde{p} = \frac{\tilde{\mu}}{\tilde{\mu}(\mathbb{X})} = \sum_{i\geq 1} \frac{J_i}{\bar{J}} \delta_{Y_i}, \tag{2}$$

where $\bar{J} := \sum_{i\geq 1} J_i = \tilde{\mu}(\mathbb{X})$. In the sequel, we will write $\tilde{p} \sim \mathrm{NRMI}(c, \rho; P)$ to denote the distribution of the so-called Normalized Random Measure with Independent Increments (NRMI) $\tilde{p}$, as first introduced in [12]. Note that $\tilde{p}$ in (2) is well defined if $\mathbb{P}(0 < \tilde{\mu}(\mathbb{X}) < \infty) = 1$ is in force, see [12] for a discussion on such an assumption and its relation with the Lévy intensity.

Being provided with d different random probability measures $\tilde{p}_1, \ldots, \tilde{p}_d$, one may enable dependence across them in the following hierarchical fashion:

$$\begin{aligned} \tilde{p}_i \mid \tilde{p}_0 &\overset{\text{iid}}{\sim} \mathrm{NRMI}(c_i, \rho_i; \tilde{p}_0) \qquad i = 1, \ldots, d \\ \tilde{p}_0 &\sim \mathrm{NRMI}(c_0, \rho_0; P_0), \end{aligned} \tag{3}$$

where P_0 is a diffuse measure on $(\mathbb{X}, \mathscr{X})$. In (3), the base measure referring to each $\tilde{p}_i$, for $i = 1, \ldots, d$, is taken to be random and equals another NRMI $\tilde{p}_0$: such a construction allows for sharing of atoms across $\tilde{p}_1, \ldots, \tilde{p}_d$. This vector of hierarchical NRMIs may be used in (1) to define the de Finetti measure Q_d.

2.2 Hierarchical Pitman–Yor Processes

A second relevant construction arises when $\tilde{p}$ is a random probability measure having distribution obtained by a suitable transformation of the distribution of a CRM. In particular, let $\mathbb{P}_\sigma$ be the probability distribution on $(\mathsf{M}_{\mathbb{X}}, \mathscr{M}_{\mathbb{X}})$ of a σ-stable CRM, with $\sigma \in (0, 1)$. For $\theta > 0$ define $\mathbb{P}_{\sigma,\theta}$ on $(\mathsf{M}_{\mathbb{X}}, \mathscr{M}_{\mathbb{X}})$ as absolutely continuous w.r.t. $\mathbb{P}_\sigma$ and such that its Radon–Nikodym derivative is

$$\frac{\mathrm{d}\mathbb{P}_{\sigma,\theta}}{\mathrm{d}\mathbb{P}_\sigma}(m) = \frac{\Gamma(\theta/\sigma)}{\sigma\, \Gamma(\theta)}\, m^{-\theta}(\mathbb{X}).$$

The resulting random measure $\tilde{\mu}_{\sigma,\theta}$ with distribution $\mathbb{P}_{\sigma,\theta}$ is not completely random, but via normalization

$$\tilde{p} = \frac{\tilde{\mu}_{\sigma,\theta}}{\tilde{\mu}_{\sigma,\theta}(\mathbb{X})} \sim \mathrm{PY}(\sigma, \theta;\, P)$$

we obtain the well-known Pitman–Yor (PY) process [11]. Correspondingly, we may define a vector of hierarchical PY processes as in (3):

$$\begin{aligned} \tilde{p} \mid \tilde{p}_0 &\overset{\mathrm{d}}{=} \mathrm{PY}(\sigma, \theta;\, \tilde{p}_0) \\ \tilde{p}_0 &\overset{\mathrm{d}}{=} \mathrm{PY}(\sigma_0, \theta_0;\, P_0) \end{aligned} \tag{4}$$

with P_0 being a non-atomic probability measure on $(\mathbb{X}, \mathscr{X})$. The theoretical analysis beyond this structure and the previous one (3) has been carried out in [1], see also [2] for some applications and [3] for a discussion of the case $d = 1$ (exchangeable hierarchical processes).

3 Prediction in Species Models

In the present section, we assume to be provided with a sample $\boldsymbol{X}_{n_i} = (X_{i,1}, \ldots, X_{i,n_i})$ of size n_i for each $i = 1, \ldots, d$, satisfying (1). The vector of random probability measures $(\tilde{p}_1, \ldots, \tilde{p}_d)$ in (4) are assumed to have a hierarchical structure, in particular we have considered hierarchies of PY processes to carry out the numerical experiments of Sect. 3.1.

Our interest here consists in predicting specific features of additional and unobserved samples from the d populations, which will be denoted by $\boldsymbol{X}_{m_i}^{(n_i)} := (X_{i,n_i+1}, \ldots, X_{i,n_i+m_i})$, as $i = 1, \ldots, d$. For the sake of illustration, we consider an additional sample of the same size m for each population, namely $m = m_1 = \ldots = m_d$. In the following numerical experiments, we concentrate our attention on two statistics which depend on the additional unobserved samples. The first one is the number of hitherto unobserved species that will be discovered in further sampling, more precisely we intend to forecast

$$K_m^{(n_i)} | \boldsymbol{X} := \sum_{r=1}^{k_{i,m}} \mathbb{1}_{\{\boldsymbol{X}\}^c}(X_{i,r}^*), \tag{5}$$

where $X_{i,1}^*, \ldots, X_{i,k_{i,m}}^*$ are the $k_{i,m}$ distinct values out of the ith additional sample $\boldsymbol{X}_m^{(n_i)}$ and $\mathbb{1}$ denotes the indicator function. Another statistic that one could be interested to predict is the following

$$S_{m,i}^{(n_i)} | \boldsymbol{X} = \sum_{r=1}^{k_{i,m}} \mathbb{1}_{\{\boldsymbol{X}\}^c}(X_{i,r}^*) \prod_{j \neq i} \mathbb{1}_{\{X_{j,n_j+1}, \ldots, X_{j,n_j+m}\}^c}(X_{i,r}^*). \tag{6}$$

$S_{m,i}^{(n_i)}$ counts the number of new and distinct observations in a future sample of size m for the ith population, which are not shared with the other $d - 1$ additional samples of size m.

The posterior distributions of $K_m^{(n_i)}$ and $S_{m,i}^{(n_i)}$ are not available in closed form, hence, one needs to resort to a simulation algorithm in order to estimate all these quantities. To solve this issue, we apply the MCMC procedure developed in [1] (see also [2]) to generate trajectories of additional samples and then estimate $K_m^{(n_i)}$ and $S_{m,i}^{(n_i)}$ on the basis of the MCMC runs.

3.1 Numerical Experiments

We consider four populations ($d = 4$), each one containing $K_i = 3{,}000$ different species chosen at random from a pool of $K = 4.000$ total species. For each population, we have chosen at random the labels of the species from the total pool of K labels, and then we have assigned to them the Zipf distribution with parameter s_i. More specifically, if $j_1, \ldots, j_{K_i}$ are the K_i species of population i, then we assign to the kth label of population i (denoted as j_k) a frequency proportional to $1/k^{s_i}$, for $k = 1, \ldots, K_i$ and $i = 1, \ldots, 4$. For the sake of illustration, we have chosen the Zipf's parameters as follows $(s_1, \ldots, s_4) = (1.1, 1.1, 1.2, 1.2)$ and we have generated a sample of size $n = 200$ for each population. We have run an MCMC sampler for a total of 35,000 iterations and a burn-in period of 15,000 iterations to predict the number of new species that will be observed in an additional sample of size m, where m varies from 20 to 200. The red curve in Fig. 1 depicts the estimated number of new species that will be observed in further sampling for the different populations, obtained applying the MCMC procedure of [1, Sect. 6.1]. The black curve represents the number of new species estimated with an oracle strategy, i.e., sampling from the true distribution generating the data. We observe that the two curves are close in all the four populations, leading us to conclude that our strategy is able to truly predict the number of new species observed in additional sampling. Figure 2 compares the prediction of $S_{m,i}^{(n_i)}$ obtained through the hierarchical PY (red curve) and through the oracle strategy (black curve), each panel corresponds to a population. We observe an accurate prediction of this statistic. It is remarkable to underline that the prediction of $S_{m,i}^{(n_i)}$ is achievable in a dependent framework only.

4 Discussion

We addressed the problem of prediction in the context of species models with multiple-samples information. We remark that similar problems were first faced in [10] under the exchangeability assumption, i.e., in the presence of only one population. In [7, 10], the authors derived tractable analytical expressions for many

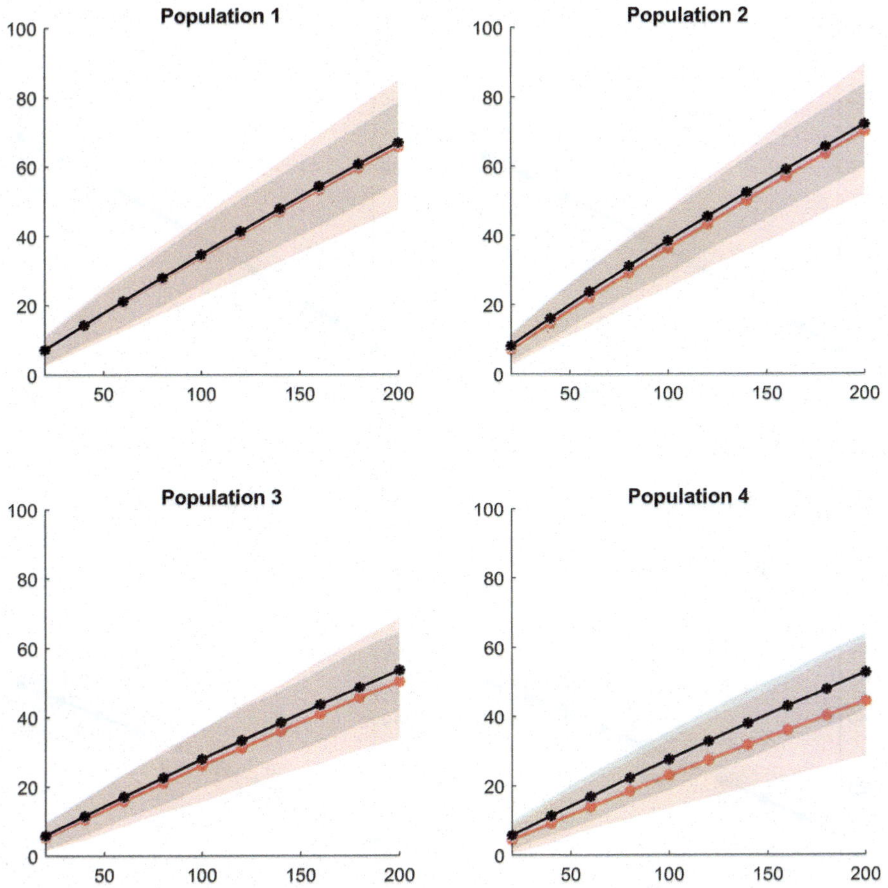

Fig. 1 Prediction of the number of new species observed in each population for different values of the additional sample size. The oracle prediction is shown in black, the estimated value is in red. Shaded bands correspond to 95% estimated credible intervals

quantities of interest when the nonparametric prior is a Pitman–Yor process. The partially exchangeable framework we have investigated here is much more involved and the posterior distributions of the two statistics (5) and (6) are not available in closed form. Therefore, we have used an MCMC sampler to estimate these quantities, implementing the prediction algorithm suggested in [1] and showing its role in the context of species sampling. It is possible to use such a procedure to predict the outcome of many other statistics depending on an additional sample of arbitrary size, e.g., the number of shared species across two or more populations. Work on this and related applications is ongoing.

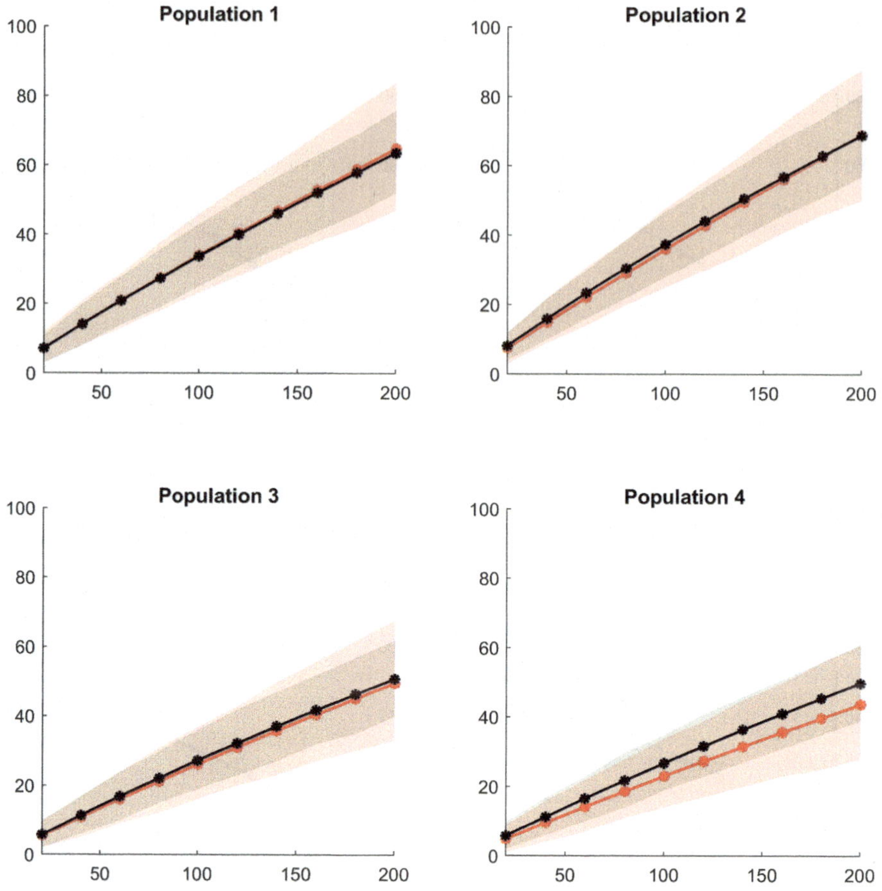

Fig. 2 Prediction of the number of new species specific of each population (i.e., not shared with the others) for different values of the additional sample size. The oracle prediction is shown in black, the estimated value is in red. Shaded bands correspond to 95% estimated credible intervals

Acknowledgments A. Lijoi and I. Prünster are supported by MIUR, PRIN Project 2015SNS29B.

References

1. Camerlenghi, F., Lijoi, A., Orbanz, P., Prünster, I.: Distribution theory for hierarchical processes. Ann. Statist. **47**, 67–92 (2019)
2. Camerlenghi, F., Lijoi, A., Prünster, I.: Bayesian prediction with multiple-sample information. J. Multivar. Anal. **156**, 18–28 (2017)
3. Camerlenghi, F., Lijoi, A., Prünster, I.: Bayesian nonparametric inference beyond the Gibbs-type framework. Scand. J. Stat. **45**, 1062–1091 (2018)

4. Daley, D.J., Vere-Jones, D.: An introduction to the theory of point processes, vol. II. Springer, New York (2008)
5. de Finetti, B.: Sur la condition d'equivalence partielle. Actualités scentiques et industrielles 5–18 (1938)
6. de Finetti, B.: Probabilismo. Logos **14**, 163–219 (1931) [Translated in *Erkenntnis* **31**, 169–223 (1989)]
7. Favaro, S., Lijoi, A., Mena, R.H., Prünster, I.: Bayesian nonparametric inference for species variety with a two parameter Poisson-Dirichlet process prior. J. R. Stat. Soc. Ser. B **71**, 993–1008 (2009)
8. Gasthaus, J., Teh, Y.W.: Improvements to the sequence memoizer. Adv. Neuronal Inf. Process. Syst. **23** (2010)
9. Kingman, J.F.C.: Completely random measures. Pac. J. Math. **21**, 59–78 (1967)
10. Lijoi, A., Mena, R.H., Prünster, I.: Bayesian nonparametric estimation of the probability of discovering a new species. Biometrika **94**, 769–786 (2007)
11. Pitman, J., Yor, M.: The two-parameter Poisson-Dirichlet distribution derived from a stable subordinator. Ann. Probab. **25**, 855–900 (1997)
12. Regazzini, E., Lijoi, A., Prünster, I.: Distributional results for means of normalized random measures with independent increments. Ann. Stat. **31**, 560–585 (2003)
13. Teh, Y.W., Jordan, M.I., Beal, M.J., Blei, D.M.: Hierarchical Dirichlet processes. J. Am. Stat. Assoc. **101**, 1566–1581 (2006)
14. Teh, Y.W., Jordan, M.I.: Hierarchical Bayesian nonparametric models with applications. Bayesian Nonparametrics, pp. 158–207. Cambridge University Press, Cambridge (2010)
15. Wood, F., Gasthaus, J., Archambeau, C., James, L.F., Teh, Y.W.: The sequence memoizer. Commun. ACM **54**, 91–98 (2011)

Algorithm for Automatic Description of Historical Series of Forecast Error in Electrical Power Grid

Gaia Ceresa, Andrea Pitto, Diego Cirio, and Nicolas Omont

Abstract The EU-FP7 iTesla project developed a toolbox that assesses dynamic security of large electric power systems in the presence of forecast uncertainties. In particular, one module extracts plausible realizations of the stochastic variables (power injections of RES Renewable Energy Sources, load power absorptions). It is built upon historical data series of hourly forecasts and realizations of the stochastic variables at HV (High-Voltage) nodes in the French transmission grid. Data reveal a large diversity of forecast error distributions: characterizing them allows to adapt the module to the data, improving the results. The algorithm here presented is aimed to automatically classify all the forecast error variables and to cluster them into smoother variables. The main steps of the algorithm are filtering of the variables with too many missing data or too low variance, outliers detection by two methods (Chebyshev inequality, quantile method), separation of unimodal variables from multimodal ones by exploiting a peak counting algorithm, Gaussian mixtures, comparison with asymmetrical distributions, multimodality index, clustering of the multimodal variables whose sum is unimodal, comparing two alternative algorithms (the former based on hierarchical clusterization, accounting for correlation and geographical closeness, and the latter on the identification of the same initial characters in the identification codes).

Keywords Data analysis · Variables classification · Clustering · Multimodality detection

G. Ceresa (✉) · A. Pitto · D. Cirio
Ricerca sul Sistema Energetico RSE S.p.A., via Rubattino 54, 20134 Milano, Italy
e-mail: gaia.ceresa@rse-web.it

A. Pitto
e-mail: andrea.pitto@rse-web.it

D. Cirio
e-mail: diego.cirio@rse-web.it

N. Omont
RTE Réseau de Transport d'Électricité, 92932 Paris la Défense Cedex, France
e-mail: nicolas.omont@rte-france.com

M. La Rocca et al. (eds.), *Nonparametric Statistics*, Springer Proceedings in Mathematics & Statistics 339, https://doi.org/10.1007/978-3-030-57306-5_12

1 Introduction

The iTesla project [1, 2], led by the French transmission system operator RTE and co-funded by the European Commission FP7, develops an approach to perform the dynamic security assessment of large power systems in an online environment, accounting for dynamic problems and for the forecast uncertainties due to renewable sources and load, associated to a variable time horizon spanning from online operation to several hours ahead of operation. The outcome of the project is a free toolbox described in [3] and available on GitHub [4]. After the end of the iTesla project, further developments of the platform have been carried out in two directions [5–7]: the choice of the suitable historical dataset to train the offline uncertainty model of the platform, and a deeper analysis of the large amounts of available historical series of forecasts and snapshots. The forecast error time series has very different profiles: some are continuous and others take discrete values, some reflect a kind of periodicity and others have sudden variations; also, the relevant distributions differ a lot. The iTesla tool can take into account this variety, thanks to a deep analysis of the forecast errors described in this paper: a classification into unimodal and multimodal variables allows to better tune the sampling module, while the clustering phase combines some subsets of multimodal variables transforming them into unimodal, obtaining a reduction of the problem dimensionality.

The novelty shown in this paper is the automatic processing of some thousands of historical series [8]. This paper starts with the algorithm description in Sect. 2, which explains the phases of raw data pre-processing, the overall descriptive statistics and, finally, the comparison of two clustering methods. Section 3 shows one application of the overall algorithm in a real dataset. Section 4 concludes.

2 Algorithm

The input is composed by a set of thousands of historical series of renewable energy sources power injections and load absorptions of the French electrical high transmission grid and one set of their forecasts done the day before; the timestamps are hourly, the time domain is at least 1 month. The variables under study are the forecast errors of active and reactive powers, computed as in Eq. 1.

$$error_{hour,node} = snapshot_{hour,node} - forecast_{hour,node} \quad \forall hour \in [hour_{min}, hour_{max}] \tag{1}$$

The algorithm for automatic description of historical series of forecast errors has three main steps:

1. Preprocessing: removal of not significant variables, detection and elimination of outliers;

2. Descriptive statistics: calculation of the first four moments, calculation of linear correlations, multimodality analysis, classification of variables;
3. Clustering: algorithm based on hierarchical clustering, clustering of variables lying in the same substation.

2.1 Preprocessing

The raw input data contain many time series that are not significant from a statistical point of view; the input series where more than 30% of timestamps are missing values, or more than 70% of timestamps are constant values, usually 0, and the series with a variance lower than 1 $MW^2(Mvar^2)$ are filtered out. The preprocessing regards both forecast and snapshot series, and then it runs also on their differences. The subsequent steps run only on the forecast error time series.

The retained variables must still be managed in order to remove the observations that have an abnormal distance from the other values in the random sample, the *outliers*. This definition leaves to the analyst the decision of which distances will be considered abnormal, and several methods can help him [9], but none of them works correctly on all of the thousands of forecast error time series. The sequel describes the process of outliers' detection and deletion that runs better on all the variety of the input series.

The tool implements the outliers detection method based on Chebyshev's Inequality: for each integrable random variable X, with finite mean μ and variance σ, it holds valid the expression:

$$P\left(|X-\mu| \geq n\sigma\right) \leq \frac{1}{n^2}. \tag{2}$$

The most used parameter in literature is $n = 3$, and a lot of tests based on *trials-and-errors* confirm the goodness of this value; the extreme values stay in the complementary set of $[\mu - 3\sigma, \mu + 3\sigma]$.

The second outliers detection method is the Quartile Method, based on the computation of first $Q1$ and third $Q3$ quartiles; the extreme values are contained in the complementary set of $[Q1 - n\,(Q3 - Q1)\,, Q3 + n\,(Q3 - Q1)]$. From literature, the most frequently used parameter is $n = 1.5$, but for the analysed forecast error time series this value detects too many extreme values, and several tests based on *trials-and-errors* select the parameter $n = 3$. After that, the MAD test runs on the detected extreme values: be X the series and X_i its elements, the outliers are those X_i that satisfy the condition

$$\frac{|X_i - median(X)|}{median\left(|X_i - \mu|\right)} > 5. \tag{3}$$

Only the operator's expertise can decide if an extreme value is an outlier or not; but it is not possible to make the resolve for some thousands of variables. In order to automate the decision, the output of both methods are compared: their resulting sets are of different size, where the smallest is a subset of the biggest, and only the last one is selected for the subsequent steps. The final check is about the cardinality of the selected set: if it is lower than 7% of the number of variable's records, their elements are classified as outliers and so removed, otherwise, they are considered as extreme values. This limit is due to the discrete variables, where the record related to an extreme value might be misinterpreted as outliers.

2.2 Descriptive Statistics

This subsection summarizes the phases of the overall statistical description.

The initial descriptive information comes from the computation of the first four moments: average, variance, skewness and kurtosis; their quantiles allow an initial classification of the variables.

The linear correlation between each pair of time series is computed through the Pearson index, in view of the future clustering phase. The analysis of the output matrix helps detect also the variables that are replicated in two different nodes due to the state estimation system.

The Multimodality Algorithm carries out the time series classification into variables with multimodal or unimodal distribution. It is composed of four consecutive steps where the first two work on the whole set of variables, and the last two run only on the variables detected as multimodal in the previous steps. Four different methods for multimodality detection are combined because none of them can guarantee the best result if applied individually on all the variables, which are characterized by large differences in the relevant distributions; each step of the algorithm improves the result of its predecessor. The workflow, shown in Fig. 1, runs once for each time series.

The first step finds the number of peaks working as follows: it generates the histogram of the variable samples with 50 equally spaced bins, each one containing at least 10 elements, both numbers decided by the rule of thumb. It considers the height of each bin and compares it with the previous one and next one: if the height of the analysed bin is higher than its neighbours, it is considered a local maximum. The *peaks* are the local maxima that stay between two lower local maxima. The result is a big set of values, most of which are not significant: it is necessary to better define the number of modes of each variable in the next step.

A Gaussian Mixture tries to fit the distribution. Two nested loops are composed: the outer runs changing every time the number of the mixture components, choosing from the number of peaks identified in the previous step down to 1. For each number of components, the inner loop runs three times the iterative Expectation–Maximization Algorithm [10], to find the better parameters (averages, covariance matrices, component proportions) of the mixture. The best fitting of the inner loop has the lowest Akaike Information Criterion (AIC) [11], while the best fitting of the outer loop has the lowest Bayesian Information Criterion (BIC) [12] ; both indices

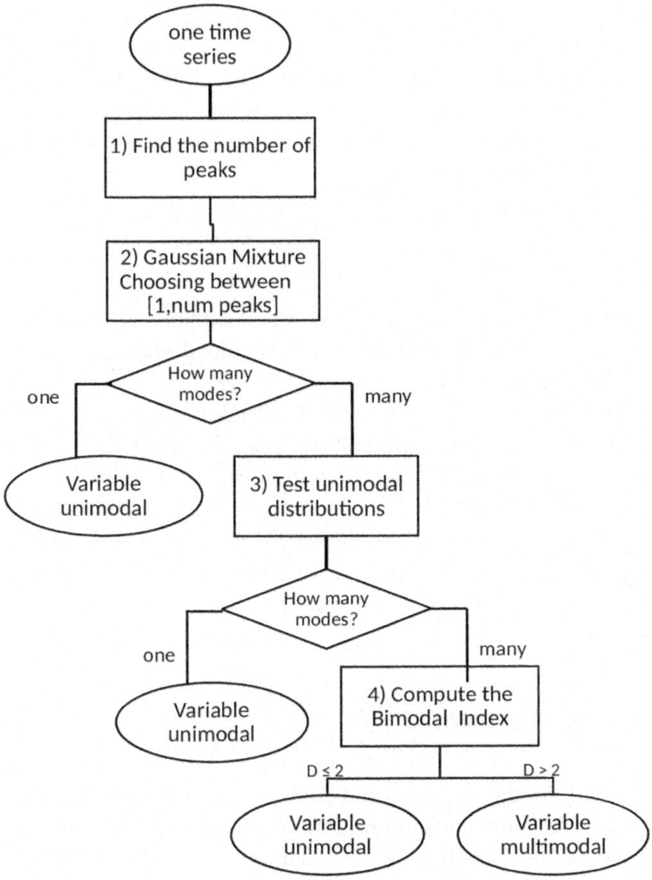

Fig. 1 Multimodality algorithm

depend on the *likelihood* function L penalized by the number k of the distribution's parameters, BIC in a heavier manner because k is multiplied by the logarithm of the length n of the series.

$$BIC = -2ln(L) + k \cdot ln(n); \qquad AIC = -2ln(L) + 2k. \tag{4}$$

A skewed and platykurtic histogram usually is approximated by a multimodal mixture, but a unimodal distribution could fit the histogram even better. Six unimodal distributions (Weibull, Logistic, Gamma, Log-Normal, Generalized extreme value, T-location scale) try to fit the variable, and the one with lowest BIC is selected and compared also with the Gaussian Mixture. If the unimodal distribution has the lowest BIC, the algorithm stops here and restarts analysing the next variable.

The variables described by a multimodal Gaussian Mixture are subjected to a final step, the computation of Ashman's D index [13], that measures the distance between the pairs of mixture components. This index is applied to the mixtures of distributions with unequal variances: let μ_1 and μ_2 the component's averages, σ_1 and σ_2 their standard deviations; a two-component mixture is unimodal if and only if

$$D = \sqrt{2}\frac{|\mu_1 - \mu_2|}{\sqrt{\sigma_1^2 + \sigma_2^2}} \leq 2. \tag{5}$$

If a mixture contains three or more modes, the Ashman's D index is computed for each pair of components, and the variable is included among the multimodal if at least one D index is higher than 2.

At this point, all the variables are classified, taking into account their averages, variances and number of peaks. All the averages are collected, and their percentiles are computed: the "low" averages stay within the percentiles [25th, 75th), while the "high" averages stay in the complementary interval. Also, the variances are classified based on their percentiles: they are "low" if they are lower than the 80th percentile, "high" if higher. The third step of classification is the number of peaks of the distribution, that can be "one" or "more than one".

2.3 *Clustering Algorithms*

Clustering the multimodal variables and combine them into fewer unimodal series is important for two reasons: the sampling module of the iTesla platform can provide a more accurate result when dealing with unimodal variables. Moreover, the dimensionality of the problem is reduced. Two algorithms, which are based on different clustering criteria, are run. The clusters are composed of two or three variables of the same type, all active or all reactive power.

The first algorithm is based on Hierarchical Clustering, shown in the left part of Fig. 2. The power absorptions or injections at two electrical nodes that are linearly correlated could be influenced by the same local phenomenon; this correlation is significant and durable if the nodes are geographically closed and subject to the same phenomena for a long time, otherwise, it could be only a random correlation. The algorithm collects six consecutive steps; at first, it computes the distances between nodes based on Pearson's index previously computed: $dist(X, Y) = 1 - |corr(X, Y)|$; after that it implements the hierarchical clustering, grouping the more correlated variables in pairs. Then, each cluster is subjected to three checks. The first looks for the *equal* variables: X and Y are *equal* if they differ for at most 1 MW (Mvar) for at least the 97% of their elements. N is the number of elements of the time series, $maxi = 3\%$ of N:

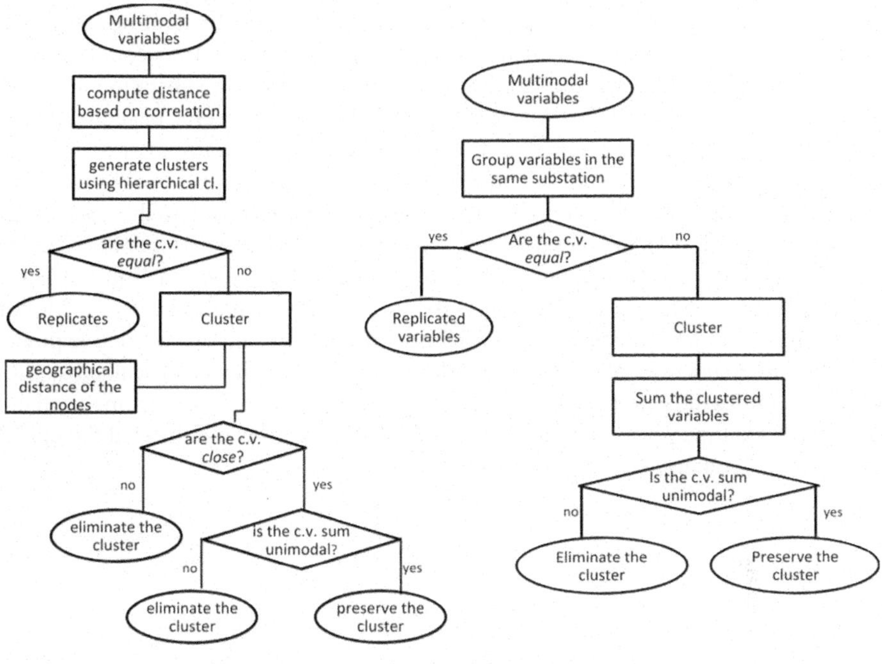

Fig. 2 Left: Algorithm based on hierarchical clustering. Right: Algorithm CVBSS

$$\begin{aligned} given\ I &= \{1, 2, \ldots, N\},\ J = \{j_1, j_2, \ldots j_{maxi}\} \subset I \\ if\ |X_i - Y_i| &< 1\ \forall i \in I \setminus J \Rightarrow X = Y \end{aligned} \tag{6}$$

1 MW(Mvar) is negligible in the forecast and snapshot time series, considering also some rounding or measurement errors and their propagation. If two variables are *equal*, they are put within the replicated variables and their cluster is eliminated. The second check is about the geographical distance, computed by considering the latitude and longitude of the nodes. Given the clustered variables X and Y, the cluster is retained only if Y belongs to the 50 nodes closest to X, selected by the nearest neighbour algorithm, where 50 is a suitable trade-off for both the very concentrated urban nodes and the distant nodes in the countryside. If a cluster is retained until here, the historical series of the involved variables are summed together: the sum is preserved only if the multimodality algorithm of Fig. 1 identifies it as unimodal. This new unimodal variable is used in the iTesla tool instead of the two or three multimodal clustered variables.

The second algorithm is Clustering Variables Belonging to the Same Substation (CVBSS), on the right side of Fig. 2. In the electrical grid, many substations contain one bus-bar that works like one electrical node if the bus coupler is closed, and it is splitted into two or three electrical nodes if the bus coupler is open. The variable associated with each node contains the measured values when the bus coupler is open, while the total substation measure is divided into two or three fictitious measures when the bus coupler is closed. The grid topology information (bus coupler status) is not available in advance, thus the forecast works well at the substation level, but it cannot predict correctly the power at the individual node level. Consequently, the forecast errors of each individual node could be large, with an irregular distribution and many peaks, but the sum of the series at the substation level usually becomes Gaussian, so the adopted strategy is to sum the variables referring to the nodes lying in the same substation, in order to obtain one variable with the smoother distribution.

The algorithm CVBSS groups the variables which refer to the same substation and which have a multimodal distribution of the forecast error. As shown in Fig. 2 left, it selects only multimodal variables, working separately on active and reactive power. It selects the pairs or triplets of variables that stay in the same electrical substation, then it identifies the *equal* variables like in Eq. 6 and it separates them from the others. The clustered time series are summed together: if the sum is multimodal the cluster is eliminated, otherwise, it is retained.

3 Case Study

This section shows one application of the algorithm on 7808 stochastic variables of the electrical French transmission grid, only active power from 1 January 2016, 00:30 to 31 January 2016, 23:30; considering that there is a subset of instants with missing values, in total, there are 737 hourly timestamps in each time series. The preprocessing phase selects the significant variables in both snapshots and forecasts, removing those with too many missing or constant values, or with a variance lower than 1MW; this phase retains 5194 variables, 67% of the input, from which the forecast error series derive as in Eq. 1. The distribution of the time series averages is Gaussian, half of the values are concentrated within the interval $[-0.27, 0.06]$ and the 25% stay out of the interval $[-1.27, 1.06]$. One variable has an average very far from the rest of the series, equal to -47.6 MW. The 80% of variables have the variance lower than 9.07 MW^2, while, in the 2% of cases, it is greater than 135 MW^2; two variables have the variance higher than 36,000 MW^2 (one has also the highest average, the other lies in the same substation). A total of 3093 nodes are combined in different manners, with some repetitions, to generate 2006 pairs of variables correlated more than 0.99; 209 variables are *equal* in pairs or triplets; 201 series are combined, with some repetition, in 127 pairs with a correlation higher than 0.9 in absolute value. Figure 3a shows the results of the application of the multimodality algorithm to the case study: each step reduces the number of multimodal variables. In the end, the algorithm finds 614 multimodal variables and 4580 unimodal ones. The variable classification, shown in Fig. 3b, is based on the values of their first two moments and on the number of their peaks. "Low averages" fall within the interval around 0, "low variances" are lower than 9.07 MW^2, the "high" levels stay in the complementary intervals, different colours refer to the number of peaks in the distributions, the height of each bin is the cardinality of each class. The major group is composed by variables

Fig. 3 Multimodality Fig. 3b and classification Fig. 3b summary; example of clustering Fig. 3c

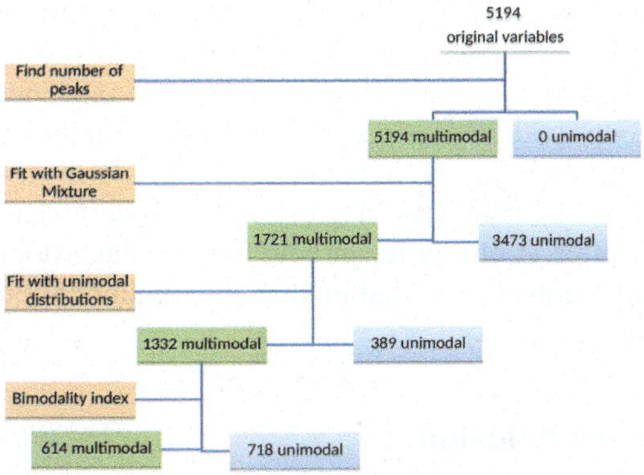

(a) Number of multimodal variables detected by each step of the algorithm in Fig. 1.

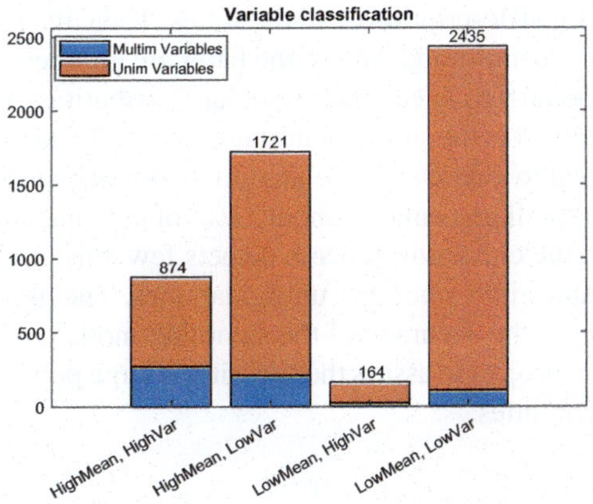

(b) Result of the classification of the variables.

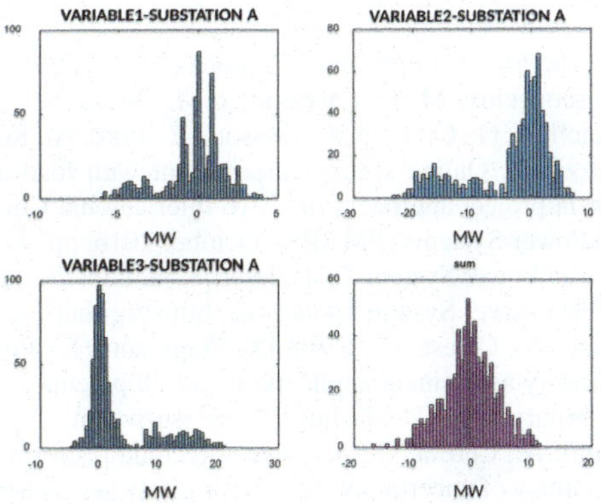

(c) Histogram of three series referring to the same substation (in blue) and of their sum (in magenta).

with low average, low variance and one peak (the desirable group); the smaller group have low average, high variance and many peaks; considering each bin of the bar diagram, the most numerous part is the one composed by unimodal variables; the groups with high variance are smaller than those with low variance. The clustering based on hierarchical method finds 16 clusters, while CVBSS generates 18 groups with 2 variables and 4 groups with three variables: the latter is preferable in this example. A cluster image is in Fig. 3c: the histograms of three variables in the same substation are reported in blue, their unimodal sum in magenta.

4 Conclusion

This paper has presented an algorithm for the automatic analysis of historical series of the forecast errors in power systems. Initially, the algorithm proposes an overall statistical description of all the series. Then, it allows to divide the unimodal variables from the multimodal ones; the latter are grouped in clusters, and then aggregated into unimodal variables with smoother distributions, because they are more suitable to be processed in the subsequent stages of the iTesla platform. The results of the algorithm applied to a case study related to the French system show that multimodal variables are a small percentage (about 12%) of the total number of variables under test. Moreover, the clustering process detects few tens of clusters which combine multimodal variables into smoother unimodal ones. The algorithm is a valuable contribution to increase the accuracy of the sampling module of the platform developed during the iTesla project to assess the security of large power systems in the presence of forecast uncertainties.

References

1. iTesla Project. https://cordis.europa.eu/project/id/283012/en
2. Vasconcelos, M.H., Carvalho, L.M., Meirinhos, J., Omont, N., Gambier-Morel, P., Jamgotchian, G., Cirio, D., Ciapessoni, E., Pitto, A., Konstantelos, I., Strbac, G., Ferraro, M., Biasuzzi, C.: Online security assessment with load and renewable generation uncertainty: the itesla project approach. In: 2016 International Conference on Probabilistic Methods Applied to Power Systems (PMAPS), October 2016, pp. 1–8 (2016)
3. iTesla Power System Tools description. https://github.com/itesla
4. iTesla Power System Tools code. https://github.com/itesla
5. Pitto, A., Ceresa, G., Cirio, D., Ciapessoni, E.: Power system uncertainty models for on-line security assessment applications: developments and applications. RSE Report no. 17001186, February 2017 (2017). http://www.rse-web.it
6. Pitto, A., Ceresa, G.: Automated techniques for analyzing historical data and improving the accuracy of uncertainty models for safety assessments of the electrical system. RSE Report no. 17007093, February 2018 (2018). http://www.rse-web.it
7. Ceresa, G., Ciapessoni, E., Cirio, D., Pitto, A., Omont, N.: Verification and upgrades of an advanced technique to model forecast uncertainties in large power systems. In: 2018 Interna-

tional Conference on Probabilistic Methods Applied to Power Systems (PMAPS), June 2018, pp. 1–8 (2018)
8. Algorithm for automatic description of historical series of forecast error in electrical power grid. https://github.com/itesla/ipst/tree/master/stat_analysis
9. Seo, S.: A review and comparison of methods for detecting outliers in univariate data sets. Master's thesis in science, University of Pittsburgh (2006)
10. Chen, Y., Gupta, M.R.: EM demystified: an expectation-maximization tutorial. UWEE technical report, UWEETR-2010-0002, February 2010 (2010)
11. Akaike, H.: Akaike's Information Criterion. Springer, Berlin, p. 25 (2011)
12. Konishi, S., Kitagawa, G.: Bayesian Information Criteria. Springer Series in Statistics, pp. 211–237. Springer, New York (2008)
13. Ashman, K., Bird, C., Zepf, S.: Detecting bimodality in astronomical datasets. Astron. J. **108**, 2348–2361 (1994)

Linear Wavelet Estimation in Regression with Additive and Multiplicative Noise

Christophe Chesneau, Junke Kou, and Fabien Navarro

Abstract In this paper, we deal with the estimation of an unknown function from a nonparametric regression model with both additive and multiplicative noises. The case of the uniform multiplicative noise is considered. We develop a projection estimator based on wavelets for this problem. We prove that it attains a fast rate of convergence under the mean integrated square error over Besov spaces. A practical extension to automatically select the truncation parameter of this estimator is discussed. A numerical study illustrates the usefulness of this extension.

Keywords Nonparametric regression · Multiplicative noise · Rates of convergence · Wavelets

1 Introduction

We consider the following unidimensional nonparametric regression model

$$Y_i = U_i f(X_i) + V_i, \quad i \in \{1, \dots, n\}, \tag{1}$$

where $f : [0, 1] \to \mathbb{R}$ is an unknown regression function, $X_1, \dots, X_n$ are n identically distributed random variables with support on $[0, 1]$, $U_1, \dots, U_n$ are n identically distributed random variables having the uniform distribution on a symmetric interval around 0 and $V_1, \dots, V_n$ are n identically distributed random variables. Moreover, it is supposed that X_i and U_i are independent, and U_i and V_i are independent for any

C. Chesneau
Université de Caen - LMNO, Caen, France
e-mail: christophe.chesneau@unicaen.fr

J. Kou
Guilin University of Electronic Technology, Guilin, China
e-mail: kjkou@guet.edu.cn

F. Navarro (✉)
CREST-ENSAI, Bruz, France
e-mail: fabien.navarro@ensai.fr

M. La Rocca et al. (eds.), *Nonparametric Statistics*, Springer Proceedings in Mathematics & Statistics 339, https://doi.org/10.1007/978-3-030-57306-5_13

$i \in \{1, \ldots, n\}$. Additional technical assumptions on the model will be formulated later. We aim to estimate the unknown function $r := f^2$ from $(X_1, Y_1), \ldots, (X_n, Y_n)$; the random vectors $(U_1, V_1), \ldots, (U_n, V_n)$ form the multiplicative-additive noise. The model (1) can be viewed as a natural extension of the standard nonparametric regression model; the main novelty is the presence of a multiplicative uniform noise that perturbed the unknown function f. Such multiplicative regression model as (1) is very popular in various application areas, particularly in signal processing (e.g., for Global Positioning System (GPS) signal detection in which not only additive noise but also multiplicative noise is encountered [1]), or in econometrics (e.g., for volatility estimation where the source of noise is multiplicative [2], also for deterministic and stochastic frontier estimation where the noise is multiplicative and both multiplicative and additive, respectively [3]). On the other hand, let us mention that some connexions exist with the so-called heteroscedastic nonparametric regression model. See, for instance, [4–6]. In particular, [4] consider the estimation of r in the heteroscedastic nonparametric regression model defined as (1) with X_1 deterministic, V_1 deterministic but unknown (it is an unknown function of X_1) and general assumptions on U_1. The form of the model is the same but the intrinsic definition is different. In this paper, we propose to estimate r with wavelet methods. Such methods have the advantage to capture the possible complexity of this unknown function. A natural linear wavelet estimator is then developed. With a suitable choice of a tuning parameter inherent of this estimator, we prove that it attains a fast rate of convergence under the mean integrated square error over Besov spaces. One drawback of this estimator is that the theoretical choice for the tuning parameter depends on a supposed unknown smoothness of r. We then provide a practical solution to this problem to choose the truncation level of our linear wavelet estimator using an adapted version of the twofold Cross-Validation (2FCV) method introduced by Nason [7]. A numerical study is performed to show the applicability of this extension.

The rest of this paper is organized as follows. In Sect. 2, we briefly present basics on wavelets and Besov balls. Additional assumptions on the model (1), the considered wavelet estimator and the main result are given in Sect. 3. Section 4 is devoted to the simulation study. The technical details for the proof of our main result are postponed in Sect. 6.

2 Basics on Wavelets and Besov Balls

For the purpose of this paper, we use the compactly supported wavelets of the Daubechies family. We present the essential below, all the details can be found in, e.g., [8, 9]. For any $j \geq 0$, we set $\Lambda_j = \{0, \ldots, 2^j - 1\}$ and, for $k \in \Lambda_j$,

$$\phi_{j,k}(x) = 2^{j/2}\phi(2^j x - k), \qquad \psi_{j,k}(x) = 2^{j/2}\psi(2^j x - k).$$

Following the methodology of [10], there exists an integer τ such that, for any integer $j_0 \geq \tau$, the collection of functions

$$\mathscr{S} = \{\phi_{j_0,k},\ k \in \Lambda_{j_0};\ \psi_{j,k};\ j \in \mathbb{N} - \{0, \ldots, j_0 - 1\},\ k \in \Lambda_j\}$$

forms an orthonormal basis of $\mathbb{L}^2([0, 1])$. Therefore, for any integer $j_0 \geq \tau$ and $h \in \mathbb{L}^2([0, 1])$, we have the following wavelet expansion:

$$h(x) = \sum_{k\in\Lambda_{j_0}} \alpha_{j_0,k}\phi_{j_0,k}(x) + \sum_{j=j_0}^{\infty} \sum_{k\in\Lambda_j} \beta_{j,k}\psi_{j,k}(x), \quad x \in [0, 1],$$

where

$$\alpha_{j_0,k} = \int_0^1 h(x)\phi_{j_0,k}(x)dx, \quad \beta_{j,k} = \int_0^1 h(x)\psi_{j,k}(x)dx,$$

Also, let us mention that $\int_0^1 \phi_{j,k}(x)dx = 2^{-j/2}$, which will be a crucial technical point in the proof. Let P_j be the orthogonal projection operator from $L^2([0, 1])$ onto the space V_j with the orthonormal basis $\{\phi_{j,k}(\cdot) = 2^{j/2}\phi(2^j \cdot -k), k \in \Lambda_j\}$. Then, for any $h \in L^2([0, 1])$, we have

$$P_j h(x) = \sum_{k\in\Lambda_j} \alpha_{j,k}\phi_{j,k}(x), \quad x \in [0, 1].$$

Besov spaces have the feature to capture a wide variety of smoothness properties in a function including spatially inhomogeneous behavior, see [11–13] for further details. Definitions of those spaces are given below. Suppose that ϕ is m regular (i.e., $\phi \in C^m$ and $|D^\alpha\phi(x)| \leq c(1 + |x|^2)^{-l}$ for each $l \in \mathbb{Z}$, with $\alpha = 0, 1, \ldots, m$). Let $h \in L^p([0, 1])$, $p, q \in [1, \infty]$ and $0 < s < m$. Then the following assertions are equivalent:

(1) $h \in B^s_{p,q}([0, 1])$; (2) $\left\{2^{js}\|P_{j+1}h - P_j h\|_p\right\} \in l_q$; (3) $\{2^{j(s-\frac{1}{p}+\frac{1}{2})}\|\beta_{j,.}\|_p\} \in l_q$. The Besov norm of h can be defined by

$$\|h\|_{B^s_{p,q}} := \|(\alpha_{\tau,.})\|_p + \|(2^{j(s-\frac{1}{p}+\frac{1}{2})}\|\beta_{j,.}\|_p)_{j\geq\tau}\|_q, \text{ where } \|\beta_{j,.}\|_p^p = \sum_{\mathbf{k}\in\Lambda_j} |\beta_{j,k}|^p.$$

3 Assumptions, Estimators, and Main Result

Technical assumptions on the model (1) are formulated below.

A.1 We suppose that $f : [0, 1] \to \mathbb{R}$ is bounded from above.
A.2 We suppose that $X_1 \sim \mathscr{U}([0, 1])$.
A.3 We suppose that $U_1 \sim \mathscr{U}([-\theta, \theta])$ with $\theta > 0$ a fixed real number.
A.4 We suppose that V_1 has a moment of order 4.
A.5 We suppose that X_i and V_i are independent for any $i \in \{1, \ldots, n\}$.

Let us observe that A.2 specifies that we consider a uniform design and that A.3 specifies that the uniform multiplicative noise is considered over a symmetric interval around 0. The assumption A.5 implies that V_i is not a function of X_i a fortiori.

We construct our linear wavelet estimators for r as follows:

$$\hat{r}_{j_0,n}(x) := \sum_{k\in\Lambda_{j_0}} \hat{\alpha}_{j_0,\mathbf{k}}\phi_{j_0,k}(x), \quad x \in [0,1], \tag{2}$$

where

$$\hat{\alpha}_{j,k} := \frac{3}{\theta^2}\left(\frac{1}{n}\sum_{i=1}^{n} Y_i^2\phi_{j,k}(X_i) - \mathbb{E}(V_1^2)2^{-j/2}\right). \tag{3}$$

The definition of $\hat{\alpha}_{j,k}$ rests on technical consideration which will be presented later. In spite of the simplicity of its construction, its performances strongly depend on the choice of level j_0. Further details on the linear wavelet estimator in a standard nonparametric regression setting can be found in [11]. Recent developments can be found in [14].

The following result determines the rates of convergence attained by $\hat{r}_{j_0,n}$ via the MISE over Besov spaces.

Proposition 1 *Consider the problem defined by* (1) *under the assumptions A.1–A.5, let* $r \in B^s_{p,q}([0,1])$ *with* $p, q \in [1,\infty)$, $s > 1/p$. *Then the linear wavelet estimator* $\hat{r}_{j_0,n}$ *with* $2^{j_*} \sim n^{\frac{1}{2s'+1}}$ *and* $s' = s - (1/p - 1/2)_+$ *satisfies*

$$\mathbb{E}\left[\int_0^1 \left(\hat{r}_{j_0,n}(x) - r(x)\right)^2 dx\right] \lesssim n^{-\frac{2s'}{2s'+1}}.$$

The level j_0 as defined in Proposition 1 is chosen to minimize as possible the MISE of $\hat{r}_{j_0,n}$ over Besov spaces. The rate of convergence $n^{-\frac{2s'}{2s'+1}}$ is not a surprise; it generally corresponds to the one obtained in the standard nonparametric regression estimation. See [6, 11, 15]. The proof of Proposition 1 is based on a suitable decomposition of the MISE and some intermediary results on the probabilistic properties of the wavelet coefficient estimator (3) (see Lemmas 1 and 2 in Sect. 6). The rest of this section is devoted to the practical aspect of the estimator (2), with alternatives on the choice of the level j_0. In particular, we propose a candidate by adapting version of the 2FCV method originally developed by Nason for choosing the threshold parameter in wavelet shrinkage [7].

4 Simulation Study

In order to illustrate the empirical performance of the proposed estimator, a numerical illustration was produced. In order to set in a realistic context, we proposed to use an automatic selection method of the estimator truncation parameter (not depending

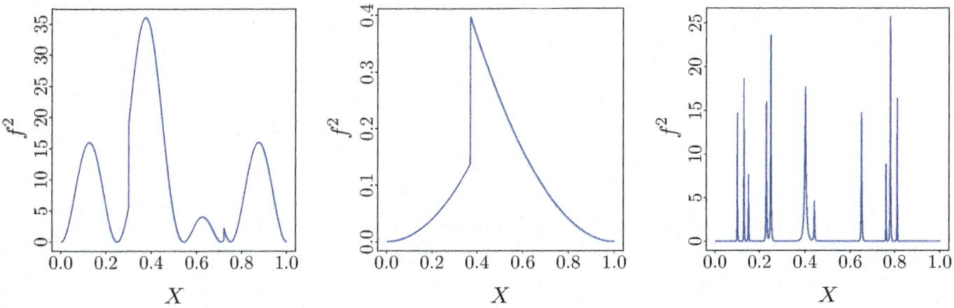

(a) *HeaviSine* (b) *Ramp* (c) *Bumps*

Fig. 1 **a–c**: The three test (squared) functions to be estimated

on the regularity of the function to be estimated). Simulations were performed using R and in particular the rwavelet package [16], available from https://github.com/fabnavarro/rwavelet.

The simulated data were generated according to (1), where $n = 4096$, X_i's are uniformly distributed on $[0, 1]$, U_i's are $\mathcal{U}([-1, 1])$ (so $\theta = 1$) and V_i are $\mathcal{N}(0, \sigma^2)$ variables and independent of X_i's with $\sigma^2 = 0.01$. Daubechies' compactly-supported wavelet with eight vanishing moments were used. We consider three standard test functions for f, commonly used in the wavelet literature (*HeaviSine, Ramp* and *Bumps*, see [17]). Recall that we wish to estimate $r = f^2$. The squared version of those functions are plotted in Fig. 1.

In the case of fixed design, the calculation of wavelet-based estimators is simple and fast, thanks to Mallat's pyramidal algorithm [9]. In the case of uniform random design, the implementation requires some changes and several strategies have been developed in the literature (see e.g., [18, 19]). For uniform design regression, [20] proposed to use and studied an approach in which the wavelet coefficients are computed by a simple application of Mallat's algorithm using the ordered Y_i's as input variables. We have followed this approach because it preserves the simplicity of calculation and the efficiency of the equispaced algorithm. In the context of wavelet regression in random design with heteroscedastic noise, [21, 22] also adopted this approach. Nason adjusted the usual 2FCV method to choose the threshold parameter in wavelet shrinkage (see [7]). His strategy was used for the selection of linear wavelet estimators by [22]. We have chosen to use this approach to select the truncation parameter j_0 of the linear estimator $\hat{r}_{j_0,n}$. More precisely, we built a collection of linear estimators $\hat{r}_{j_0,n}$, $j_0 = 0, 1, \ldots, \log 2(n) - 1$ (by successively adding whole resolution levels of wavelet coefficients), and select the best among this collection by minimizing a 2FCV criterion denoted by $2\mathrm{FCV}(j_0)$. The resulting estimator of the truncation level is denoted by $\hat{j}_0$ and the corresponding estimator of r by $\hat{r}_{\hat{j}_0,n}$ (see [22, 23] for more details).

For a single experiment, and for each of the three test functions, with a sample size $n = 4096$, we display the observations and the unknown function r in Fig. 2a. A sample of three estimators from the collection is also shown in the Fig. reffig:singleb.

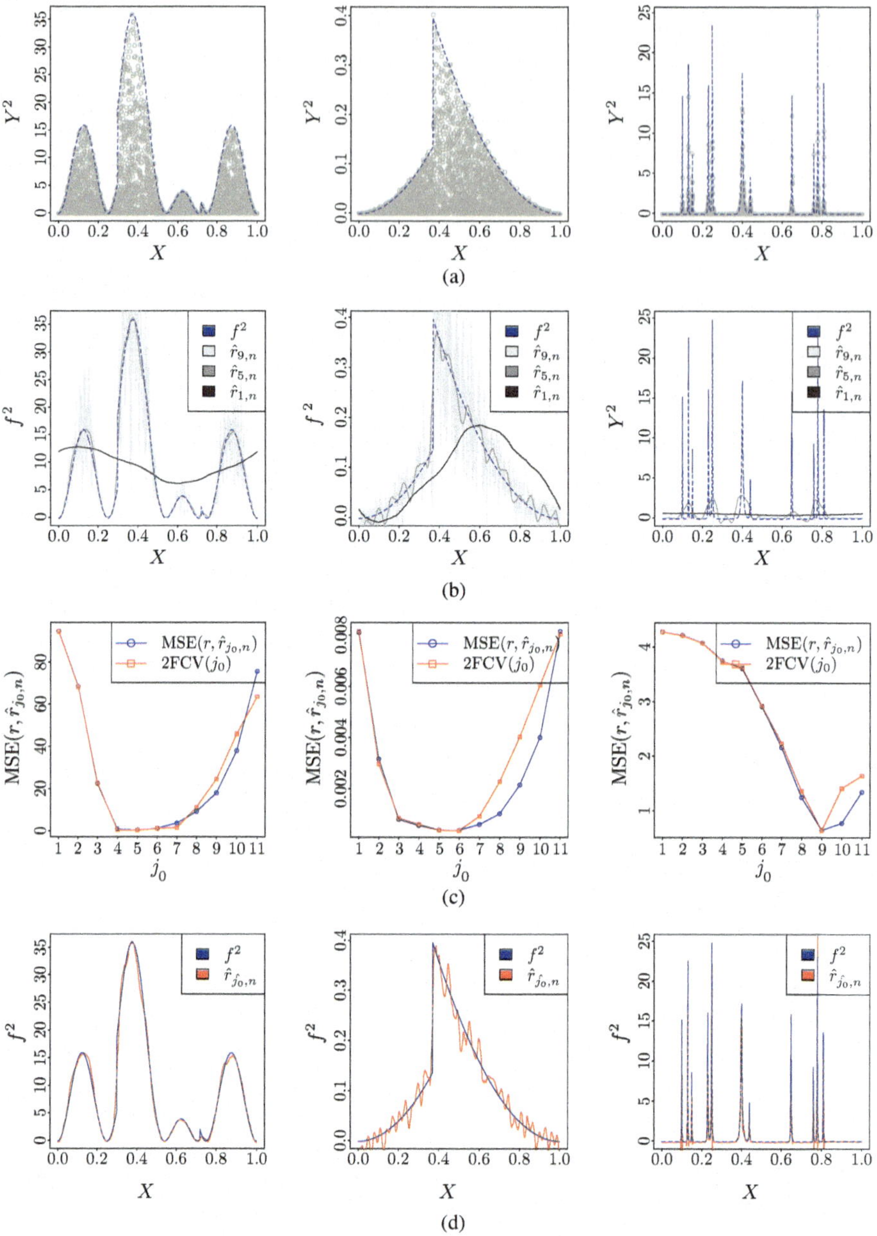

Fig. 2 **a**: Noisy observations (X, Y^2). **b**: Sample of the model collection. **c**: Graph of the MSE (blue) against j_0 and (re-scaled) 2FCV criterion. **d**: Typical estimations from one simulation with $n = 4096$. Blue lines indicate the true functions, red lines correspond to the estimators $\hat{r}_{\hat{j}_0,n}$

Graphs of the curves associated with the selection criterion (i.e. 2FCV(j_0)) are also displayed in Fig. 2c. In order to be able to evaluate the performance of this criterion, the Mean Square Error curves (i.e., MSE($\hat{r}_{j_0,n}, r) = \frac{1}{n}\sum_{i=1}^{n}(r(X_i) - \hat{r}_{j_0,n}(X_i))^2$)) are also shown (in blue). We denote by j_0^*, the parameter selected by minimizing this quantity. It can be observed that 2FCV(j_0) gives very reliable estimate for the MSE($\hat{r}_{j_0,n}, r$), and in turn, also a high-quality estimate of the optimal model. Indeed, in this case, the method allows to find the oracle of the collection (i.e., that obtained by assuming the regularity of the function to be estimated known) for the three signals.

5 Conclusion

In this paper, we develop a simple wavelet methodology for the problem of estimating an unknown function subject to additive and multiplicative noises. Focusing on a uniform multiplicative noise, we construct a linear wavelet estimator that attains a fast rate of convergence. Then some extensions of the estimator are presented, with a numerical study showing the usefulness of the method.

A possible extension of this work would be to consider a more general assumption on the distribution of the multiplicative noise. Another possible extension would be to construct another wavelet estimation procedure involving thresholding of the wavelet coefficient estimators and also dependence on the observations, as in [24] for the additive noise only. These aspects need further investigations that we leave for future work.

6 Proofs

To prove Proposition 1, we use the following two lemmas.

Lemma 1 *Let* $j \geq \tau$, $k \in \Lambda_j$, $\hat{\alpha}_{j,k}$ *be* (3). *Then, under A.1–A.5, we have*

$$\mathbb{E}[\hat{\alpha}_{j,k}] = \alpha_{j,k}.$$

Proof of Lemma 1. Using the independence assumptions on the random variables, A.1–A.5 with $\mathbb{E}[U_1] = 0$, observe that

$$\mathbb{E}\left[U_1 V_1 f(X_1)\phi_{j,k}(X_1)\right] = \mathbb{E}[U_1]\mathbb{E}[V_1]\mathbb{E}\left[f(X_1)\phi_{j,k}(X_1)\right] = 0$$

and

$$\mathbb{E}\left[V_1^2\phi_{j,k}(X_1)\right] = \mathbb{E}[V_1^2]\mathbb{E}\left[\phi_{j,k}(X_1)\right] = \mathbb{E}[V_1^2]\int_0^1 \phi_{j,k}(x)dx = \mathbb{E}[V_1^2]2^{-j/2}.$$

Therefore, using similar mathematical arguments with $\mathbb{E}\left[U_1^2\right] = \frac{\theta^2}{3}$, we have

$$\begin{aligned}
\mathbb{E}[\hat{\alpha}_{j,k}] &= \frac{3}{\theta^2}\mathbb{E}\left[\frac{1}{n}\sum_{i=1}^{n} Y_i^2\phi_{j,k}(X_i) - \mathbb{E}[V_1^2]2^{-j/2}\right] \\
&= \frac{3}{\theta^2}\left(\mathbb{E}\left[Y_1^2\phi_{j,k}(X_1)\right] - \mathbb{E}[V_1^2]2^{-j/2}\right) \\
&= \frac{3}{\theta^2}\Big(\mathbb{E}\left[U_1^2 r(X_1)\phi_{j,k}(X_1)\right] + 2\mathbb{E}\left[U_1V_1 f(X_1)\phi_{j,k}(X_1)\right] + \mathbb{E}\left[V_1^2\phi_{j,k}(X_1)\right] \\
&\qquad - \mathbb{E}\left[V_1^2\phi_{j,k}(X_1)\right]\Big) \\
&= \frac{3}{\theta^2}\mathbb{E}\left[U_1^2\right]\mathbb{E}\left[r(X_1)\phi_{j,k}(X_1)\right] = \int_0^1 r(x)\phi_{j,k}(x)dx = \alpha_{j,k}.
\end{aligned}$$

Lemma 1 is proved. □

Lemma 2 *Let $j \geq \tau$ such that $2^j \leq n$, $k \in \Lambda_j$, $\hat{\alpha}_{j,k}$ be* (3). *Then, under (A.A.1)–(A.5),*

$$\mathbb{E}\left[(\hat{\alpha}_{j,k} - \alpha_{j,k})^2\right] \lesssim \frac{1}{n}.$$

Proof of Lemma 2. Owing to Lemma 1, we have $\mathbb{E}[\hat{\alpha}_{j,k}] = \alpha_{j,k}$. Therefore

$$\begin{aligned}
\mathbb{E}[(\hat{\alpha}_{j,k} - \alpha_{j,k})^2] = Var\left[\hat{\alpha}_{j,k}\right] &= \frac{9}{\theta^4}Var\left[\frac{1}{n}\sum_{i=1}^{n} Y_i^2\phi_{j,k}(X_i) - \mathbb{E}[V_1^2]2^{-j/2}\right] \\
&= \frac{9}{\theta^4}Var\left[\frac{1}{n}\sum_{i=1}^{n} Y_i^2\phi_{j,k}(X_i)\right] \\
&= \frac{9}{\theta^4}\frac{1}{n}Var\left[Y_1^2\phi_{j,k}(X_1)\right] \lesssim \frac{1}{n}E\left[Y_1^4\phi_{j,k}^2(X_1)\right] \\
&\lesssim \frac{1}{n}\left[\mathbb{E}\left[U_1^4 f^4(X_1)\phi_{j,k}^2(X_1)\right] + \mathbb{E}\left[V_1^4\phi_{j,k}^2(X_1)\right]\right] \\
&= \frac{1}{n}\left[\mathbb{E}\left[U_1^4\right]\mathbb{E}\left[f^4(X_1)\phi_{j,k}^2(X_1)\right] + \mathbb{E}\left[V_1^4\phi_{j,k}^2(X_1)\right]\right]. \qquad (4)
\end{aligned}$$

By A.1 and $\mathbb{E}\left[\phi_{j,k}^2(X_1)\right] = \int_0^1 \phi_{j,k}^2(x)dx = 1$, we have $\mathbb{E}\left[f^4(X_1)\phi_{j,k}^2(X_1)\right] \lesssim 1$. On the other hand, by A.4 and A.5, we have

$$\mathbb{E}\left[V_1^4\phi_{j,k}^2(X_1)\right] = \mathbb{E}\left[V_1^4\right]\mathbb{E}\left[\phi_{j,k}^2(X_1)\right] = \mathbb{E}\left[V_1^4\right] \lesssim 1$$

Thus, all the terms in the brackets of (4) are bounded from above. This ends the proof of Lemma 2. □

Proof of Proposition 1 **from Lemmas** 1 **and** 2. The main lines of the proof use standard arguments (see, for instance, [11]). The key result remains Lemma 2 above and a suitable choice for j_0 which balance the bias and the rest term of term. More precisely, by the definition of projector, we have

$$\mathbb{E}\left[\int_0^1 \left|\hat{r}_{j_0,n}(x) - r(x)\right|^2 dx\right] = \mathbb{E}\left[\left\|\hat{r}_{j_0,n} - P_{j_*}r\right\|_2^2\right] + \left\|P_{j_*}r - r\right\|_2^2. \tag{5}$$

The orthonormality of the wavelet basis gives

$$\mathbb{E}\left[\left\|\hat{r}_{j_0,n} - P_{j_*}r\right\|_2^2\right] = \mathbb{E}\left[\left\|\sum_{k\in\Lambda_{j_*}} (\hat{\alpha}_{j_*,k} - \alpha_{j_*,k})\phi_{j_*,k}\right\|_2^2\right] = \sum_{k\in\Lambda_{j_*}} \mathbb{E}[(\hat{\alpha}_{j_*,k} - \alpha_{j_*,k})^2].$$

According to Lemma 2, $|\Lambda_{j_*}| \sim 2^{j_*}$ and $2^{j_*} \sim n^{\frac{1}{2s'+1}}$,

$$\mathbb{E}\left[\left\|\hat{r}_{j_0,n} - P_{j_*}r\right\|_2^2\right] \lesssim \frac{2^{j_0}}{n} \lesssim n^{-\frac{2s'}{2s'+1}}. \tag{6}$$

When $p \geq 2$, $s' = s$. By Hölder inequality and $r \in B^s_{p,q}([0,1])$,

$$\|P_{j_0}r - r\|_2^2 \lesssim \|P_{j_0}r - r\|_p^2 \lesssim 2^{-2j_0 s} \lesssim n^{-\frac{2s}{2s+1}}.$$

When $1 \leq p < 2$ and $s > 1/p$, $B^s_{p,q}([0,1]) \subseteq B^{s'}_{2,\infty}([0,1])$

$$\|P_{j_0}r - r\|_2^2 \lesssim \sum_{j=j_0}^{\infty} 2^{-2js'} \lesssim 2^{-2j_0 s'} \lesssim n^{-\frac{2s'}{2s'+1}}.$$

Therefore, in both cases,

$$\|P_{j_0}r - r\|_2^2 \lesssim n^{-\frac{2s'}{2s'+1}}. \tag{7}$$

By (5), (6) and (7), we obtain

$$\mathbb{E}\left[\int_0^1 \left|\hat{r}_{j_0,n}(x) - r(x)\right|^2 dx\right] \lesssim n^{-\frac{2s'}{2s'+1}}.$$

Proposition 1 is proved. □

References

1. Huang, P., Pi, Y., Progri, I.: Gps signal detection under multiplicative and additive noise. J. Navig. **66**(4), 479–500 (2013)
2. Härdle, W., Tsybakov, A.: Local polynomial estimators of the volatility function in nonparametric autoregression. J. Econ. **81**(1), 223–242 (1997)
3. Simar, L., Wilson, P.W.: Statistical inference in nonparametric frontier models: the state of the art. J. Product. Anal. **13**(1), 49–78 (2000)
4. Tony Cai, T., Wang, L., et al.: Adaptive variance function estimation in heteroscedastic nonparametric regression. Ann. Stat. **36**(5), 2025–2054 (2008)
5. Chichignoud, M.: Minimax and minimax adaptive estimation in multiplicative regression: locally Bayesian approach. Probab. Theory Relat. Fields **153**(3–4), 543–586 (2012)
6. Comte, F.: Estimation non-paramétrique. Spartacus-IDH (2015)
7. Nason, G.P.: Wavelet shrinkage using cross-validation. J. R. Stat. Soc. Ser. B (Methodol.) 463–479 (1996)
8. Daubechies, I.: Ten Lectures On Wavelets, vol. 61. SIAM (1992)
9. Mallat, S.: A Wavelet Tour of Signal Processing: The Sparse Way. Academic (2008)
10. Cohen, A., Daubechies, I., Vial, P.: Wavelets on the interval and fast wavelet transforms. Appl. Comput. Harmon. Anal. (1993)
11. Härdle, W., Kerkyacharian, G., Picard, D., Tsybakov, A.: Wavelets, Approximation, And Statistical Applications, vol. 129. Springer Science & Business Media (2012)
12. Meyer, Y.: Wavelets and Operators, vol. 1. Cambridge University Press, Cambridge (1992)
13. Triebel, H.: Theory of function spaces ii. Bull. Am. Math. Soc. **31**, 119–125 (1994)
14. Chaubey, Y.P., Chesneau, C., Doosti, H.: Adaptive wavelet estimation of a density from mixtures under multiplicative censoring. Statistics **49**(3), 638–659 (2015)
15. Tsybakov, A.B.: Introduction to nonparametric estimation. Revised and extended from the 2004 french original. Translated by vladimir zaiats (2009)
16. Navarro, F., Chesneau, C.: R package rwavelet: wavelet analysis (Version 0.1.0) (2018)
17. Donoho, D.L., Johnstone, I.M., Kerkyacharian, G., Picard, D.: Wavelet shrinkage: asymptopia? J. R. Stat. Soc. Ser. B (Methodol.), pp. 301–369 (1995)
18. Tony Cai, T., Brown, L.D., et al.: Wavelet shrinkage for nonequispaced samples. Ann. Stat. **26**(5), 1783–1799 (1998)
19. Hall, P., Turlach, B.A., et al.: Interpolation methods for nonlinear wavelet regression with irregularly spaced design. Ann. Stat. **25**(5), 1912–1925 (1997)
20. Tony Cai, T., Brown, L.D.: Wavelet estimation for samples with random uniform design. Stat. Probab. Lett. **42**(3), 313–321 (1999)
21. Kulik, R., Raimondo, M., et al.: Wavelet regression in random design with heteroscedastic dependent errors. Ann. Stat. **37**(6A), 3396–3430 (2009)
22. Navarro, F., Saumard, A.: Slope heuristics and v-fold model selection in heteroscedastic regression using strongly localized bases. ESAIM: Probab. Stat. **21**, 412–451 (2017)
23. Navarro, F., Saumard, A.: Efficiency of the v-fold model selection for localized bases. In: Bertail, P., Blanke, D., Cornillon, P.-A., Matzner-Løber, E. (eds.) Nonparametric Statistics, pp. 53–68. Springer International Publishing, Cham (2018)
24. Chesneau, C., Fadili, J., Maillot, B.: Adaptive estimation of an additive regression function from weakly dependent data. J. Multivar. Anal. **133**, 77–94 (2015)

Speeding up Algebraic-Based Sampling via Permutations

Francesca Romana Crucinio and Roberto Fontana

Abstract Algebraic sampling methods are a powerful tool to perform hypothesis tests on conditional spaces. We analyse the link of the sampling method introduced in [6] with permutation tests and we exploit this link to build a two-step sampling procedure to perform two-sample comparisons for non-negative discrete exponential families. We thus establish a link between standard permutation and algebraic-statistics-based sampling. The proposed method reduces the dimension of the space on which the MCMC sampling is performed by introducing a second step in which a standard Monte Carlo sampling is performed. The advantages of this dimension reduction are verified through a simulation study, showing that the proposed approach grants convergence in the least time and has the lowest mean squared error.

Keywords Conditional tests · Discrete exponential families · Markov basis · Markov chain monte carlo

1 Introduction

Consider two samples $\mathbf{Y}_1$ and $\mathbf{Y}_2$ of size n_1 and n_2, respectively, coming from some non-negative discrete exponential family with natural parameter $\psi(\cdot)$, base measure $H(\cdot)$ and normalising constant $G(\cdot)$

$$f(y \mid \mu_i) = G(\mu_i)H(y)\exp\{y \cdot \psi(\mu_i)\} \qquad i = 1, 2.$$

We are interested in conditional tests that exploit the joint sample $\mathbf{Y} = (\mathbf{Y}_1, \mathbf{Y}_2)$ of size $N = n_1 + n_2$ to test

F. R. Crucinio (✉)
Department of Statistics, University of Warwick, Coventry CV4 7AL, UK
e-mail: F.Crucinio@Warwick.ac.uk

R. Fontana
Dipartimento di Scienze Matematiche, Politecnico di Torino, Corso Duca degli Abruzzi 24, 10129 Torino, Italy
e-mail: roberto.fontana@polito.it

M. La Rocca et al. (eds.), *Nonparametric Statistics*, Springer Proceedings in Mathematics & Statistics 339, https://doi.org/10.1007/978-3-030-57306-5_14

$$H_0 : \mu = \mu_1 = \mu_2 \quad \text{against} \quad H_1 : \mu_1 \gtrless \mu_2. \tag{1}$$

Specifically, there exists a uniformly most powerful unbiased (UMPU) procedure performed conditionally on the sum of the entries of the pooled sample $T = \sum_{i=1}^{N} Y_i$ [10]. Moreover, T is a sufficient statistic for the nuisance parameter of the test, the population constant β_0, if we assume the standard one-way ANOVA model for the means $\psi(\mu_i) = \beta_0 + \beta_i,\ i = 1, 2$ [11].

The test statistic adopted in UMPU tests is $U = \sum_{i=1}^{n_1} Y_i$ and its conditional distribution given T under H_0 in (1) is

$$f_U(u \mid T = t) = \frac{\sum\limits_{\mathbf{y}_1 \in \mathscr{F}_{n_1,u}} \prod\limits_{i=1}^{n_1} H(y_i) \cdot \sum\limits_{\mathbf{y}_2 \in \mathscr{F}_{n_2,t-u}} \prod\limits_{i=n_1+1}^{n_1+n_2} H(y_i)}{\sum\limits_{u=0}^{t} \sum\limits_{\mathbf{y}_1 \in \mathscr{F}_{n_1,u}} \prod\limits_{i=1}^{n_1} H(y_i) \cdot \sum\limits_{\mathbf{y}_2 \in \mathscr{F}_{n_2,t-u}} \prod\limits_{i=n_1+1}^{n_1+n_2} H(y_i)}, \tag{2}$$

where H is the base measure of the non-negative discrete exponential family f. We denote by $\mathscr{F}_{n,x}$ the set of non-negative integer vectors of length n with sum of entries equal to x.

In order to perform the test (1), we can either find the critical values for any given risk of type I error or, alternatively, compute the p-value corresponding to the observed value u_{obs} of U. Unfortunately, the distribution (2) can rarely be computed in closed form. In most cases, it is necessary to approximate (2) through Markov Chain Monte Carlo (MCMC).

MCMC sampling methods suffer two major drawbacks in the discrete setting: the construction of the Markov basis needed to build the chain is computationally expensive and the chain may mix slowly [8]. The idea of speeding-up the MCMC sampling is therefore not new in the Algebraic Statistics literature. In [7], the first drawback is addressed in the case of bounded contingency tables, instead of computing the entire Markov basis in an initial step, sets of local moves that connect each table in the reference set with a set of neighbouring tables are studied. A similar approach for bounded two-way contingency tables under the independence model with positive bounds is presented in [13]. A hybrid scheme using MCMC and sequential importance sampling able to address both drawbacks has been proposed in [8]. We propose a strategy that exploits the structure of the sample space for UMPU tests and does not require sequential importance sampling but only standard independent Monte Carlo sampling.

2 Markov Chain Monte Carlo Samplings

As a consequence of the conditioning on $T = \sum_{i=1}^{n_1+n_2} Y_i$, the sample space to be inspected under H_0, is the *fibre* of non-negative integer vectors of size $N = n_1 + n_2$ and with entries which add up to t

$$\mathscr{F}_{N,t} = \{(Y_1, \ldots, Y_N) \in \mathbb{N}^N : \sum_{i=1}^{N} Y_i = \mathbf{1}_N^T \mathbf{Y} = t\}, \tag{3}$$

where $\mathbf{1}_N = (1, \ldots, 1)$ is the vector of length N with all entries equal to 1.

The distribution we are interested in is the cumulative distribution function of the test statistic $U = \sum_{i=1}^{n_1} Y_i$ given $T = \sum_{i=1}^{N} Y_i = t$ under H_0 as specified in (1)

$$F_U(u \mid \mathscr{F}_{N,t}) = \mathbb{P}(U(\mathbf{y}) \leq u \mid \mathbf{y} \in \mathscr{F}_{N,t}) = \sum_{\mathbf{y} \in \mathscr{F}_{N,t}} \mathbb{I}_{(U(\mathbf{y}) \leq u)}(\mathbf{y}) f(\mathbf{y} \mid \mu), \tag{4}$$

where $U(\mathbf{y}) = \sum_{i=1}^{n_1} y_i$ and $\mathbb{I}_{(U(\mathbf{y}) \leq u)}(\mathbf{y})$ is 1 if $U(\mathbf{y}) \leq u$ and 0 otherwise and $f(\mathbf{y} \mid \mu) = \prod_{i=1}^{N} f(y_i \mid \mu)$ with a slight abuse of notation.

In the following, we describe two MCMC algorithms to sample from $\mathscr{F}_{N,t}$. The first one samples *vectors* $\mathbf{y} \in \mathscr{F}_{N,t}$, while the second one samples *orbits* of permutations $\pi \subseteq \mathscr{F}_{N,t}$. Both MCMC algorithms make use of a Markov basis, a set of moves allowing to build a connected Markov chain over $\mathscr{F}_{N,t}$ using only simple additions/subtractions [6]. Specifically, a Markov basis for a matrix A is a finite set of moves $\{\mathbf{m}_1, \ldots, \mathbf{m}_K\}$ such that

1. $\mathbf{m}_i$ belongs to the integer kernel of A, $1 \leq i \leq K$;
2. every pair of elements $\mathbf{x}, \mathbf{y}$ in $\mathscr{F}_{N,t}$ is connected by a path formed by a sequence $(\mathbf{m}, \varepsilon)$ of moves $\mathbf{m}$ and signs $\varepsilon = \pm 1$, and this path is contained in $\mathscr{F}_{N,t}$.

Markov bases can be found analytically [5, 6] or using the algebraic software `4ti2` [1].

2.1 MCMC—Vector-Based

The first MCMC we consider is an adaptation of the algorithm used in [2, 3, 6] for the fibre $\mathscr{F}_{N,t}$.

An example of the Markov chain over $\mathscr{F}_{N,t}$ is shown in Fig. 1a for $N = 3$ and $t = 6$. Each vertex of the graph represents a vector $\mathbf{y} \in \mathscr{F}_{N,t}$ and each edge represents an applicable move in the Markov basis. The number of states (i.e. vectors) and the number of edges is given by

$$|V| = \binom{t + N - 1}{N - 1}$$

$$|E| = 2(N - 1)\binom{t - 1}{N - 1} + \sum_{z=1}^{N-1} (N - z)\binom{t - 1}{N - 1 - z}\binom{N - 1}{z - 1}\binom{2N - 2}{z},$$

respectively [5].

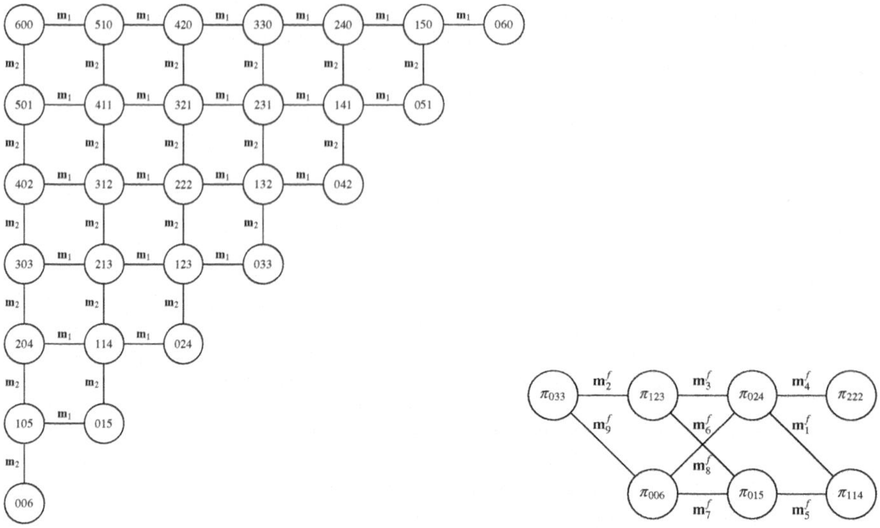

(a) Graph on the fiber $\mathcal{F}_{3,6}$.

(b) Graph on the orbits of $\mathcal{F}_{3,6}$.

Fig. 1 Vector-based and orbit-based parametrisation of the fibre $\mathcal{F}_{N,t}$

The target distribution of the Markov chain is the probability of sampling $\mathbf{y} \in \mathcal{F}_{N,t}$ under H_0 as specified in (1)

$$f(\mathbf{y} \mid \mu) = \prod_{i=1}^{N} f(y_i \mid \mu) = G(\mu)^N \exp\{\psi(\mu)t\} \prod_{i=1}^{N} H(y_i) \propto \prod_{i=1}^{N} H(y_i).$$

The estimator used to approximate (4) is the indicator function $\mathbb{I}_{(U(\mathbf{y}) \leq u)}(\mathbf{y})$, where the $\mathbf{y}$s are the sampled vectors.

2.2 MCMC—Orbit-Based

The second algorithm is built by exploiting the link of $\mathcal{F}_{N,t}$ with orbits of permutations π. Clearly, if $\mathbf{y} \in \mathcal{F}_{N,t}$, every permutation of $\mathbf{y}$ is an element of $\mathcal{F}_{N,t}$ too. Moreover, different orbits do not intersect. Therefore the orbits of permutations $\pi \subseteq \mathcal{F}_{N,t}$ form a partition of $\mathcal{F}_{N,t}$.

This partition is particularly interesting, as the elements which belong to the same orbit have the same probability of being sampled from $\mathcal{F}_{N,t}$, [12].

Therefore, it is possible to devise a two-step sampling procedure:

Step 1: Sample one orbit π from the set of orbits $\pi \subseteq \mathcal{F}_{N,t}$.
Step 2: Sample uniformly from π.

The first step can be performed through a MCMC algorithm similar to the one described in Sect. 2.1 with target distribution the probability of sampling $\mathbf{y}$ in orbit π

$$\sum_{\mathbf{y}\in\pi} f(\mathbf{y} \mid \mu),$$

while the second one corresponds to a standard Monte Carlo sampling from π.

The number of orbits of permutation π contained in the fibre is given by part(t, N) [5], with part defined in [9, 14]. The values of the partition function can be computed using the recurrence

$$|O| = \text{part}(t, N) = \text{part}(t, N-1) + \text{part}(t-N, N)$$

and depend on both the sample size N and the sum of entries t.

To perform Step 1, we parametrise the fibre $\mathscr{F}_{N,t}$ in such a way that all the vectors in the same permutation orbit are mapped to the same element. To do so, we consider a frequency-based representation. In this representation the orbit $\pi_{(0,2,4)} \subseteq \mathscr{F}_{3,6}$ is mapped into $\mathbf{f}_\pi = (1, 0, 1, 0, 1, 0, 0)$. In this notation, vectors $(0, 4, 2)$ and $(2, 0, 4)$, which belong to the same orbit, correspond to the same frequency vector.

The target distribution in the frequency-based parametrisation is

$$\sum_{\mathbf{y}\in\pi} f(\mathbf{y} \mid \mu) = \#\pi \cdot C \prod_{j=0}^{t} H(j)^{f_j} \propto \frac{N!}{f_0! \cdot \ldots \cdot f_t!} \prod_{j=0}^{t} H(j)^{f_j},$$

with $\#\pi$ being the number of distinct elements in the orbit π.

Because Step 2 corresponds to a standard permutation sampling, we consider the distribution of U given T over one orbit π, i.e. the usual permutation cdf,

$$F_U(u \mid \pi) = \mathbb{P}(U(\mathbf{y}) \le u \mid \mathbf{y} \in \pi) = \frac{1}{\#\pi} \sum_{\mathbf{y}\in\pi} \mathbb{I}_{(U(\mathbf{y})\le u)}(\mathbf{y}). \tag{5}$$

3 Comparison of Vector-Based and Orbit-Based MCMC

Dividing the sampling procedure into the two steps described in Sect. 2.2 has a clear computational advantage: Step 2 corresponds to a standard Monte Carlo sampling from the orbit π, which is faster than performing an MCMC sampling. On the other hand, Step 1 performs an MCMC sampling over the set of orbits π contained in $\mathscr{F}_{N,t}$, whose cardinality is smaller than that of the set of vectors in $\mathscr{F}_{N,t}$:

$$|V| = \binom{t+N-1}{N-1} > \text{part}(t, N) = |O| \qquad \text{for } t, N > 1.$$

Table 1 Ratio between the number of orbits $|O|$ and the number of vectors $|V|$ in $\mathscr{F}_{N,t}$ for several values of N and t

N \ t	5	10	15	20	30	50	100
5	0.056	0.030	0.022	0.018	0.015	0.012	0.010
10	0.004	$4.5 \cdot 10^{-4}$	$1.3 \cdot 10^{-4}$	$5.3 \cdot 10^{-5}$	$1.7 \cdot 10^{-5}$	$5.0 \cdot 10^{-6}$	$1.5 \cdot 10^{-6}$
15	$6.0 \cdot 10^{-4}$	$2.1 \cdot 10^{-5}$	$2.3 \cdot 10^{-6}$	$4.4 \cdot 10^{-7}$	$4.4 \cdot 10^{-8}$	$2.9 \cdot 10^{-9}$	$1.4 \cdot 10^{-10}$
20	$1.6 \cdot 10^{-4}$	$2.1 \cdot 10^{-6}$	$9.5 \cdot 10^{-8}$	$9.1 \cdot 10^{-9}$	$2.9 \cdot 10^{-10}$	$3.9 \cdot 10^{-12}$	$2.0 \cdot 10^{-14}$
50	$2.2 \cdot 10^{-6}$	$6.7 \cdot 10^{-10}$	$1.1 \cdot 10^{-12}$	$5.4 \cdot 10^{-15}$	$1.0 \cdot 10^{-18}$	$4.0 \cdot 10^{-24}$	$2.8 \cdot 10^{-32}$

Table 1 shows the ratios between the cardinality of $\pi \subseteq \mathscr{F}_{N,t}$ and the cardinality of $\mathbf{y} \in \mathscr{F}_{N,t}$ for values of N and t between 5 and 100. Even for moderately sized samples, the number of orbits contained in $\mathscr{F}_{N,t}$ is about two orders of magnitude smaller than the number of vectors (e.g. for $N = 5$ and $t = 5$ $|O|/|V| = 5.6 \cdot 10^{-2}$).

Hence, if we keep the number of iterations fixed and compare the number of vectors inspected by the two algorithms, the orbit-based algorithm gives the highest value, namely the number of iterations $\times$ the number of distinct vectors in the orbit sampled at iteration i.

We show how the reduction in the dimension of the space explored by the MCMC algorithm improves convergence and accuracy with respect to the truth (4) through the simulation study in the next section.

4 Simulation Study

Assume that the two samples $\mathbf{Y}_1$ and $\mathbf{Y}_2$ are Poisson distributed with mean μ_1 and μ_2, respectively. In this case, the exact distribution (4) under H_0 is the binomial distribution

$$F_U(u \mid \mathscr{F}_{N,t}) = \mathbb{P}(\text{Binomial}(t, \theta_0) \leq u) = \sum_{k=0}^{u} \binom{t}{k} \theta_0^k (1-\theta_0)^{t-k}, \tag{6}$$

with $\theta_0 = n_1/(n_1 + n_2)$ [10, 11].

We compare the *exact* conditional cdf above with the *approximated* cdfs given by the vector-based algorithm, the orbit-based algorithm and the standard permutation cdf over the orbit of the observed pooled vector (this is the limit case of the orbit-based algorithm when only the observed orbit is sampled). A preliminary simulation study is presented in [4].

We consider 9 scenarios built taking three sample sizes (n_1, n_2) and, for each sample size, three different couples of population means (μ_1, μ_2).

The comparisons performed are two: first, we check the convergence behaviour on a fixed runtime (15 s) for both algorithms; then we compare their accuracy through the mean squared error (MSE).

4.1 Convergence Comparison

To compare how fast the two MCMC procedures converge to the true distribution (4), we draw one random sample $\mathbf{y}_{obs}$ for each scenario above and we run both algorithms for 15 s. Figure 2 shows four examples of the behaviour of the two MCMC procedures which are representative of the nine scenarios.

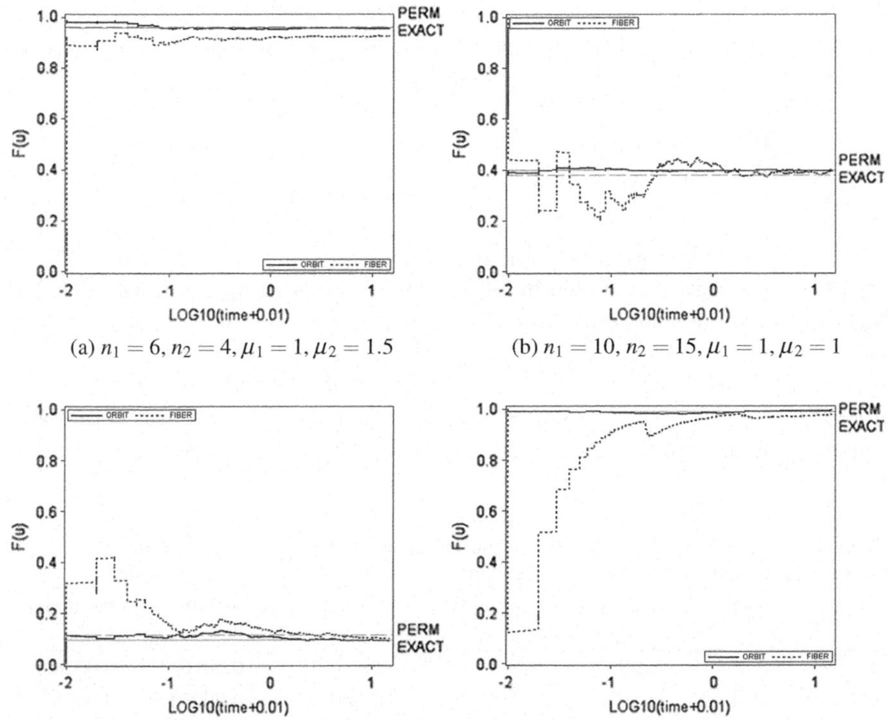

(a) $n_1 = 6, n_2 = 4, \mu_1 = 1, \mu_2 = 1.5$

(b) $n_1 = 10, n_2 = 15, \mu_1 = 1, \mu_2 = 1$

(c) $n_1 = 10, n_2 = 15, \mu_1 = 1, \mu_2 = 1.5$

(d) $n_1 = 30, n_2 = 20, \mu_1 = 1, \mu_2 = 1$

Fig. 2 Comparison of the convergence to the exact value (solid horizontal line) in 15 s for the vector-based algorithm (dashed line) and the orbit-based algorithm (solid line). The Monte Carlo permutation estimate of $F_U(u \mid \mathscr{F}_{N,t})$ (dashed horizontal line) is reported too. The number of Monte Carlo permutations per orbit is 5, 000. The plots show the estimates achieved as functions of the log-time

The orbit-based algorithm is very fast and achieves good convergence in ~ 0.1 seconds. On the contrary, the vector-based algorithm is much less efficient, in fact, its convergence to the exact value is not always satisfactory even after 15 s (Fig. 2a, b).

Remark 1 It would be possible to further reduce the computational time required by the orbit-based algorithm by exploiting one of the key features of this new approach, namely the possibility of sampling from each orbit independently. The Monte Carlo samplings in Step 2 could be made in *parallel*: once the chain reaches an orbit π the Monte Carlo sampling over π can be performed while the chain keeps on moving on the set of orbits.

4.2 Accuracy Comparison

For each scenario, we randomly generate 1,000 samples through which we compute the MSE of the distribution estimated by the three procedures under study

$$\mathrm{MSE} = \frac{1}{1000} \sum_{j=1}^{1000} \left(\sum_{k=0}^{u_j} \binom{t_j}{k} \theta_0^k (1-\theta_0)^{t_j-k} - \hat{F}_U(u_j \mid \mathscr{F}_{N,t_j}) \right)^2 .$$

Both MCMC algorithms are run for 15 s with no burn-in steps. The resulting MSE for the nine scenarios is shown in Table 2. As a further comparison, we report the MSE given by the standard Monte Carlo permutation sampling over the observed orbit.

The orbit-based algorithm always give the smallest MSE apart from scenario $n_1 = 20$, $n_2 = 30$, $\mu_1 = 1$, $\mu_2 = 1$, where the standard Monte Carlo permutation sampling has the smallest MSE. Table 3 shows the ratio between the MSE of the

Table 2 Mean Squared Error (MSE) for the vector-based, the orbit-based and the Monte Carlo permutation sampling. Both MCMC algorithms were run for 15 s with no burn-in steps

n_1	n_2	μ_1	μ_2	Orbit-based	Vector-based	Permutation
6	4	1	1	0.00012	0.0016	0.00284
6	4	1	1.5	0.00012	0.00083	0.00212
6	4	1	2	0.00016	0.00043	0.00221
10	15	1	1	0.00034	0.00131	0.00077
10	15	1	1.5	0.00009	0.00046	0.00074
10	15	1	2	0.00007	0.00017	0.00057
20	30	1	1	0.00069	0.00132	0.00036
20	30	1	1.5	0.00006	0.00053	0.00027
20	30	1	2	0.00001	0.00011	0.00009

Table 3 Ratio between the MSE of the vector-based and the MSE of the orbit-based algorithm (column 5) and ratio between the MSE of the standard Monte Carlo permutation sampling and the MSE of the orbit-based algorithm (column 6).

n_1	n_2	μ_1	μ_2	MSE vector/MSE orbit	MSE perm/MSE orbit
6	4	1	1	12.82	22.7
6	4	1	1.5	6.98	17.79
6	4	1	2	2.67	13.82
10	15	1	1	3.9	2.31
10	15	1	1.5	4.96	8
10	15	1	2	2.45	8.27
20	30	1	1	1.9	0.52
20	30	1	1.5	9.03	4.58
20	30	1	2	15.9	12.77

Table 4 Number of iterations for 15 s

Scenario				N. iterations		Ratio
n_1	n_2	μ_1	μ_2	Orbit	Vector	Vector/Orbit
6	4	1	1	23,977	53,842	2.25
6	4	1	1.5	24,169	53,210	2.20
6	4	1	2	24,560	57,382	2.34
10	15	1	1	11,950	52,504	4.39
10	15	1	1.5	11,564	54,836	4.74
10	15	1	2	7326	53,492	7.30
20	30	1	1	4675	45,576	9.75
20	30	1	1.5	3174	44,817	14.12
20	30	1	2	2572	48,003	18.66

vector-based algorithm and the MSE of the orbit-based algorithm (column 5) and the ratio between the MSE of the standard Monte Carlo permutation sampling and the MSE of the orbit-based algorithm (column 6). The MSE of the vector-based algorithm is at least 1.9 times bigger than that of the orbit-based algorithm, while the MSE of the standard Monte Carlo permutation sampling can be 22.7 times bigger than that of the orbit-based algorithm (scenario 1).

The number of iterations made by the vector-based and the orbit-based algorithms in the allocated 15 s are reported in Table 4. The orbit-based algorithm performs better than the vector-based one even if the number of iterations made is lower: in 15 s, the ratio between the numbers of iterations increases from twice to almost 19 times. Despite this difference in the number of iterations, the orbit-based algorithm always achieves lower MSE than the vector-based algorithm.

5 Conclusions

The orbit-based algorithm grants a faster convergence to the exact distribution if compared to the standard MCMC algorithm proposed in [6]. At the same time, it gives more reliable estimates by decreasing the MSE. This simulation-based observation can be proved by comparing the variance of the estimators used by the two algorithms (the indicator function in (4) and the permutation cdf (5) respectively) [5].

When permutation-invariant statistics are used, the orbit-based algorithm is dramatically simplified. In this case, it is only necessary to walk among orbits of permutations without performing the second-step sampling and thus the reduction in computational time is significant.

Finally, it is worth noting that the MCMC sampling procedure based on orbits of permutations establishes a link between standard permutation and algebraic-statistics-based sampling that, to the best of our knowledge, has not been previously noted.

A preliminary version of this work has been presented at the 4th ISNPS conference in June 2018 in Salerno, Italy. Extensions of the present work include hypothesis testing for $K > 2$ groups and data fitting [5].

Acknowledgments R. Fontana acknowledges that the present research has been partially supported by MIUR grant Dipartimenti di Eccellenza 2018–2022 (E11G18000350001). The authors thank the anonymous referee for his/her helpful comments.

References

1. 4ti2 team. 4ti2 version 1.6.7. A software package for algebraic, geometric and combinatorial problems on linear spaces (2015). www.4ti2.de
2. Aoki, S., Hara, H., Takemura, A.: Markov Bases in Algebraic Statistics. Springer Series in Statistics. Springer, New York (2012)
3. Aoki, S., Takemura, A.: Markov Chain Monte Carlo tests for designed experiments. J. Stat. Plan. Inference **140**(3), 817–830 (2010)
4. Crucinio, F.R., Fontana, R.: Comparison of conditional tests on Poisson data. In: Statistics and Data Science: Proceedings of the Conference of the Italian Statistical Society, pp. 333–338. Firenze University Press (2017)
5. Fontana R., Crucinio F. R.: Orbit-based conditional tests. A link between permutations and Markov bases. J. Stat. Plan. Inference 205:(23–33), 2020
6. Diaconis, P., Sturmfels, B.: Algebraic algorithms for sampling from conditional distributions. Ann. Stat. **26**(1), 363–397 (1998)
7. Dobra, A.: Dynamic Markov bases. J. Comput. Graph. Stat. **21**(2), 496–517 (2012)
8. Kahle, D., Yoshida, R., Garcia-Puente, L.: Hybrid schemes for exact conditional inference in discrete exponential families. Ann. Inst. Stat. Math. **70**(5), 983–1011 (2018)
9. Kunz, M.: Partitions and their lattices. ArXiv Mathematics e-prints, April 2006 (2006). arXiv:0604203.pdf. Accessed 20 Apr 2017
10. Lehmann, E.L., Romano, J.P.: Testing Statistical Hypotheses. Springer Texts in Statistics. Springer, New York (2006)
11. McCullagh, P., Nelder, J.A.: Generalized Linear Models, 2nd edn. Chapman & Hall/CRC Monographs on Statistics & Applied Probability. Taylor & Francis (1989)

12. Pesarin, F., Salmaso, L.: Permutation Tests for Complex Data: Theory, Applications and Software. Wiley (2010)
13. Rapallo, F., Yoshida, R.: Markov bases and subbases for bounded contingency tables. Ann. Inst. Stat. Math. **62**(4), 785–805 (2010)
14. Wilf, H.S.: Lectures on integer partitions (2000). https://www.math.upenn.edu/~wilf/PIMS/PIMSLectures.pdf. Accessed 20 Apr 2017

Obstacle Problems for Nonlocal Operators: A Brief Overview

Donatella Danielli, Arshak Petrosyan, and Camelia A. Pop

Abstract In this note, we give a brief overview of obstacle problems for nonlocal operators, focusing on the applications to financial mathematics. The class of nonlocal operators that we consider can be viewed as infinitesimal generators of non-Gaussian asset price models, such as Variance Gamma Processes and Regular Lévy Processes of Exponential type. In this context, we analyze the existence, uniqueness, and regularity of viscosity solutions to obstacle problems which correspond to prices of perpetual and finite expiry American options.

Keywords Obstacle problem · Nonlocal operators · Lévy processes · American options · Viscosity solutions · Existence and uniqueness

2010 Mathematics Subject Classification. Primary 35R35 · Secondary 60G51 · 91G80

1 Introduction

The purpose of this note is to give a brief overview of obstacle problems for nonlocal operators, focusing on the applications to financial mathematics. Natural classes of nonlocal operators are infinitesimal generators of *Lévy processes*. We recall that a Lévy process $\{X(t)\}_{t\geq 0}$ defined on a filtered probability space $(\Omega, \mathcal{F}, \{\mathcal{F}_t\}_{t\geq 0}, \mathbb{P})$ is a random process that is stochastically continuous and has stationary and independent increments. More precisely, $\{X(t)\}_{t\geq 0}$ is a Lévy process if:

D. Danielli (✉) · A. Petrosyan
Department of Mathematics, Purdue University, West Lafayette, IN 47907, USA
e-mail: danielli@math.purdue.edu

A. Petrosyan
e-mail: arshak@math.purdue.edu

C. A. Pop
School of Mathematics, University of Minnesota, Minneapolis, MN 55455, USA
e-mail: capop@umn.edu

M. La Rocca et al. (eds.), *Nonparametric Statistics*, Springer Proceedings in Mathematics & Statistics 339, https://doi.org/10.1007/978-3-030-57306-5_15

1. $X(0) = 0$ with probability 1;
2. For all $0 \leq t_1 < t_2 < \cdots < t_n$, $X(t_1)$, $X(t_2) - X(t_1), \ldots, X(t_n) - X(t_{n-1})$ are independent;
3. For all $0 \leq s < t < \infty$, the probability distribution of $X(t) - X(s)$ is the same as the one of $X(t-s)$;
4. For all $\varepsilon > 0$, we have that

$$\lim_{t \downarrow 0} \mathbb{P}(|X(t)| > \varepsilon) = 0.$$

We begin the introduction with Sect. 1.1 where we give representations of Lévy processes using the Lévy–Khintchine formula and the Lévy–Itô decomposition. We continue in Sect. 1.2 to describe the connection to nonlocal (integro-differential) operators and we present in Sect. 1.3 more general stochastic equations, which give rise to a wider class of nonlocal operators. In Sect. 1.4, we give a brief introduction to obstacle problems and we summarize in Sect. 1.5 previous results obtained in the literature.

1.1 Representations of Lévy Processes

Our starting point is the *Lévy–Khintchine formula* [1, Corollary 2.4.20], which shows that, for all $t \geq 0$ and $\xi \in \mathbb{R}^n$, we have

$$\mathbb{E}\left[e^{i\xi \cdot X(t)}\right] = e^{t\psi(\xi)}, \tag{1.1}$$

where the characteristic exponent $\psi(\xi)$ is given by

$$\psi(\xi) = -\frac{1}{2}\xi \cdot A\xi + ib \cdot \xi + \int_{\mathbb{R}^n \setminus \{0\}} \left(e^{i\xi \cdot y} - 1 - i\xi \cdot y\chi_{|y|<1}\right) \nu(dy). \tag{1.2}$$

Here, A is a $n \times n$-dimensional, symmetric, positive-semidefinite matrix, $b \in \mathbb{R}^n$ and ν is a Lévy measure on $\mathbb{R}^n \setminus \{0\}$, i.e., it satisfies

$$\int_{\mathbb{R}^n \setminus \{0\}} \min\{1, |y|^2\} \, \nu(dy) < \infty.$$

When $A \equiv 0$ and $\nu \equiv 0$, that is $\mathbb{E}\left[e^{i\xi \cdot X(t)}\right] = e^{itb \cdot \xi}$, the process $X(t) = tb$ is deterministic motion on a straight line, with velocity of motion, or *drift*, b. If instead $A \equiv 0$, but $\nu \not\equiv 0$ has finite variation, that is it satisfies

$$\int_{\mathbb{R}^n \setminus \{0\}} \min\{1, |y|\} \, \nu(dy) < \infty, \tag{1.3}$$

then we can rewrite the characteristic exponent (1.2) as

$$\psi(\xi) = ib' \cdot \xi + \int_{\mathbb{R}^n \setminus \{0\}} \left(e^{i\xi \cdot y} - 1\right) \nu(dy).$$

The simplest possible case is when $\nu = \lambda \delta_h$, where $\lambda > 0$ and δ_h is the Dirac mass concentrated at $h \in \mathbb{R}^n \setminus \{0\}$. If we let $X(t) = b't + N(t)$, then the process $\{N(t)\}_{t \geq 0}$ is such that

$$\mathbb{E}\left[e^{i\xi \cdot N(t)}\right] = \exp\left[\lambda t \left(e^{i\xi \cdot h} - 1\right)\right],$$

and therefore, $\{N(t)\}_{t \geq 0}$ is a Poisson process of intensity λ taking values in $\{mh,\ m \in \mathbb{N}\}$. The physical interpretation is that $\{X(t)\}_{t \geq 0}$ follows the path of a straight line with drift b' and has jump discontinuities of size $|h|$. The time between two consecutive jumps are independent random variables exponentially distributed with parameter λ.

The next step is to take $\nu = \sum_{j=1}^{m} \lambda_j \delta_{h_j}$, with $m \in \mathbb{N}, \lambda_j > 0,\ h_j \in \mathbb{R}^n \setminus \{0\},\ 1 \leq j \leq m$. In this instance, we can write $\{X(t)\}_{t \geq 0}$ as

$$X(t) = b't + \sum_{j=1}^{m} N_j(t),$$

where the $\{N_j(t)\}_{t \geq 0},\ 1 \leq j \leq m$, are independent Poisson processes with intensity λ_j taking values in $\{mh_j,\ m \in \mathbb{N}\}$. The path is still deterministic with drift b' and has jumps of size in $\{|h_1|, \ldots, |h_m|\}$ occurring at exponentially distributed random times. When we let m tend to ∞ in a suitable sense, or more generally, when the Lévy measure ν is of finite variation, that is, condition (1.3) holds, we can write

$$X(t) = b't + \sum_{0 \leq s \leq t} \Delta X(s),$$

where $\Delta X(s) = X(s) - X(s-)$ is the jump at time s. Instead of dealing with jumps directly, it is more convenient to count the number of jumps that belong to a set A up to time t. To this end, for a Borel set $A \subseteq \mathbb{R}^n \setminus \{0\}$ and $t \geq 0$, we define the random Poisson measure with intensity ν

$$N(t, A) = \#\{0 \leq s \leq t \mid \Delta X(s) \in A\},$$

which allows us to write

$$\sum_{0 \leq s \leq t} \Delta X(s) = \int_{\mathbb{R}^n \setminus \{0\}} x N(t, dx).$$

However, in the most general case, the Lévy measure μ may not satisfy the finite variation condition (1.3) and to deal with the accumulation of small jumps, we make use of the compensated Poisson measure:

$$\widetilde{N}(dt, dx) = N(dt, dx) - dt\ \nu(dx).$$

Finally, in case of a general Lévy measure ν and of a diffusion matrix A, one has the *Lévy–Itô decomposition* [1, Theorem 2.4.16]:

$$X(t) = DW(t) + bt + \int_{0<|x|<1} x\widetilde{N}(t, dx) + \int_{|x|\geq 1} xN(t, dx), \tag{1.4}$$

where D is a $n \times n$-dimensional matrix such that $DD^T = A$, and $\{W(t)\}_{t\geq 0}$ is a n-dimensional Brownian motion.

1.2 Connections to Integro-Differential Operators

At this point, we want to explore the connection between stochastic processes and integro-differential operators. Using the fact that any Lévy process is a Markov process, by defining

$$T_t f(x) := \mathbb{E}\left[f(x + X(t))\right], \quad \forall x \in \mathbb{R}^n,$$

we obtain that $\{T_t\}_{t\geq 0}$ defines a one-parameter semigroup of linear operators on the Banach space of bounded continuous functions, $C(\mathbb{R}^n)$. One can think of the semigroup $\{T_t\}_{t\geq 0}$ as a tool to give a deterministic, macroscopic description of the Lévy process as an average of microscopic random dynamics. The infinitesimal generator corresponding to the semigroup semigroup $\{T_t\}_{t\geq 0}$ is defined formally by

$$Lf(x) = \lim_{t\downarrow 0} \frac{T_t f(x) - f(x)}{t},$$

and takes the form

$$Lf(x) = \frac{1}{2}\mathrm{tr}(AD^2 f) + b \cdot \nabla f(x) + \int_{\mathbb{R}^n\setminus\{0\}} \left[f(x+y) - f(x) - y \cdot \nabla f(x)\chi_{|y|<1}(y)\right] \nu(dy).$$

Under suitable regularity assumptions that allow us to apply Itô's rule [1, Theorem 4.4.7] to solutions to the parabolic differential equation $u_t = Lu$ on $(0, \infty) \times \mathbb{R}^n$, with initial condition $u(0, \cdot) = f$ on $\mathbb{R}^n$, we obtain that $u(t, x) = T_t f(x)$, for all $t \geq 0$ and $x \in \mathbb{R}^n$, and so $T_t = e^{tL}$.

We can also establish a connection between the infinitesimal generator L of the process $\{X(t)\}_{t\geq 0}$ and the characteristic exponent $\psi(\xi)$ appearing in the Lévy–Khintchine formula (1.1). Viewed as a pseudo-differential operator [6, 27], the symbol of the infinitesimal generator L is the characteristic exponent (1.2) appearing in identity (1.1). In our survey, we will be concerned with generalizations of symbols that contain only a drift and a nonlocal term (the second-order diffusion term is removed). This gives rise to mathematical challenges in the study of the regularity of solutions when the drift term dominates the nonlocal component—the so-called *supercritical regime*. This property is often encountered in financial models for stock prices, such as Variance Gamma and Regular Lévy Processes of Exponential Type described in greater detail in Sect. 2.

1.3 Stochastic Integro-Differential Equations

More generally than the infinitesimal generators of Lévy processes, in this survey we are specifically concerned with *nonlocal* operators that are infinitesimal generators of strong Markov processes, which can be written as solutions to stochastic integro-differential equations of the form:

$$dX(t) = b(X(t-))dt + \int_{\mathbb{R}^n\setminus\{0\}} F(X(t-), y)\tilde{N}(dt, dy), \quad t > 0. \tag{1.5}$$

Here, $\tilde{N}(dt, dy)$ is a compensated Poisson random measure with intensity measure $d\nu$, as defined in Sect. 1.1, and b and F satisfy suitable conditions, which we describe in detail in Sect. 3. Our conditions ensure, by [1, Theorem 6.2.9], that for any initial condition $X^x(0) = x \in \mathbb{R}^n$, there exists a unique strong solution $\{X^x(t)\}_{t\geq 0}$ to equation (1.5) with *càdlàg* paths a.s. The process $\{X^x(t)\}_{t\geq 0}$ satisfies the strong Markov property, and therefore, it is uniquely determined by its infinitesimal generator

$$Lu(x) = b \cdot \nabla u(x) + \int_{\mathbb{R}^n\setminus\{0\}} (u(x + F(x, y)) - u(x) - F(x, y) \cdot \nabla u(x)) \; \nu(dy) \tag{1.6}$$

for all $u \in C^2(\mathbb{R}^n)$ (this denotes all functions with bounded and continuous derivatives up to and including order 2 in $\mathbb{R}^n$). The term *nonlocal* refers to the fact that the value of $Lu(x)$ depends on the whole solution u and not only on its behavior nearby the point x. A typical example of a nonlocal integro-differential operator is the fractional Laplacian $(-\Delta)^s$, with $s \in (0, 1)$, which is defined on the Fourier transform side by the formula

$$\widehat{(-\Delta)^s u}(\xi) = |\xi|^{2s}\hat{u}(\xi),$$

or, equivalently, by the pointwise representation

$$(-\Delta)^s u(x) = \gamma(n, s)\; p.v. \int_{\mathbb{R}^n} \frac{2u(x) - u(x + y) - u(x - y)}{|x|^{n+2s}}\, dy,$$

γ being a normalization constant depending only on n and s. The fractional Laplacian $(-\Delta)^s$ is the infinitesimal generator of the symmetric $2s$-*stable Lévy process* with characteristic exponent in the Lévy–Khintchine formula given by $\psi(\xi) = |\xi|^{2s}$.

1.4 Obstacle Problems

In recent years, there has been a resurgence of interest in the study of nonlocal operators, motivated by applications. In fact, such operators and the associated integro-differential equations naturally arise in a variety of contexts, ranging from temperature control to linear elasticity, from fluid dynamics to financial mathematics. To describe the latter application in more detail, we assume that

$$S(t) = e^{X(t)} \tag{1.7}$$

models an asset price process, where $\{X(t)\}_{t\geq 0}$ is a solution to the stochastic equation (1.5). We let $\varphi : \mathbb{R}^n \to \mathbb{R}$ be the payoff function of an American option (i.e., a profit of $\varphi(s)$ is generated when exercising the option at time t and the stock level is $s = S(t)$). Without loss of generality, we can assume that the payoff can be written as a function of $\{X(t)\}_{t\geq 0}$. We recall that, unlike the European option, in the American option framework the holder has the right to exercise at any date prior to maturity, and not only at the expiry date. Hence, the value of the American option with expiry date T can be written as

$$v(t, x) = \sup\ \mathbb{E}[e^{-rt}\varphi(X(\theta))|X(t) = x], \qquad \text{for all } (t, x) \in (0, T) \times \mathbb{R}^n,$$

where the supremum is taken over all stopping times θ bounded by $T - t$, and we assume that the expectation is taken under a risk-neutral probability measure and r is the risk-free interest rate. Letting τ be the first time that the stochastic process $\{X(t)\}_{t\geq 0}$ enters the *exercise region* $\{v = \varphi\}$, and assuming that the value function $u(t, x)$ is regular enough, probabilistic arguments ensure that the stopped process $\{e^{-rt\wedge\tau}v(t \wedge \tau, X(t \wedge \tau))\}_{t\geq 0}$ is a martingale, which is equivalent to the equality

$$\partial_t v + Lv - rv = 0, \qquad \text{for all } (t, x) \in \{v > \varphi\}. \tag{1.8}$$

In general, however, the discounted option price process $\{e^{-rt}v(t, X(t))\}_{t\geq 0}$ is a supermartingale, which translates into the inequality

$$\partial_t v + Lv - rv \leq 0, \qquad \text{for all } (t, x) \in (0, T) \times \mathbb{R}^n. \tag{1.9}$$

Combining equations (1.8) and (1.9) together with the property that $v \geq \varphi$ gives us that the value function v is a solution to the *evolution obstacle problem*:

$$\min\{-\partial_t v - Lv + rv, v - \varphi\} = 0, \qquad \text{for all } (t, x) \in (0, T) \times \mathbb{R}^n, \tag{1.10}$$

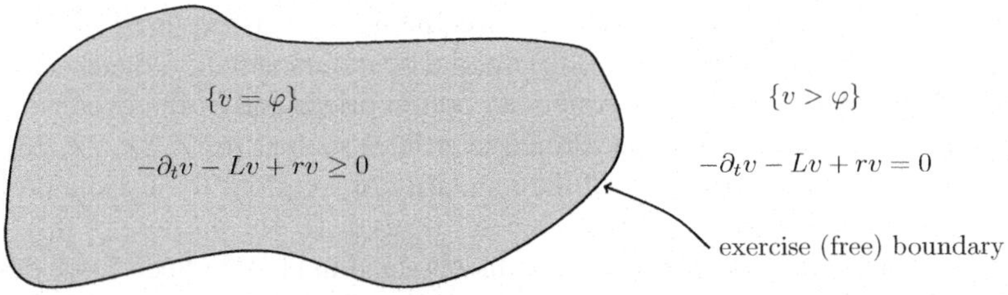

Fig. 1 A schematic description of the complementarity conditions for the evolution obstacle problem at a time slice t. The exercise region $\{v = \varphi\}$ is represented by the gray area, and the remaining region is the continuation region $\{v > \varphi\}$

where L is the infinitesimal generator of $\{X(t)\}_{t\geq 0}$. The strong Markov property of $\{X(t)\}_{t\geq 0}$ implies that the exercise decision at any time t depends only on t and $X(t)$. Therefore, for each t there exist an *exercise region* $\{v = \varphi\}$, in which one should exercise the option, and a *continuation region* $\{v > \varphi\}$, in which one instead should wait. The *exercise boundary* is the interface separating the two. See Fig. 1 for a schematic representation. We briefly mention here that in the case of perpetual American option, when the option has a infinite expiration time, the value function depends only on the current value of the process $\{X(t)\}_{t\geq 0}$ and is a solution to a stationary obstacle problem. We refer to Sect. 3 for further details.

1.5 Review of Literature and Outline of the Survey

If the underlying stochastic process is Brownian motion, then the infinitesimal generator of the underlying process is $L = \Delta$ and u will satisfy the classical obstacle problem, which is by now very well understood [8–10, 20]. However, Brownian motion falls short in some respects:

1. Stock prices do not move continuously, which prompts us to consider models that allow jumps in small time intervals;
2. Empirical studies of stock price returns indicate distributions with heavy tails, which are not compatible with a Gaussian model.

For these reasons, it becomes necessary to study jump diffusion processes, whose infinitesimal generator is an integro-differential operator of the form (1.4). Such type of operators was introduced in finance by the Nobel Prize winner Merton [25]. The novel element, which reflects the presence of jumps, is the integral term. Its presence leads to new theoretical and numerical issues. Since no closed-form solutions are known in general for the American option, it becomes important to determine the regularity of the exercise boundary, which in turn is closely related to the behavior of the value function.

In the framework of jump diffusion models with a non-degenerate diffusion matrix, regularity of the value function and efficient numerical schemes were studied in [2, 3, 5, 23], and regularity of the free boundary was explored in [4]. Using methods from the theory of pseudo-differential operators and the Wiener–Hopf factorization, qualitative studies of American option prices and of the exercise region under pure-jump models were performed in articles such as [6, 7, 21, 22, 26].

Our work continues the study of the regularity of solutions to obstacle problems for nonlocal operators with (possibly supercritical) drift. The purpose of this note is to give an overview of the regularity results obtained in [17]. In Sect. 2, we describe two examples of stochastic processes of interest in mathematical finance to which our results apply. In Sect. 3, we state the problem precisely, and provide the statements of our main results. Finally, in Sect. 4, we indicate some future directions.

2 Motivating Examples

In this section, we assume $n = 1$ and that the asset price process can be written as in (1.7). Moreover, r denotes the risk-free interest rate. It is crucially important to ensure that the discounted asset price process $\{e^{-rt} S(t)\}_{t\geq 0}$ is a martingale in order to obtain an arbitrage-free market. Assume that $\{X(t)\}_{t\geq 0}$ is a one-dimensional Lévy process that satisfies the stochastic equation:

$$dX(t) = b\,dt + \int_{\mathbb{R}^n} y\,\widetilde{N}(dt, dy), \quad \forall t > 0, \tag{2.1}$$

where b is a real constant and $\widetilde{N}(dt, dy)$ is a compensated Poisson random measure with Lévy measure $\nu(dy)$. Using [1, Theorem 5.2.4 and Corollary 5.2.2], a sufficient condition that guarantees that the discounted asset price process $\{e^{-rt+X(t)}\}_{t\geq 0}$ is a martingale is:

$$\int_{|x|\geq 1} e^x\,\nu(dx) < \infty \quad \text{and} \quad -r + \psi(-i) \;= 0, \tag{2.2}$$

where $\psi(\xi)$ denotes the characteristic exponent of the Lévy process $\{X(t)\}_{t\geq 0}$, that is,

$$\psi(\xi) = ib\xi + \int_{\mathbb{R}\setminus\{0\}} (e^{ix\xi} - 1 - ix\xi)\,\nu(dx). \tag{2.3}$$

Examples in mathematical finance to which our results apply include the *Variance Gamma Process* [24] and *Regular Lévy Processes of Exponential type* (RLPE) [6].

When the jump part of the nonlocal operator L corresponding to the integral term in the characteristic exponent (2.3) has sublinear growth as $|\xi| \to \infty$, we say that the drift term $b \cdot \nabla$ corresponding to $ib \cdot \xi$ in the characteristic exponent (2.3) is *supercritical*. An example of a nonlocal operator with supercritical drift is the Variance Gamma Process and a subcollection of Regular Lévy Processes of Exponential type described below.

2.1 Variance Gamma Process

Following [14, Identity (6)], the Variance Gamma Process $\{X(t)\}_{t\geq 0}$ with parameters ν, σ, and θ has Lévy measure given by

$$\nu(dx) = \frac{1}{\nu|x|}\left(e^{-\frac{|x|}{\eta_p}}\mathbf{1}_{\{x>0\}} + e^{-\frac{|x|}{\eta_n}}\mathbf{1}_{\{x<0\}}\right)dx,$$

where $\eta_p > \eta_n$ are the roots of the equation $x^2 - \theta\nu x - \sigma^2\nu/2 = 0$, and ν, σ, θ are positive constants. From [14, Identity (4)], we have that the characteristic exponent of the Variance Gamma Process with constant drift $b \in \mathbb{R}$, $\{X(t) + bt\}_{t\geq 0}$, has the expression:

$$\psi_{\mathrm{VG}}(\xi) = \frac{1}{\nu}\ln\left(1 - i\theta\nu\xi + \frac{1}{2}\sigma^2\nu\xi^2\right) + ib\xi, \quad \forall\xi \in \mathbb{C},$$

and so the infinitesimal generator of $\{X(t) + bt\}_{t\geq 0}$ is given by

$$L = \frac{1}{\nu}\ln(1 - \theta\nu\nabla - \frac{1}{2}\sigma^2\nu\Delta) + b\cdot\nabla,$$

which is a sum of a pseudo-differential operator of order less than any $s > 0$ and one of order 1. When $\eta_p < 1$ and $r = \psi_{VG}(-i)$, condition (2.2) is satisfied and the discounted asset price process $\{e^{-rt+X(t)}\}_{t\geq 0}$ is a martingale. Thus, applying the results in Sect. 3 to the Variance Gamma Process $\{X(t)\}_{t\geq 0}$ with constant drift b, we obtain that the prices of perpetual and finite expiry American options with bounded and Lipschitz payoffs are Lipschitz functions in the spatial variable. Given that the nonlocal component of the infinitesimal generator L has order less than any $s > 0$, this may be the optimal regularity of solutions that we can expect.

2.2 Regular Lévy Processes of Exponential Type

Following [6, Chap. 3], for parameters $\lambda_- < 0 < \lambda_+$, a Lévy process is said to be of exponential type $[\lambda_-, \lambda_+]$ if it has a Lévy measure $\nu(dx)$ such that

$$\int_{-\infty}^{-1} e^{-\lambda_+ x}\nu(dx) + \int_{1}^{\infty} e^{-\lambda_- x}\nu(dx) < \infty.$$

Regular Lévy Processes of Exponential type $[\lambda_-, \lambda_+]$ and order ν are non-Gaussian Lévy processes of exponential type $[\lambda_-, \lambda_+]$ such that, in a neighborhood of zero, the Lévy measure can be represented as $\nu(dx) = f(x)\,dx$, where the density $f(x)$ satisfies the property that

$$|f(x) - c|x|^{-\nu-1}| \le C|x|^{-\nu'-1}, \quad \forall\, |x| \le 1,$$

for constants $\nu' < \nu$, $c > 0$, and $C > 0$. Our results apply to RLPE type $[\lambda_-, \lambda_+]$, when we choose the parameters $\lambda_- \le -1$ and $\lambda_+ \ge 1$. The class of RLPE include the CGMY/KoBoL processes introduced in [14]. Following [14, Eq. (7)], CGMY/KoBoL processes are characterized by a Lévy measure of the form

$$\nu(dx) = \frac{C}{|x|^{1+Y}}\left(e^{-G|x|}\mathbf{1}_{\{x<0\}} + e^{-M|x|}\mathbf{1}_{\{x>0\}}\right)dx,$$

where the parameters $C > 0$, $G, M \ge 0$, and $Y < 2$. Our results apply to CGMY/KoBoL processes, when we choose the parameters $G, M > 1$ and $Y < 2$, or $G, M \ge 1$ and $0 < Y < 2$.

3 Statements of the Main Results

In this section, we provide the statements of our main results. Complete proofs can be found in [17], where these results have originally appeared.

We begin by listing the required assumptions on the measure $\nu(dx)$ and the coefficients $b(x)$ and $F(x, y)$ appearing in the operator (1.6):

1. There is a positive constant K such that for all $x_1, x_2 \in \mathbb{R}^n$, we have

$$\int_{\mathbb{R}^n\setminus\{O\}} |F(x_1, y) - F(x_2, y)|^2\, d\nu(y) \le K|x_1 - x_2|^2,$$
$$\sup_{z\in B_{|y|}} |F(x, z)| \le \rho(y), \quad \forall x, y \in \mathbb{R}^n,$$
$$\int_{\mathbb{R}^n\setminus\{O\}} (|y| \vee \rho(y))^2\, \nu(dy) \le K,$$

   where $\rho : \mathbb{R}^n \to [0, \infty)$ is a measurable function.
2. The coefficient $b : \mathbb{R}^n \to \mathbb{R}^n$ is bounded and Lipschitz continuous, i.e., $b \in C^{0,1}(\mathbb{R}^n)$.
3. For the stationary problem, we assume that $F(x, y) = F(y)$ (independent of x).

3.1 Stationary Obstacle Problem

We consider the obstacle problem

$$\min\{-Lv + cv - f, v - \varphi\} = 0 \quad \text{on } \mathbb{R}^n, \tag{3.1}$$

where L is the infinitesimal generator of the unique strong solution $\{X^x(t)\}_{t\geq 0}$ to the stochastic equation (1.5), with initial condition $X^x(0) = x$. We explicitly remark here that, in the applications in Sect. 2, one chooses $c \equiv r$, the risk-free interest rate. Solutions to the obstacle problem (3.1) are constructed using the *stochastic representation formula* of the value function:

$$v(x) := \sup\{v(x; \tau) : \tau \in \mathcal{T}\}.$$

where $\mathcal{T}$ is the set of stopping times and

$$v(x;\tau) := \mathbb{E}\left[e^{-\int_0^\tau c(X^x(s))\,ds}\varphi(X^x(\tau)) + \int_0^\tau e^{-\int_0^t c(X^x(s))\,ds} f(X^x(t))\,dt\right], \quad \forall \tau \in \mathcal{T}.$$

In order to state our results, we need to introduce the relevant function spaces. We denote by $C(\mathbb{R}^n)$ the space of bounded continuous functions $u : \mathbb{R}^n \to \mathbb{R}$ such that

$$\|u\|_{C(\mathbb{R}^n)} := \sup_{x\in\mathbb{R}^n} |u(x)| < \infty.$$

For all $\alpha \in (0, 1]$, a function $u : \mathbb{R}^n \to \mathbb{R}$ belongs to the Hölder space of functions $C^{0,\alpha}(\mathbb{R}^n)$ if

$$\|u\|_{C^{0,\alpha}(\mathbb{R}^n)} := \|u\|_{C(\mathbb{R}^n)} + [u]_{C^{0,\alpha}(\mathbb{R}^n)} < \infty,$$

where, as usual, we define

$$[u]_{C^{0,\alpha}(\mathbb{R}^n)} := \sup_{x_1,x_2\in\mathbb{R}^n, x_1\neq x_2} \frac{|u(x_1) - u(x_2)|}{|x_1 - x_2|^\alpha}.$$

When $\alpha \in (0, 1)$, we denote for brevity $C^\alpha(\mathbb{R}^n) := C^{0,\alpha}(\mathbb{R}^n)$. Our first result concerns the regularity of the value function.

Theorem 3.1 *Let $c, \varphi, f : \mathbb{R}^n \to \mathbb{R}$ be bounded Lipschitz continuous functions, and assume that there is a constant $c_0 > 0$ such that $c(x) \geq c_0 > 0$, $\forall x \in \mathbb{R}^n$. Then the following hold:*

(i) (Hölder continuity) *There is a constant $\alpha = \alpha([b]_{C^{0,1}(\mathbb{R}^n)}, c_0) \in (0, 1)$, such that the value function $v \in C^\alpha(\mathbb{R}^n)$.*
(ii) (Lipschitz continuity) *If in addition we have that*

$$c_0 \geq [b]_{C^{0,1}(\mathbb{R}^n)}, \tag{3.2}$$

then the value function $v \in C^{0,1}(\mathbb{R}^n)$.

The proof of Theorem 3.1 hinges on the stochastic representation of solutions and on the continuity of the strong solutions to the SDE with respect to the initial conditions. To proceed, we introduce the notion of viscosity solution, which gives an intrinsic definition of a solution which is local in nature but does not assume a priori any regularity, except for continuity.

Definition 3.2 Let $v \in C(\mathbb{R}^n)$. We say that v is a *viscosity subsolution (supersolution)* to the stationary obstacle problem if, for all $u \in C^2(\mathbb{R}^n)$ such that $v - u$ has a global max (min) at $x_0 \in \mathbb{R}^n$ and $u(x_0) = v(x_0)$, then

$$\min\{-Lu(x_0) + c(x_0)u(x_0) - f(x_0), u(x_0) - \varphi(x_0)\} \leq (\geq)\, 0. \tag{3.3}$$

We say that v is a *viscosity solution* if it is both a sub- and supersolution.

Next, we show that the value function is the unique solution to (3.1).

Theorem 3.3 (Existence) *Assume in addition*

$$\int_{\mathbb{R}^n \setminus \{O\}} |F(y)|^{2\alpha}\, \nu(dy) < \infty$$

where $\alpha \in (0, 1)$ is the constant appearing in Theorem 3.1. Then the value function v is a viscosity solution to the stationary obstacle problem.

Theorem 3.4 (Uniqueness) *Suppose that $c, f, \varphi \in C(\mathbb{R}^n)$ and c is a positive function. If the stationary obstacle problem has a viscosity solution, then it is unique.*

We remark that a sufficient condition on the Lévy measure to ensure that perpetual American put option prices are Lipschitz continuous, but not continuously differentiable, is provided in [6, Theorem 5.4, p. 133]. However, the condition is in terms of the Wiener–Hopf factorization for the characteristic exponent of the Lévy process, and it is difficult to find a concrete example for which it holds. Since in our case the order of the nonlocal operator is strictly less than the order of the drift component, and there is no second-order term, the issue of regularity of solutions is quite delicate.

The proof of the existence result hinges in a crucial way on a *Dynamic Programming Principle*. In order to state it precisely, we need the following definition.

Definition 3.5 For all $r > 0$ and $x \in \mathbb{R}^n$, we let

$$\tau_r := \inf\{t \geq 0 : X^x(t) \notin B_r(x)\},$$

where $B_r(x)$ denoted the open Euclidean ball of radius $r > 0$ centered at $x \in \mathbb{R}^n$.

Theorem 3.6 (Dynamic Programming Principle) *The value function $v(x)$ satisfies:*

$$v(x) = \sup\{v(x; r, \tau) : \tau \leq \tau_r\}, \quad \forall r > 0,$$

where we define

$$v(x;r,\tau) := \mathbb{E}\left[e^{-\int_0^\tau c(X^x(s))\,ds}\left(\varphi(X^x(\tau))\mathbf{1}_{\{\tau<\tau_r\}} + v(X^x(\tau))\mathbf{1}_{\{\tau=\tau_r\}}\right)\right]$$
$$+\mathbb{E}\left[\int_0^{\tau\wedge\tau_r} e^{-\int_0^t c(X^x(s))\,ds} f(X^x(t))\,dt\right].$$

Uniqueness is proved instead with the aid of the following theorem.

Theorem 3.7 (Comparison principle) *Suppose that the assumptions of the uniqueness theorem hold. If u and v are a viscosity subsolution and supersolution to the stationary obstacle problem, respectively, then $u \leq v$.*

In financial terms, comparison principles simply translate into arbitrage inequalities: if the terminal payoff of an American option dominates the terminal payoff of another one, then their values should verify the same inequality.

3.2 Evolution Obstacle Problem

The evolution obstacle problem is given by

$$\begin{cases} \min\{-\partial_t v - Lv + cv - f, v - \varphi\} = 0 & \text{on } [0,T)\times\mathbb{R}^n, \\ v(T,\cdot) = g & \text{on } \mathbb{R}^n, \end{cases} \tag{3.4}$$

with the compatibility condition

$$g \geq \varphi(T,\cdot) \quad \text{on} \quad \mathbb{R}^n. \tag{3.5}$$

The treatment of this problem is very similar to the stationary case. For the sake of brevity, we confine ourselves to mentioning here that the main new difficulty is to establish regularity in the time variable. This is done with the aid of the following result concerning the continuity properties of $\{X(t)\}_{t\geq 0}$, which in turn is a consequence of Doob's Martingale Inequality.

Lemma 3.8 *There is a positive constant $C = C(\|b\|_{C^{0,1}(\mathbb{R}^n)}, K)$ such that*

$$\mathbb{E}\left[\max_{s\in[0,t]}\left|X^{x_1}(s) - X^{x_2}(s)\right|^2\right] \leq C|x_1 - x_2|^2 e^{Ct}, \quad \forall x_1, x_2 \in \mathbb{R}^n,\ t \geq 0,$$
$$\mathbb{E}\left[\max_{r\in[s,t]}\left|X^x(r) - X^x(s)\right|^2\right] \leq C|t-s| \vee |t-s|^2, \quad \forall x \in \mathbb{R}^n,\ 0 \leq s < t.$$

The use of this lemma also allows to relax the assumptions on the coefficients in that we no longer require condition (3.2) to hold and we can allow the jump size $F(x,y)$ to be a function of the current state x of the process.

The relevant function spaces, in the evolution case, are as follows. For all $T > 0$, we denote by $C_t^{\frac{1}{2}} C_x^{0,1}([0, T] \times \mathbb{R}^n)$ the space of functions $u : [0, T] \times \mathbb{R}^n \to \mathbb{R}$ such that

$$\|u\|_{C_t^{\frac{1}{2}} C_x^{0,1}([0,T]\times\mathbb{R}^n)} := \|u\|_{C([0,T]\times\mathbb{R}^n)} + \sup_{\substack{t_1,t_2\in[0,T],t_1\neq t_2 \\ x_1,x_2\in\mathbb{R}^n,x_1\neq x_2}} \frac{|u(t_1, x_1) - u(t_2, x_2)|}{|t_1 - t_2|^{\frac{1}{2}} + |x_1 - x_2|} < \infty,$$

and we let $C_t^1 C_x^2([0, T] \times \mathbb{R}^n)$ denote the space of functions $u : [0, T] \times \mathbb{R}^n \to \mathbb{R}$ such that the first-order derivative in the time variable and the second-order derivatives in the spatial variables are continuous and bounded. Let $\mathcal{T}_t$ denote the set of stopping times $\tau \in \mathcal{T}$ bounded by t, for all $t \geq 0$. Solutions to problem (3.4) are constructed using the stochastic representation formula,

$$v(t, x) := \sup\{v(t, x; \tau) : \tau \in \mathcal{T}_{T-t}\}, \tag{3.6}$$

where we define

$$\begin{aligned} v(t, x; \tau) := \mathbb{E}\left[e^{-\int_0^\tau c(t+s,X^x(s))\,ds}\varphi(t + \tau, X^x(\tau))\mathbf{1}_{\{\tau<T-t\}}\right] \\ +\mathbb{E}\left[e^{-\int_0^\tau c(t+s,X^x(s))\,ds}g(X^x(T - t))\mathbf{1}_{\{\tau=T-t\}}\right] \\ +\mathbb{E}\left[\int_0^\tau e^{-\int_0^s c(t+r,X^x(r))\,dr} f(t + s, X^x(s))\,ds\right], \end{aligned} \tag{3.7}$$

for all $(t, x) \in [0, T]\times \in \mathbb{R}^n$.

Proposition 3.9 (Regularity) *Suppose that c, φ, f belong to $C^{0,1}([0, T] \times \mathbb{R}^n)$, the final condition g is in $C^{0,1}(\mathbb{R}^n)$, and the compatibility condition* (3.5) *holds. Then the value function v defined in* (3.4) *belongs to $C_t^{\frac{1}{2}} C_x^{0,1}([0, T] \times \mathbb{R}^n)$.*

We next define a notion of viscosity solution for the evolution obstacle problem (3.4) extending that of its stationary analogue for Eq. (3.1) similarly to the ideas described in [15, Sect. 8]:

Definition 3.10 (*Viscosity solutions*) Let $v \in C(\mathbb{R}^n)$. We say that v is a viscosity subsolution (supersolution) to the evolution obstacle problem (3.4) if

$$v(T, \cdot) \leq (\geq) g, \tag{3.8}$$

and, for all $u \in C_t^1 C_x^2([0, T] \times \mathbb{R}^n)$ such that $v - u$ has a global max (min) at $(t_0, x_0) \in [0, T) \times \mathbb{R}^n$ and $u(t_0, x_0) = v(t_0, x_0)$, we have that

$$\min\{-\partial_t u(t_0, x_0) - Lu(t_0, x_0) + c(t_0, x_0)u(t_0, x_0) - f(t_0, x_0), u(t_0, x_0) - \varphi(t_0, x_0)\} \leq (\geq)\, 0. \tag{3.9}$$

We say that v is a viscosity solution to Eq. (3.4) if it is both a sub- and supersolution.

We then have the following theorems regarding the existence and uniqueness of viscosity solutions.

Theorem 3.11 (Existence) *Suppose that the hypotheses of Proposition 3.9 hold. Then the value function v defined in* (3.6) *is a viscosity solution to the evolution obstacle problem* (3.4).

Theorem 3.12 (Uniqueness) *Suppose that g belongs to $C(\mathbb{R}^n)$, c, f, φ are in $C([0, T] \times \mathbb{R}^n)$, the compatibility condition* (3.5) *holds, and*

$$\lim_{y \to O} F(x, y) = 0, \quad \forall x \in \mathbb{R}^n. \tag{3.10}$$

If the obstacle problem (3.4) *has a viscosity solution, then it is unique.*

4 Concluding Remarks

We conclude this note by observing that optimal regularity of solutions and the regularity of the free boundary are completely unexplored for the classes of operators we consider. In this connection, we mention that some of the most powerful techniques to investigate these issues for nonlocal operators are based on an extension approach à la Caffarelli–Silvestre [13], see e.g., [12, 16, 18, 19]. However, there are now methods not relying on an extension procedure (such as the one developed by Caffarelli et al. [11]), but those appear to be limited to a class of operators of positive fractional order.

References

1. Applebaum, D.: Lévy Processes and Stochastic Calculus. Cambridge Studies in Advanced Mathematics, vol. 116, 2nd edn. Cambridge University Press, Cambridge (2009)
2. Bayraktar, E.: A proof of the smoothness of the finite time horizon American put options for jump diffusions. SIAM J. Control. Optim. **48**, 51–572 (2009)
3. Bayraktar, E., Xing, H.: On the perpetual American put options for level dependent volatility models with jumps. Quant. Financ. **11**, 335–341 (2011).
4. Bayraktar, E., Xing, H.: Analysis of the optimal exercise boundary of American options for jump diffusions. SIAM J. Math. Anal. **41**, 825–860 (2009)
5. Bayraktar, E., Xing, H.: Regularity of the optimal stopping problem for jump diffusions. SIAM J. Control. Optim. **50**(3), 1337–1357 (2012)
6. Boyarchenko, S.I., Levendorskiĭ, S.Z.: Non-Gaussian Merton-Black-Scholes Theory. World Scientific, River Edge (2002)
7. Caffarelli, L.A.: Optimal stopping in Lévy models for nonmonotone discontinuous payoffs. SIAM J. Control. Optim. **49**, 2062–2082 (2011)
8. Caffarelli, L.A.: The regularity of free boundaries in higher dimensions. Acta Math. **139**, 155–184 (1977)

9. Caffarelli, L.A., Ros-Oton, X., Serra, J.: Compactness methods in free boundary problems. Commun. Partial Differ. Equ. **5**, 427–448 (1980)
10. Caffarelli, L.A., Ros-Oton, X., Serra, J.: The obstacle problem revisited. J. Fourier Anal. Appl. **4**, 383–402 (1998)
11. Caffarelli, L.A., Ros-Oton, X., Serra, J.: Obstacle problems for integro-differential operators: regularity of solutions and free boundaries. Invent. Math. **208**, 1155–1211 (2017)
12. Caffarelli, L.A., Salsa, S., Silvestre, L.: Regularity estimates for the solution and the free boundary of the obstacle problem for the fractional Laplacian. Invent. Math. **171**, 425–461 (2008)
13. Caffarelli, L.A., Silvestre, L.: An extension problem related to the fractional Laplacian. Commun. Partial Differ. Equ. **32**(7–9), 1245–1260 (2007)
14. Carr, P., Geman, H., Madan, D.B., Yor, M.: The fine structure of asset returns: an empirical investigation. J. Bus. **75**(2), 305–333 (2002)
15. Crandall, M.G., Ishii, H., Lions, P.-L.: User's guide to viscosity solutions of second order partial differential equations. Bull. Am. Math. Soc. (N.S.) **27**, 1–67 (1992)
16. Danielli, D., Garofalo, N., Petrosyan, A., To, T.: Optimal regularity and the free boundary in the parabolic Signorini problem. Mem. Am. Math. Soc. **249**(1181) (2017).
17. Danielli, D., Petrosyan, A., Pop, C.A.: Obstacle problems for nonlocal operators. In: New Developments in the Analysis of Nonlocal Operators. Contemporary Mathematics, vol. 723, pp. 191–214. American Mathematical Society, Providence (2019)
18. Garofalo, N., Petrosyan, A.: Some new monotonicity formulas and the singular set in the lower dimensional obstacle problem. Invent. Math. **177**, 415–461 (2009)
19. Garofalo, N., Petrosyan, A., Pop, C.A., Smit Vega Garcia, M.: Regularity of the free boundary for the obstacle problem for the fractional Laplacian with drift. Ann. Inst. H. Poincaré Anal. Non Linéaire **34**(3), 533–570 (2017)
20. Laurence, P., Salsa, S.: Regularity of the free boundary of an American option on several assets. Commun. Pure Appl. Math. **62**, 969–994 (2009)
21. Levendorskiĭ, S.Z.: Early exercise boundary and option prices in Lévy driven models. Quant. Financ. **4**, 525–547 (2004)
22. Madan, D.B., Seneta, E.: Pricing of the American put under Lévy processes. Int. J. Theor. Appl. Financ. **7**, 303–335 (2004)
23. Madan, D.B., Seneta, E.: American and European options in multi-factor jump-diffusion models, near expiry. Financ. Stoch. **12**, 541–560 (2008)
24. Madan, D.B., Seneta, E.: The Variance Gamma (V.G.) model for share market returns. J. Bus. **63**, 511–524 (1990)
25. Merton, R.: Option pricing when underlying stock returns are discontinuous. J. Financ. Econ. **3**, 125–144 (1976)
26. Mordecki, E.: Optimal stopping and perpetual options for Lévy processes. Financ. Stoch. **6**(4), 473–493 (2002)
27. Taylor, M.E.: Partial differential equations II. Qualitative studies of linear equations. In: Applied Mathematical Sciences, vol. 116. Springer, New York (2011)

Low and High Resonance Components Restoration in Multichannel Data

Daniela De Canditiis and Italia De Feis

Abstract A technique for the restoration of low resonance component and high resonance component of K independently measured signals is presented. The definition of low and high resonance component is given by the Rational Dilatation Wavelet Transform (RADWT), a particular kind of finite frame that provides sparse representation of functions with different oscillations persistence. It is assumed that the signals are measured simultaneously on several independent channels and in each channel the underlying signal is the sum of two components: the low resonance component and the high resonance component, both sharing some common characteristic between the channels. Components restoration is performed by means of the lasso-type penalty and backfitting algorithm. Numerical experiments show the performance of the proposed method in different synthetic scenarios highlighting the advantage of estimating the two components separately rather than together.

Keywords RADWT · Lasso regression · Multichannel signals

1 Introduction

The problem of recovering multiple signals recorded in different channels under the assumption that they share some common characteristics is very frequent in various fields of application, for example, biology, neuroscience, and information technology.

In this paper, we deal with the problem of recovering the low resonance component and the high resonance component of K simultaneous measured signals. This is very useful for the analysis of EEG data, as explained in [4]. Specifically, we hypothesize to have K channels and the signal measured by each of them is the sum of two components: a low resonance component and a high resonance component; the first

D. De Canditiis (✉)
IAC-CNR, via dei Taurini 19, Rome, Italy
e-mail: d.decanditiis@iac.cnr.it

I. De Feis
IAC-CNR, via Piuetro Castellino 111, Naples, Italy
e-mail: d.defeis@iac.cnr.it

M. La Rocca et al. (eds.), *Nonparametric Statistics*, Springer Proceedings
in Mathematics & Statistics 339, https://doi.org/10.1007/978-3-030-57306-5_16

being common to all the channels, as a grand mean; the second being channel specific but sharing some common characteristics among the channels. The definition of low and high resonance component will be given through the introduction of the RADWT which is a modern and fast computational tool for analyzing a very general class of signals and will be clarified later in Sect. 2. Here, however, we want to stress that the aim is not to recover the compound signal in each channel, as discussed in [4], but, rather, to reconstruct each of the two components separately. Obviously, this goal returns for free the reconstruction of the compound signal in each channel, but it has the advantage over the technique proposed in [4] to better reconstruct the components rather than their sum. This fact can be useful in some studies, such as those presented in [1], where the good reconstruction of the channel-specific effect allows a better understanding of the underlying phenomenon.

The proposal resembles the Morphological Component Analysis (MCA), an active line of research in image processing. The MCA is a quite new method which allows us to separate features contained in an image when these features present different morphological aspects, see [7].

The problem discussed in this paper is the equivalent in the field of signal processing. In fact, the hypothesis of work is that in each channel the signal is a mixture of two components *morphologically* different from each other and the goal is to separate them.

The remainder of the paper is organized as follows. Section 2 describes the data model we are considering with the working hypothesis. Section 3 presents and discusses the estimation procedure within the paradigm of Lasso procedures, enlightening the connections with the procedure proposed in [4]. Finally, Sect. 4 shows the empirical performance through some numerical experiments.

2 Statistical Model

Consider the following data model:

$$y^{(k)} = c + u^{(k)} + \varepsilon^{(k)} \qquad k = 1, \ldots, K \quad \text{and} \quad \varepsilon^{(k)} \sim N(0, \sigma^2 I), \tag{1}$$

where vector $y^{(k)}$ represents n-equispaced observations of function $c(t) + u^{(k)}(t)$ over the equispaced grid design $t_1 < t_2 < \cdots < t_n$ for each channel $k = 1, \ldots, K$, i.e., $y^{(k)} \in \mathrm{R}^{n\times 1}$. In this contribution, the goal is to reconstruct the two signals $c(t)$ and $\mathrm{u}^{(k)}(t)$ separately in each channel and not their sum as in [4]. From the practical point of view, the aim is to reconstruct two deterministic vectors c and $u^{(k)} \in \mathrm{R}^{n\times 1}$ given the data in (1) in each channel. We stress that, from the triangular inequality, one has $\|c + u - (\hat{c} + \hat{u})\| \leq \|c - \hat{c}\| + \|u - \hat{u}\|$, hence the task of reconstructing each of the two components gives for free the task of reconstructing their sum. We do not hypothesize functions $c(t)$ and $u^{(k)}(t)$ which belong to some functional Sobolev space $H^s_{p,q}[a, b]$ as it is usually done in functional nonparametric regression setting, instead we let these functions to be much more general and we restrict our attention to their finite-dimensional representation. Since many physiological and physical signals are not only non-stationary but also exhibit a mixture of oscillatory and non-oscillatory transient behaviors (for example, speech, stock-market, biomedical EEG, etc.) we suppose that each signal in each channel is the sum of two *morphologically* different signals, a "high resonance" component and a "low resonance" component. By a high resonance component, we mean a signal consisting of multiple simultaneous sustained oscillations; in contrast, by a low resonance component, we mean a signal consisting of non-oscillatory transients of unspecified shape and duration. We stress that the high and low resonance component of a signal can not be extracted from its high- and low-frequencies components in a time-scale decomposition, but they can be well represented by a high Q-factor RADWT and a low Q-factor RADWT, respectively, as very well explained in [6]. Hence, in this contribution we use two different RADWT to sparsely represent the two different components.

The RADWT is a normalized tight frame[1] of $L_2(R)$ defined as $\left\{(\frac{q}{p})^{k/2}\psi\left((\frac{q}{p})^k t + \frac{sp}{q}l\right)\right\}_{k,l\in Z}$ where ψ is a wavelet function and (p, q, s) is a triplet of integer parameters which gives the time-scale characteristic of the frame. In particular, the ratio $q/p > 1$ is closely related to the scale (or frequency) dilatation factor, the parameter $s \geq 1$ is closely related to the time dilatation factor, and $\frac{p}{s(q-p)} \geq 1$ is the redundant factor. The Q-factor depends on these parameters although there is not an explicit formula. In a particular setting, the dilatation factor q/p between 1 and 2 and $s > 1$ gives a RADWT with high Q-factor, while setting $s = 1$ we obtain a low Q-factor RADWT with time-scale characteristic similar to the dyadic wavelet transform. When $q = 2$, $p = 1$, and $s = 1$ the frame reduces to the classical wavelet basis. Given a finite energy signal x of length n, the finite representation of the RADWT transform is a matrix $W \in R^{n\times d}$ with $d \geq n$ (the higher the Q-factor the higher the redundancy d) such that $WW^t = I_n$. This matrix represents the finite frame operator, being $W^t x$ the analysis operation and $W(W^t x)$ the synthesis operation. See [2] for details on fast analysis and synthesis schemes obtained by a sequence of proper down-sampling operations (downsample of q and upsample of p) and fast Fourier transforms.

Let $\Psi \in \mathrm{R}^{n\times d_1}$ be the finite matrix representation of the low Q-factor analysis filter and let $\Phi \in \mathrm{R}^{n\times d_2}$ be the finite matrix representation of the high Q-factor analysis filter (the synthesis operators being just the transpose matrices), then our working hypothesis is the following:

(H1) signal c is sparse in Ψ, i.e., setting $\alpha_0 = \Psi^t c$ we have that $\left|S_0^\alpha\right| = \left|\{j : \alpha_{0_j} \neq 0\}\right| << d_1$;
(H2) signals $u^{(k)}$ have a **jointly** sparse representation in Φ, i.e., setting $\beta_0^{(k)} = \Phi^t u^{(k)}$ and $S_0^{(k),\beta} = \{j : \beta_{0_j}^{(k)} \neq 0\}$ we have that $S_0^{(1),\beta} = \cdots = S_0^{(K),\beta}$, with the common cardinality denoted by $\left|S_0^\beta\right| << d_2$.
(H3) the columns of matrices Ψ and Φ are normalized to have norm 1.

[1]A collection of functions $\{w_i\}$ of $L_2(R)$ forms a frame if exist two constants c_l and c_r such that $c_l\|f\|^2 \leq \sum_i < f, w_i >^2 \leq c_r\|f\|^2$ for all $f \in L_2(R)$. The frame is tight if $c_l = c_r$.

3 Estimation

Model in (1) can be rewritten as a linear model in terms of RADWT coefficients as follows:

$$\begin{cases} y^{(1)} &= \Psi\alpha + \Phi\beta^{(1)} + \varepsilon^{(1)} \\ y^{(2)} &= \Psi\alpha + \Phi\beta^{(2)} + \varepsilon^{(2)} \\ \quad\vdots & \\ y^{(K)} &= \Psi\alpha + \Phi\beta^{(K)} + \varepsilon^{(K)}. \end{cases} \tag{2}$$

With some basic linear algebra transformation, problem (2) can be reformulates as follows:

$$y = \begin{bmatrix} y^{(1)} \\ y^{(2)} \\ \vdots \\ y^{(K)} \end{bmatrix} = (\Psi \otimes 1_k)\alpha + (\Phi \otimes I_K) \begin{bmatrix} \beta^{(1)} \\ \beta^{(2)} \\ \vdots \\ \beta^{(K)} \end{bmatrix} + \begin{bmatrix} \varepsilon^{(1)} \\ \varepsilon^{(2)} \\ \vdots \\ \varepsilon^{(K)} \end{bmatrix} = X_1\alpha + X_2\beta + \varepsilon, \tag{3}$$

where y is a column vector of nK response variables, 1_K is a $(K \times 1)$ array of 1, I_K is the identity matrix of dimension K, and ε is a nK-variate Gaussian random column vector with zero mean and covariance matrix $\sigma^2 I_{n\times K}$. For completeness, we express the design matrices explicitly:

$$\Psi \otimes 1_K = \begin{bmatrix} \Psi \\ \Psi \\ \cdots \\ \Psi \end{bmatrix} \quad \text{and} \quad \Phi \otimes I_K = \begin{bmatrix} \Phi & 0 & \cdots & 0 \\ 0 & \Phi & \cdots & 0 \\ & \cdots & & \\ 0 & 0 & \cdots & \Phi \end{bmatrix}. \tag{4}$$

Vectors α and β are unknown regression coefficients of length d_1 and Kd_2, respectively. Under the working hypothesis (H1), we expect the coefficients of the common part α to be sparse into the dictionary Ψ, while under the working hypothesis (H2), we expect the coefficients of the channel-specific effects β to be *grouped* sparse into the dictionary Φ, i.e., for all $j = 1, \ldots, d_2$ we have $\beta_j^{(k)} = 0$, for all $k = 1, \ldots, K$ or $\beta_j^{(k)} \neq 0$ for all $k = 1, \ldots, K$. This observation provides the following non-overlapping group structure for vector β

$$\{1, 2 \ldots, Kd_2\} = G_1 \cup G_2 \cup \cdots \cup G_{d_2}, \quad \text{with} \quad G_j = \{\beta_j^{(1)}, \beta_j^{(2)}, \ldots \beta_j^{(K)}\}. \tag{5}$$

While in [4], this problem has been approached by a "global" technique, i.e., a technique for recovering a single vector $\theta = (\alpha^t \ \beta^t)^t$; in this work we propose a different technique that aims to recover α and β separately. We stress that this is not an alternative algorithm to solve the same problem, but instead a different problem, i.e., reconstructing each component and not their sum. More specifically, in [4] the solution is expressed by the following:

$$\hat{\theta} = \text{argmin}_{\theta \in R^{(d1+Kd_2)\times 1}} \left\{ \frac{1}{nK} ||y - X\theta||_2^2 + \lambda \|\alpha\|_1 + \lambda \sum_{j=1}^{d_2} ||\beta_{G_j}\|_2 \right\}, \quad (6)$$

with design matrix $X = [X_1\ X_2]$ of dimension $nK \times d_1 + kd_2$. In this paper, the perspective is quite different and we look at model (3) as an additive model in which we are interested in recovering the two components: $X_1\alpha$ and $X_2\beta$. The literature on additive models is very extensive and surely [5] is one of the most complete references on that subject. In [5], it is explained how a natural approach to this problem is the *backfitting* technique, which consists in cyclically updating one of the two components using the partial residual obtained with the other component fixed. Specifically, if we suppose to know the high resonance components $u^{(k)} = \Phi\beta^{(k)}$, then we can evaluate for each channel k the partial residual $z_1^{(k)} = y^{(k)} - \Phi\beta^k$ and estimate the common low resonance component by

$$\hat{\alpha} = \text{argmin}_{\alpha \in R^{d_1 \times 1}} \left\{ \frac{1}{nK} \|z_1 - X_1\alpha\|_2^2 + \lambda \|\alpha\|_1 \right\}, \quad (7)$$

where $z_1 \in R^{nK\times 1}$ is the concatenation of the partial residuals $z_1^{(k)}$, for each $k = 1, \ldots, K$.

Analogously, if we suppose to know the low resonance component $c = \Psi\alpha$, then we can evaluate for each channel k the partial residual $z_2^{(k)} = y^{(k)} - \Psi\alpha$, and estimate the channel-specific high resonance components by

$$\hat{\beta} = \text{argmin}_{\beta \in R^{Kd_2 \times 1}} \left\{ \frac{1}{nK} \|z_2 - X_2\beta\|_2^2 + \lambda \sum_{j=1}^{d_2} ||\beta_{G_j}\|_2 \right\} \quad (8)$$

where $z_2 \in R^{nK\times 1}$ is the concatenation of partial residuals $z_2^{(k)}$, for each $k = 1, \ldots, K$. Problems (7) and (8) are both convex and can be solved by fast algorithms. Specifically, problem (7) is a classical LASSO regression problem and can be solved by a coordinate descendent algorithm, while problem (8) is a grouped LASSO problem and can be solved by a group descendent algorithm as the one proposed in [3]. Summarizing, we propose the following algorithm

- INPUT: λ, $y^{(k)}$, Φ, Ψ
- initialize $\hat{\beta}_0^k = 0$, for all $k = 1, \ldots, K$
- repeat until convergence for $l = 0, 2, \ldots$
  - update partial residual $z_1^{(k)} = y^{(k)} - \Phi\hat{\beta}_l^{(k)}$, for all $k = 1, \ldots, K$

  - solve problem (7) to obtain $\hat{\alpha}_{l+1}$
  - update partial residual $z_2^{(k)} = y^{(k)} - \Psi\hat{\alpha}_l$, for all $k = 1, \dots, K$
  - solve problem (8) to obtain $\hat{\beta}_{l+1}$

- OUTPUT: $\hat{\alpha}$ and $\hat{\beta} = \left(\left(\hat{\beta}^{(1)}\right)^t, \left(\hat{\beta}^{(2)}\right)^t, \dots, \left(\hat{\beta}^{(K)}\right)^t\right)^t$

In this contribution, the convergence is established if a maximum number of iterations is reached or solution improvement is negligible. The unknown components are finally obtained by the synthesis operation $\hat{c} = \Psi\hat{\alpha}$ and $\hat{u}^{(k)} = \Phi\hat{\beta}^{(k)}$, for $k = 1, \dots, K$.

4 Numerical Experiments

The aim of this section is to demonstrate, at least under the model hypothesis, the advantage of using the proposed backfitting technique with respect to solving the single problem in (6) as done in [4]. The delicate point is the choice of the regularization parameter λ that can greatly affect the results of both procedures. For this reason, recalling from the theory that the optimal λ is of order $\sim log(dimension)/size$, we fix $\lambda = log(d_1 + kd_2)/nK$ in all our experiments.

We generated data according to model (2) using three channels ($K = 3$) and $n = 256$ observations in each channel. Matrix Ψ was generated using the following choice $p_{low} = 1$, $q_{low} = 2$, $s_{low} = 1$ with 4 levels of decomposition ($d_1 = 496$), and matrix Φ was generated using $p_{high} = 8$, $q_{high} = 9$, $s_{high} = 3$ with 10 levels of decomposition ($d_2 = 695$). These matrices represent RADWT with Q-factor almost 1 and 5, respectively. We considered three scenarios with different sparsity level:

Scenario 1: low sparsity, corresponding to $|S_\alpha| = 24$ and $\left|S_\beta\right| = 24$;
Scenario 2: medium sparsity, corresponding to $|S_\alpha| = 12$ and $\left|S_\beta\right| = 12$;
Scenario 3: high sparsity, corresponding to $|S_\alpha| = 6$ and $\left|S_\beta\right| = 6$;

and for each scenario we used three signal to noise ratios (SNR): 1.5, 3, 6, defined as

$$SNR = \frac{\frac{1}{K}\sum_{i=1}^{K} Var(\Psi\alpha + \Phi\beta^{(k)})}{\sigma^2}.$$

The numerical setting mimics the one presented in [4], as well as the following performance indexes:

- Root Mean Square Error (RMSE) defined as

$$RMSE = \sqrt{\frac{1}{n}\sum_{i=1}^{n}\left(\hat{f}^{(k)}(t_i) - f^{(k)}(t_i)\right)^2}, \quad k = 1, \dots, K;$$

  with $f^{(k)} = c + u^{(k)}$ and $\hat{f}^{(k)} = \Psi\hat{\alpha} + \Phi\hat{\beta}^{(k)}$ as its estimate;

- Root Mean Square Error for the low resonance component (RMSE_{low}) defined as

$$RMSE_{low} = \sqrt{\frac{1}{n}\sum_{i=1}^{n}\left(\hat{c}(t_i) - c(t_i)\right)^2};$$

- Root Mean Square Error for the high resonance component (RMSE_{high}) defined as

$$RMSE_{high} = \sqrt{\frac{1}{n}\sum_{i=1}^{n}\left(\hat{u}^{(k)}(t_i) - u^{(k)}(t_i)\right)^2}, \ k = 1, \ldots, K;$$

RMSE_{low} and RMSE_{high} aim at evaluating a component wise accuracy.

Tables 1, 2, and 3 report results obtained for Scenario 1, 2, and 3, respectively. In these tables, `backfitting` refers to the proposed technique, while `multi-c` refers to the one proposed in [4]. We note that in each scenario the estimate of the two components is better using the `backfitting` technique with respect to the `multi-c` technique, while the estimate of the compound signal is very similar being possible that some errors are compensated in the sum for the triangular inequality. Finally, we observe that, as expected, for all types of scenarios the error improves when the SNR increases.

We conclude this contribution observing that the proposed technique is very interesting and deserves further investigations both from an applicative point of view as well as from a theoretical perspective.

Table 1 Average values (standard deviation between parentheses) of RMSE, RMSE_{low} (the same for each channel), and RMSE_{high} based on 10 simulations with different noise realizations. Experiment carried out on Scenario 1 with SNR=1.5, 3, and 6

	RMSE		RMSE_{low}		RMSE_{high}	
	multi-c	backfitting	multi-c	backfitting	multi-c	backfitting
SNR=1.5						
`ch1`	0.2051 (0.0031)	0.2048 (0.0029)	0.1952 (0.0016)	0.1870 (0.0015)	0.2185 (0.0131)	0.2104 (0.0113)
`ch2`	0.2171 (0.0113)	0.2167 (0.0112)	0.1952 (0.0016)	0.1870 (0.0015)	0.2386 (0.0093)	0.2304 (0.0109)
`ch3`	0.2189 (0.0102)	0.2188 (0.0101)	0.1952 (0.0016)	0.1870 (0.0015)	0.2404 (0.0066)	0.2342 (0.0055)

(continued)

Table 1 (continued)

	RMSE		$RMSE_{low}$		$RMSE_{high}$	
	multi-c	backfitting	multi-c	backfitting	multi-c	backfitting
SNR=3						
ch1	0.1487 (0.0050)	0.1481 (0.0050)	0.1627 (0.0125)	0.1531 (0.0105)	0.1890 (0.0086)	0.1788 (0.0088)
ch2	0.1587 (0.0057)	0.1579 (0.0056)	0.1627 (0.0125)	0.1531 (0.0105)	0.2013 (0.0122)	0.1911 (0.0111)
ch3	0.1557 (0.0045)	0.1553 (0.0045)	0.1627 (0.0125)	0.1531 (0.0105)	0.2000 (0.0078)	0.1911 (0.0078)
SNR=6						
ch1	0.1185 (0.0057)	0.1178 (0.0057)	0.1439 (0.0157)	0.1346 (0.0133)	0.1712 (0.0112)	0.1597 (0.0089)
ch2	0.1327 (0.0063)	0.1322 (0.0065)	0.1439 (0.0157)	0.1346 (0.0133)	0.1932 (0.0109)	0.1828 (0.0101)
ch3	0.1421 (0.0117)	0.1415 (0.0115)	0.1439 (0.0157)	0.1346 (0.0133)	0.1988 (0.0189)	0.1885 (0.0179)

Table 2 Average values (standard deviation between parentheses) of RMSE, $RMSE_{low}$ (the same for each channel), and $RMSE_{high}$ based on 10 simulations with different noise realizations. Experiment carried out on Scenario 2 with SNR=1.5, 3, and 6

	RMSE		$RMSE_{low}$		$RMSE_{high}$	
	multi-c	backfitting	multi-c	backfitting	multi-c	backfitting
SNR=1.5						
ch1	0.1641 (0.0038)	0.1641 (0.0036)	0.1172 (0.0081)	0.1163 (0.0081)	0.1485 (0.0076)	0.1478 (0.0078)
ch2	0.1835 (0.0064)	0.1834 (0.0063)	0.1172 (0.0081)	0.1163 (0.0081)	0.1777 (0.0060)	0.1773 (0.0061)
ch3	0.1719 (0.0100)	0.1716 (0.0100)	0.1172 (0.0081)	0.1163 (0.0081)	0.1624 (0.0140)	0.1619 (0.0138)
SNR=3						
ch1	0.1061 (0.0035)	0.1057 (0.0035)	0.0865 (0.0054)	0.0861 (0.0055)	0.0938 (0.0070)	0.0933 (0.0066)
ch2	0.1113 (0.0027)	0.1110 (0.0027)	0.0865 (0.0054)	0.0861 (0.0055)	0.1083 (0.0059)	0.1077 (0.0051)
ch3	0.1116 (0.0095)	0.1113 (0.0094)	0.0865 (0.0054)	0.0861 (0.0055)	0.1082 (0.0130)	0.1074 (0.0129)
SNR=6						
ch1	0.0849 (0.0022)	0.0848 (0.0022)	0.0606 (0.0045)	0.0605 (0.0045)	0.0827 (0.0039)	0.0824 (0.0038)
ch2	0.0978 (0.0039)	0.0977 (0.0039)	0.0606 (0.0045)	0.0605 (0.0045)	0.1049 (0.0049)	0.1046 (0.0050)
ch3	0.0856 (0.0071)	0.0855 (0.0071)	0.0606 (0.0045)	0.0605 (0.0045)	0.0876 (0.0079)	0.0875 (0.0078)

Table 3 Average values (standard deviation between parentheses) of RMSE, RMSE_{low} (the same for each channel), and RMSE_{high} based on 10 simulations with different noise realizations. Experiment carried out on Scenario 3 with SNR=1.5, 3, and 6

	RMSE		RMSE_{low}		RMSE_{high}	
	multi-c	backfitting	multi-c	backfitting	multi-c	backfitting
SNR=1.5						
ch1	0.0386 (0.0012)	0.0386 (0.0012)	0.0263 (0.0023)	0.0263 (0.0023)	0.0282 (0.0007)	0.0282 (0.0007)
ch2	0.0500 (0.0015)	0.0500 (0.0015)	0.0263 (0.0023)	0.0263 (0.0023)	0.0424 (0.0017)	0.0425 (0.0016)
ch3	0.0478 (0.0008)	0.0478 (0.0008)	0.0263 (0.0023)	0.0263 (0.0023)	0.0399 (0.0015)	0.0399 (0.0015)
SNR=3						
ch1	0.0392 (0.0016)	0.0392 (0.0016)	0.0257 (0.0021)	0.0257 (0.0021)	0.0295 (0.0004)	0.0295 (0.0004)
ch2	0.0495 (0.0013)	0.0495 (0.0013)	0.0257 (0.0021)	0.0257 (0.0021)	0.0422 (0.0013)	0.0422 (0.0013)
ch3	0.0497 (0.0013)	0.0497 (0.0013)	0.0257 (0.0021)	0.0257 (0.0021)	0.0425 (0.0004)	0.0425 (0.0004)
SNR=6						
ch1	0.0391 (0.0012)	0.0391 (0.0012)	0.0258 (0.0019)	0.0258 (0.0019)	0.0293 (0.0002)	0.0293 (0.0002)
ch2	0.0498 (0.0012)	0.0498 (0.0012)	0.0258 (0.0019)	0.0258 (0.0019)	0.0425 (0.0011)	0.0425 (0.0011)
ch3	0.0496 (0.0009)	0.0496 (0.0009)	0.0258 (0.0019)	0.0258 (0.0019)	0.0423 (0.0001)	0.0423 (0.0001)

References

1. Barros, A.K., Rosipal, R., Girolami, M., Dorffner, G., Ohnishi, N.: Extraction of sleep-spindles from the electroencephalogram (EEG). In: Malmgren, H., Borga, M., Niklasson, L. (eds.) Artificial Neural Networks Medicine and Biology, Perspectives in Neural Computing, pp. 125–130. Springer, London (2000)
2. Bayram, I., Selesnick, I.W.: Frequency-domain design of overcomplete rational-dilation wavelet transform. IEEE Trans. Signal Process. **57**(8), 2957–2972 (2009)
3. Breheny, P., Huang, J.: Group descent algorithms for nonconvex penalized linear and logistic regression models with grouped predictors. Stat. Comput. **25**(2), 173–187 (2015)
4. De Canditiis, D., De Feis, I.: Simultaneous nonparametric regression in RADWT dictionaries. In: Comput. Stat. Data Anal. (2019). https://doi.org/10.1016/j.csda.2018.11.003
5. Hastie, T., Tibshirani, R.: Genaralized Additive Modcels. Chapman & Hall, London (1990)
6. Selesnick, I.W.: Resonance-based signal decomposition: a new sparsity-enabled signal analysis. Signal Process. **91**(12), 2793–2809 (2011)
7. Starck, J.-L., Moudden, Y., Bobin, J., Elad, M., Donoho, D.L.: Morphological component analysis. In: Proceedings of SPIE 5914, Wavelets XI, 59140Q, 17 September 2005 (2005). https://doi.org/10.1117/12.615237

Kernel Circular Deconvolution Density Estimation

Marco Di Marzio, Stefania Fensore, Agnese Panzera, and Charles C. Taylor

Abstract We consider the problem of nonparametrically estimating a circular density from data contaminated by angular measurement errors. Specifically, we obtain a kernel-type estimator with weight functions that are reminiscent of deconvolution kernels. Here, differently from the Euclidean setting, discrete Fourier coefficients are involved rather than characteristic functions. We provide some simulation results along with a real data application.

Keywords Circular kernels · Deconvolution · Fourier coeffcients · Measurement errors · Movements of ants

1 Introduction

Circular data arise when the sample space is described by a unit circle. By comparison with a linear scale, the main features of circular observations are that the beginning and the end of the measurement scale coincide, and their common location is called the origin (or zero direction) which is usually chosen arbitrarily. Once the origin and the direction of rotation have been chosen, any circular observation can be measured by an angle ranging, in radians, from 0 to 2π. Circular data often arise in biology, meteorology and geology; other examples include phenomena that are periodic in

M. Di Marzio (✉) · S. Fensore
University of Chieti-Pescara, Chieti, Italy
e-mail: marco.dimarzio@unich.it

S. Fensore
e-mail: stefania.fensore@unich.it

A. Panzera
University of Florence, Florence, Italy
e-mail: agnese.panzera@unifi.it

C. C. Taylor
University of Leeds, Leeds, England
e-mail: charles@maths.leeds.ac.uk

M. La Rocca et al. (eds.), *Nonparametric Statistics*, Springer Proceedings
in Mathematics & Statistics 339, https://doi.org/10.1007/978-3-030-57306-5_17

time. For comprehensive accounts of statistics for circular data see, for example, [7] and [8].

In this paper, we discuss the problem of nonparametrically estimating a circular density when data are observed with error. Specifically, here we consider the case of measurement errors described by i.i.d. circular random variables. This problem has been studied by [6], who proposed an estimator constructed as a truncated development of the density which is estimated by trigonometric basis in which the theoretical coefficients are replaced by empirical ones. Then, using a model selection procedure, [1] derived an adaptive penalized contrast estimator, and [9] proposed an orthogonal series estimator which is optimal in the minimax sense.

In the Euclidean setting, the problem of estimating a density in the context of errors-in-variables has been widely investigated. The most popular method is a nonparametric one based on kernel-type estimators. A kernel density estimator in the case of homoscedastic, classical measurement errors with known distribution has been introduced by [10]. Kernel density estimation with a different type of measurement error, named Berkson error, has been considered in [2]. A further estimator for the case of heteroscedastic, classical measurement error has been proposed by [5] who also considers the case of unknown error density. For this latter problem see, among others, [4]. An exhaustive treatment of density estimation with errors-in-variables and related topics is provided by [3]. In the directional setting, the kernel-based methods for errors-in-variables problems seem to be substantially unexplored. In this article, we propose to extend the Euclidean approach to the estimation of a circular density in the case of classical, homoscedastic measurement errors being circular random variables with known distribution.

After recalling in Sect. 2, some preliminaries about Fourier series and nonparametric estimation of circular densities in the error-free case, in Section 3 we discuss the extension of the kernel-type density estimator to the case where variables are observed with errors. Then, in Sect. 4 we present some simulation results, and in Sect. 5 we conclude with a real data application.

2 Preliminaries

In this section, we provide some basic facts about Fourier series representation of circular densities and recall the definition of the circular kernel density estimator.

2.1 Trigonometric Moments and Fourier Series

Let Q be a circular random variable and denote by f_Q its probability density function. Due to the periodic nature of Q, its distribution is the same as the distribution of $Q + 2\pi$; this implies that the characteristic function of Q, which is

$$\varphi_Q(\ell) = E[e^{i\ell Q}] = \int_0^{2\pi} e^{i\ell q} f_Q(q)dq,$$

is defined only for integer ℓs. Moreover, for any $\ell \in \mathbb{Z}$, one has

$$|\varphi_Q(\ell)| \leq 1, \quad \varphi_Q(0) = 1, \quad \bar{\varphi}_Q(\ell) = \varphi_Q(-\ell),$$

where $\bar{\varphi}_Q(\cdot)$ is the complex conjugate of $\varphi_Q(\cdot)$. Notice that the complex numbers $\{\varphi_Q(\ell), \ell \in \mathbb{Z}\}$ are the coefficients in the Fourier series representation (in complex form) of f_Q and correspond to the *trigonometric moments* of Q about the mean direction, i.e. letting

$$\alpha_\ell = E[\cos(\ell Q)], \quad \beta_\ell = E[\sin(\ell Q)],$$

it holds that $\varphi_Q(\ell) = \alpha_\ell + i\beta_\ell$; clearly, for any $\ell \in \mathbb{Z}$,

$$\alpha_{-\ell} = \alpha_\ell, \quad \beta_{-\ell} = -\beta_\ell, \quad |\alpha_\ell| \leq 1, \quad |\beta_\ell| \leq 1.$$

Then, assuming that f_Q is a square integrable function on $[0, 2\pi)$, for $q \in [0, 2\pi)$, one can recover $f_Q(q)$ from the Fourier coefficients using the expansion

$$f_Q(q) = \frac{1}{2\pi} \sum_{\ell=-\infty}^{\infty} \varphi_Q(\ell) e^{-i\ell q} = \frac{1}{2\pi} \left\{ 1 + 2 \sum_{\ell=1}^{\infty} (\alpha_\ell \cos(\ell q) + \beta_\ell \sin(\ell q)) \right\}. \tag{1}$$

Equation (1) is analogous to the inversion formula for characteristic functions of real-valued random variable. In the Euclidean setting, the smoothness of a density can be determined by the rate of decay of the Fourier transforms: a polynomial decay characterizes *ordinary smooth* functions, while an exponential decay characterizes *supersmooth* ones. Analogously, for a circular density the smoothness can be defined according to the rate of decay of the coefficients in its Fourier series representation.

We recall that for a wrapped circular distribution, the trigonometric moment of order $\ell \in \mathbb{Z}$ corresponds to the value of the characteristic function of the unwrapped random variable, say φ_X, at (integer) ℓ, i.e. $\varphi_Q(\ell) = \varphi_X(\ell)$.

Examples of supersmooth densities include the densities of wrapped Normal and wrapped Cauchy distribution; conversely, the wrapped Laplace and the wrapped Gamma densities are examples of ordinary smooth ones. See Fig. 1 for some examples of wrapped distributions and their Fourier coefficients.

2.2 *Circular Density Estimation in the Error-Free Case*

Given a random sample of angles $\Theta_1, \ldots, \Theta_n$ from an unknown circular density f_Θ, the kernel estimator of f_Θ at $\theta \in [0, 2\pi)$ is defined as

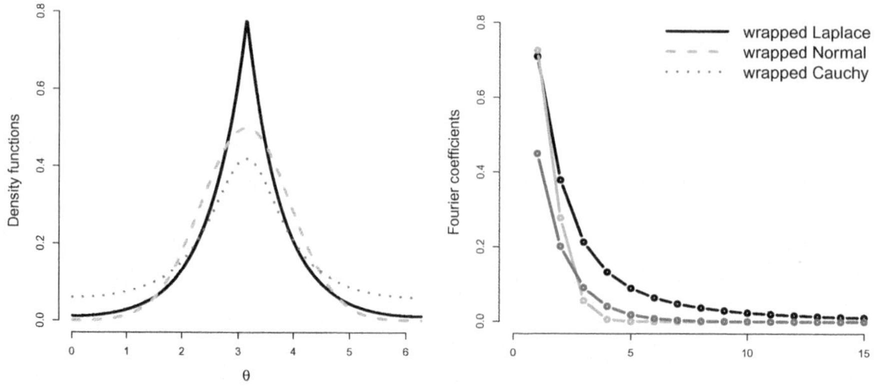

Fig. 1 Examples of wrapped densities sharing the values of mean and variance of their unwrapped versions (left) and corresponding Fourier coefficients (right)

$$\hat{f}_\Theta(\theta;\kappa) = \frac{1}{n}\sum_{i=1}^{n} K_\kappa(\Theta_i - \theta),$$

where K_κ is a circular kernel, i.e. a periodic, unimodal, symmetric density function with concentration parameter $\kappa > 0$, which admits a convergent Fourier series representation as follows

$$K_\kappa(\theta) = \frac{1 + 2\sum_{\ell=1}^{\infty} \gamma_\ell(\kappa)\cos(\ell\theta)}{2\pi}.$$

Notice that, by comparison with Equation (1), due to the symmetry, the Fourier coefficients of K_κ satisfy $\beta_\ell = 0$ and $\alpha_\ell = \gamma_\ell(\kappa)$ for any ℓ. It is well known that the choice of the kernel is generally not crucial. This means that in our case it suffices to select any symmetric circular density function able to arbitrarily concentrate its whole mass around zero by increasing the value of the concentration parameter κ. Also, note that circular data have a periodic range, whereas in the Euclidean case the presence of boundaries of the sample space could require ad hoc, shape-designed kernels. Classical examples of circular kernels are the von Mises density with $\gamma_\ell(\kappa) = \mathcal{I}_\ell(\kappa)/\mathcal{I}_0(\kappa)$, where $\mathcal{I}_\ell(\kappa)$ is the modified Bessel function of order ℓ; the Wrapped Normal and Wrapped Cauchy densities where $\gamma_\ell(\kappa) = \kappa^{\ell^2}$ and $\gamma_\ell(\kappa) = \kappa^\ell$, respectively. As in the linear setting, the role of the kernel function is to emphasize, in the estimation process, the contribution of the observations which are in a neighbourhood of the estimation point. Here, κ controls the width of that neighbourhood in such a way that its role is the inverse of the square of the bandwidth in the linear case, in the sense that smaller values of κ give wider neighbourhoods.

3 Kernel Density Estimator in the Errors-in-Variables Case

We consider the problem of estimating the density of a circular random variable Θ which is observed with error. In particular, we deal with the classical, homoscedastic measurement error case where we wish to estimate the density f_Θ of Θ but we observe independent copies of the circular random variable

$$\Phi = (\Theta + \varepsilon)\text{mod}(2\pi),$$

where ε is a random angle independent of Θ, whose density f_ε is assumed to be a known circular density symmetric around zero. Notice that the density f_Φ of Φ is the *circular convolution* of f_Θ and f_ε, i.e. for $\theta \in [0, 2\pi)$,

$$f_\Phi(\theta) = \int_0^{2\pi} f_\Theta(\omega) f_\varepsilon(\theta - \omega) d\omega, \tag{2}$$

so, the estimation of f_Θ reduces to a circular *deconvolution* density problem. Similarly to the Euclidean case, equation (2) implies that, for $\ell \in \mathbb{Z}$,

$$\varphi_\Phi(\ell) = \varphi_\Theta(\ell)\varphi_\varepsilon(\ell),$$

then, a naive estimator of f_Θ at $\theta \in [0, 2\pi)$ could be

$$\tilde{f}_\Theta(\theta) = \frac{1}{2\pi} \sum_{\ell=-\infty}^{\infty} \frac{\hat{\varphi}_\Phi(\ell)}{\varphi_\varepsilon(\ell)} e^{-i\ell\theta}, \tag{3}$$

where $\hat{\varphi}_\Phi(\ell) = \frac{1}{n}\sum_{j=1}^{n} e^{i\ell\Phi_j}$ is the empirical version of $\varphi_\Phi(\ell)$. Now, since

$$\int_{-\pi}^{\pi} \left(f_\Theta(\theta) - \tilde{f}_\Theta(\theta)\right) d\theta = \frac{1}{2\pi} \sum_{\ell=-\infty}^{\infty} \left(\varphi_\Theta(-\ell) - \frac{\hat{\varphi}_\Phi(-\ell)}{\varphi_\varepsilon(-\ell)}\right)\left(\varphi_\Theta(\ell) - \frac{\hat{\varphi}_\Phi(\ell)}{\varphi_\varepsilon(\ell)}\right)$$

we have that rapid decays of $\varphi_\varepsilon(\ell)$ lead to big discrepancies between $f_\Theta(\theta)$ and $\tilde{f}_\Theta(\theta)$ even in correspondence of small discrepancies between $\varphi_\Theta(\ell)$ and $\hat{\varphi}_\Phi(\ell)$. Therefore, in order to regularize estimator (3), a possible remedy is to introduce the characteristic function of a circular kernel K_κ, say $\varphi_{K_\kappa}(\ell)$, as a damping factor, i.e.

$$\begin{aligned}\tilde{f}_\Theta(\theta;\kappa) &= \frac{1}{2\pi}\sum_{\ell=-\infty}^{\infty}\frac{\hat{\varphi}_\Phi(\ell)}{\varphi_\varepsilon(\ell)}\varphi_{K_\kappa}(\ell)e^{-i\ell\theta}\\ &= \frac{1}{2\pi}\sum_{\ell=-\infty}^{\infty}\frac{\varphi_{K_\kappa}(\ell)}{\varphi_\varepsilon(\ell)}\frac{1}{n}\sum_{j=1}^{n}e^{i\ell\Phi_j}e^{-i\ell\theta}\\ &= \frac{1}{2\pi}\frac{1}{n}\sum_{\ell=-\infty}^{\infty}\sum_{j=1}^{n}\frac{\varphi_{K_\kappa}(\ell)}{\varphi_\varepsilon(\ell)}e^{-i\ell(\theta-\Phi_j)},\end{aligned}$$

which leads to the following circular deconvolution estimator of $f_\Theta(\theta)$ at $\theta \in [0, 2\pi)$

$$\tilde{f}_\Theta(\theta;\kappa) = \frac{1}{2\pi}\frac{1}{n}\sum_{j=1}^{n}\left(1 + 2\sum_{\ell=1}^{\infty}\frac{\gamma_\ell(\kappa)}{\lambda_\ell(\kappa_\varepsilon)}\cos(\ell(\theta-\Phi_j))\right), \tag{4}$$

where $\gamma_\ell(\kappa)$ and $\lambda_\ell(\kappa_\varepsilon)$, respectively, are the ℓth coefficients in the Fourier series representation of K_κ and f_ε. Also, in order to guarantee that estimator (4) is well defined, we assume that *a*) the error density is an infinitely divisible distribution, i.e. it has nonvanishing coefficients $|\lambda_\ell(\kappa_\varepsilon)| > 0$ for any integer ℓ, and *b*) the kernel K_κ and $\tilde{f}_\Theta(\cdot;\kappa)$ are square integrable functions, i.e using the Parseval's identity,

$$\frac{1}{2\pi}\left(1 + 2\sum_{\ell=1}^{\infty}\gamma_\ell^2(\kappa)\right) < \infty \quad \text{and} \quad \frac{1}{2\pi}\left(1 + 2\sum_{\ell=1}^{\infty}\frac{\gamma_\ell^2(\kappa)}{\lambda_\ell^2(\kappa_\varepsilon)}\right) < \infty.$$

The way in which the rate of decay of the $\lambda_\ell(\kappa_\varepsilon)$s affects the performance of the estimator will be shown in the simulation experiments in the next section.

Notice that estimator (4) is suitable for the case where the ε_is are homoscedastic errors with known distribution. The Euclidean version designed for the case of heteroscedastic errors has been studied by [5] who also introduce a modified estimator for the problem where the distribution of the errors is unknown, but replicated observations are available.

4 Simulations

In this section, we compare the performance of the deconvolution estimator and the standard kernel density estimate in a simulation setting. In particular, we consider the von Mises density (vM) with mean direction and concentration, respectively, equal to π and 2 as the target density f_Θ, and the wrapped Laplace (wL), wrapped Normal (wN) or wrapped Cauchy (wC) with zero mean direction and different values of the concentration parameter as the error density f_ε. Notice that the concentration parameter takes non-negative real values for both vM and wL but with opposite meaning in that, for wL, lower values of the concentration parameter give higher concentration, whereas for wN and wC the concentration parameter ranges from 0 to 1 with the concentration increasing with the value of the parameter.

Let $\nu_\ell(\kappa_\Theta)$ be the ℓth Fourier coefficient of f_Θ, for $\ell \in \mathbb{N}$. We consider the noise-to-signal ratio (NSR), defined as the ratio between the *circular variance* of ε and that of Θ, which can be expressed in terms of Fourier coefficients as $\{1 - \lambda_2(\kappa_\varepsilon)\}/\{1 - \nu_2(\kappa_\Theta)\}$. Specifically, we consider three different settings corresponding to a NSR ranging from 16% to 47%, which are shown in Fig. 2, where for ease of presentation the target density has mean zero. In particular, for $\ell \in \mathbb{N}$, we have $\nu_\ell(\kappa_\Theta) = \mathcal{I}_\ell(\kappa_\Theta)/\mathcal{I}_0(\kappa_\Theta)$, while $\lambda_\ell(\kappa_\varepsilon) = \kappa_\varepsilon^{\ell^2}$, $\lambda_\ell(\kappa_\varepsilon) = \kappa_\varepsilon^{-2}/(\ell^2 + \kappa_\varepsilon^{-2})$, and $\lambda_\ell(\kappa_\varepsilon) = \kappa_\varepsilon^{\ell}$, respectively, give the wN, the wL and wC as the error distributions.

We generate 500 samples of size $n = 100, 200$ and 400 and compare the estimators in terms of averaged integrated squared error (AISE). In particular, we calculate the ratio $AISE_{dec}/AISE_{kde}$, where dec and kde, respectively, stand for $\tilde{f}_\Theta(\theta; \kappa)$ and $\hat{f}_\Theta(\theta; \kappa)$. The smoothing parameter κ has been selected by using least squares cross-validation. The results are summarized in Table 1 and Fig. 3. It can be seen that the deconvolution estimator outperforms the standard one especially when the NSR is moderate or the error density is ordinary smooth.

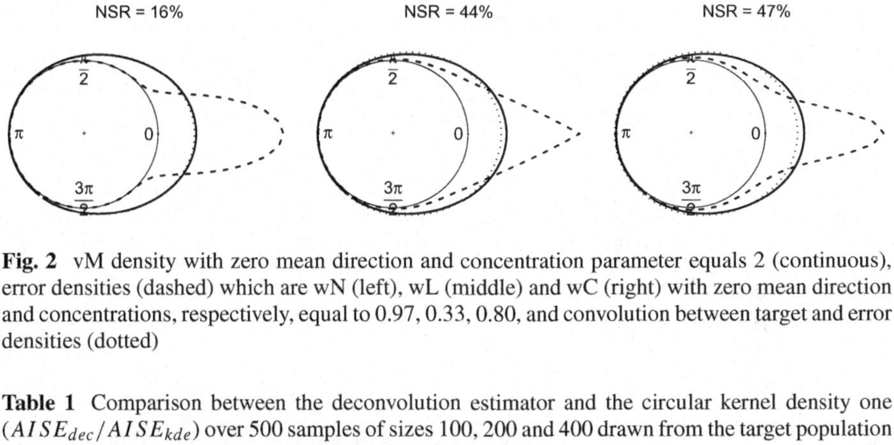

Fig. 2 vM density with zero mean direction and concentration parameter equals 2 (continuous), error densities (dashed) which are wN (left), wL (middle) and wC (right) with zero mean direction and concentrations, respectively, equal to 0.97, 0.33, 0.80, and convolution between target and error densities (dotted)

Table 1 Comparison between the deconvolution estimator and the circular kernel density one ($AISE_{dec}/AISE_{kde}$) over 500 samples of sizes 100, 200 and 400 drawn from the target population contaminated by noise obtained by different error populations

NSR	Target density	Error density	$n = 100$	$n = 200$	$n = 400$
16%	vM(π, 2)	wN(0, 0.97)	0.755	0.782	0.769
44%	vM(π, 2)	wL(0, 0.33)	0.866	0.857	0.839
47%	vM(π, 2)	wC(0, 0.80)	0.966	1.015	1.085

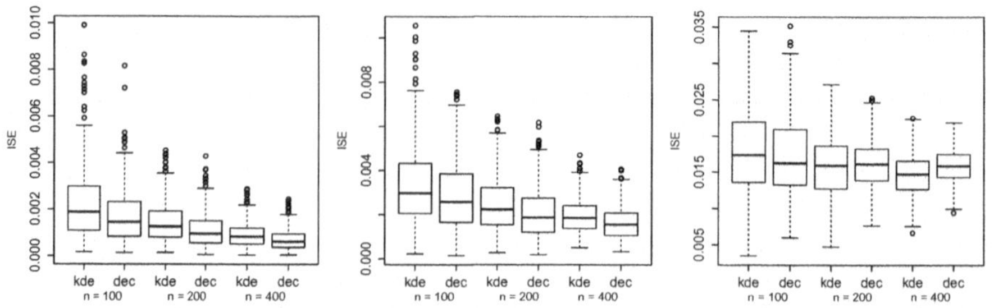

Fig. 3 Comparison between the deconvolution estimator and the circular kernel density one in terms of integrated squared errors (ISE) over 500 contaminated samples of sizes 100, 200 and 400 with a NSR equals to 16% (left), 44% (middle) and 47% (right)

5 Real Data Example

We consider the classical dataset described by [7] concerning the directions chosen by 100 ants in response to an evenly illuminated black target placed at π. [7] showed that classical parametric models, like von Mises, are not suited for these data. However, he concluded them for a unimodal population. A nonparametric approach has been suggested by [6], who, in the context of errors-in-variables modelling, revealed some evidence about multimodality. His approach is based on orthogonal trigonometric series. The rationale behind the errors-in-variables hypothesis is that, due to the typical jerky movement of the insect, the point where the ant intersects the circle can be treated as indirect observation of the direction chosen by the ant.

We compare the standard circular kernel density estimator with our deconvolution one. Specifically, we have assumed a wrapped Laplace error with zero mean and

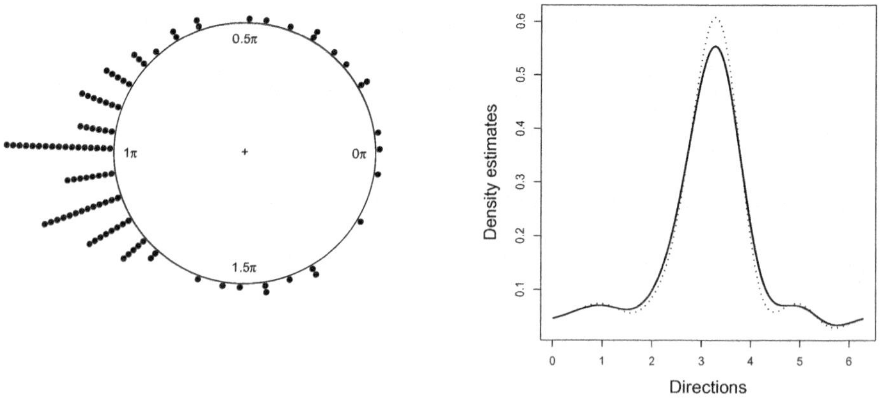

Fig. 4 Ants data (left) and kernel density estimate (continuous) and deconvolution estimate (dotted) of the directions of ants (right)

concentration equal to 0.2, employing a wrapped Normal weight function whose smoothing parameter has been selected by least squares cross-validation. As it can be seen in Fig. 4 the proposed deconvolution estimator reveals the presence of three modes more distinctly than the standard method.

References

1. Comte, F., Taupin, M.L.: Adaptive density deconvolution for circular data. Prépublication MAP5 2003-10 report, Université Paris Descartes (2003)
2. Delaigle, A.: Nonparametric density estimation from data with a mixture of Berkson and classical errors. Canad. J. Statist. **35**, 89–104 (2007)
3. Delaigle, A.: Nonparametric kernel methods with errors-in-variables: constructing estimators, computing them, and avoiding common mistakes. Aust. N. Z. J. Stat. **56**, 105–124 (2014)
4. Delaigle, A., Hall, P., Meister, A.: On Deconvolution with repeated measurements. Ann. Statist. **36**, 665–685 (2008)
5. Delaigle, A., Meister, A.: Density estimation with heteroscedastic error. Bernoulli **14**, 562–579 (2008)
6. Efromovich, S.: Density Estimation for the Case of Supersmooth Measurement Error. J. Amer. Statist. Assoc. **92**, 526–535 (1997)
7. Fisher, N.I.: Statistical analysis of circular data. Cambridge University Press (1993)
8. Jammalamadaka, S.R., SenGupta, A.: Topics in Circular Statistics. World Scientific (2001)
9. Jhoannes, J., Schwarz, M.: Adaptive circular deconvolution by model selection under unknown error distribution. Bernoulli **19**, 1576–1611 (2013)
10. Stefanski, L.A., Carroll, R.J.: Deconvoluting kernel density estimators. Statistics **21**, 169–184 (1990)

Asymptotic for Relative Frequency When Population Is Driven by Arbitrary Unknown Evolution

Silvano Fiorin

Abstract Strongly consistent estimates are shown, via relative frequency, for the probability of *white balls* inside a dichotomous urn when such a probability is an arbitrary unknown continuous time-dependent function over a bounded time interval. The asymptotic behaviour of relative frequency is studied in a nonstationary context using a Riemann-Dini type theorem for strong law of large numbers of random variables with arbitrarily different expectations; furthermore, the theoretical results concerning the strong law of large numbers can be applied for estimating the mean function of an unknown form of a general nonstationary process.

1 Introduction

Several different areas of statistics deal with an urn model including *white* and *black* balls with probability p and $1-p$, respectively. In this very classic context a time-dependent component is introduced, and p is replaced with $p_0(t)$ which denotes a time varying quantity $0 \le p_0(t) \le 1$ in such a way that at any instant $t \in [0, T]$ only one observation is taken from the corresponding urn with probability $p_0(t)$ and the random variable $Y(t)$ is obtained such that $P(Y(t) = 1) = p_0(t)$, $P(Y(t) = 0) = 1 - p_0(t)$, $E(Y(t)) = p_0(t)\ \forall t \in [0, T]$, defining the nonstationary process

$$Y = \{Y(t) : t \in [0, T]\} \tag{1}$$

with mean function $E(Y(t)) = p_0(t)$. The description of the above model is specified introducing some reasonable assumptions:

A 1 we assume continuity for the usually unknown mean function $p_0 : [0, T] \mapsto [0, 1]$;

S. Fiorin (✉)
Dipartimento di Scienze Statistiche, via C. Battisti, 241-35121 Padova, Italy
e-mail: fiorin@stat.unipd.it

M. La Rocca et al. (eds.), *Nonparametric Statistics*, Springer Proceedings in Mathematics & Statistics 339, https://doi.org/10.1007/978-3-030-57306-5_18

A 2 for any fixed pair of instants $t_1, t_2 \in [0, T]$ the independence is assumed for the random variables $Y(t_1)$ and $Y(t_2)$.

This assumption is introduced in order to apply the Rajchman Theorem (see [5]) or the classical results concerning Strong Law of Large Numbers (SLLN) (see [4]). Namely, only pairwise uncorrelation is requested for $Y(t_1)$ and $Y(t_2)$ but, it can be easily checked in this case, the uncorrelation implies independence; furthermore, independence is here a very mild condition; in fact, we may suppose that the total number of white and black balls in the urn is big enough that the knowledge of $Y(t_1) = 1$ or $Y(t_1) = 0$ does not produce a meaningful modification of the probability distribution for $Y(t_2)$.

The main purpose of this paper consists of a double aim:

1. to study the asymptotic behaviour of relative frequency in a nonstationary context;
2. to estimate the unknown function p_0, i.e. the mean function $p_0(t) = E(Y(t))$ of the nonstationary process (1), which is an arbitrary continuous map form $[0, T]$ into $[0, 1]$.

The urn evolution has effects concerning sampling; for instance, if the observations number n is big enough, a not slight time interval will be needed in order to receive the n observations which surely are not values taken by the same random variable. Then, for the sake of simplification, we assume that any r.v. $Y(t)$ may be observed at most only one time. The point of view we adopt is then characterized by a strong nonstationarity and the consistent estimation for the mean $m(t_0)$ at a fixed time t_0 may appear as a very hard objective.

An approach to estimation for the mean function $m(\cdot)$ of a nonstationary process was given by M. B. Priestley (see [10] in page 587 and [11] in page 140) when the form of m is known and the case is suggested of a polynomial function in t. Viceversa: *with no information on the form of m we obviously cannot construct a consistent estimate of it.* The approach here adopted is quite different from classical methods of time series analysis; the only information available for m is the continuity property over $[0, T]$, and no approximation of m is introduced by continuous functions of a known form. The estimation technique involves the process (1) which is a specified case of nonstationarity but the theoretical results given in the last section hold true for a general nonstationary process. The case (1) is only a concrete example of a process having no regularity properties; nevertheless, the continuity for the mean function m is a reasonable and not restrictive assumption which denotes compatibility with a context of an arbitrary but not brutal evolution for the composition of the urn.

Concerning estimation problem for the mean function $m(\cdot)$ of a nonstationary process, some well-known approach is available in the literature as, for instance, *the smoothing spline estimation* by [13] or *nonparametric regression estimation* as in [7] and [9]. These classical approaches, following the sieves technique, need the first k_n functions belonging to a base inside a vector space and the usual assumptions involved for the smooth function $m(\cdot)$ are concerning the derivatives m', m'' and so on. Thus the estimation procedure developed in this paper may be seen as an alternative method; only continuity is adopted for $m(\cdot)$ and the use of sieves technique is omitted.

The answer to above arguments is the relative frequency

$$\frac{1}{n}\sum_{j=1}^{n} Y(t_j) \tag{2}$$

where $\{t_j : j = 1, \ldots, n\}$ are the first n observation times of a sequence $\{t_j : j \geq 1\} \subset [0, T]$ and the main purpose is that of getting consistent estimations of $m(t) = p_0(t)$ via almost sure convergence for the sequence (2). The SLLN is then the theoretical tool needed in the below analysis, but the classical approach based on the zero-mean r.v.'s $(Y(t_j) - p_0(t_j))$, i.e.

$$\frac{1}{n}\sum_{j=1}^{n}(Y(t_j) - p_0(t_j)) \to 0 \text{ a.s.} \tag{3}$$

is not enough; in fact, we need convergence for (2) with the not zero-mean r.v.'s $Y(t_j)$. This argument, investigated by Fiorin [8] is now improved with the help of new results given in Sect. (5).

Nevertheless, the application of usual SLLN for studying the asymptotic behaviour of (2) is not a trivial step and several problems arise concerning the process (1). The family of r.v.'s $\{Y(t_j) : j \geq 1\}$ is not a stationary process and then we have no possibility of applying the classical ergodic theory (see, for instance, Chap. 3 in [2]) based on a stationary probability distribution over R^{∞} and on a measure-preserving transformation. Analogously the generalizations of ergodic theory such as Dunford and Schwartz pointwise ergodic theorem (see [6] in page 675) or Chacon and Ornstein theorem [3] cannot be applied to our problem. Also law of large numbers for random functions cannot be adapted to the above problem; taking, for instance, the Ranga Raw law for $D[0, 1]$ valued r.v.'s [12], the main argument is given by the observable trajectories inside the Skorohod space $D[0, 1]$ of functions with discontinuities only for the first kind; thus the trajectories of process (1), including any arbitrary function taking only values 0 and 1, are not a random element into $D[0, T]$. Moreover let us observe that, because of the discontinuity at any point t, the observation of any trajectory over all the interval $[0, T]$, and then any law of large numbers based on trajectories, are a too hard purpose. Consequently, the asymptotic arguments are concerning the sequence (2), where the number of observed r.v.'s $Y(t_j)$ tends to infinity.

The convergence of (2) is studied via the sequence $\{E(Y(t_j)) = p_0(t_j) : j \geq 1\}$ and permutations (i.e. bijections) $\pi : N \to N$; in fact, if a permutation π is introduced, the possible almost sure limit of

$$\frac{1}{n}\sum_{j=1}^{n} Y(t_{\pi(j)}) \tag{4}$$

is depending on π. If $\{P^0_{\pi n}\}$ is a sequence of probability measures, where each $P^0_{\pi n}$ assigns mass $\frac{1}{n}$ to each point $\{p_0(t_{\pi(j)}) : j = 1, \ldots, n\}$, then the *weak* or *vague* convergence for the sequence $\{P^0_{\pi n}\}$ to a probability measure P^0 implies almost sure convergence of (4) to the limit $\int_0^1 I(v) dP^0(v)$ where $I(v)$ is the identity map over $[0, 1]$ and P^0 depends on the sequence $\{Y(t_j) : j \geq 1\}$ and on permutation π. All the below analysis is based on the possibility of finding a permutation π in such a way that the convergence of (4) is driven to a limit $\int_0^1 I(v) dP^0(v)$ where P^0 is a previously chosen probability measure over $[0, 1]$; under a theoretical point of view this is a result for SLLN (4) which is the analogous of the well-known Riemann-Dini Theorem for real simply convergent (but not absolutely convergent) series. Under the operative point of view the strongly consistent estimates are the result of an experimental design based on choosing

(I) the sequence of observation times $\{t_j : j \geq 1\} \subset [0, T]$;
(II) the permutation $\{t_{\pi(j)} : j \geq 1\}$.

2 Convergence Elements

If the observation times $\{t_j : j \geq 1\}$ are given jointly with the observable r.v.'s $\{Y(t_j) : j \geq 1\}$, an intuitive approach for studying the almost sure convergence for (2) is suggested by the elementary equality

$$\frac{1}{n}\sum_{j=1}^{n} Y(t_j) = \frac{1}{n}\sum_{j=1}^{n} (Y(t_j) - E(Y(t_j))) + \frac{1}{n}\sum_{j=1}^{n} E(Y(t_j)); \tag{5}$$

if the $Y(t_j)$'s are pairwise uncorrelated and their second moments have a common bound (see [5]) then the a.s. convergence to 0 for $\frac{1}{n}\sum_{j=1}^{n}(Y(t_j) - E(Y(t_j)))$ jointly with the convergence to a limit L for the deterministic sequence

$$\frac{1}{n}\sum_{j=1}^{n} E(Y(t_j)) \tag{6}$$

imply that (2) is a.s. convergent to the limit L. Thus the argument of below analysis is the possible convergence to some limit L for the sequence (6). Then writing (6) as an integral

$$\frac{1}{n}\sum_{j=1}^{n} E(Y(t_j)) = \int_0^1 I(x) dP_n(x), \tag{7}$$

where $I(x)$ is the identity map and P_n is the probability measure which assigns the weight $\frac{1}{n}$ to each point $\{E(Y(t_j)) : j = 1, \ldots, n\}$, and the argument of below analysis is the possible limit for the sequence of integrals (7) adopting the technique of *weak* or *vague* convergence for the sequence of measures P_n's; in fact, by definition of weak convergence for measures, we have that if the P_n's are weakly convergent to P then

$$\lim_{n\to\infty} \int_0^1 I(x)dP_n(x) = \int_0^1 I(x)dP(x). \tag{8}$$

Nevertheless, the weak convergence for P_n's is not so easy to obtain since the expectations $\{E(Y(t_j)) : j \geq 1\}$ define an arbitrary deterministic sequence and then weak convergence is achieved via permutations.

3 A General SLLN via Permutations

A permutation is any bijection $\pi : N \to N$ defined over the naturals N in such a way that the sequence of random variables is introduced

$$\{Y(t_{\pi(j)}) : j \geq 1\} \text{ with expectations } \{E(Y(t_{\pi(j)})) : j \geq 1\} \tag{9}$$

and thus, for any assigned natural n, $P_{\pi n}$ is defined as the probability measure giving mass $\frac{1}{n}$ to each point $\{E((Y(t_{\pi(j)})) : j = 1, \ldots, n\}$. The main theoretical result shows the technique of finding a permutation π such that the sequence $P_{\pi n}$ is weakly convergent to an assigned probability measure P. For a rigorous proof of below statement see Theorem 7 in [8].

Theorem 1 *For any assigned sequence of constants* $\{E(Y(t_j)) : j \geq 1\} \subset [0, 1]$ *there exists a class* $\mathcal{M}$ *of probability measures (over* $[0, 1]$*) such that for each given* $P \in \mathcal{M}$ *a corresponding permutation can be constructed such that the sequence* $P_{\pi n}$ *is weakly (or vaguely) convergent to* P *and then*

$$\lim_{n\to\infty} \int_0^1 I(x)dP_{\pi n}(x) = \int_0^1 I(x)dP(x) \textit{ and}$$

$$\lim_{n\to\infty} \frac{1}{n} \sum_{j=1}^{n} Y(t_{\pi(j)}) = \int_0^1 I(x)dP(x) \textit{ almost surely.}$$

Some comments and remarks may help to clarify the meaning of above result:

(a) The final goal is not only the construction of a permutation π making the $P_{\pi n}$'s a weakly convergent sequence, but also that of driving convergence to a chosen limit measure belonging to class $\mathcal{M}$.
(b) The definition of class $\mathcal{M}$ is, of course, a central and rather technical argument: for details and a rigorous treatment see the construction leading to Definition 6 in [8].

(c) The main theorem may appear as an analogous of the well-known Riemann-Dini theorem for convergent real series: both the proofs are clearly involving permutations, but the technique adopted in proving the above main result is a constructive one.

(d) The above result is a generalization of the classical SLLN concerning a sequence of r.v.'s Y_j having a common finite expectation $\mu = E(Y_j), \forall j \geq 1$. By the elementary equality

$$\frac{1}{n}\sum_{j=1}^{n} Y(t_j) = \frac{1}{n}\sum_{j=1}^{n}(Y(t_j) - E(Y(t_j))) + \frac{1}{n}\sum_{j=1}^{n} E(Y(t_j))$$

and if the convergence holds true:

$$\lim_{n\to\infty} \frac{1}{n}\sum_{j=1}^{n}(Y(t_j) - E(Y(t_j))) = 0 \text{ a.s.}$$

an easy direct comparison is possible:

1. in the standard case, when $E(Y(t_j)) = \mu, \forall j \geq 1$, we trivially have

$$\frac{1}{n}\sum_{j=1}^{n} E(Y(t_j)) = \mu, \forall n.$$

This means that for each n the weight 1 is assigned to value μ and then the probability measure $P_{\pi n} = \delta_\mu$ are invariant with respect to any given permutation π and the $P_{\pi n}$'s are weakly convergent to measure $P = \delta_\mu$.

2. In the general case, when expectations $\{E(Y(t_j)) : j \geq 1\} \subset [0, 1]$ are arbitrarily different values,

$$\frac{1}{n}\sum_{j=1}^{n} E(Y(t_{\pi(j)})) = \int_0^1 I(x) dP_{\pi n}$$

depends on the sequence $\{E(Y(t_j)) : j \geq 1\}$ and π, and the technique based on weak convergence for $P_{\pi n}$'s is a generalization of the standard case.

Moreover, the limit for SLLN is written as an integral $\int_0^1 I(x)dP(x)$, i.e. as an expectation with respect to the probability measure P which is the weak limit of $P_{\pi n}$'s; thus P is defined through π, independently of probability distribution of r.v.'s $Y(t_j)$.

Finally, let us observe that the main theorem cannot be directly applied for finding π because the proof technique is fully based on the knowledge of values $E(Y(t_j))$'s which are the estimation object.

4 Estimating $E(Y(t))$

Let us choose as observation times any sequence $\{t_j : j \geq 1\}$ which is dense into $[0, T]$ and thus Theorem 1 can be applied to $\{t_j : j \geq 1\}$; because of the density of t_j's the class $\mathscr{M}$ of the weak limit measures contains all the absolutely continuous probability measures over $[0, T]$. Thus $P_U \in \mathscr{M}$ where P_U denotes the uniform probability measure over $[0, T]$ having density

$$f_U(t) = \frac{1}{T} \; \forall\, t \in [0, T]$$

and, applying the main theorem, a permutation π can be found such that $P_{\pi n}$, which assigns weight $\frac{1}{n}$ to each point $\{t_{\pi(j)} : j = 1, \ldots, n\}$, is weakly convergent to P_U. The continuity of the unknown function $p_0(t) = E(Y(t))$ for each $t \in [0, T]$ keeps weak convergence for the induced measures over $[0, 1]$: then $p_0(P_{\pi n})$ is weakly convergent to $p_0(P_U)$, where $p_0(P_{\pi n})$ assigns weight $\frac{1}{n}$ to each point

$$\{p_0(t_{\pi(j)}) = E(Y(t_{\pi(j)})) : j = 1, \ldots, n\}$$

and then, by the mean value theorem for integrals, the limits hold true:

$$\lim_{n\to\infty} \frac{1}{n} \sum_{j=1}^{n} E(Y(t_{\pi(j)})) = \lim_{n\to\infty} \int_0^1 I(x) dp_0(P_{\pi n}) =$$

$$= \int_0^1 I(x) dp_0(P_U) = \int_0^T p_0(t) dP_U = \frac{1}{T} \int_0^T p_0(t) dt = p_0(\underline{t})$$

for some points $\underline{t} \in [0, T]$, and

$$\lim_{n\to\infty} \frac{1}{n} \sum_{j=1}^{n} Y(t_{\pi(j)}) = p_0(\underline{t}) \text{ almost surely.}$$

An analogous version of above result holds true for any assigned interval $(a, b] \subset [0, T]$. Using the same above permutation π such that the $P_{\pi n}$'s are weakly convergent to P_U over $[0, T]$, for each n fixed, we collect inside the set $\{t_{\pi(j)} : j = 1, \ldots, n\}$ all the $t_{\pi(j)}$'s falling into $(a, b]$, i.e. the set is defined

$$A(\pi, n, (a, b]) = \{t_{\pi(j)} \in (a, b] : j = 1, \ldots, n\}$$

and if $n(a, b]$ denotes its cardinality, the following statement holds true (see Theorem 4 in [8] for a complete proof).

Theorem 2 *The sequence of r.v.'s*

$$\frac{1}{n(a,b]} \sum_{t_{\pi(j)} \in A(\pi,n,(a,b])} Y(t_{\pi(j)}),$$

when $n \to \infty$, is a strongly consistent estimate of $p_0(\underline{t})$ for some points $\underline{t} \in [a,b]$.

5 Remarks

1. Theorem (2) may be applied, at the same time, to several different subintervals of $[0, T]$; for instance, to all the subintervals belonging to a finite partition of $(0, T]$.
2. The policy of choosing the observation times $\{t_j : j \geq 1\}$ as a dense subset of $[0, T]$ is a technique which is common to several areas of statistical inference. In this context it can be easily checked that
   (a) this choice derives directly from evolution of the nonstationary process $\{Y(t) : t \in [0, T]\}$; in fact at most only one observation is possible for any r.v. $Y(t)$. Thus to increase the number of observations implies to choose new t_j's and their density in $[0, T]$ ensures a good knowledge of the process.
   (b) The density of t_j's makes necessary the use of permutations; in fact, the sequence $\frac{1}{n}\sum_{j=1}^{n} Y(t_j)$ has no meaning if a permutation is not assigned for choosing the t_j's. But the choice of π, as it was shown above, has a deep effect in terms of measures $P_{\pi n}$ and of convergence.

References

1. Ash, R.B., Doleans-Dade, C.A.: Probability & measure theory, 2nd edn. Academic Press, London (2000)
2. Ash, R.B., Gardner, M.F.: Topics in stishastic processer, 2nd edn. Academic Press, New York (1975)
3. Chacon, R.V., Ornstein, D.S.: A general ergodic theorem Illinois. J. Math **4**(2), 153–160 (1960)
4. Chandra, T.K.: Laws of large numbers. Narosa Publishing House, New Delhi (2012)
5. Chung, K.L.: A course in probability theory, 3rd edn. Academic Press, London (2001)
6. Dunford, N., Schwartz, J.T.: Linear operators part I, general theory. Wiley Classic, New York (1988)
7. Eubank, R.L.: Spline smoothing and nonparametric regression. Marrcel Dekker, New York (1988)
8. Fiorin, S.: *Asymptotic for relative frequency when population is driven by arbitrary evolution*, eprint arXiv:1709.06313 (2017)
9. Gyorfi, L., Kohler, M, Krzyzak, A. Walk, H.: *A distribution-free theory of nonparametric regression*. Springer, New York (2002)
10. Priestley, M.B.: *Spectral analysis and time series*, vol 1, univariate series. Academic Press, London (1981)

11. Priestley, M.B.: Non-linear and non-stationary time series analysis. AcaDEMIC Press, London (1988)
12. Ranga Rao, R.: The law of large numbers for $D[0, 1]$-valued random variables. Theor. Prob. Appl. **8**, 70–74 (1963)
13. Wahba, G.: Spline models for the observational data. S.I.A.M, Philadelphia PA (1990)

Semantic Keywords Clustering to Optimize Text Ads Campaigns

Pietro Fodra, Emmanuel Pasquet, Bruno Goutorbe, Guillaume Mohr, and Matthieu Cornec

Abstract In this paper, we describe how to use some well-known machine learning tools to make groups of textual queries of similar semantic meaning. Such a clusterization can be used to improve the performances of bidding algorithms for online advertising, by mutualizing the signal gathered by text ads displayed on result pages of search queries which share a similar meaning. Indeed, search engines organize auctions wherein participants bid on selected search terms on which they wish to display an ad. Generalist e-commerce companies such as Cdiscount bid simultaneously on millions of terms that reflect the diversity of their catalog of products, according to the expected profits associated with the ads. Methods to estimate these expected returns suffer from a sparsity of data, since most of the keywords have little or no historical signal. Grouping them and exploiting information on the most frequent keywords (short tail) to infer information on the less frequent ones (long tail), allow to anticipate the user behavior by semantics and improve the bidding strategy. The plan is the following: pre-process the keywords by stemming, choose an e-commerce training corpus for the Word2Vec model, train it, and perform an embedding into a euclidean space where we can cluster keywords thanks to a K-means algorithm. We validate our approach on a sub-sample of the keywords for which they have anon-

P. Fodra (✉)
Chief of Data Scoring, Cdiscount, France
e-mail: pietro.fodra@cdiscount.com

E. Pasquet
Data Scientist, Cdiscount, France
e-mail: emmanuel.pasquet@cdiscount.com

B. Goutorbe
Chief of Data Science, Cdiscount, France
e-mail: bruno.goutorbe@cdiscount.com

G. Mohr
Chief of Data Traffic, Cdiscount, France
e-mail: guillaume.mohr@cdiscount.com

M. Cornec
Cdiscount, France

M. La Rocca et al. (eds.), *Nonparametric Statistics*, Springer Proceedings in Mathematics & Statistics 339, https://doi.org/10.1007/978-3-030-57306-5_19

semantic distance available. Finally, all the keywords in the same cluster share the same bid, which is computed aggregating the cluster historical signal.

Keywords NLP · SEA · Semantic clustering

1 Introduction

For large e-commerce websites, visibility is crucial and largely depends on how Internet users find the website through the most common online search engines. In order to appear in the search engine result page, three options are usually available.

1. *The free way* (SEO): the search engine matches the user query to a page of the website. This option, despite being free, has the disadvantage of being uncontrollable: the search engine is entirely responsible of the matching (if and which page) and there is no way to improve the result in a short amount of time.
2. *Text Ads*: for each user query, the search engine creates an auction where e-commerce actors participate to get the best position in the result page and increase the chances of their link being clicked. Each participant creates an ad (a rich message and a link) and associates it to a set of keywords, for which he can specify the maximum amount of money he is willing to pay for each click (max cost-per-click). Then the search engine chooses, according to the overall quality of the ad and the bid amount, which ads to show, and the position of each one on the result page. These ads render as the textual content of the ad, equipped with a re-direction link exactly as for free results, but with a small extra label ("Ad," for example) on the left. They usually appear on the top and on the bottom of the result page.
3. *Product List Ads*: similar to textual ads, but rendering as a priced image usually displayed on the top of the page, before all the textual results. For this type of ads, even if triggered by a textual query, the participants bid directly on the product (and not on keywords), while the search engine is usually responsible for the keyword-product match.

In this work, we will talk about how *semantic clustering of keywords can help to improve performances of textual ads campaigns* (option 2). We will address this problem by means of Machine Learning and Natural Language Processing (NLP) tools as Word2Vec, clustering techniques, and text processing.

In Sect. 2 we will formalize the bidding problem and explain in detail why semantic clustering helps to improve the bidding strategy. In Sect. 3 we will explore the Word2Vec embedding, focusing on the metric nature of the landing space, and how to validate the model by introducing a different distance (based on behavior) on a subset of frequent keywords. Section 4 is devoted to the semantic clustering using K-Means, while in Sect. 5 we present a practical application and the result of an A/B testing.

2 Formalization of the Bidding Problem

Before detailing the problem, let us fix some notation.

Definition 1 We take an agent participating to search engine auctions where he bids to show textual ads for $\mathbb{K}$ a (finite) set of *keywords* (despite the name keywords, keywords can be more than one words). We denote by $k \in \mathbb{K}$ a generic element of this set and define $\mathbb{W}$ as the set of all words contained in all the keywords in $\mathbb{K}$.

Definition 2 We define a *clusterization* of keywords as a partition of $\mathbb{K}$: a cluster C is a non-empty subset $C \subseteq \mathbb{K}$ of keywords sharing some property (not necessarily an ad group), with the property that for any two clusters $C_i, C_j, C_i \cap C_j = \emptyset$, and $\sum_i C_i = \mathbb{K}$.

Definition 3 For each keyword k, we define ϕ_k as the maximum cost-per-click (abbr. max CPC) the agent is willing to pay for a click, while we will denote by ϕ_C the max CPC associated to a cluster if all the keywords $k \in C$ share the same bid.

Definition 4 For each keyword k, we define two random variables X_k and V_k representing the total amount of money spent on k during the time interval $[0, T]$ and the total revenue associated to keyword: both random sequences are controlled by the max CPC bid ϕ_k. Notice that increasing the bid ϕ_k would make both the cost X_k and the revenue V_k increase, even though the cost usually explodes much faster than the revenue (saturation effect).

The bidding problem, for which we assume for simplicity that the control Φ is constant throughout $[0, T]$, is to maximize:

$$\max_{\Phi \in \mathcal{A}} \mathbb{E}\left[\mathcal{U}\left(\sum_{k \in \mathbb{K}} V_k\right)\right], \quad \text{under} \sum_k X_k = X, \tag{1}$$

where X is the total budget of the agent for all his keywords, $\mathcal{A}$ the set of admissible strategies and $\mathcal{U}$ a convex monotone utility function. We will focus on a problem which comes before (1): in order to maximize the agent utility function, we need a reliable model for the random variables $(V_k)_k$.

Definition 5 We use the following decomposition of the revenue:

$$V_k = \nu_k N_k, \tag{2}$$

where ν_k is an uncontrolled (independent from Φ) variable representing the revenue associated to each click (0 when a click is not followed by a purchase, which is most of the time), and N_k (controlled) is the number of clicks during $[0, T]$.

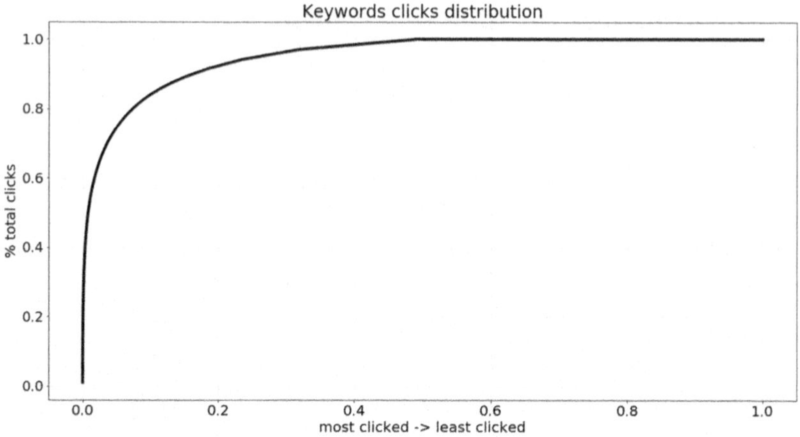

Fig. 1 Example of keywords clicks distribution: in this curve 20% of the keywords account for more than 90% of the total clicks and roughly half of the keywords have no signal at all (flat tail). For very large accounts with a lot of keywords, this curve can be much steeper

However, even for $(\nu_k)_k$ i.i.d., estimating the average value per click can be a challenge. In fact, as shown in (Fig. 1), *most of the keywords have little or no historical signal* since they are only clicked few times during their lifetime; however, since there are a lot of them, the impact of ignoring these keywords can be dramatic. The main idea of this paper is to exploit semantic similarity (which is independent from Φ) to group them into clusters having a much stronger historical signal, and provide an average estimator for the cluster only. Instead of waiting or forcing as in multi-armed bandit theory (see [2]) for rare keywords to accumulate signal over time, we use their short-tail neighbors. Given clusters $(C_i)_{i=1}^n$, the problem (1) becomes

$$\max_{\Phi \in \mathcal{A}'} \mathbb{E}\left[\mathcal{U}\left(\sum_{i=1}^{n} V_{C_i}\right)\right], \quad \text{under} \sum_{i=1}^{n} X_{C_i} = X, \tag{3}$$

where X_{C_i} and V_{C_i} are the global cost and revenue of a whole cluster. Notice that a strategy $\Phi \in \mathcal{A}'$ if all the keywords in the same cluster share the same bid.

3 The Word2Vec Modeling

The first step is producing an embedding of $\mathbb{W}$ into a Euclidean metric space whose distance respects the semantic distance among *words*.

In order to find this embedding, we use the Word2Vec model introduced by the seminal paper [6] and [7], implemented by the Python package `gensim` (see [9]). The Word2Vec model embeds the one-hot-encoding representation of $\mathbb{W}$, which is a space of very-high dimension (as large as the number of keywords, which in our case is more than one million), to a smaller Euclidean space, usually of dimension between 100 and 500, thanks to a neural network whose hidden layer transformation matrix is used as coordinates for the embedding. This technique works pretty well and is able to capture the linear correlation among vectors: the classic example provided by the paper takes into account the relationship

$$\psi(king) - \psi(queen) = \psi(man) - \psi(woman), \tag{4}$$

which captures the semantics among the four terms involved.

However, even though the vector space structure of the landing space $\mathbb{R}^d$ is used to express semantic differences among couples (as in the king to queen = man to woman example), the metric used to compare words is not the one induced by the vector space structure (the norm), but the cosine one.

Definition 6 We define $d_\Theta : \mathbb{W}^2 \to [0, 1]$ as

$$d_\Theta(w_1, w_2) = \frac{1}{2}\left(1 - \frac{< w_1, w_2 >}{\|w_1\| \, \|w_2\|}\right) \in [0, 1] \tag{5}$$

as the semantic distance between words induced by the Word2Vec embedding.

3.1 *The Behavioral Distance*

Word2Vec is an unsupervised model whose performances are not easily measurable. That is why we validate our approach on a keywords subset for which another distance is available.

Definition 7 We define a subset $\mathbb{K}^* \subset \mathbb{K}$ (roughly 10%) of search terms which are particular frequent in our internal search engine (the one allowing to explore the website) and define a behavioral distance on $\mathbb{K}^*$ ([3]) as

$$d_J(k_1, k_2) = 1 - \frac{\pi(k_1) \cap \pi(k_2)}{\pi(k_1) \cup \pi(k_2)} \in [0, 1], \; \forall k_1, k_2 \in \mathbb{K}^*, \tag{6}$$

where $\pi(k)$ is the set of all the products that have been clicked on the internal engine after searching for k.

It is worth noticing that this distance is not available on the whole $\mathbb{K}$ since most of the elements of $\mathbb{K}$ have never been searched . This distance is commonly known as the Jaccard distance . For example, $d_J(k_1, k_2) = 1$ if the two queries lead to no common product, while $d_J(k_1, k_2) = 0$ if the user clicks are exactly identical for the two queries. We call this distance behavioral, in contrast with the semantic one, since it depends only on the user behavior and does not take into account the keyword meaning.

3.2 The Model Training and the Semantic Distance

In order to train the model, we need to define a training corpus $\mathbb{D}$ (a collection of documents). We may use as corpus the collection of all our search terms $\mathbb{K}$, i.e., $\mathbb{D} = \mathbb{K}$, but in this case, we would lose semantic information due to the documents being very short (keywords are often less than 5 words) and we would not have enough context to train the model.

That is why we have chosen the *internal product description catalog*, which provides a very rich and specific corpus, i.e., $\mathbb{D}$ is the collection of all the product descriptions, for all the products in our catalog. In order to have a faster implementation, we have chosen to train a single model for keywords in a given category only with the descriptions of products in the same category: since training time does not scale linearly, this allows us to break a big problem into smaller problems (one for each category) and still perform well.

So far, we have defined an Euclidean embedding of $\mathbb{W}$, but our goal is to find a distance between keywords, not words.

Definition 8 We define the weight associated to a word w as the inverse document frequency ([8])

$$\alpha_w = \frac{1}{\sum_{D \in \mathbb{D}} \mathbb{I}_{w \in \Omega(D)}}, \qquad w \in \mathbb{W}, \tag{7}$$

where $\mathbb{D}$ is the corpus used to train the Word2Vec model.

Definition 9 We define

$$d_S(k_1, k_2) = d_\Theta \left(\sum_{w \in \Omega(k_1)} \alpha_w \psi_w, \sum_{z \in \Omega(k_2)} \alpha_z \psi_z \right), \quad \forall k_1, k_2 \in \mathbb{K}, \tag{8}$$

where $\Omega(k)$ is the set of all the words contained in the keyword k (e.g., $\Omega(\text{hello world}) = \{\text{hello, world}\}$).

The semantic distance d_S is the Euclidean distance between the baricenters of the two keywords, where the weight of each words decreases with the word frequency in the corpus $\mathbb{D}$. This allows us to reduce the weight of the word *telephone* in its category, since its presence does not help to semantically distinguish two keywords.

Once we have defined these two distances, we can compare them on $\mathbb{K}^* \subset \mathbb{K}$ by taking the correlation value between the entries of the matrices $\{d_J(k_i, k_j)\}_{ij}$ and $\{d_S(k_i, k_j)\}_{ij}$. This metric allows to optimize the model hyper-parameters, as the embedding dimension p, the extension of the training corpus, or the measure the impact of text pre-processing. Here is our conclusions.

corpus	no tf-idf	tf-idf
product catalogs	19.3%	35.2%
product catalogs + generic	19.1%	34.8%

Fig. 2 Correlation between behavioral and semantic distance for $d = 300$ on the telephone department depending on the training corpus and the use of the idf normalization (8): while adding a generic catalog does not improve the results, and the idf normalization is essential to obtain a good performance

- *General purpose corpus*: we add a general purpose corpus (e.g., WaCkY [1]) to the training phase, which leads to a much slower training without a significant performance improvement. We conclude that the product description corpus is sufficiently rich for our purpose.
- *Pre-processing*: we have improved the model performance by stemming words, which allows us to normalize and reduce the keywords space, as well as removing special characters and converting Latin numbers to Arabic ones to have a homogeneous text treatment. We have massively relied on the Python package `nltk` ([5]).
- *Embedding dimensions*: we have found that $p = 300$ is a good compromise between learning speed and performances.
- Weighting by the IDF weights significantly improves the quality of the metric (Fig. 2).

4 Keyword Clusterization

Thanks to the embedding described in the previous section, we can use the metrics induced by d_S to cluster all the keywords. However, since the number of keywords is relatively large, we would like to use K-means ([4]) to minimize the fitting time of the clustering model. However, we need to take care of some details.

We recall that the distance d_S is the cosine distance between baricenters in the Euclidean embedding space of the Word2Vec model. If we used a K-means on keywords using those baricenters as Euclidean coordinates, we would be using the norm distance to create clusters, and not the cosine one (which defines similarities in the W2V model). To overcome this problem, it is enough to, for each K-means iteration (Fig. 3):

1. project the whole embedding space $\mathbb{R}^d - \{0\}$ to the sphere S^d dividing by the norm;
2. use the euclidean distance on the projected coordinates to compute centroids (approximation of the sphere chord distance);
3. project the new centroids on the sphere.

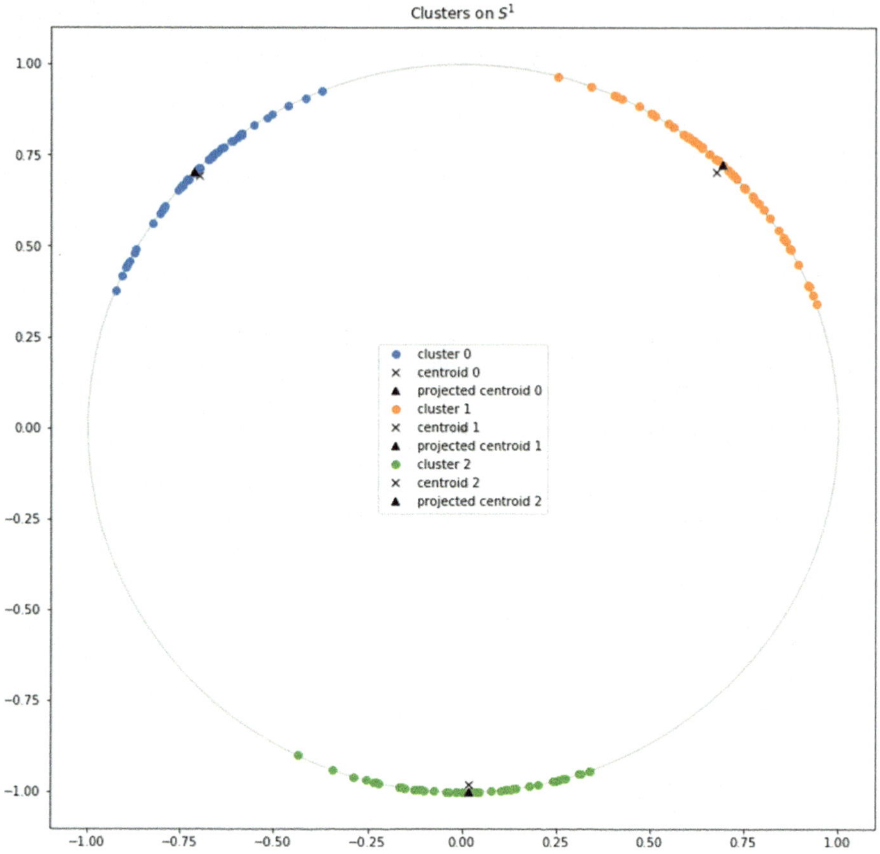

Fig. 3 Example of K-means on S^1: the projection of the cluster centroids guarantees that centroids are still on the sphere

5 A/B Testing and Conclusions

We used the keyword clusterization defined in the previous section to create ad groups whose keywords are forced to share the same bid (max CPC). How bids are determined is out of the scope of this paper; however, the idea is that bidder sees a group of keywords in the same ad group as a unique one, aggregating all the history into a unique shared signal. We have finally build an A/B test where the bidder bids as usual (one bid per keyword) in the A part, and a bid per ad group on the B part.

The result of the A/B encourages our researches to go further: while the A part outperforms the B part on a relatively small of very short-tail keywords, the B part proves to be better on the long tail, leading to an overall tie. This is mainly due to the purely semantic clusterization: some very generic (and often short keyword) have so much signal that they would deserve to have a special treatment. Generic keywords

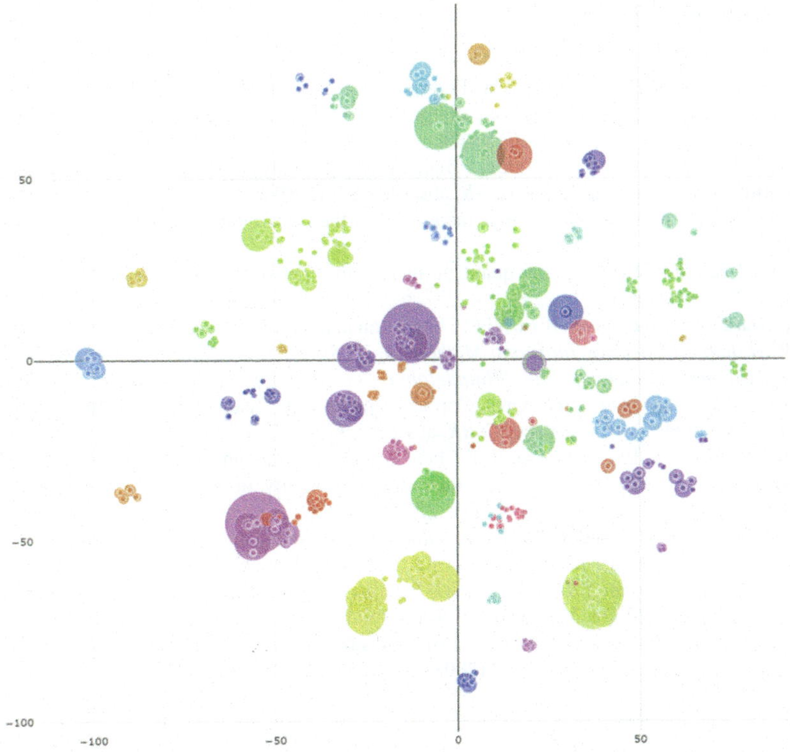

Fig. 4 Graphical representation of clusters via TSNE [10] on the telephones category. Colors represent clusters, which are computed before TSNE, while the circle size the number of clicks for a given keyword in log-scale

(e.g., *telephone*) and specific keywords (e.g., *telephone brand color*) have different behavior, even if their semantics can be similar: that is the way, further work will be dedicated to a keyword tagging allowing to order keywords according to their level of specification. This could be done by analyzing lexical property (as for the keyword length, for example) or their previous behavior if available.

Another development axis is the clustering technique; instead of partitioning the keyword space, we can identify neighbors for each keyword and use smoothing techniques (as a kernel density or a KNN) to smooth the historical signal coming from each keyword. This approach would allow us to introduce weights into the smoothing densities depending on the keyword behavior.

References

1. Baroni, Marco., Bernardini, Silvia., Ferraresi, Adriano, Zanchetta, Eros: The wacky wide web: a collection of very large linguistically processed web-crawled corpora. Language Resources and Evaluation **43**(3), 209–226 (2009)
2. Sébastien Bubeck and Nicolò Cesa-Bianchi. Regret analysis of stochastic and nonstochastic multi-armed bandit problems. *CoRR*, abs/1204.5721, 2012
3. Sven Kosub. A note on the triangle inequality for the jaccard distance. *CoRR*, abs/1612.02696, 2016
4. Lloyd, Stuart P.: Least squares quantization in pcm. IEEE Transactions on Information Theory **28**, 129–137 (1982)
5. Edward Loper and Steven Bird. Nltk: The natural language toolkit. In *In Proceedings of the ACL Workshop on Effective Tools and Methodologies for Teaching Natural Language Processing and Computational Linguistics. Philadelphia: Association for Computational Linguistics*, 2002
6. Tomas Mikolov, Kai Chen, Greg Corrado, and Jeffrey Dean. Efficient estimation of word representations in vector space. *CoRR*, abs/1301.3781, 2013
7. Tomas Mikolov, Ilya Sutskever, Kai Chen, Greg S Corrado, and Jeff Dean. Distributed representations of words and phrases and their compositionality. In *Advances in neural information processing systems*, pages 3111–3119, 2013
8. Juan Ramos et al. Using tf-idf to determine word relevance in document queries
9. Radim Řehůřek and Petr Sojka. Software framework for topic modelling with large corpora. In *Proceedings of the LREC 2010 Workshop on New Challenges for NLP Frameworks*, pages 45–50, Valletta, Malta, May 2010. ELRA. http://is.muni.cz/publication/884893/en
10. Van der Maaten, Laurens, Hinton, Geoffrey: Visualizing data using t-SNE. Journal of Machine Learning Research **9**, 2579–2605 (2008)

A Note on Robust Estimation of the Extremal Index

M. Ivette Gomes, Miranda Cristina, and Manuela Souto de Miranda

Abstract Many examples in the most diverse fields of application show the need for statistical methods of analysis of extremes of dependent data. A crucial issue that appears when there is dependency is the reliable estimation of the *extremal index* (EI), a parameter related to the clustering of large events. The most popular EI-estimators, like the blocks' EI-estimators, are very sensitive to anomalous cluster sizes and exhibit a high bias. The need for robust versions of such EI-estimators is the main topic under discussion in this paper.

Keywords Dependent sequences · Monte-Carlo simulation · Robust semi-parametric estimation · Statistics of extremes

1 Introductory Notes

The *extremal index* (EI), denoted by θ, is a parameter of extreme events related to the clustering of exceedances of high thresholds. In the semi-parametric estimation of this parameter, we have to cope with problems similar to those that appear in the estimation of the primary parameter of extreme events, the *extreme value index* (EVI), here denoted by ξ, related to the tail heaviness: increasing bias, as the threshold decreases and a high variance for high thresholds. See [14] for a recent overview on the topic of univariate statistical *extreme value theory* (EVT).

We generally assume to be working with a strictly stationary sequence of *random variables* (RVs), $\{X_n\}_{n\geq 1}$, from a *cumulative distribution function* (CDF) F,

M. I. Gomes (✉)
CEAUL and DEIO, Faculdade de Ciências, Universidade de Lisboa, Lisboa, Portugal
e-mail: ivette.gomes@fc.ul.pt

M. Cristina
ISCA and CIDMA, Universidade de Aveiro, Aveiro, Portugal
e-mail: cristina.miranda@ua.pt

M. Souto de Miranda
CIDMA, Universidade de Aveiro, Aveiro, Portugal
e-mail: manuela.souto@ua.pt

M. La Rocca et al. (eds.), *Nonparametric Statistics*, Springer Proceedings in Mathematics & Statistics 339, https://doi.org/10.1007/978-3-030-57306-5_20

under general asymptotic and long-range dependence restrictions, like the long-range dependence condition **D** ([24]) and the local dependence condition **D**" ([23]). Let $\{X_{i:n}\}_{n\geq 1}$, $1 \leq i \leq n$, denote the associated sequences of ascending order statistics.

The stationary sequence $\{X_n\}_{n\geq 1}$ is said to have an EI, θ $(0 < \theta \leq 1)$, if, for all $\tau > 0$, we can find a sequence of levels $u_n = u_n(\tau)$ such that with $\{Y_n\}_{n\geq 1}$ the associated *independent, identically distributed* (IID) sequence (*i.e.*, an IID sequence from the same CDF, F),

$$\mathbb{P}(Y_{n:n} \leq u_n) = F^n(u_n) \underset{n\to\infty}{\longrightarrow} e^{-\tau} \quad \text{and} \quad \mathbb{P}(X_{n:n} \leq u_n) \underset{n\to\infty}{\longrightarrow} e^{-\theta\tau}.$$

There is thus a *'shrinkage'* of maximum values, but the limiting CDF of the maximum, $X_{n:n}$, linearly normalized, is still an *extreme value* (EV) CDF, with a functional form of the type

$$\mathrm{EV}_\xi(x) = \begin{cases} \exp\{-(1+\xi x)^{-1/\xi}\},\ 1+\xi x > 0, & \text{if } \xi \neq 0 \\ \exp(-\exp(-x)),\ x \in \mathbb{R}, & \text{if } \xi = 0. \end{cases}$$

Under the two mixing conditions **D** and **D**" (see [32]), the EI can also be defined as

$$\theta = \frac{1}{\text{limiting mean size of clusters}} = \lim_{n\to\infty} P(X_2 \leq u_n | X_1 > u_n),$$

with

$$u_n: \quad F(u_n) = 1 - \tau/n + o(1/n), \quad \text{as } n \to \infty, \ \text{with} \tau > 0, \text{fixed.} \qquad (1)$$

The very simple m-dependent (m-DEP) processes are used here for illustration. Those processes, with an EI given by $\theta = 1/m$, are based on IID Fréchet(ξ) RVs Y_i, $i \geq 1$, from a CDF $\Phi_\xi^{1/m}$, with $\Phi_\xi(x) = \exp\big(-x^{-1/\xi}\big)$, $x \geq 0$, the standard Fréchet CDF. They are then built upon the relation $X_i = \max_{1\leq j \leq i+m-1} Y_j, i \geq 1$. To enhance the clustering of high values (with an asymptotic mean size equal to m), we present Fig. 1.

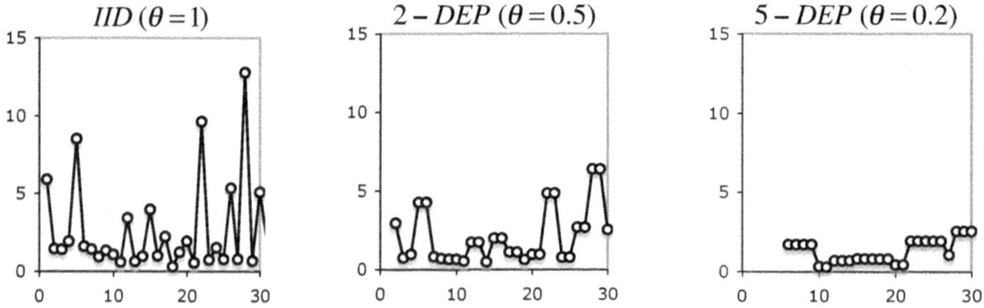

Fig. 1 Sample paths of an IID (*left*), 2-DEP (*center*) and 5-DEP (*right*) processes from the same underlying Fréchet($\Phi_{\xi=1}$), but with EIs, respectively, equal to 1, 0.5, and 0.2

Notice the richness of these processes regarding clustering of exceedances: there is a 'shrinkage' of maximum values, together with larger and larger 'clusters' of exceedances of high values, as θ decreases. Indeed, serial dependence leads to large values occurring close in time and forming clusters.

The scope of the article is the following: In Sect. 2, we deal with the EI-estimation, giving primordial emphasis to the blocks' estimator, since it is perhaps the most widely known EI-estimator. Robust versions of the blocks' EI-estimators are discussed in Sect. 3. Such an approach provides also a bias reduction, particularly in the presence of anomalous observations. A Monte-Carlo (MC) simulation study is described in Sect. 4, in the framework of m-DEP processes. The proposed robust version of the blocks' EI-estimators is compared with other EI-estimators in the literature. Finally, in Sect. 5, a few overall comments are put forward.

2 Extremal Index Estimation

The traditional estimators of θ differ mainly in the approaches and definitions used for identifying the clusters of exceedances (see, among others, [12, 19, 20, 22, 26, 31, 33]). The most relevant approaches in the literature are (a) the blocks estimator, where the sample is partitioned into b blocks and exceedances of high levels are identified and counted in each block that has at least one exceedance; (b) the runs estimator, for which the occurrence of a first exceedance determines the beginning of a cluster. Other estimators have been recommended in the literature, like an improved version of the block's suggested in [33], the intervals estimator in [12], the k-gaps estimator (see [34] or [35], among others), or the Nandagopalan estimator (see [23] and [15], also among others). Herein we focus on the blocks' estimator. The main goal is to improve its robustness within the family of m-DEP processes.

Consider a sequence of high levels $u_n = u_n(\tau)$ such that (1) holds, and a sequence r_n, such that $r_n \in \mathbb{N}$ and $r_n = o(n)$ as $n \to \infty$, *i.e.*, $r_n/n \to 0$, as $n \to \infty$. Let $b_n = \lfloor n/r_n \rfloor$, where $\lfloor x \rfloor$ denotes the integer part of x, and take the partition of a sample with size n into b_n adjacent disjoint blocks, all with size r_n. The number of times that $\{X_n\}$ exceeds a fixed level u_n is counted by the point process $N_n\{u_n(\tau)\}$. A cluster of exceedances is defined by the number of exceedances within a block in which there is at least one exceedance. Note that, according to this definition, the blocks of observations without any exceedance are ignored.

In limit, it was proved in [21] that under a broad Δ condition, the number of exceedances N_n converges to a compound Poisson process with multiplicities equal to the dimension of the clusters. Moreover, clusters' size distribution is given by

$$\pi_n(j) = \mathbb{P}\left[\sum_{i=1}^{r_n} \mathbb{I}_{X_i>u_n} = j \,|\, \sum_{i=1}^{r_n} \mathbb{I}_{X_i>u_n} > 0\right], \quad j = 1, 2, \ldots$$

where $\mathbb{I}_A$ stands for the indicator function of A. If the limit exists when $n \to \infty$, the distribution of the clusters' size associated with the compound Poisson process is $\pi = \lim_{n\to\infty} \pi_n$. In general, π is not known and can be diverse. Nevertheless, whenever it exists, and under the aforementioned dependence condition, the limiting mean coincides with the inverse of the EI, *i.e.*, the EI can be expressed as $\theta^{-1} = \lim_{n\to\infty} \sum_{j\geq 1} j\pi_n(j)$. Finally, the blocks' estimator is defined by the inverse of the mean number of exceedances *per* cluster, *i.e.*, by

$$\hat{\theta}_B = (N_n/Z_n)^{-1} = Z_n/N_n, \tag{2}$$

where Z_n denotes the number of blocks that contain at least one exceedance, *i.e.*, the number of clusters by the definition of cluster. In the present paper, we consider the blocks' estimator (B) in a different but equivalent form of the one presented in [31]. The estimator is defined by

$$\widehat{\theta}_{\mathrm{W}} = -\frac{\log\left(\frac{1}{k}\sum_{i=1}^{k} \mathbb{I}_{M_{(i-1)r,ir} \leq u_n}\right)}{\frac{1}{k}\sum_{i=1}^{rk} \mathbb{I}_{X_i > u_n}}, \tag{3}$$

with $M_{s,r} = \max_{s<i\leq r} X_i$, for $0 \leq s < r$. The estimator $\widehat{\theta}_{\mathrm{W}}$ is a consistent and asymptotically normal EI-estimator, but with a second-order asymptotic behavior better than the one of the EI-estimator in (2).

The blocks' estimator has a simple interpretation and important asymptotic properties, but it is based on the mean and it is very sensitive to an anomalous cluster's size. The occurrence of just one atypical cluster size is enough to produce a disastrous estimate for θ. Thus, our main goal is to investigate procedures that can improve the robustness of the blocks' estimator.

3 Introducing Robustness

Dealing with robustness in statistical EVT seems to be an apparent contradiction. Indeed, the main robust proposals were conceived for down-weighting extreme observations and in EV–analysis, those extreme observations are the most interesting ones. In fact the conjugation is unusual and challenging but has been successfully exploited in papers like [10] or [37] or [5], among others. Most of those papers are devoted to the estimation of Pareto-type parameters and the robust estimation of a positive EVI. As far as we know, the EI-estimation has not been treated from a robust point of view.

3.1 Parametric Distribution of the Limiting Cluster Size

Since in limit the inverse of the EI represents the cluster mean dimension, we investigate the more appropriate robust procedure for estimating that mean. At a first glance this seems to be trivial, but it deserves particular attention. A robust estimator must exhibit a good performance under the assumed statistical model in spite of not being optimum and, simultaneously, it must produce reliable estimates if real data show small departures from the assumptions. See [18] and [16], where the main contributions for the systematization of robust statistics can be found.

Hampel's approach considers the distribution of any estimator under $\mathcal{F}(\Xi)$, the family of all possible probability distributions defined in the sample space Ξ and for which the estimator is defined. It is within this framework that some fundamental tools of robustness were defined: the influence function (IF) of an estimator, the definition of a robust estimator when it has a bounded IF, or the development of robust M-estimators (see, e.g., [16]), which are proportional to their bounded IF. We have adopted the robust approach for dealing with the limiting distribution π of the cluster dimension, *i.e.,* we have considered π as $\pi(\theta)$ in the broad family $\mathcal{F}$ of distributions. According to [21], $\pi(\theta)$ is unknown and can be diverse. Some authors assumed specific distributions in their work, such as the Poisson model. Herein we assume such a neighborhood approach, considering that the true distribution of the limiting cluster size belongs to a neighborhood of the Poisson family. Such an assumption was chosen specifically for the B-estimator, since it counts the number of exceedances *per* cluster, and the Poisson process is perhaps the most used in modeling counting processes.

3.2 Robust Estimators

In general, we expect to have small or very small cluster sizes. Their mean is thus strongly affected by the occurrence of clusters with atypical dimensions. Robust estimators can control the effect of anomalous data and they have a good performance in a neighborhood of the assumed model. Nevertheless, the most popular and efficient robust estimators were conceived and are computationally implemented for dealing with considerable sample sizes, weighting tails usually with symmetric models, namely, the Normal model. Thus, the selection of a robust mean estimator deserves some concerns, particularly because robust estimators, in general, are not explicitly defined and so, computational components play a decisive role in the results.

There is a great collection of robust location estimators whose properties are well studied. Most of them are included in the broad family of M-estimators, which generalizes the class of *maximum likelihood* (ML)-estimators. Currently, the most used robust estimators for location are perhaps the MM-estimators, a subfamily of the M-estimators that combines efficiency with a high breakdown point (another important measure of robustness), but their computational setup is prepared for the Normal distribution, and they are not adequate for the Poisson model. Dealing with *generalized linear models* (GLM), there are robust estimators developed for the logistic regression, for the Poisson regression, for the Normal and the Gamma error terms. Those estimators are implemented in statistical packages for the R environment (see [30]), like the popular *robustbase* or the *robmixglm* (see [25], and [3] for documentation). They have been tested by researchers in computational statistics and they are used by a wide community. We intend to use known and verified computational procedures, aiming to simplify the procedures for data analysts. The proposals herein presented keep thus that main goal in mind.

Due to the great number of blocks without exceedances, the probability of occurrence of the zero value could be poorly modeled by the Poisson model. The assumption of a Poisson distribution with parameter λ for the number of exceedances *per* block, N_0, including those without exceedances, would imply that $\lambda > 1$ and $\mathbb{P}[N_0 = 0] < \exp(-1) \approx 0.37$, which seems unrealistic when dealing with exceedances of very high thresholds. So, robust estimators prepared for the Poisson model should not be directly applied and it is necessary to consider robust estimators that can deal with a great number of zero observations. More precisely, the mean cluster size will be estimated assuming a GLM framework and two different models in the neighborhoods of the Poisson family, namely, a hurdle Poisson and a mixture of GLMs with Poisson error terms.

First consider the robust hurdle model. The model was suggested by Heritier ([17]) as a possible way of dealing with an excess of zeros in count data. It consists of two functionally independent processes: the first is a binary process generating the zero values, while the second is conditional on the first one, according to a zero truncated Poisson distribution. The model was studied in detail in [27]; also [4] presented the link, variance, and deviance functions for the zero truncated Poisson. The Poisson hurdle model is defined by

$$\mathbb{P}[Y_i = y_i] = \begin{cases} 1 - p(\boldsymbol{x}_i), & y_i = 0, \\ p(\boldsymbol{x}_i)\frac{\exp[-\lambda(\boldsymbol{u}_i)][\lambda(\boldsymbol{u}_i)]^{y_i}}{y_i![1-\exp[\lambda(\boldsymbol{u}_i)]]}, & y_i = 1, 2, \ldots, \end{cases} \tag{4}$$

where y_i denotes the counts, $\boldsymbol{x}_i \in \mathbb{R}^p$, $\boldsymbol{u}_i \in \mathbb{R}^{\tilde{p}}$. In the framework of the GLM, the first part of the model assumes a logistic model for $p(\boldsymbol{x}_i)$, with

$$logit[p(\boldsymbol{x}_i)] = \log\left[\frac{p(\boldsymbol{x}_i)}{1 - p(\boldsymbol{x}_i)}\right] = \boldsymbol{x}_i^T\boldsymbol{\alpha}, \quad (\boldsymbol{\alpha} \in \mathbb{R}^p),$$

and the second part considers a log-linear model for $\lambda(\boldsymbol{u}_i)$ conditionally on $p(\boldsymbol{x}_i)$, with $\log[\lambda(\boldsymbol{u}_i)] = \boldsymbol{u}_i^T\boldsymbol{\gamma}$, $(\boldsymbol{\gamma} \in \mathbb{R}^{\tilde{p}})$. The log-likelihood of the hurdle model can be written in the form $l(\boldsymbol{\alpha}, \boldsymbol{\gamma}; \mathrm{y}) = l(\boldsymbol{\alpha}; \mathrm{y}) + l(\boldsymbol{\gamma}; \mathrm{y})$, establishing the orthogonality of the parameters $\boldsymbol{\alpha}$ and $\boldsymbol{\gamma}$. This allows the independent estimation of the two parts of the model. Note also that for obtaining the mean cluster size estimate only the second part is necessary. Robust estimators for the coefficients of the hurdle model are investigated in [9]. For the logistic component of the model robust counterparts are available, like the methods proposed in [7] and [8] or those suggested by [6]. They are implemented in the *robustbase* package. In [9] the authors generalize the work in [7] to the truncated Poisson distribution: for a GLM with covariates $\boldsymbol{x}_i$ and unknown parameter $\boldsymbol{\beta}$, they use a robust M-estimator which is the solution of

$$\sum_{i=1}^{n} \psi(y_i, \boldsymbol{\mu}_i) = \sum_{i=1}^{n} \left[\psi_c(r_i)\omega(\boldsymbol{x}_i) \frac{1}{\sqrt{\nu_{\mu_i}}} \boldsymbol{\mu}_i^T - a(\boldsymbol{\beta}) \right] = \mathbf{0}, \tag{5}$$

where ψ_c denotes the Huber function, which will control the effect of anomalous residuals, being c the tuning constant that will regulate the degree of robustness/efficiency of the estimator. Moreover, $r_i = (y_i - \mu_i)/\sqrt{\nu_{\mu_i}}$ are the Pearson residuals, with $\nu_{\mu_i} = \mathbb{V}[Y_i|\boldsymbol{x}_i]$, $\omega(\boldsymbol{x}_i)$ are weights that will control anomalous covariates observations, $\mu_i = \mathbb{E}[Y_i|\boldsymbol{x}_i]$ and $a(\boldsymbol{\beta})$ is a correction term that ensures Fisher's consistency. Robust estimators defined by ψ-functions, as in the intermediate term of (5), are called Mallows-type estimators. When $\omega(\boldsymbol{x}_i) = 1$ for all i, the estimator becomes the Huber estimator. That is, the adequate case in the present study, since for the mean cluster size estimation only the constant term estimate is taken. In [17], it is recommended a c-value between $c = 1.3$ and $c = 1.8$, and we have used $c = 1.6$. The properties of the robust estimator result from general M-estimation theory, namely, their influence function is $\mathrm{IF}(y; \psi, \pi) = \mathcal{M}(\psi, \pi)\psi(y, \mu)$, where

$$\mathcal{M}(\psi, \pi) = -\mathbb{E}\left[\frac{\partial}{\partial \boldsymbol{\beta} \psi(y, \mu)} \right]$$

and their asymptotic covariance matrix is

$$\mathcal{M}(\psi, \pi)^{-1} Q(\psi, \pi) \mathcal{M}(\psi, \pi)^{-T},$$

with $Q(\psi, \pi) = \mathbb{E}[\psi(y, \mu)\psi(y, \mu)^T]$. Cantoni and Zedini (see [9]) concretized the form of matrices $\mathcal{M}(\psi, \pi)$ and $Q(\psi, \pi)$ for the truncated Poisson and they deducted robust estimators from the asymptotic covariance matrix of the corresponding Mallows quasi-likelihood estimators.

The truncated Poisson component of the hurdle model can be alternatively estimated with an MT-estimator (see [36]). MT-estimators are another subfamily of M-estimators that consider a variance stabilizing transformation in the response variable and a redescending ψ function in (5), (instead of ψ_c). The aforementioned general properties also follow from M-estimation theory. The computational process for obtaining MT-estimates in simulation studies has been more complex and more time consuming than for computing robust Mallows-type estimates. The obtained results were very similar, and so we focus only on the former process.

Consider now the second approach referred to above, which assumes a mixture model of GLMs by considering potential outliers coming from an overdispersed GLM as in [1], namely,

$$g(\mu_i|c_i, \lambda_i) = \begin{cases} \boldsymbol{x}_i^T \boldsymbol{\beta}, & c_i = 1, \\ \boldsymbol{x}_i^T \boldsymbol{\beta} + \lambda_i, & c_i = 2, \end{cases} \tag{6}$$

where $c_i = 1$ stands for the standard model belonging to the exponential family, and $c_i = 2$ groups potential outliers considering a random effect $\lambda_i \sim N(0, \tau^2)$ and assuming mixture proportions p_1 and p_2 ($p_1 + p_2 = 1$) fixed over $\boldsymbol{x}_i$. The estimates are obtained by fitting the model with EM (*expectation-maximization*) optimization methods, particularly, the GEM algorithm and the quasi-Newton methods. The inclusion of the λ random effect can accommodate discordant observations, allowing good estimates for the parameters in the standard model component. From this point of view it is a robust estimation procedure, in spite of not being defined through a particular robust estimators family. The methodological support is explained in [2] and computations were performed using the *robmixglm* package and its estimating function with the same name. Once again, results were very similar to those achieved with Cantoni and Zeidini proposal in [9]. In the following, we have decided to present only the results achieved by the latter suggestion. Recall that the above computational procedures are related to the estimation of the constant term of a GLM with a link function $g(\mu) = \log(\mu)$. The obtained constant term estimate, $\tilde{\lambda}$, needs thus to be transformed to $\hat{\lambda} = \exp(\tilde{\lambda})$, for obtaining the mean cluster size estimate. So, the robust version of the EI blocks' estimator is defined by

$$\hat{\theta}_{\text{Rob}} = 1/\hat{\lambda}. \tag{7}$$

The main steps of the computational procedures through the *robustbase* package are summarized in the following algorithm.

Algorithm

Step 1. Use the function *glmrob* in the R-package *robustbase*, inserting the observed clusters' size as observations of the response variable (N_n^*). Consider the linear predictor as a constant term of a regression without any other regressor.

Step 2. Select the following options in the *glmrob* function: family="Poisson", method="Mqle", weights.on.x= "none", control=glmrobMqe.control(tcc= c), with c a value between 1.3 and 1.8, to get a robust estimate $\tilde{\lambda}^*$ of the constant term.

Step 3. Transform $\tilde{\lambda}^*$ by the inverse of the link function, obtaining the mean cluster size estimate $\hat{\lambda} = \exp \tilde{\lambda}^*$.

Step 4. Compute the EI–estimate $\hat{\theta}_{\text{Rob}}$, already defined in (7).

4 A Simulation Study

In the first part of the simulation study, we aim to compare $\hat{\theta}_{\text{Rob}}$ (computed by the hurdle model and the aforementioned algorithm for $c = 1.6$) with other EI-estimators, namely, the traditional blocks estimator $\hat{\theta}_{\text{B}}$ in (2), its improved version $\hat{\theta}_{\text{W}}$ in (3), the interval estimator $\hat{\theta}_{\text{Int}}$ in [12], the runs estimator $\hat{\theta}_{\text{Runs}}$ in [23] (see also [15]) and the k-gaps estimator $\hat{\theta}_{\text{Gap}}$ in [34]. A comparison of different EI-estimators, done through MC simulations, can be found in [11]. Next, we introduced contamination in the samples for analyzing the effect on the estimates and the advantages of the robust proposal.

The performance of all methods was evaluated regarding the simulated values in terms of the estimated expected mean value (or equivalently, bias) and in terms of the estimated mean squared error (MSE). All computations were developed with R software. For the alternatives $\hat{\theta}_{\text{Int}}$ and $\hat{\theta}_{\text{Runs}}$ the package *extRemes* ([13]) was used and for the $\hat{\theta}_{\text{Gap}}$ estimator we used the *revdbayes* package (see [28] and a last version of package documentation in [29]). Notice that in previous studies we have compared the three robust procedures cited above in Sect. 3, namely, the robust approach of the hurdle model in (4), with Huber estimators, the robust approach of the hurdle model, again in (4), but with MT-estimators and the mixture of GLMs in (6), and the results were similar, either with or without contamination. Thus, the results presented in this section for $\hat{\theta}_{\text{Rob}}$ were computed through the algorithm written above, at the end of Sect. 3, with $c = 1.6$.

4.1 Simulation Study Design

Observations were simulated from a standard Fréchet model with CDF $\Phi_\xi(x) = \exp\left(-x^{-1/\xi}\right), x > 0,\ \xi > 0$. In the present study we consider m-DEP sequences. This type of structures verifies the limit conditions imposed by the theory and the EI can be straightforwardly computed. Originally, we have assumed different EI-values, namely, $\theta = 0.5, 0.2$, and 0.1. Those θ values, respectively, represent the expectations $\lambda = 2, 5$, and 10, in the Poisson model, and $\theta = 1/m$, with $m = 2, 5$, and 10 in the m-DEP structures. The simulation is illustrated for $\theta = 0.2$, a sample size $n = 2000$ and for 500 replications. We have used blocks determined by three different partitions, associated with a number of blocks $b = 100, 150, 200$. The performance of the estimators for each sample was evaluated considering 30 thresholds corresponding to upper sample quantiles from 0.80 up to 0.99.

The advantages of the robust version were evaluated by comparing the results obtained under the previous conditions with those obtained after introducing contamination in the samples. We have contaminated the same samples used before in a deterministic way. To guarantee an anomalous cluster size although not inducing changes in the extremal index value, it has been necessary to generate a number of sequencial exceedences so it produces an outlier in the N_0 sequence. After ordering observations in each sample, $(x_1, \ldots, x_n)$, the central values around the median were thus replaced, in a tiny percentage (1.2%), by the corresponding value of the order statistic $x_{n:n}$, assuring in this way an outlier over cluster dimension in every sample, independently of the exceedance value that determined the atypical cluster size. With this type of contamination one can observe how the estimates can be affected by the presence of just one discordant value.

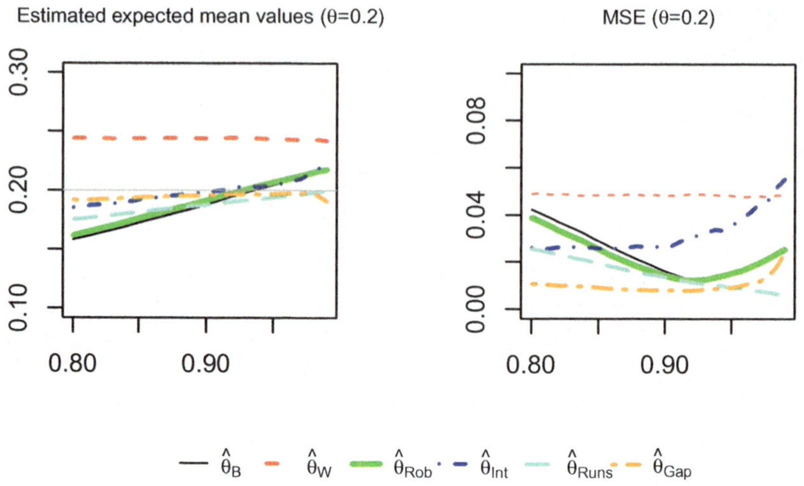

Fig. 2 Estimated mean values (*left*) and MSEs (*right*) of $\hat{\theta}_B$, $\hat{\theta}_W$, $\hat{\theta}_{Rob}$, $\hat{\theta}_{\text{Int}}$, $\hat{\theta}_{\text{Runs}}$, and $\hat{\theta}_{\text{Gap}}$ in a 5-DEP structure ($\theta = 0.2$) for the sample **without contamination**

4.1.1 Illustration

We next provide an illustration of the performed studies with a 5-DEP structure. We compared the EI-estimators in terms of estimated mean cluster size (for bias) and estimated MSE, considering $b = 100$ disjoint blocks. Figure 2 shows the results using samples without contamination. The robust version of the blocks' estimator $\hat{\theta}_{\text{Rob}}$ produced good results, very close to the original version of the estimator. Comparing with other estimators globally, considering both bias and MSE, the k-gaps estimator had the best performance, followed by the runs estimator.

The scenery is very different when contamination is introduced (see Fig. 3). The k-gaps and the runs estimator lose the advantages since their bias increases for high quantiles, as well as their MSEs. The three versions of the block' estimator performed better, particularly, the robust blocks' estimator which globally had the best performance among all, observing simultaneously bias and variability.

5 Final Comments

- A robust version of the blocks' EI-estimator has been presented, considering the limit distribution of the cluster dimension in the neighborhood of a Poisson model. Such an approach allows the truncation associated with the definition of clusters of exceedances. Then, the limiting mean cluster size can be estimated through the constant term of a GLM, namely, using a truncated Poisson.

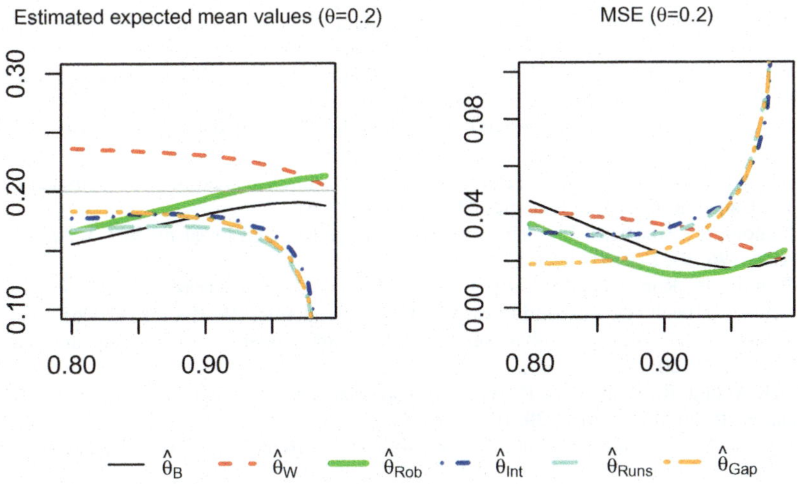

Fig. 3 Estimated mean values (*left*) and MSEs (*right*) of $\hat{\theta}_{\mathrm{B}}$, $\hat{\theta}_{\mathrm{W}}$, $\hat{\theta}_{\mathrm{Rob}}$, $\hat{\theta}_{\mathrm{Int}}$, $\hat{\theta}_{\mathrm{Runs}}$, and $\hat{\theta}_{\mathrm{Gap}}$ in a 5-DEP structure ($\theta = 0.2$) for the **contaminated sample**

- We have paid attention to robust methods whose computational procedures are available and tested, in order to facilitate their potential use by data analysts. In the present comparative study, the robustness was integrated in the process by assuming a hurdle model and using Huber M-estimators. Other robust estimators could have been considered, which justifies a future and deeper investigation.
- Compared with other EI-estimators and without contaminated samples, the robust proposal performance was similar to the traditional blocks' estimator, and the *k*-gaps estimator produced the best results. With contaminated samples and under the simulated conditions, the robust version had the best performance among all the considered estimators, in what respects both bias and variability.
- Further work is required in investigating robust procedures for other models with known theoretical EI.

Acknowledgements Research partially supported by National Funds through Fundação para a Ciência e a Tecnologia (FCT), within projects UID/MAT/00006/2019 (CEA/UL) and UID/MAT/04106/2019 (CIDMA).

References

1. Aitkin, M.: A general maximum likelihood analysis of overdispersion in generalized linear models. Stat. Comput. **6**, 251–262 (1996)
2. Beath, K.: A mixture-based approach to robust analysis of generalised linear models. Journal of Applied Statist. **45**, 2256–2268 (2017)

3. Beath, K.: robmixglm—Robust Generalized Linear Models (GLM) using Mixtures. R package version 1.0-2. https://CRAN.R-project.org/package=robmixglm (2018)
4. Barry, S., Welsh, A.: Generalized additive modelling and zero inflated count data. Ecological Modelling **157**, 179–188 (2002)
5. Beran, J., Schell, D.: On robust tail index estimation. Comput. Statist. & Data Anal. **56**, 3430–3443 (2012)
6. Bianco, A., Ben, M., Yohai, V.: Robust estimation for linear regression with asymmetric errors. The Canadian Journal of Statistics **33**, 511–528 (2005)
7. Cantoni, E., Ronchetti, E.: Robust inference for generalized linear models. J. Amer. Statist. Assoc. **96**, 1022–1030 (2001)
8. Cantoni, E., Ronchetti, E.: A robust approach for skewed and heavy-tailed outcomes in the analysis of health care expenditures. J. of Health Economics **25**, 198–213 (2006)
9. Cantoni, E., Zedini, A.: A robust version of the hurdle model. J. Statist. Plann. and Infer. **141**, 1214–1223 (2011)
10. Dell'Aquila, R., Embrechts, P.: Extremes and robustness: a contradiction? Financial Markets and Portfolio Management **20**, 103–118 (2006)
11. Ferreira, M.: Heuristic tools for the estimation of the extremal index: a comparison of methods. Revstat—Statist. J. **16**, 115-13 (2018)
12. Ferro, C.A.T., Segers, J.: Inference for clusters of extreme values. J. Royal Statist. Soc., Series B **65**, 545-556 (2003)
13. Gilleland, E., Katz, R.: extRemes 2.0: An extreme value analysis package in R. J. of Statistical Software **72**, 1-39 (2016)
14. Gomes, M.I., Guillou, A.: Extreme value theory and statistics of univariate extremes: a review. International Statistical Review **83**(2), 263–292 (2015)
15. Gomes, M.I., Hall, A., Miranda, C.: Subsampling techniques and the Jackknife methodology in the estimation of the extremal index. Comput. Statist. & Data Anal. **52**, 2022–2041 (2008)
16. Hampel, F.R., Ronchetti, E.M., Rousseeuw, P.J., Stahel, W.A.: Robust Statistics: The Approach based on Influence Functions. John Wiley, New York (1986)
17. Heritier, S., Cantoni, E., Copt, S., Victoria-Feser, M.-P.: Robust Methods in Biostatistics. John Wiley & Sons, United Kingdom (2009)
18. Huber, P.: Robust Statistics. John Wiley, New York (1981)
19. Hsing, T.: Estimating the parameters of rare events. Stoch. Proc. and Appl. **37**, 117–139 (1991)
20. Hsing, T.: Extremal index estimation for a weakly dependent stationary sequence. The Ann. of Statist. **21**, 2043–2071 (1993)
21. Hsing, T., Hüsler, J., Leadbetter, M.R.: On the exceedance point process for a stationary sequence. Probab. Th. and Rel. Fields **78**, 97–112 (1988)
22. Laurini, F., Tawn, J.: New estimators for the extremal index and other cluster characteristics. Extremes **6**, 189–211 (2003)
23. Leadbetter, M.R., Nandagopalan, S.: On exceedance point processes for stationary sequences under mild oscillation restrictions. In Hüsler, J. and R.-D. Reiss (eds.), Extreme Value Theory, Springer-Verlag, pp. 69–80 (1989)
24. Leadbetter, M.R., Lindgren, G., Rootzén, H.: Extremes and Related Properties of Random Sequences and Series. Springer Verlag, New York (1983)
25. Maechler, M., Rousseeuw, P., Croux, C., Todorov, V., Ruckstuhl, A., Salibian-Barrera, M., Verbeke, T., Koller, M., Conceicao, E., Palma, M.: robustbase—Basic Robust Statistics R package version 0.93-5. http://CRAN.R-project.org/package=robustbase (2019)
26. Markovich, N.: Modeling clusters of extreme values. Extremes **17**, 97–125 (2014)
27. Min, Y., Agresti, A.: Modeling nonnegative data with clumping at zero: a survey. J. of the Iranian Statistical Society **1**, 7–33 (2002)
28. Northrop, P.: An efficient semiparametric maxima estimator of the extremal index. Extremes **18**, 585–603 (2015)
29. Northrop, P.: revdbayes: Ratio-of-Uniforms Sampling for Bayesian Extreme Value Analysis version 1.3.4. https://cran.r-project.org/web/packages/revdbayes/revdbayes.pdf (2019)

30. R Core Team. R: A Language and Environment for Statistical Computing. R Foundation for Statistical Computing. Vienna, Austria (2014)
31. Robert, C.Y.: Inference for the limiting cluster size distribution of extreme values. The Ann. of Statist. **37**, 271–310 (2009)
32. Robinson, M., Tawn, J.: Extremal analysis of processes sampled at different frequencies. J. R. Statist. Soc. B. **62**, 117–135 (2000)
33. Smith, R.L., Weissman, I.: Estimating the extremal index. J. Royal Statist. Soc., Series B **56**, 41–55 (1994)
34. Süveges, M.: Likelihood estimation of the extremal index. Extremes **10**, 41–55 (2007)
35. Süveges, M., Davison, A.: Model misspecification in peaks over threshold analysis. The Annals of Applied Statistics **4**, 203–221 (2010)
36. Valdora, M., Yohai, V.: Robust estimators for generalized linear models. J. Statist. Plann. and Infer. **146**, 31–48 (2014)
37. Vandewalle, B., Beirlant, J., Christmann, A., Hubert, M.: A robust estimator for the tail index of Pareto-type distributions. Comput. Statist. & Data Anal. **51**, 6252–6268 (2007)

Multivariate Permutation Tests for Ordered Categorical Data

Huiting Huang, Fortunato Pesarin, Rosa Arboretti, and Riccardo Ceccato

Abstract The main goal of this article is to compare whether different groups with ordinal responses on the same measurement scale satisfy stochastic dominance and monotonic stochastic ordering. In the literature, the majority of inferential approaches to settle the univariate case are proposed within the likelihood framework. These solutions have very nice characterizations under their stringent assumptions. However, when the set of alternatives lie in a positive orthant with more than four dimensions, it is quite difficult to achieve proper inferences. Further, it is known that testing for stochastic dominance in multivariate cases by likelihood approach is much more difficult than the univariate case. This paper intends to discuss the problem within the conditionality principle of inference through the permutation testing approach and the nonparametric combination (NPC) of dependent permutation tests. The NPC approach based on permutation theory is generally appropriate to suitably find exact good solutions to this kind of problems. Moreover, some solutions for a typical medical example are provided.

Keywords Multivariate permutation testing · Stochastic dominance · Conditional inference · Nonparametric combination

---

H. Huang (✉) · F. Pesarin
Department of Statistical Sciences, University of Padova, Padova, Italy
e-mail: huiting.huang@studenti.unipd.it

F. Pesarin
e-mail: pesarin@stat.unipd.it

R. Arboretti
Department of Civil, Environmental and Architrctural Engineering,
University of Padova, Padova, Italy
e-mail: rosa.arboretti@unipd.it

R. Ceccato
Department of Management and Engineering, University of Padova, Padova, Italy
e-mail: ceccato@gest.unipd.it

M. La Rocca et al. (eds.), *Nonparametric Statistics*, Springer Proceedings
in Mathematics & Statistics 339, https://doi.org/10.1007/978-3-030-57306-5_21

1 Introduction

Ordered categorical data are frequently encountered in many research and decision-making fields. For instance, records from patients under different treatments in clinical experiments, feedbacks of questionnaire in social sciences, data on some questions about feeling, thought or opinion collected in a natural way in psychology, quality examination of products in marketing and technology, etc. Taking clinical trails as a guide, we intend to find if results of cure plans satisfy stochastic ordering and to find the best treatment among cure plans. Problems of comparing whether different groups with ordinal responses on the same measurement scale satisfy stochastic dominance ($C = 2$) is our principal interest. Thereby, we intend to provide tests of hypotheses with ordinal responses especially by testing for stochastic dominance since that for stochastic ordering, ($C > 2$), is obtained as a combination of $C - 1$ dominance partial tests. This is known to be a rather difficult problem. Many approaches are proposed in the literature to settle it within likelihood frameworks. [14] proposed an iterative procedure with censored data which is based on a pair-wise algorithm to find the asymptotic MLE's of Kaplan–Meier form. [25] introduced numerical approximation of MLE's of two V-dimensional distributions under stochastic ordering. [27] derived the null asymptotic distribution for the likelihood ratio test statistic for some testing procedures. Testing procedures based on maximum likelihood estimates of odds ratios have been considered by [2, 3] and others. Moreover, Kateri and Agresti (2013), in place of traditional frequentist methods, applied a Bayesian approach to test if the structure of an association between the response variable and the explanatory variable in two samples is ordinal. When available, likelihood-based solutions within their stringent assumptions are provided with known inferential properties. In general, however, it is quite difficult to obtain proper testing inference, especially for the multivariate case. Multivariate case is much more difficult to be analyzed within likelihood frameworks than the univariate one. In such a setting, the number of underlying nuisance parameters and/or that of observed variables can often be much larger than sample sizes. So, unless clearly justified assumptions allowing for considerable reduction of underlying complexity, the most intriguing of which is when one pseudo-parameter is expressed as a function of many underlying nuisance parameters, no correct general testing solution is possible within that approach.

Our approach to this kind of problems is within the conditionality principle of inference [13], where the conditioning is with respect to a set of sufficient statistics in the null hypothesis as usually the pooled observed data is. That is, by using the permutation testing theory and the nonparametric combination (NPC) of dependent permutation tests [4–8, 18–23]. When the underlying population distribution is unknown, nonparametric permutation methods might become a necessity. This is especially true when the number of categories and/or that of underlying nuisance parameters are not very small. [16] studied the testing for marginal inhomogeneity and direction-independent marginal order under the global permutation tests. [15] utilized χ^2-P statistic with small sample size under the permutation approach. The NPC approach is a general methodology for multivariate problems, especially, for stochastic dominance and stochastic ordering. The NPC testing solution performs [24] Union-Intersection (UI) approach when an equivalent set of sub-problems is properly carried out.

In principle, the exact calculations of required testing distributions are obtained through complete enumeration of all data permutations. This, however, becomes impossible in practice when the cardinality of permutation spaces are large. To this end, a conditional Monte Carlo procedure was suggested to practically obtain their estimations, at any desired degree of accuracy ([18, 20, 22]). Main NPC routines are achieved in MATLAB, R, Python, StatXact, SAS, etc.

The rest of the paper is organized as follows. Section 2 introduces a typical real example. Section 3 discusses the two-sample basic problem under unidimensional and multidimensional cases. Section 4 studies approaches for stochastic ordering restriction in C-sample designs. Solutions to the example are in Sect. 5. Some concluding remarks are in Sect. 6.

2 A Typical Medical Example

Let us consider the example in Table 1 from Chuang-Stein and Agresti (1997), also reported by [1, 2, 12, 26]. It regards a unidimensional survey on subarachnoid hemorrhage measured by Glasgow outcome scale, where 210 patients received a Placebo, 190 received a Low dose, 207 a Medium dose, and 195 a High dose. Response data, related to the extent of trauma, measured on the same ordinal scale, are classified according to $C = 4$ doses of a treatment, $\{Placebo, Low, Medium, High\}$, with outcome classified in $K = 5$ ordered categories $\{Death, Vegetative\ state, Major\ disability, Minor\ disability, Good\ recovery\}$.

Based on our intuition, but also in accordance with quoted authors, patients taking Placebo are expected to achieve lower treatment effect than those taking Low dose, patients taking Low dose have lower effects than those with Medium dose, and so forth. Therefore, it is expected that patients exhibit monotonically non-decreasing responses X as the dose increases. Thus, it is required to test whether there is a monotonic stochastic ordering on related response data. Formally, the hypotheses to consider are $H_0 : X_P \overset{d}{=} X_L \overset{d}{=} X_M \overset{d}{=} X_H$ against $H_1 : X_P \overset{d}{\preceq} X_L \overset{d}{\preceq} X_M \overset{d}{\preceq} X_H$ with

Table 1 Dose and Extent of trauma due to subarachnoid hemorrhage

Treatment	Death	Veget	Major	Minor	Recov	Total
Placebo	59	25	46	48	32	210
Low	48	21	44	47	30	190
Medium	44	14	54	64	31	207
High	43	4	49	58	41	195
Total	194	64	193	217	134	802

at least one strict inequality. If responses were quantitative, this problem is also termed of isotonic regression. Defining the cumulative distribution function for responses X at ordered categories $c_1 \prec \ldots \prec c_K$ as $F_X(c_k) = \Pr\{X \preceq c_k\}$, namely, the hypotheses are equivalently expressed as $H_0 : \{F_{X_P} = F_{X_M} = F_{X_L} = F_{X_H}\}$ against $H_1 : \{F_{X_P} \geq F_{X_M} \geq F_{X_L} \geq F_{X_H}\}$, with at least one strict inequality.

With clear meaning of the symbols, the rationale for this formulation resides in that if, according to increasing doses, non-decreasing treatment effects δ occur at latent variables Y, i.e., $\delta_h \leq \delta_j, 1 \leq h < j \leq C$, then latent responses should behave as $Y_h = (Y + \delta_h) \overset{d}{\leq} Y_j = (Y + \delta_j)$.

The related testing problem has a rather difficult solution within the likelihood-ratio theory, which with categorical data in addition presents quite a serious difficulty: even for moderate number of cells it is recognized to be not unique ([10–12, 26, 27]; etc.). Moreover, to get a solution, important supplementary options, difficult to justify in terms of the real problem under study, are required. This difficulty mostly consists in that the set of alternatives is restricted to lie in the $(C - 1) \times (K - 1)$-*Dimensional positive orthant* where the likelihood cannot be maximized under H_0 by ordinary methods of maximization.

Our solution does firstly consider the setting of two treatments, and then, according to [24] UI and Jonckheere–Terpstra's approaches, by a breakdown of the hypotheses into $C - 1$ pairs of sub-hypotheses. Later, all resulting dependent partial tests are combined by a NPC method.

3 The Two-Sample Basic Problem

Let us firstly consider the two-sample basic case, where data are in a $2 \times K$ table and the specific hypotheses are expressed as $H_0 : X_1 \overset{d}{=} X_2 \equiv \{F_1(c_k) = F_2(c_k), k = 1, \ldots K\}$ against $H_1 : X_1 \overset{d}{\prec} X_2 \equiv \{F_1(c_k) \geq F_2(c_k), k = 1, \ldots K\}$ with at least one strict inequality. The related testing problem can be equivalently set as $H_0 : \bigcap_{k=1}^{K-1} [F_1(c_k) = F_2(c_k)]$ against the set of *restricted alternatives* $H_1 : \bigcup_{k=1}^{K-1} [F_1(c_k) > F_2(c_k)]$.

It is worth noting that i) according to [24] the problem is equivalently broken-down into $K - 1$ *one-sided sub-problems*; ii) H_1 defines a *multi-one-sided* set of alternatives; iii) since under both H_0 and H_1 it is $F_1(c_K) = F_2(c_K) = 1$, category c_K is not considered; iv) the global solution requires the joint comparison of $K - 1$ random relative frequencies: $\hat{F}_1(c_k) - \hat{F}_2(c_k), k = 1, \ldots, K - 1$.

Since the number of unknown nuisance parameters to take care in any $2 \times K$ testing process is $2 \times K - 1$ and the likelihood is to be maximized in the $(K - 1)$-dimensional positive orthant, indeed a very difficult task especially when $K > 4$, our approach is to stay within the conditional principle of inference.

The conditioning should be on a set of sufficient statistics in the null hypothesis for the unknown underlying common distribution F. To this end, let $p_F(X)$ be the underlying likelihood related to F, and let the two independent samples of IID data, respectively, sized n_1 and n_2, be $\mathbf{X}_1 = (X_{11}, \ldots, X_{1n_1})$ and $\mathbf{X}_2 = (X_{21}, \ldots, X_{2n_1})$. So the data set is $\mathbf{X} = (\mathbf{X}_1, \mathbf{X}_2)$, whose joint likelihood is $p_F(\mathbf{X}) = \prod_{i=1}^{n_1} p_{F_1}(X_{1i}) \prod_{i=1}^{n_2} p_{F_2}(X_{2i})$. In null hypothesis, it is assumed that there is no difference between two distributions, namely, $F_1 = F_2 = F$. Thus, the joint null likelihood $p_F(\mathbf{X}) = \prod_{j=1}^{2} \prod_{i=1}^{n_j} p_F(X_{ji})$ is invariable with respect to any permutation $\mathbf{X}^*$ of the observed pooled data $\mathbf{X} = (\mathbf{X}_1 \biguplus \mathbf{X}_2)$, where $\biguplus$ is the symbol for pooling two data sets. This shows that data under H_0 are exchangeable, i.e., permutable. Moreover, under H_0 pooled data $\mathbf{X}$ are always a set of sufficient statistics for any underlying distribution F [18–20, 22]; so, any information on parameters defining F is wholly contained in $\mathbf{X}$. The set of all permutations $\mathbf{X}^*$ of $\mathbf{X}$ is indicated with $\Pi(\mathbf{X})$. It is worth noting that $\Pi(\mathbf{X}) = \Pi(\mathbf{X}^*)$, i.e., the set of permutations of $\mathbf{X}$ coincides, $\forall \mathbf{X}^* \in \Pi(\mathbf{X})$, with that of $\mathbf{X}^*$. Of course, under the alternative H_1 the above invariable property does not work, because the two distributions are different by assumption: indeed $\mathbf{X}_1$ is sufficient for F_1 and $\mathbf{X}_2$ is sufficient for F_2 and so pooled data are not exchangeable.

The act of conditioning on a set of sufficient statistics for F in H_0 entails that any conditional inference is independent of the underlying population distribution F. This conditioning gives rise to the following fundamental property:

Let $(\mathcal{X}, \mathcal{A}, F)$ be the probability space related to data X, then sufficiency of $\mathbf{X}$ *for underlying F, under H_0, implies that the null conditional probability of any event $A \in \mathcal{A}$, given* $\mathbf{X}$, *is independent of F, i.e., $Pr\{\mathbf{X}^* \in A; F \mid \mathbf{X}\} = Pr\{\mathbf{X}^* \in A \mid \mathbf{X}\} = P[A \mid \mathbf{X}]$.*

Three relevant consequences of this property are c1) under H_0 all M permutations $\mathbf{X}^*$ of $\mathbf{X}$ are equally likely; c2) so $P[A \mid \mathbf{X}] = \#(\mathbf{X}^* \in A)/M$, where $\#(\cdot)$ is the number of elements of $\Pi(\mathbf{X})$ that satisfy condition $(\cdot)$, i.e., $P[A \mid \mathbf{X}]$ is properly a count ratio; c3) if $\mathbf{T} = (T_1, \ldots, T_S)^\top$ is a vector of $S \geq 1$ permutation statistics (e.g., tests) and $\varphi : \mathcal{R}^S \rightarrow \mathcal{R}^1$ is any measurable function, then the conditional null distribution of φ is independent of F; indeed,

$$\begin{aligned} Pr\{\varphi(T_1^*, \ldots, T_S^*) \leq z; F \mid \mathbf{X}\} &= Pr\{\varphi(T_1^*, \ldots, T_S^*) \leq z \mid \mathbf{X}\} \\ &= Pr[\varphi_{\mathbf{T}}^{-1}(z) \mid \mathbf{X}] = \frac{\#[\mathbf{X}^* \in \varphi_{\mathbf{T}}^{-1}(z)]}{M}, \end{aligned} \tag{1}$$

since, due to measurability of φ, $\forall z \in \mathcal{R}^1$, it is $\varphi_{\mathbf{T}}^{-1}(z) \in \mathcal{A}$.

It is worth noting that (c3) is the central property for deducing and justifying the NPC of dependent permutation tests. Also worth noting is (i) the conditional probability $P[A \mid \mathbf{X}]$ has always an *objective existence*; (ii) the conditional null distribution of φ is independent of all dependence parameters underlying $\mathbf{T}$; (iii) to characterize sufficiency of $\mathbf{X}$ in H_0, permutation tests require the existence of a likelihood $p_F(\mathbf{X}) > 0$, not its calculability; (iv) when $\mathbf{X}$ is minimal sufficient for F, it makes no sense to work outside the permutation testing principle [20]; (v) permutation tests are nonparametric, distribution-free, and intrinsically robust.

For operating with categorical data, to the usual contingency table we prefer using its unit-by-unit representation: $\mathbf{X} = \{X(i), i = 1, \ldots, n; n_1, n_2\}$, with $n = n_1 + n_2$, where it is intended that the first n_1 data belong to the first sample and the rest belong to the second. Such a representation, is one-to-one related with the table for unidimensional variables, as with the example; it is much more efficient for multidimensional variables, where working with contingency tables becomes more and more difficult, up to impossibility, as the number of variables increases. Using that representation, a random permutation $\mathbf{X}^* \in \Pi(\mathbf{X})$ can be expressed as $\mathbf{X}^* = \{X(u_i^*), i = 1, \ldots, n; n_1, n_2\}$, where $\mathbf{u}^* \in \Pi(\mathbf{u})$ is a random permutation of the unit labels $\mathbf{u} = \{1, \ldots, n\}$. The corresponding permuted table is calculate as $\{f_{jk}^* = \#(X_{ji}^* \in c_k), k = 1, \ldots, K, j = 1, 2\}$. Obviously, the marginal frequencies are permutation invariable quantities since $f_{\cdot k} = f_{1k} + f_{2k} = f_{1k}^* + f_{2k}^* = f_{\cdot k}^*$, $k = 1, \ldots K$. Similarly, the cumulative marginal frequencies are also invariable: $N_{\cdot k} = N_{1k} + N_{2k} = N_{\cdot k}^*$ with $N_{jk} = \sum_{s \le k} f_{js}$.

3.1 The 2 × K One-Dimensional Case

We start with the two-samples one-dimensional problem. For the case of $C = 2$, the related stochastic dominance testing problem becomes $H_0 : F_1 = F_2 \equiv \bigcap_{k=1}^{K-1}[F_1(c_k) = F_2(c_k)]$ against $H_1 : F_1 > F_2 \equiv \bigcup_{k=1}^{K-1}[F_1(c_k) > F_2(c_k)]$, whose global analysis requires the joint comparison of $K - 1$ differences of random frequencies: $\hat{F}_1(c_k) - \hat{F}_2(c_k), k = 1, \ldots, K - 1$. Since the crucial point for that joint analysis is the proper handling of all underlying dependences, to attain general solutions we must work within the UI-NPC of related dependent permutation tests because, due to c3) (see Sect. 3), the estimation of dependence coefficients is not required since NPC works independently of such dependences, how complex these are.

Accordingly, the $K - 1$ partial test statistics are

$$T_k^* = C(n_1, n_2) \cdot [\hat{F}_{1k}^* - \hat{F}_{2k}^*]\left[\bar{F}_{\cdot k}(1 - \bar{F}_{\cdot k})\right]^{-\frac{1}{2}}, \quad k = 1, \ldots, K - 1, \qquad (2)$$

where $\hat{F}_{jk}^* = \hat{F}_j^*(c_k) = N_{jk}^*/n_j, j = 1, 2; \bar{F}_{\cdot k} = N_{\cdot k}/n$ are permutation and marginal empirical distribution functions (EDFs); N_{1k}^* and $N_{2k}^*, k = 1, \ldots, K - 1$ are permutation cumulative frequencies obtained from the permuted table $\{f_{jk}^*, k = 1, \ldots, K, j = 1, 2\}$.

It is worth noting that (i) EDFs $\hat{F}_{jk}$ are maximum likelihood unbiased estimates of population CDFs $F_j(c_k), k = 1, \ldots, K - 1, j = 1, 2$; (ii) each partial tests T_k^* is a reformulation of Fisher's exact probability test and so it is a *best conditional test*; (iii) large values of each partial test T_k^* are significant against its related null sub-hypothesis H_{1k}; (iv) the $K - 1$ partial tests are positively dependent; (v) for computation of T_k^*, 0 is assigned to expressions with the form 0/0; (vi) $C(n_1, n_2) = [n_1 n_2 (n - 1)/n^2]^{1/2}$ is a permutation constant not dependent on k; (vii) for increasing

Table 2 Representation of the conditional Monte Carlo method in multivariate tests

$\mathbf{X}$	$\mathbf{X}_1^*$		$\mathbf{X}_r^*$		$\mathbf{X}_R^*$
T_1^o	T_{11}^*	...	T_{1r}^*	...	T_{1R}^*
$\vdots$	$\vdots$		$\vdots$		$\vdots$
T_K^o	T_{K1}^*	...	T_{Kr}^*	...	T_{KR}^*
		$\downarrow$			
T_ψ^o	$T_{\psi 1}^*$	...	$T_{\psi r}^*$	...	$T_{\psi R}^*$

sample sizes, each T_k^* under H_0 converges to the standardized normal distribution: $T_k^* \xrightarrow{d} \mathcal{N}(0, 1)$.

According to the approach discussed in [18, 19, 22], the global testing solution can be obtained by their UI-NPC while using any admissible combining function. The simplest admissible combination is by the direct sum of partial tests:

$$T_{AD}^* = \sum_{k=1}^{K-1} T_k^* = C(n_1, n_2) \cdot \sum_{k=1}^{K-1} [\hat{F}_{1k}^* - \hat{F}_{2k}^*] \left[\bar{F}_{\cdot k}(1 - \bar{F}_{\cdot k})\right]^{-\frac{1}{2}}. \tag{3}$$

Such a solution looks like the discrete version of Anderson–Darling goodness-of-fit type test for multi-one-sided alternatives. It is worth noting that (i) each partial test is unbiased and so T_{AD}^* is unbiased; (ii) at least one partial test is consistent and so T_{AD}^* is consistent; (iii) T_{AD}^* is an admissible combination of partial best tests and so provided with *good power behavior*. Of course, by using other admissible combining functions one can obtain other *good solutions*, none of which, however, being uniformly better than any other.

The corresponding p-value-like statistics can be written as $\lambda_{AD} = \Pr\{T_{AD}^* \geq T_{AD}^o \mid \mathbf{X}\}$, where $T_{AD}^o = T_{AD}(\mathbf{X})$ is the observed value of T_{AD} on pooled data $\mathbf{X}$. So, remembering that p-value-like statistics play the role of tests whose common critical value is α, if $\lambda_{AD} \leq \alpha$, the null hypothesis is rejected at significance level $\alpha > 0$.

Consider the representation displayed in Table 2. It corresponds to the NPC procedure for a general problem with K partial tests, R random permutations, and combining function ψ.

Under H_0, the sub-matrix $\{T_{kr}^*\}_{K \times R}$ simulates the K-dimensional null distribution of K partial permutation tests. The sub-vector $\{T_{\psi r}^*\}_R$ simulates the null permutation distribution of combined test T_ψ.

Thus, the statistic $\hat{\lambda}_\psi = \#(T_\psi^* \geq T_\psi^o)/R$ gives an unbiased and, as R diverges, a strongly consistent estimate of the p-value statistic λ_ψ of T_ψ.

Under H_1, at least one T_k^o presents larger observed values than in H_0; so, if the combining function ψ is non-decreasing in each argument, the p-value statistic satisfies the relation: $\hat{\lambda}_{\psi;H_1} \overset{d}{\leq} \hat{\lambda}_{\psi;H_0}$ uniformly for every data set $\mathbf{X}$ and every underlying distribution F. Hence, the latter justifies that H_0 is rejected when $\hat{\lambda}_\psi \leq \alpha$; moreover, it can be proved that T_ψ is provided with the unbiasedness and consistency properties. Details and proofs for these and other properties are in [18–20, 22].

3.2 The 2 x K Multidimensional Case

In the general multidimensional case, let us start from two-sample V-dimensional problem, $V \geq 2$. The formulation of testing for multidimensional hypotheses are $H_0 : \mathbf{X}_1 \overset{d}{=} \mathbf{X}_2$ against $H_1 : \mathbf{X}_1 \overset{d}{\prec} \mathbf{X}_2$. The hypotheses H_0 and H_1, according to [24] are assumed to be equivalently broken-down into $K \geq 2$ sub-hypotheses, $H_0 \equiv \bigcap_{k=1}^{K} H_{0k}$ and $H_1 \equiv \bigcup_{k=1}^{K} H_{1k}$. Thus, with V dimensional ordinal data and K ordered categories for each variable, the hypotheses are equivalently written as $\bigcap_{v=1}^{V} \bigcap_{k=1}^{K-1} [F_{1v}(c_k) = F_{2v}(c_k)]$ and $\bigcup_{v=1}^{V} \bigcup_{k=1}^{K-1} [F_{1v}(c_k) > F_{2v}(c_k)]$, respectively. Thus, for variable $v = 1, \ldots, V$, partial test is T^*_{ADv} according to Sect. 3.1. Since all these partial tests are standardized and so, sharing the same asymptotic null distribution, for their combination we can proceed with their direct sum. This provides for the V-dimensional extension of Anderson–Darling test for multi-one-sided alternatives:

$$T^*_{AD} = \sum_{v=1}^{V} T^*_{ADv} = C(n_1, n_2) \cdot \sum_{v=1}^{V} \sum_{k=1}^{K-1} [\hat{F}^*_{1vk} - \hat{F}^*_{2vk}] \left[\bar{F}_{\cdot vk}(1 - \bar{F}_{\cdot vk})\right]^{-\frac{1}{2}}. \quad (4)$$

It is worth noting that, now, with symbol $\mathbf{X}$ it is represented the V-dimensional variable and the pooled sample data matrix, the context generally suffices avoiding misunderstandings. Of course, the V-dimensional T^*_{AD} enjoys the same *good* properties as the unidimensional. In place of the direct combination of V partial tests T^*_{ADv}, i.e., one Anderson–Darling test for each variable, it is possible to think of a more general combination like, for instance, $T^*_\psi = \psi(T^*_{AD1}, \ldots, T^*_{ADV})$. The most commonly used combining functions ψ are Fisher's $T_F = -2\sum_v \log(\lambda^*_{ADv})$, or Liptak's $T^*_L = \sum_v \Phi^{-1}(1 - \lambda^*_{ADv})$, where λ^*_{ADv} is the p-value statistic of T^*_{ADv} and $\Phi(\cdot)^{-1}$ is the inverse standard normal CDF. Since in T^*_{AD} all summands are well defined, it is also of some interest to observe that the double summation can equivalently be computed as $\sum_k \sum_v$.

4 The C-sample Stochastic Ordering Problem

Considering the Jonckheere–Terpstra idea, the $C \times K$ table can be broken-down into $(C - 1)$ sub-tables. Accordingly, the testing problem is broken-down into $(C - 1)$ sub-problems each based on a $2 \times K$ sub-table. To be specific, for any $j \in \{1, \ldots, C - 1\}$, we divide the data set into two pooled pseudo-groups, where the first pseudo-group is obtained by pooling data of the first j ordered groups and the second by pooling the rest. Thus, the procedure considers the first pooled pseudo-group as $\mathbf{Y}_{1(j)} = \mathbf{X}_1 \biguplus \ldots \biguplus \mathbf{X}_j$ and the second as $\mathbf{Y}_{2(j)} = \mathbf{X}_{j+1} \biguplus \ldots \biguplus \mathbf{X}_C$, $j = 1, \ldots, C-1$, where $\mathbf{X}_j = \{X_{ji}, i = 1, \ldots, n_j\}$ is the data set in the jth group.

In the null hypothesis H_0, related pooled variables satisfy the relationships $Y_{1(j)} \overset{d}{=} Y_{2(j)}$, $j = 1, \ldots, C-1$, thus, data from every pair of pseudo-groups are exchangeable. In the alternative H_1, as for at least one j the relation inequality $X_j \overset{d}{\prec} X_{j+1}$, $1 \leq j \leq C-1$ is strict, the corresponding stochastic dominance between each pair of pseudo-groups $Y_{1(j)} \overset{d}{\prec} Y_{2(j)}$ is true for all $j \leq C-1$. Therefore, the hypotheses for monotonic stochastic ordering problem can be equivalently written as $H_0 : \{\bigcap_{j=1}^{C-1}(Y_{1(j)} \overset{d}{=} Y_{2(j)})\}$ and $H_1 : \{\bigcup_{j=1}^{C-1}(Y_{1(j)} \overset{d}{\prec} Y_{2(j)})\}$, emphasizing a break-down into a set of $C-1$ sub-hypotheses. For each sub-problem we can consider the test:

$$T^*_{AD(j)} = C(n_{1(j)}, n_{2(j)}) \cdot \sum_{k=1}^{K-1} \left[\hat{F}^*_{1(j)k} - \hat{F}^*_{2(j)k}\right] [\bar{F}_{\cdot(j)k}(1 - \bar{F}_{\cdot(j)k})]^{-\frac{1}{2}}, \ j = 1, \ldots, C-1, \tag{5}$$

where $n_{1(j)} = n_1 + \ldots + n_j$, $n_{2(j)} = n - n_{1(j)}$; the permutation relative frequencies are $\hat{F}^*_{l(j)k} = \#(X^*_{l(j)} \preceq c_k)/n_{l(j)}$, $l = 1, 2$; the marginal relative frequencies are $\bar{F}_{\cdot(j)k} = [\#(X^*_{1(j)} \preceq c_k) + \#(X^*_{2(j)} \preceq c_k)]/n$; partial tests $T^*_{AD(j)}$ are positively dependent; and $C(n_{1(j)}, n_{2(j)})$ are the permutation k-invariable constants. So the global problem is solved by combining the $C-1$ partial tests within the UI-NPC as, for instance, by

$$T^*_{AD} = \sum_{j=1}^{C-1} T^*_{AD(j)}. \tag{6}$$

According to our experience, except for the direct, the most suitable combining functions for this problem are Fisher's and Liptak's. Since in the stochastic ordering alternative all $C-1$ partial tests contain a positive non-centrality quantity, i.e., all lie in their respective sub-alternatives, Tippett's combination is less sensitive than others.

Of course, if $V > 1$ variables were involved, the multivariate stochastic ordering solution would require one stochastic ordering partial test for each variable $v = 1, \ldots, V$. So, with clear meanings of the symbols, the global test, by direct combination, is

$$T^*_{AD,V} = \sum_{j=1}^{C-1} C(n_{1(j)}, n_{2(j)}) \cdot \sum_{v=1}^{V} \sum_{k=1}^{K-1} \left[\hat{F}^*_{1v(j)k} - \hat{F}^*_{2v(j)k}\right] [\bar{F}_{\cdot v(j)k}(1 - \bar{F}_{\cdot v(j)k})]^{-\frac{1}{2}}. \tag{7}$$

Table 3 p-values based on UI-NPC approach

	$T^*_{(1)}$	$T^*_{(2)}$	$T^*_{(3)}$	T''_D	T''_F	T''_L	T''_T
$\hat{\lambda}_{AD(j)}$	0.0141	0.0025	0.0074	0.0017	0.0015	0.0012	0.0068
$\hat{\lambda}_{W(j)}$	0.0131	0.0021	0.0076	0.0010	0.0012	0.0010	0.0053
$\hat{\lambda}_{M(j)}$	0.0144	0.0024	0.0062	0.0011	0.0014	0.0011	0.0068

5 Solution of Medical Example

The analyses of the data from medical example, based on $R = 100\,000$ random permutations, for tests: Anderson–Darling T^*_{AD}, on scores T^*_W, and on mid-ranks T_M, and their combination functions: T''_D direct, T''_F Fisher's, T''_L Liptak's, and T''_T Tippett's are shown in Table 3. Note that (i) W scores are assigned to ordering integer numbers as ($w_1 = 1, w_2 = 2, w_3 = 3, w_4 = 4, w_5 = 5$); (ii) since small p-value statistics are evidence for H_1, Fisher's, Liptak's, and Tippett's are non-increasing functions of partial p-values. The p-values based on UI-NPC method are

Results in Table 3 clearly show that the p-values based on four different combination functions T''_D, T''_F, T''_L, and T''_T, all reject the null hypothesis at significance level $\alpha = 0.01$ of monotonic stochastic ordering among the $C = 4$ doses. So the inferential conclusion is that patients present non-decreasing responses as the dose increases.

It is worth noting that the three combined p-value statistics T''_D, T''_F, and T''_L differ only slightly in the fourth digit. This means that related tests are all suitable for testing unidimensional dominance and stochastic ordering alternatives. In our case, if the stochastic ordering alternative is true, it is also jointly true by construction for all $C - 1$ partial tests $T^*_{(j)}$. So, Tippett's T''_T differs from other combination functions because its power behavior is mostly sensitive when only one partial test lies in the alternative. Due to too many ties in the data set, test with rank transformations was not considered.

Since all p-values statistics related to $T_{AD(3)}$ are $< 0.05/3$, by simple Bonferroni's rule it results that subjects taking High dose exhibit significantly lower responses than those taking lower doses.

6 Concluding Remarks

The basic idea in this paper is to test for stochastic ordering restrictions with multivariate ordered categorical data through a suitable combination of a set of partial tests by UI-NPC approach based within the permutation theory. Such problems have quite difficult solutions within the likelihood ratio theory which, when available, have nice characterizations under their usually too stringent assumptions.

The UI-NPC approach is within the conditionality principle of inference, where the conditioning is with respect to a set of sufficient statistics in the null hypothesis like the pooled observed data. So, it is based on the permutation testing approach and the NPC of dependent permutation tests. The NPC approach shows a good general power behavior, it is rather efficient and less demanding in terms of underlying assumptions comparing to parametric competitors when these exist and are available.

References

1. Agresti, A., Coull, B.A.: Order-restricted inference for monotone trend alternatives in contingency tables. Comput. Stat. Data Anal. **28**, 139–155 (1998)
2. Agresti, A., Coull, B.A.: The analysis of contingency tables under inequality constraints. J. Stat. Plann. Infer. **107**, 45–73 (2002)
3. Agresti, A., Mehta, C.R., Patel, N.R.: Exact inference for contingency tables with ordered categories. J. Am. Stat. Assoc. **85**, 453–458 (1990)
4. Arboretti, G.R., Bonnini, S.: Moment-based multivariate permutation tests for ordinal categorical data. J. Nonparametr. Stat. **20**, 383–393 (2008)
5. Arboretti, G.R., Bonnini, S.: Some new results on univariate and multivariate permutation tests for ordinal categorical variables under restricted alternatives. Stat. Methods Appl. **18**, 221–236 (2009)
6. Basso, D., Pesarin, F., Salmaso, L., Solari, A.: Permutation tests for stochatic ordering and ANOVA: theory and applications in R. Springer, New York (2009)
7. Bazyari, A., Pesarin, F.: Parametric and permutation testing for multivariate monotonic alternatives. Stat. Comput. **23**, 639–652 (2013)
8. Bonnini, S., Prodi, N., Salmaso, L., Visentin, C.: Permutation Approaches for Stochastic Ordering. Commun. Stat. - Theory Methods. **43**, 2227–2235 (2014)
9. Chuang-Stein, C., Agresti, A.: A review of tests for detecting a monotone dose-response relationship with ordinal response data. Stat. Med. **16**, 2599–2618 (1997)
10. Cohen, A., Kemperman, J.H.B., Sackrowitz, H.B.: Properties of likelihood inference for order restricted models. J. Multivar. Anal. **72**, 50–77 (2000)
11. Cohen, A., Madigan, D., Sackrowitz, H.B.: Effective directed tests for models with ordered categorical data. Aust. N. Z. J. Stat. **45**, 285–300 (2003)
12. Colombi, R., Forcina, A.: Testing order restrictions in contingency tables. Metrika **79**, 73–90 (2016)
13. Cox, D.R., Hinkley, D.V.: Theoretical Statistics. Chapman & Hall, London (1974)
14. Feltz, C.J., Dykstra, R.L.: Maximum likelihood estimation of the survival functions of N stochastically ordered random variables. J. Am. Stat. Assoc. **80**, 1012–1019 (1985)
15. Gökpinar, F., Gökpinar, E., Bayrak, H.: Permutation approach for ordinal preference data. Commun. Stat. Simul. Comput. **46**, 2321–2332 (2017)
16. Jelizarow, M., Cieza, A., Mansmann, U.: Global permutation tests for multivariate ordinal data: alternatives, test statistics and the null dilemma. J. Royal Stat. Soc. C. **64**, 191–213 (2015)
17. Kateri, M., Agresti, A.: Bayesian inference about odds ratio structure in ordinal contingency tables. Environmetrics **24**, 281–288 (2013)
18. Pesarin, F.: Multivariate Permutation Tests: With Applications to Biostatistics. Wiley, Chichester (2001)
19. Pesarin, F.: Permutation test: Multivariate. In Encyclopedia of Statistical Sciences. John Wiley & Sons Inc, New York (2006)
20. Pesarin, F.: Some elementary theory of permutation tests. Commun. Stat. Theory Meth. **44**, 4880–4892 (2015)
21. Pesarin, F., Salmaso, L.: Permutation tests for univariate and multivariate ordered categorical data. Austrian. J. Stat. **35**, 315–324 (2006)
22. Pesarin, F., Salmaso, L.: Permutation tests for complex data: theory, applications and software. Wiley, Chichester (2010)

23. Pesarin, F., Salmaso, L., Carrozzo, E., Arboretti, R.: Union-intersection permutation solution for two-sample equivalence testing. Stat. Comput. **26**, 693–701 (2016)
24. Roy, S.N.: On a heuristic method of test construction and its use in multivariate analysis. Ann. Math. Stat. **24**, 220–238 (1953)
25. Sampson, A.R., Whitaker, L.R.: Estimation of multivariate distributions under stochastic ordering. J. Am. Stat. Assoc. **84**, 541–548 (1989)
26. Silvapulle, M.J., Sen, P.K.: Constrained Statistical Inference: Inequality. Order and Shape Restrictions. Wiley, New York (2005)
27. Wang, Y.: A likelihood ratio test against stochastic ordering in several populations. J. Am. Stat. Assoc. **91**, 1676–1683 (1996)

Smooth Nonparametric Survival Analysis

Dimitrios Ioannides and Dimitrios Bagkavos

Abstract This research proposes the local polynomial smoothing of the Kaplan–Meier estimate under the fixed design setting. This allows the development of estimates of the distribution function (equivalently the survival function) and its derivatives under the random right censoring model. The asymptotic properties of the estimate, including its asymptotic normality are all established herein.

Keywords Kaplan–Meier · Local polynomial fitting · Censoring

1 Introduction

The present research proposes the combination of the Kaplan–Meier estimate with the local polynomial fitting technique. The result is an estimate of the distribution function and its derivatives for discretized (binned) data, under the right censorship model.

The motivation behind this research is two fold. One aspect is that the original version of the Kaplan–Meier estimate comes with some significant limitations. Perhaps the most important is that it produces a step function. This contradicts the quite plausible assumption of continuity and smoothness of the distribution and survival functions. Subsequently, this limits the scope of the estimate's application, especially for inferential purposes where differentiability plays a key role. Another aspect which prompted the present research is that the literature seems to be rather thin on boundary aware kernel estimates of the density function and its derivatives under the right censorship model. However, these quantities are quite useful in bandwidth selection, estimation of the slope, curvature, or mode of a population among many other applications.

D. Ioannides
Department of Economics, University of Macedonia, Egnatias 156, 540 06 Macedonia, Greece

D. Bagkavos (✉)
Department of Mathematics, University of Ioannina, 45100 Ioannina, Greece
e-mail: dimitrios.bagkavos@gmail.com

M. La Rocca et al. (eds.), *Nonparametric Statistics*, Springer Proceedings in Mathematics & Statistics 339, https://doi.org/10.1007/978-3-030-57306-5_22

With the purpose to address all aforementioned points, this work combines the Kaplan–Meier estimate of the distribution function which intrinsically admits right-censored data and the local polynomial fitting principle which allows estimation of distribution function derivatives of any arbitrary order. The benefit of this approach is that additionally to filling these gaps, it produces distribution / survival function estimates with asymptotically smaller mean squared error compared to the Kaplan–Meier estimate.

The proposed estimates together with the necessary notation and its asymptotically equivalent form are introduced in Section 2. Their asymptotic properties together with quantification of their asymptotic distribution are discussed in Section 3. All proofs are given in Section 4.

2 Local Linear Estimation of the Distribution Function and Its Derivatives

Let $T_1, T_2, \dots, T_n$ be a sample of i.i.d. survival times censored on the right by i.i.d. random variables $U_1, U_2, \dots, U_n$, which are independent from the T_i's. Let f_T be the common probability density and F_T the distribution function of the T_i's. Denote with H the distribution function of the U_i's. Typically the randomly right-censored observed data are denoted by the pairs (X_i, δ_i), $i = 1, 2, \dots, n$ with $X_i = \min\{T_i,\ U_i\}$ and $\delta_i = \mathbf{1}_{\{T_i \le U_i\}}$, where $\mathbf{1}_{\{\cdot\}}$ is the indicator random variable of the event $\{\cdot\}$. The distribution function of X_i's satisfies $1 - F = (1 - F_T)(1 - H)$. It is assumed that estimation happens in the interval $[0, M]$ where M satisfies the relationship

$$M = \sup\{x : 1 - F(x) > \varepsilon\} \text{ for a small } \varepsilon > 0.$$

We are interested in estimating the distribution function $F_T(x)$ and its derivatives of any arbitrary order. An immediate byproduct of obtaining an estimate of $F_T(x)$ is its use in estimating the survival function $S_T(x) = 1 - F_T(x)$. The classical nonparametric estimate of F_T, [11], is given by

$$\hat{F}_S(x) = \begin{cases} 0, & 0 \le x \le Z_1, \\ 1 - \prod_{i=1}^{k-1} \left(\frac{n-i}{n-i+1}\right)^{\Lambda_i}, & Z_{k-1} < x \le Z_k,\ k = 2, \dots, n, \\ 1, & x > Z_n \end{cases} \tag{1}$$

where (Z_i, Λ_i) are the ordered X_i's, along with their censoring indicators δ_i, $i = 1, \dots, n$. According to the standard local polynomial principle, first, partition the interval $[0, M]$ into g disjoint subintervals $\{I_j,\ j = 1 \dots g\}$ of equal length b. Denote with $x_j = (j - \frac{1}{2})b$, $j = 1, \dots, g$, the center of the interval I_j. Essentially b can be determined by an optimal histogram bin width selection rule.

Denote with $\sigma^2(x_i)$ the variance of $\hat{F}_S(x_i)$ at x_i and let ε_i, $i = 1, \dots, g$ be independent random vectors with mean 0 and variance 1. Also, set $m(x_i) = F_T(x_i)$. Since

$\hat{F}_S(x_i)$ is an asymptotically unbiased estimate of $F_T(x)$ it can be used as the response to the local nonparametric regression problem

$$\hat{F}_S(x_i) = m(x_i) + \sigma(x_i)\varepsilon_i, \quad i = 1, \ldots, g.$$

Using the data $\{\hat{F}_S(x_i), x_i\}$, $i = 1, \ldots, g$, the regression function m can be approximated locally in a nearby point x_0 such that $|x - x_0| \le \varepsilon$ for an arbitrarily small ε, by a Taylor expansion

$$m(x) = \sum_{k=0}^{p} \frac{m^{(k)}(x_0)}{k!}(x - x_0)^k + R_k(x),$$

with $R_k(x)$ being the Lagrange remainder term. Set $K_h(u) = h^{-1}K(u/h)$. Here K is a kernel function, usually a symmetric density, assumed to be supported on a symmetric and compact interval. h denotes the bandwidth which controls the spread of the kernel. Define the ith kernel moment by

$$\mu_i(K) \equiv \mu_i = \int_{-\infty}^{+\infty} u^i K(u)\, du, \quad i = 0, 1, \ldots, \nu + 1.$$

It is assumed throughout that K satisfies $\mu_0 = 1$, $\mu_1 = 0$, and $\mu_2 < +\infty$. Also, let $\beta_k = m^{(k)}/k!$, $k = 0, \ldots, p$. The estimates of β_k, say $\hat{\beta}_k$ will result by solving the optimization problem

$$\min_{\beta_k, k=0,\ldots,p} \sum_{j=1}^{g} \left\{ \hat{F}_S(x_j) - \sum_{k=0}^{p} \beta_k (x_j - x)^k \right\}^2 K_h(x_j - x). \tag{2}$$

According to [5], the optimal order of the local polynomial to use in (2) depends on the order of the derivative being estimated and is given by $p = \nu + 1$. This yields the solution

$$\hat{\beta}_\nu = \sum_{i=1}^{g} K_\nu\left(\frac{x_i - x}{h}\right) \hat{F}_S(x_i), \quad \nu = 0, 1, 2, \ldots \tag{3}$$

where

$$K_\nu(u) = e_{\nu+1}^T S^{-1}(1, hu, \ldots, (hu)^\nu, (hu)^{\nu+1})^T h^{-1} K(u).$$

$e_{\nu+1}^T$ denotes a vector with $\nu + 2$ elements with 1 in the $(\nu + 1)$th position and zeros elsewhere. S is the $(\nu + 2) \times (\nu + 2)$ matrix $(S_{n,j+l})_{0 \le j,l \le \nu+1}$ with

$$S_{n,l}(x) = \sum_{i=1}^{g} K_h(x_i - x)(x_i - x)^l, \quad l = 0, 1, \ldots, 2\nu + 2.$$

Thus, $F_T^{(\nu)}(x)$ is estimated by $\hat{F}_L^{(\nu)}(x) = \nu!\hat{\beta}_\nu$.

In the definition of K_ν, the role of $e_{\nu+1}^T S^{-1}(1, hu, \ldots, (hu)^\nu, (hu)^{\nu+1})^T h^{-1}$ is to automatically reinstate the kernel mass falling outside the region of estimation back in so as to correct the estimate at the boundary. In the interior this factor equals to 1 and the estimate defaults to a regular kernel estimate. To see this, assume without loss of generality that K is supported on $[-1, 1]$ and let $0 < c < 1$ so that $x = ch \in [0, h)$ is a boundary point. Correspondingly, in the interior we have $x = ch$, $c > 1$ so that $x \in [h, M - h]$. Define

$$\mu_{i,c} = \int_{-c}^{+\infty} u^i K(u)\, du, \; i = 0, 1, \ldots, 2\nu + 2.$$

In the interior where $c > 1$, $\mu_{i,c} = \mu_i$. Let $S_c = (\mu_{i+j,c})_{0 \le i,j \le \nu+1}$. From the proof of Theorem 1 in [1],

$$S_{n,l}(x) = \sum_{i=1}^{g} K_h(x_i - x)(x_i - x)^l = b^{-1}h^{l+1}\mu_{l,c}(1 + o(1)), \; l = 0, \ldots 2\nu + 2. \tag{4}$$

Then it is easy to see that in the interior we have

$$\hat{F}_L^{(\nu)}(x) = \frac{\nu!}{h^{\nu+1}} \sum_{i=1}^{g} K_{\nu,c}^* \left(\frac{x_i - x}{h}\right) \hat{F}_S(x_i)(1 + o(1)),$$

where

$$K_{\nu,c}^*(u) = e_{\nu+1}^T S_c^{-1}(1, u, \ldots, u^\nu, u^{\nu+1})^T b^{-1} K(u) I_{\{-c,+\infty\}}(u),$$

and for $c > 1$, $K_\nu(u) = h^{-(\nu+1)} K_{\nu,c}^*(u)(1 + o(1))$. In order to facilitate the theoretical study of $\hat{F}_L^{(\nu)}(x)$ it is worth defining the following equivalent formulation of the estimate. For fixed j and for $k \in \{1, \ldots, g\}$ set

$$c_{kj} = \mathbf{1}_{[x_k - \frac{b}{2}, x_k + \frac{b}{2}]}(X_j, \delta_j = 1).$$

Since the $X_1, X_2, \ldots, X_n$ are i.i.d., using the strong law of large numbers yields

$$n^{-1}b^{-1} \sum_{j=1}^{n} c_{ij} \overset{a.s.}{\to} b^{-1} \int_{x_i - \frac{b}{2}}^{x_i + \frac{b}{2}} f_T(y)(1 - H(y))\, dy =$$

$$\simeq b^{-1} b f_T(x_i)(1 - H(x_i)) = f_T(x_i)(1 - H(x_i)). \tag{5}$$

Thus, dividing the empirical estimate of $f_T(x_i)(1 - H(x_i))$ by an estimate of $1 - H(x_i)$ yields an estimate of $f_T(x_i)$. Following [13], by reversing the intuitive role played by T_i and U_i, $1 - H(x)$ can be estimated by the (sightly modified) Kaplan–Meier estimator,

$$1-\hat{H}(x)=\begin{cases}1, & 0\le x\le Z_1,\\ \prod_{i=1}^{k-1}\left(\frac{n-i+1}{n-i+2}\right)^{1-\Lambda_i}, & Z_{k-1}<x\le Z_k, k=2,\ldots,n,\\ \prod_{i=1}^{n}\left(\frac{n-i+1}{n-i+2}\right)^{1-\Lambda_i}, & Z_n<x.\end{cases}$$

Thus, from (5), for fixed i, an empirical estimate of $bf_T(x_i)$ at the ith bin center, denoted by $\hat{f}_T(x_i)$ is defined by

$$\hat{f}_T(x_i)=\frac{1}{n}\sum_{j=1}^{n}\frac{c_{ij}}{1-\hat{H}(x_i)}\simeq bf_T(x_i).$$

Let

$$W_\nu^*(u)=\int_{-\infty}^{u}K_\nu(t)\,dt \text{ and } W_{\nu,c}^*(u)=\int_{-\infty}^{u}K_{\nu,c}^*(t)\,dt.$$

Following [15] and [10], $\hat{f}_T(x_i)$ can be used to approximate the jump of the Kaplan–Meier estimate at x_i. As a consequence $\hat{F}_L^{(\nu)}(x)$ can be approximated as

$$\begin{aligned}\hat{F}_L^{(\nu)}(x) &\simeq \nu!\sum_{i=1}^{g}W_\nu\left(\frac{x_i-x}{h}\right)\hat{f}_T(x_i)\\ &\equiv \frac{\nu!}{h^\nu}\sum_{i=1}^{g}W_{\nu,c}^*\left(\frac{x_i-x}{h}\right)\hat{f}_T(x_i)(1+o(1)).\end{aligned}$$

An obvious estimate of the survival function $S_T(x)$ is $\hat{S}_T(x)=1-\hat{F}_L^{(0)}(x)$. Additional applications include using $\hat{F}_L^{(\nu)}(x)$ (for $\nu>1$) in plug-in bandwidth selection rules in estimation of population characteristics etc. For all these it is important to establish the theoretical properties and the asymptotic distribution of $F_T^{(\nu)}(x)$. These are discussed next.

3 Asymptotic Properties

Denote with $b_{L,c}(x)$ and $\sigma_{L,c}^2(x)$ the bias and variance of $\hat{F}_L^{(\nu)}(x)$ using bandwidth h_ν at the boundary point $x=ch_\nu$. The notation h_ν instead of the simpler form h is used henceforth so as to emphasize the fact that different bandwidth should be used according to the order of the derivative being estimated. Let $b_L(x)$ and $\sigma_L^2(x)$ denote correspondingly the bias and variance in the interior. Also, let $\mu_i(K_\nu^*)$, $\mu_{i,c}(K_{\nu,c}^*)$ denote the ith kernel moment of K_ν^* in the interior and $K_{\nu,c}^*$ in the boundary. The asymptotic properties of $\hat{F}_L^{(\nu)}(x)$ are summarized in the next theorem.

Theorem 1 *Assume that for $l=0,\ldots,\nu+1$, $K^{(l)}$ is bounded, absolutely integrable, with finite second moments and F_T is $l+2$ times differentiable. Assume* *also that as* $n \to +\infty$, $h_\nu \to 0$, $nh_\nu^{2\nu} \to +\infty$ *and* $b/h_\nu \to 0$. *Then, the asymptotic bias and variance of* $\hat{F}_L^{(\nu)}(x)$ *are given by*

$$b_{L(,c)}(x) = h_\nu^2 \frac{\nu!}{(\nu+2)!}\mu_{\nu+2(,c)}(K^*_{\nu(,c)})F_T^{(\nu+2)}(x) + o(h_\nu^2),$$

$$\sigma^2_{L(,c)}(x) = \frac{(\nu!)^2}{nh_\nu^{2\nu}}\Bigg[G(x) - 2h_\nu g(x)\int tK^*_{\nu(,c)}(s)W^*_{\nu(,c)}(s)\,ds$$
$$-\left\{F_T^{(\nu)}(x) + h_\nu^2\nu!((\nu+2)!)^{-1}\mu_{\nu+2,c}(K^*_{\nu(,c)})F_T^{(\nu+2)}(x)\right\}^2\Bigg]$$
$$+ O(n^{-1}h_\nu^{2\nu}) + o(h_\nu^4),$$

where

$$g(x) = f_T(x)(1-H(x))^{-1},\quad G(x) = \int_0^x g(t)\,dt,\quad W^*_{\nu(,c)}(s) = \int_{-\infty}^s K^*_{\nu(,c)}(u)\,du.$$

Further,

$$\hat{F}_L^{(\nu)}(x) \sim N\left(F_T^{(\nu)}(x) + b_{L(,c)}(x), \sigma^2_{L(,c)}(x)\right).$$

Remark 1 *The asymptotic properties of* $\hat{F}_L^{(\nu)}(x)$ *in Theorem 1 show that the estimate automatically achieves boundary corrections. In the interior the estimate behaves like a conventional kernel estimate, e.g., the survival function estimate of [10].*

Remark 2 *Theorem 1 also implies that the derivative order leaves the bias rate of convergence unaffected. Further the second term of the variance expression is negative and this indicates that kernel smoothing improves the estimate's variance compared to the variance of* $\hat{F}_T(x)$ *by a second-order effect.*

Remark 3 *Theorem 1 also shows that random right censoring does affect the variance leading term because of the survival function* $1 - H(x)$ *in the denominator. As a result it is expected that the censored data estimate to be more variable in practice than its complete sample counterpart.*

4 Proofs and Auxiliary Lemmas

Let

$$\tilde{f}_T(x_i) = \frac{1}{n}\sum_{j=1}^n \frac{c_{ij}}{1-H(x_i)}.$$

From [12], $\sup_x |\hat{H}(x) - H(x)| = O_p(n^{-1/2})$ and thus

$$\frac{1}{1-\hat{H}(x_i)} = \frac{1}{1-H(x_i)+H(x_i)-\hat{H}(x_i)} = \frac{1}{1-H(x_i)} \frac{1}{1+\frac{H(x_i)-\hat{H}(x_i)}{1-H(x_i)}}$$
$$= \frac{1}{1-H(x_i)} \sum_{k=0}^{+\infty} (-1)^k \left(\frac{H(x_i)-\hat{H}(x_i)}{1-H(x_i)} \right)^k = \frac{1}{1-H(x_i)} \left\{ 1 + O_p\left(n^{-1/2}\right) \right\}.$$

Therefore, $\hat{f}_T(x_i)$ can be approximated asymptotically by $\tilde{f}_T(x_i)$ with negligible error. For this reason, we equivalently prove theorem 1 for estimator

$$\tilde{F}_L^{(\nu)}(x) = \frac{\nu!}{h^\nu} \sum_{i=1}^{g} W_{\nu,c}^* \left(\frac{x_i - x}{h} \right) \tilde{f}_T(x_i)(1+o(1)).$$

Lemma 1 *Assume that F_T is twice differentiable, continuous and that $b = o(h)$, then, as $n \to \infty$,*

$$\mathbb{E}c_{kj} = \mathbb{E}c_{kj}^2 = \mathbb{E}c_{kj}^3 = \mathbb{E}c_{kj}^4 = bf_T(x_k)(1-H(x_k)) + o(b), \tag{6}$$
$$\mathbb{E}(c_{kj}c_{kr}) = \mathbb{E}(c_{kj}c_{kr})^2 = \mathbb{E}(c_{kj}^2 c_{kr}) = \mathbb{E}(c_{kj}^3 c_{kr}) = b^2 f_T^2(x_k)(1-H(x_k))^2 + o(b^2) \text{ for } r \neq j, \tag{7}$$
$$\mathbb{E}(c_{kj}c_{kr}c_{kl}) = \mathbb{E}(c_{kj}^2 c_{kr}c_{kl}) = b^3 f_T^3(x_k)(1-H(x_k))^3 + o(b^3) \text{ for } r \neq j \neq l, \tag{8}$$
$$\mathbb{E}(c_{kj}c_{kr}c_{kl}c_{kt}) = b^4 f_T^4(x_k)(1-H(x_k))^4 + o(b^4) \text{ for } r \neq j \neq l \neq t, \tag{9}$$

where all r, j, l, t above are between 1 and g.

Proof First note that conditioning on $X_j = y$ and $\delta_j = 1$, for fixed k and j,

$$\mathbb{E}\left\{ \mathbf{1}_{[x_k-\frac{b}{2}, x_k+\frac{b}{2}]}(X_j, \delta_j = 1) \right\} = \int_{x_k-\frac{b}{2}}^{x_k+\frac{b}{2}} f_T(y)(1-H(y))\, dy$$
$$= bf_T(x_k)(1-H(x_k)) + o(b). \tag{10}$$

Now, using

$$\mathbb{E}(c_{kj}c_{kr}) = \mathbb{E}\mathbf{1}_{[x_k-\frac{b}{2}, x_k+\frac{b}{2}]}(X_j, \delta_j = 1)\mathbb{E}\mathbf{1}_{[x_k-\frac{b}{2}, x_k+\frac{b}{2}]}(X_r, \delta_r = 1)$$
$$= b^2 f_T^2(x_k)(1-H(x_k))^2 + o(b),$$

together with the fact that

$$\mathbb{E}c_{kj}^2 c_{kr} = \mathbb{E}c_{kj}^2 \mathbb{E}c_{kr} = \mathbb{E}c_{kj}\mathbb{E}c_{kr} = (\mathbb{E}c_{kj})^2 = \mathbb{E}(c_{kj}c_{kr})^2 = \mathbb{E}c_{kj}^3 c_{kr}$$

completes the proof of (7). The proofs of (6), (8), and (9) follow similarly.

4.1 Proof of Theorem 1

The proof of the theorem is based on a combination of lemma 1 with straightforward algebra and well-known results. Thus only a sketch is provided here and only for the boundary case $x = ch_\nu, 0 < c < 1$ as the result for the interior follows by letting $c \to +\infty$. Combining (3) and (4), for $\nu = 0, 1, 2, \ldots$, and by Lemma 1,

$$\mathbb{E}\tilde{F}_L^{(\nu)}(x) = \frac{\nu!}{nh_\nu^\nu}\sum_{i=1}^{g} W_{\nu,c}^*\left(\frac{x_i - x}{h_\nu}\right)\frac{bf_T(x_i)(1 - H(x_i))}{1 - H(x_i)}(1 + o(b)).$$

By lemma 2 of [1], we have

$$\left|\sum_{i=1}^{g} W_{\nu,c}^*\left(\frac{x_i - x}{h_\nu}\right)bf_T(x_i) - \int W_{\nu,c}^*\left(\frac{u - x}{h_\nu}\right)f_T(u)\,du\right| \leq \frac{b^2}{4}\int\left(W_{\nu,c}^*\left(\frac{u - x}{h_\nu}\right)f_T(u)\right)''du. \quad (11)$$

Then, by applying integration by parts, performing the change of variable $u - x = sh_\nu$, Taylor expanding around x and using the boundary conditions

$$\int_{-c}^{+\infty} u^q K_{\nu,c}^*(u)\,du = \delta_{\nu,q},\ 0 \leq \nu, q \leq p$$

where $\delta_{\nu,q}$ is Kronecker's delta, which establishes the bias expression. The variance is treated similarly by combining Lemma 1 and approximating the sums by integrals based on lemma 2 of [1]. Now, for $\nu = 0, 1, \ldots$ set

$$\tilde{F}_L^{(\nu)}(x) = W = \sum_{j=1}^{n} W_j,\quad W_j = \frac{\nu!}{nh_\nu^\nu}\sum_{i=1}^{g} W_{\nu,c}^*\left(\frac{x_i - x}{h_\nu}\right)\frac{c_{ij}}{1 - H(x_i)}.$$

Note that the random variable W_j depends only on the pair (X_j, δ_j) and thus

$$\tilde{F}_L^{(\nu)}(x) = W = \sum_{j=1}^{n} W_j$$

is a sum of independent random variables. Hence, the asymptotic normality of $\tilde{F}_L^{(\nu)}(x)$ will result by the application of the Lyapunov Central Limit Theorem (Theorem 4.9 in [14]) according to which a sufficient condition for

$$\frac{W - \mathbb{E}W}{\sqrt{\mathbb{V}\mathrm{ar}(W)}} \xrightarrow{d} N(0, 1)$$

to hold is

$$\lim_{n\to\infty} \mathbb{V}\mathrm{ar}(W)^{-3/2} \sum_{i=1}^{n} \mathbb{E}|W_i - \mathbb{E}W_i|^3 = 0. \tag{12}$$

To verify the condition, first note that fixing j and using the nonnegativity of W_j in the first step below, using an approximation similar to (11) in the third step, in the fourth step the change of variable $u - x = th_\nu$, subsequently expanding $f_T(x + th_\nu)$ in Taylor series around x and by the assumption that K_ν, and therefore its integral over its support is bounded, yields

$$\begin{aligned}
\sum_{j=1}^{n} \mathbb{E}|W_j - \mathbb{E}W_j|^3 &\le n(8\mathbb{E}(|W_j|^3) + 8|\mathbb{E}(W_j)|^3) \le 16n|\mathbb{E}(W_j)|^3 \\
&= n\frac{(\nu!)^3}{n^3h_\nu^{3\nu}} \left\{ \sum_{i=1}^{g} W_{\nu,c}^* \left(\frac{x_i - x}{h_\nu} \right) \frac{\mathbb{E}c_{ij}}{1 - H(x_i)} \right\}^3 \\
&= \frac{(\nu!)^3}{n^2h_\nu^{3\nu}} \left\{ \sum_{i=1}^{g} W_{\nu,c}^* \left(\frac{x_i - x}{h_\nu} \right) bf_T(x_i)(1 + o(b)) \right\}^3 \\
&\le \frac{(\nu!)^3}{n^2h_\nu^{3\nu}} \left\{ \int W_{\nu,c}^* \left(\frac{u - x}{h_\nu} \right) f_T(u)(1 + o(b))\, du \right\}^3 \\
&= \frac{(\nu!)^3}{n^2h_\nu^{3\nu}} \left\{ h_\nu \int W_{\nu,c}^*(t) f_T(x + th_\nu)(1 + o(b))\, dt \right\}^3 \\
&= O(n^{-2}h_\nu^{-(3\nu-3)}).
\end{aligned} \tag{13}$$

Also, fixing j in the second step below, using in the fourth step twice an approximation similar to (11), applying the change of variable $u - x = th_\nu$ and subsequently Taylor expanding $f_T(x + th_\nu)(1 - H(x + th_\nu))^{-1}$ and $f_T^2(x + th_\nu)$ around x and using (as in obtaining (13)) the fact that $W_{\nu,c}^*$ is bounded, the variance of W becomes

$$\begin{aligned}
\mathrm{Var}(W) &= \frac{(\nu!)^2}{n^2h_\nu^{2\nu}} \mathrm{Var} \left\{ \sum_{j=1}^{n} \sum_{i=1}^{g} W_{\nu,c}^* \left(\frac{x_i - x}{h_\nu} \right) \frac{c_{ij}}{1 - H(x_i)} \right\} \\
&\le \frac{(\nu!)^2}{nh_\nu^{2\nu}} \sum_{i=1}^{g} W_{\nu,c}^* \left(\frac{x_i - x}{h_\nu} \right)^2 \frac{\left\{ \mathbb{E}c_{ij}^2 - (\mathbb{E}c_{ij})^2 \right\}}{(1 - H(x_i))^2} \\
&= \frac{(\nu!)^2}{nh_\nu^{2\nu}} \sum_{i=1}^{g} W_{\nu,c}^* \left(\frac{x_i - x}{h_\nu} \right)^2 \frac{bf_T(x_i)(1 - H(x_i)) - \{bf_T(x_i)(1 - H(x_i))\}^2}{(1 - H(x_i))^2} \\
&\quad \times (1 + o(b)) \\
&= \frac{(\nu!)^2}{nh_\nu^{2\nu}} \int W_{\nu,c}^* \left(\frac{u - x}{h_\nu} \right)^2 \left\{ \frac{f_T(u)}{1 - H(u)} - bf_T^2(u) \right\} du (1 + o(b)) \\
&= O(n^{-1}h_\nu^{-(2\nu-2)})(1 + o(b)).
\end{aligned} \tag{14}$$

Using (13) and (14) back to (12) yields

$$\lim_{n\to\infty} \mathbb{V}\mathrm{ar}(W)^{-3/2}\sum_{i=1}^{n}|W_i - \mathbb{E}W_i|^3 = O\left(n^{3/2}h_\nu^{3(2\nu-2)/2}n^{-2}h_\nu^{-3\nu+3}\right) = O\left(n^{-1/2}\right)$$

which verifies the condition and finishes the proof.

Acknowledgements The authors deeply thank an anonymous referee for the valuable comments and suggestions provided, leading in significantly improving this article.

References

1. Bagkavos, D., Patil, P.N.: Local polynomial fitting in failure rate estimation. IEEE Transactions on Reliability **56**, 126–163 (2008)
2. Bagkavos, D.: Local linear hazard rate estimation and bandwidth selection. Ann. Inst. Stat. Math. **63**, 1019–1046 (2011)
3. Cheng, M.-Y., Peng, L.: Regression modeling for nonparametric estimation of distribution and quantile functions. Statist. Sinica. **12**, 1043–1060 (2002)
4. Cheng, M.-Y., Fan, J., Marron, J.S.: On automatic boundary corrections. Ann. Statist. **25**, 1691–1708 (1997)
5. Fan, J., Gijbels, I.: Local polynomial modeling and its applications. Chapman and Hall, London (1996)
6. Jones, M.C., Sheather, S.J.: Using non-stochastic terms to advantage in kernel-based estimation of integrated squared density derivatives. Statistics and Probability Letters **11**, 511–514 (1991)
7. Huang, Z., Maesono, Y.: Edgeworth expansion for kernel estimators of a distribution function. Bulletin of informatics and cybernetics **10**, 1–10 (2014)
8. Hall, P., Sheather, S., Jones, M.C., Marron, J.S.: On optimal data based bandwidth selection in kernel density estimation. Biometrika **78**, 521–530 (1991)
9. Garcia-Soidán, P.H., González-Manteiga, W., Prada-Sánchez, J.M.: Edgeworth expansions for nonparametric distribution estimation with applications. J. Statist. Plann. Inference **65**, 213–231 (1997)
10. Gulati, S., Padgett, W.J.: Families of smooth confidence bands for the survival function under the general random censorship model. Lifetime Data Anal. **2**, 349–362 (1996)
11. Kaplan, E.L., Meier, P.: Nonparametric estimation from incomplete observations. J. Amer. Statist. Assoc. **53**, 457–481 (1958)
12. Karunamuni, R., Yang, S.: Weak and strong uniform consistency rates of kernel density estimates for randomly censored data. Can. J. Statist. **19**, 349–359 (1991)
13. Marron, J.S., Padgett, W.J.: Asymptotically optimal bandwidth selection for kernel density estimators from randomly right censored samples. Ann. Statist. **15**, 1520–1535 (1987)
14. Petrov, V.: Limit Theorems of Probability Theory: sequences of independent random variables. Oxford University Press, New York (1985)
15. Susarla, V., Tsai, W. Y.and Van Ryzin, J.(1984). A Buckley–James-type estimator for the mean with censored data, *Biometrika*, **71** 624–625

Density Estimation Using Multiscale Local Polynomial Transforms

Maarten Jansen

Abstract The estimation of a density function with an unknown number of singularities or discontinuities is a typical example of a multiscale problem, with data observed at nonequispaced locations. The data are analyzed through a multiscale local polynomial transform (MLPT), which can be seen as a slightly overcomplete, non-dyadic alternative for a wavelet transform, equipped with the benefits from a local polynomial smoothing procedure. In particular, the multiscale transform adopts a sequence of kernel bandwidths in the local polynomial smoothing as resolution level-dependent, user-controlled scales. The MLPT analysis leads to a reformulation of the problem as a variable selection in a sparse, high-dimensional regression model with exponentially distributed responses. The variable selection is realized by the optimization of the l1-regularized maximum likelihood, where the regularization parameter acts as a threshold. Fine-tuning of the threshold requires the optimization of an information criterion such as AIC. This paper develops discussions on results in [9].

Keywords Wavelets · Sparsity · Local polynomials · Kernel · Smoothing

1 Introduction

Due to its natural intermittency, the estimation of a non-uniform density can be described as a nonequispaced multiscale problem, especially when the density contains singularities. Indeed, when the number and the locations of the singularities remain unknown, then the estimation procedure is deemed to go through all possible combinations of locations and intersingular distances. Also, since a given bandwidth in a kernel-based method may be too small in a region of low intensity and too large in a region of high intensity, a local choice of the bandwidth can be considered as an instance of multiscale processing, where the bandwidth is seen as a notion of scale.

M. Jansen (✉)
Université libre de Bruxelles, Bruxelles, Belgium
e-mail: maarten.jansen@ulb.ac.be

M. La Rocca et al. (eds.), *Nonparametric Statistics*, Springer Proceedings in Mathematics & Statistics 339, https://doi.org/10.1007/978-3-030-57306-5_23

A popular class of multiscale methods in smoothing and density estimation is based on a wavelet analysis of the data. The classical wavelet approach for density estimation [3, 6] requires an evaluation of the wavelet basis functions in the observed data or otherwise a binning of the data into fine scale intervals, defined by equispaced knots on which the wavelet transform can be constructed. The preprocessing for the equispaced (and possibly dyadic) wavelet analysis may induce some loss of details about the exact values of the observations.

This paper works with a family of multiscale transforms constructed on nonequispaced knots. With these constructions and taking the observations as knots, no information is lost at this stage of the analysis. The construction of wavelet transforms on irregular point sets is based on the lifting scheme [11, 12]. Given the transformation matrix that maps a wavelet approximation at one scale onto the approximation and offsets at the next coarser scale, the lifting scheme factorizes this matrix into a product of simpler, readily invertible operations. Based on the lifting factorization, there exist two main directions in the design of wavelets on irregular point sets. The first direction consists of the factorization of basis functions that are known to be refinable, to serve as approximation basis, termed scaling basis in a wavelet analysis. The wavelet basis for the offsets between successive scales is then constructed within the lifting factorization of the refinement equation, taking into account typical design objectives such as vanishing moments and control of variance inflation. Examples of such existing refinable functions are B-spline functions defined on nested grids of knots [8]. The second approach for the construction of wavelets on irregular point sets does not factorize a scheme into lifting steps. Instead, it uses an interpolating or smoothing scheme as a basic tool in the construction of a lifting step from scratch. Using interpolating polynomials leads to the Deslauriers-Dubuc refinement scheme [2, 4]. To this refinement scheme, a wavelet transform can be associated by adding a single lifting step, designed for vanishing moments and control of variance inflation, as in the case of factorized refinement schemes. This paper follows the second approach, using local polynomial smoothing [5, Chapter 3] as a basic tool in a lifting scheme. For reasons explained in Sect. 2, the resulting Multiscale Local Polynomial Transform (MLPT) is no longer a wavelet transform in the strict sense, as it must be slightly overcomplete. Then, in Sect. 3, the density estimation problem is reformulated in a way that it can easily be handled by an MLPT. Section 4 discusses aspects of sparse selection and estimation in the MLPT domain for data from a density estimation problem. In Sect. 5, the sparse selection is finetuned, using information criteria and defining the degrees of freedom in this context. Finally, Sect. 6 presents some preliminary simulation results and further outlook.

2 The Multiscale Local Polynomial Transform (MLPT)

Let $\boldsymbol{Y}$ be a sample vector from the additive model $Y_i = f(x_i) + \sigma_i Z_i$, where the covariates x_i may be non-equidistant and the noise Z_i may be correlated. The underlying function, $f(x)$, is assumed to be approximated at resolution level J by a linear combination of basis functions $\varphi_{J,k}(x)$, in

$$f_J(x) = \sum_{k=0}^{n_J-1} \varphi_{J,k}(x) s_{J,k} = \boldsymbol{\Phi}_J(x) \boldsymbol{s}_J,$$

where $\Phi_J(x)$ is a row vector containing the basis functions. The choice of coefficients $\boldsymbol{s}_J$ is postponed to the moment when the basis functions are specified. At this moment, one could think of a least squares projection as one of the possibilities.

The Multiscale Local Polynomial Transform (MLPT) [7] finds the sparse coefficient vector $\boldsymbol{v}$ in $\boldsymbol{s}_J = \mathbf{X}\boldsymbol{v}$, using a linear operation $\boldsymbol{v} = \widetilde{\mathbf{X}}\boldsymbol{s}_J$. Just like in wavelet decomposition, the coefficient vector of several subvectors $\boldsymbol{v} = [\boldsymbol{s}_L^T \; \boldsymbol{d}_L^T \; \boldsymbol{d}_{L+1}^T \; \dots \; \boldsymbol{d}_{J-1}^T]^T$, leading to the following basis transformation

$$\Phi_J(x)\boldsymbol{s}_J = \Phi_J(x)\mathbf{X}\boldsymbol{v} = \Phi_L(x)\boldsymbol{s}_L + \sum_{j=L}^{J-1} \Psi_j(x)\boldsymbol{d}_j,$$

where we introduced $\Phi_L(x)$ and $\Psi_j(x)$ for the submatrices of the transformed basis $\Phi_J(x)\mathbf{X}$, corresponding to the subvectors of the coefficient vector $\boldsymbol{v}$. The detail vectors $\boldsymbol{d}_j$ are associated to successive resolution levels through the decomposition algorithm, corresponding to the analysis matrix $\widetilde{\mathbf{X}}$,
for $j = J-1, J-2, \dots, L$

- **Subsamplings**, i.e., keep a subset of the current approximation vector, $\boldsymbol{s}_{j+e,e} = \mathbf{J}_j \boldsymbol{s}_{j+1}$, with $\mathbf{J}_j$ a $n_j \times n_{j+1}$ submatrix of the identity matrix.
- **Prediction**, i.e., compute the detail coefficients at scale j as offsets from a prediction based on the subsample.
  $\boldsymbol{d}_j = \boldsymbol{s}_{j+1} - \mathbf{P}_j \boldsymbol{s}_{j+1,e}$
- **Update**, the remaining approximation coefficients. The idea is that $\boldsymbol{s}_j$ can be interpreted as smoothed, filtered, or averaged values of $\boldsymbol{s}_{j+1}$.
  $\boldsymbol{s}_j = \boldsymbol{s}_{j+1,e} + \mathbf{U}_j \boldsymbol{d}_j$

Before elaborating on the different steps of this decomposition, we develop the inverse transform $\mathbf{X}$ by straightforwardly undoing the two lifting steps in reverse order.
for $j = L, L+1, \dots, J-1$

- **Undo update**, using $\boldsymbol{s}_{j+1,e} = \boldsymbol{s}_j - \mathbf{U}_j \boldsymbol{d}_j$.
- **Undo prediction**, using $\boldsymbol{s}_{j+1} = \boldsymbol{d}_j + \mathbf{P}_j \boldsymbol{s}_{j+1,e}$.

2.1 Local Polynomial Smoothing as Prediction

The transform in this paper adopts a smoothing operation as prediction, thus incorporating the covariate values as parameters of the analysis. As an example, the Nadaraya–Watson kernel prediction leads to

$$P_{j;k,\ell} = \frac{K\left(\frac{x_{j+1,k}-x_{j,\ell}}{h_{j+1}}\right)}{\sum_{l=1}^{n_j} K\left(\frac{x_{j+1,k}-x_{j,l}}{h_{j+1}}\right)}.$$

In this expression, $K(u)$ denotes a kernel function, i.e., a positive function with integral 1. The parameter h_{j+1} is the bandwidth. While in (uniscale) kernel smoothing this is a smoothing parameter, aiming at optimal balance between bias and variance in the estimation, it acts as a user-controlled scale parameter in a Multiscale Kernel Transform (MKT). This is in contrast to a discrete wavelet transform, where the scale is inherently fixed to be dyadic, i.e., the scale at level j is twice the scale at level $j + 1$. In an MKT, and also in the forthcoming MLPT, the scale can be chosen in a data adaptive way, taking the irregularity of the covariate grid into account. For instance, when the covariates can be considered as ordered independent realizations from a uniform density, it is recommended that the scale is taken to be $h_j = h_0 \log(n_j)/n_j$ [10]. The scales at fine resolution levels are then a bit larger, allowing them cope up with the non-equidistance of the covariates.

A slightly more complex prediction, adopted in this paper, is based on local polynomial smoothing. It fills the kth row of $\mathbf{P}_j$ with $\boldsymbol{P}(x_{j+1,k})$, where the row vector $\boldsymbol{P}_j(x)$ is given by

$$\boldsymbol{P}_j(x) = \boldsymbol{X}^{(\tilde{p})}(x)\left(\mathbf{X}_j^{(\tilde{p})^T}\mathbf{W}_j(x)\mathbf{X}_j^{(\tilde{p})}\right)^{-1},$$

with the row vector of power functions, $\boldsymbol{X}^{(\tilde{p})}(x) = [1 \; x \; \ldots \; x^{\tilde{p}-1}]$ and the corresponding Vandermonde matrix at resolution level j, $\mathbf{X}_j^{(\tilde{p})} = [\mathbf{1} \; \boldsymbol{x}_j \; \ldots \; \boldsymbol{x}_j^{\tilde{p}-1}]$. The diagonal matrix of weight functions is given by $(\mathbf{W}_j)_{\ell\ell}(x) = K\left(\frac{x-x_{j,\ell}}{h_j}\right)$.

The prediction matrix has dimension $n_{j+1} \times n_j$. This expansive or redundant prediction is in contrast to lifting schemes for critically downsampled wavelet transform, such as the Deslauriers–Dubuc or B-spline refinement schemes. In these schemes, the prediction step takes the form $\boldsymbol{d}_j = \boldsymbol{s}_{j+1,o} - \mathbf{P}_j\boldsymbol{s}_{j+1,e}$, where $\boldsymbol{s}_{j+1,o} = \mathbf{J}_j^c\boldsymbol{s}_{j+1}$, with $\mathbf{J}_j^c$ the $(n_{j+1} - n_j) \times n_{j+1}$ subsampling operation, complementary to $\mathbf{J}_j$. In the MLPT, a critical downsampling with $\mathbf{J}_j$ and $\mathbf{J}_j^c$ would lead to fractal-like basis functions [7]. The omission of the complementary subsampling leads to slight redundancy, where n data points are transformed into roughly $2n$ MLPT coefficients, at least if n_j is approximately half of n_{j+1} at each scale. The corresponding scheme is known in signal processing literature as the Laplacian pyramid [1]. With an output size of $2n$, the MLPT is less redundant than the non-decimated wavelet transform (cycle spinning, stationary wavelet transform) which produces outputs of size $n\log(n)$. Nevertheless, the inverse MLPT shares with the non-decimated wavelet transform an additional smoothing occurring in the reconstruction after processing. This is because processed coefficients are unlikely to be exact decompositions of an existing data vector. The reconstruction thus involves some sort of projection.

2.2 The Update Lifting Step

The second lifting step, the update $\mathbf{U}_j$, serves multiple goals, leading to a combination of design conditions [8]. An important objective, especially in the context of density estimation, is to make sure that all functions in $\Psi_j(x)$ have zero integral. When $f_j(x) = \Phi_L(x)\boldsymbol{s}_L + \sum_{j=L}^{J-1} \Psi_j(x)\boldsymbol{d}_j$, then any processing that modifies the detail coefficients $\boldsymbol{d}_j$, e.g., using thresholding, preserves the integral of $f_j(x)$, which is interesting if we want to impose that $\int_{-\infty}^{\infty} f_j(x)dx = 1$ for an estimation or approximation of a density function. Another important goal of the update, leading to additional design conditions, is to control the variance propagation throughout the transformation. This prevents the noise from a single observation from proceeding unchanged all the way to coarse scales.

2.3 The MLPT Frame

Examples of MLPT functions are depicted in Fig. 1. It should be noted that these functions are defined on an irregular grid of knots. Nothing of the grid irregularity is reflected in the approximation and detail functions $\Phi_L(x)$ and $\Psi_j(x)$. Also, as the detail functions form an overcomplete set, they are not basis functions in the strict sense. Instead, the set of $\Phi_L(x)$ and $\Psi_j(x)$ for $j = L, L+1, \ldots, J-1$ is called a frame.

Unlike in a B-spline wavelet decomposition, observations in the knots are valid fine scale approximation coefficients [9]. More precisely, the approximation

$$f_J(x) = \sum_{i=1}^{n} f(x_i)\varphi_{J,i}(x),$$

has a convergence rate equal to that of least squares projection. This property is important when incorporating a MLPT model into the regression formulation of the problem of the density estimation problem in Sect. 3.

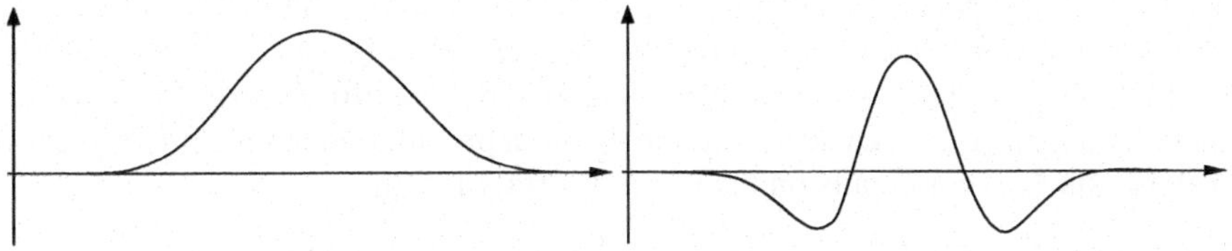

Fig. 1 Left panel: approximation function, i.e., one element of $\Phi_L(x)$. Right panel: detail or offset function, i.e., one element of $\Psi_j(x)$. It holds that $\int_{-\infty}^{\infty} \Psi_j(x)dx = \mathbf{0}_j^T$

2.4 The MLPT on Highly Irregular Grids

The regression formulation of the density estimation problem in Sect. 3 will lead to regression on highly irregular grids, that is, grids that are far more irregular than ordered observations from a random variable. On these grids, it is not possible to operate at fine scales, even if these scales are a bit wider than in the equidistant case, as discussed in Sect. 2.1. In order to cope with the irregularity, the fine scales would be so wide that fine details are lost, and no asymptotic result would be possible. An alternative solution, adopted here, is to work with dyadic scales, but only processing coefficients that have sufficient nearby neighbors within the current scale. Coefficients in sparsely sampled neighborhoods are forwarded to coarser scales. The implementation of such a scheme requires the introduction of a smooth transition between active and non-active areas at each scale [9].

More precisely, the reconstruction from the local polynomial prediction $\boldsymbol{s}_{j+1} = \boldsymbol{d}_j + \mathbf{P}_j \boldsymbol{s}_{j+1,e}$, is replaced by a weighted form

$$\boldsymbol{s}_{j+1} = \mathbf{Q}_{j+1}\left(\mathbf{P}_j \widetilde{\boldsymbol{s}}_j + \boldsymbol{d}_j\right) + (\mathbf{I}_{j+1} - \mathbf{Q}_{j+1})\widetilde{\mathbf{J}}_j^T \widetilde{\boldsymbol{s}}_j. \tag{1}$$

The diagonal matrix $\mathbf{Q}_{j+1}$ has values between 0 and 1. The value is 0 when a coefficient is not surrounded by enough neighbors to apply a regular local polynomial prediction $\mathbf{P}_j$, and it gradually (not suddenly, that is) tends to one in areas with sufficiently dense observations to apply proper polynomial prediction.

3 A Regression Model for Density Estimation

Let X be a sample of independent realization from an unknown density $f_X(x)$ on a bounded interval, which we take, without loss of generality, to be $[0, 1]$. The density function has an unknown number of singularities, i.e., points $x_0 \in [0, 1]$ where $\lim_{x \to x_0} f_X(x) = \infty$, as well as other discontinuities.

As in [9], we consider the spacings $\Delta X_{n;i} = X_{(n;i)} - X_{(n;i-1)}$, i.e., the differences between the successive ordered observations $X_{(n;i)}$. Then, by the mean value theorem, we have that there exists a value $\overline{\xi}_{n;i} \in [X_{(n;i-1)}, X_{(n;i)}]$ for which $f_X(\overline{\xi}_{n;i})\Delta X_{n;i} = \Delta U_{n;i}$, where $\Delta U_{n;i} = U_{(n;i)} - U_{(n;i-1)} = F_X(X_{(n;i)}) - F_X(X_{(n;i-1)})$. Unfortunately, the value of $\overline{\xi}_{n;i}$ cannot be used as such in the subsequent asymptotic result, due to technical issues in the proof. Nevertheless, for a fairly free choice of $\xi_{n;i} \in [X_{(n;i-1)}, X_{(n;i)}]$, close to $\overline{\xi}_{n;i}$, the theorem provides nonparametric regression of $\Delta X_{n;i}$ on $\xi_{n;i}$. For details on the proof, we refer to [9].

Theorem 1 *Let $f_X(x)$ be an almost everywhere twice continuously differentiable density function on $x \in [0, 1]$. Define $A_{M,\delta} \subset [0, 1]$ as the set where $f_X(x) \geq \delta$ and $f_X'(x) \leq M$, with δ, M arbitrary, strictly positive real numbers. Then there exist values $\xi_{n;i} \in [X_{(n;i-1)}, X_{(n;i)}]$, so that with probability one, for all intervals $[X_{(n;i-1)}, X_{(n;i)}] \subset A_{M,\delta}$,* *the value of* $f_X(\xi_{n;i})(n+1)\Delta X_{n;i}$*, given the covariate* $\xi_{n;i}$*, converges in distribution to an exponential random variable, i.e.,*

$$f_X(\xi_{n;i})(n+1)\Delta X_{n;i}|\xi_{n;i} \stackrel{d}{\to} D \sim \exp(1), \, a.s.$$

We thus consider a model with exponentially distributed response variable $Y_i = (n+1)\Delta X_{n;i}|\xi_{n;i}$ and the vector of parameters $\theta_i = f_X(\xi_{n;i}) = 1/\mu_i$ with $\mu_i = E(Y_i)$, for which we propose a sparse MLPT model $\boldsymbol{\theta} = \mathbf{X}\boldsymbol{\beta}$, with $\mathbf{X}$ the inverse MLPT matrix defined on the knots in $\boldsymbol{\xi}$.

The formulation of the density estimation problem as a sparse regression model induces no binning or any other loss of information. On the contrary, the information on the values of X_i is duplicated: a first, approximative copy can be found in the covariate values $\xi_{n;i}$. A second copy defines the design matrix. The duplication prevents loss of information when in subsequent steps some sort of binning is performed on the response variables.

4 Sparse Variable Selection and Estimation in the Exponential Regression model

For the i.i.d. exponential responses $\boldsymbol{Y} \sim \exp(|\boldsymbol{\theta}|)$ with $\boldsymbol{\theta} = \mathbf{X}\boldsymbol{\beta}$, and $\mu_i = 1/\theta_i$, the score is given by $\nabla \log L(\boldsymbol{\theta}; \boldsymbol{Y}) = \mathbf{X}^T(\boldsymbol{Y} - \boldsymbol{\mu})$, so that the maximum ℓ_1 regularized log-likelihood estimator $\widehat{\boldsymbol{\beta}} = \arg\max_{\boldsymbol{\beta}} \log L(\boldsymbol{\beta}) - \lambda\|\boldsymbol{\beta}\|_1$ can be found by solving the Karush–Kuhn–Tucker (KKT) conditions

$$\begin{array}{ll} \mathbf{X}_j^T(\boldsymbol{Y}-\boldsymbol{\mu}) = \lambda \mathrm{sign}(\beta_j) & \text{if } \beta_j \neq 0, \\ \left|\mathbf{X}_j^T(\boldsymbol{Y}-\boldsymbol{\mu})\right| < \lambda & \text{if } \beta_j = 0. \end{array}$$

Even if we knew which components of $\boldsymbol{\beta}$ were nonzero, the KKT would still be highly nonlinear. This is in contrast to the additive normal model, where $\boldsymbol{\mu} = \mathbf{X}\boldsymbol{\beta}$. The estimator *given the selection* then follows from a shrunk least squares solution. Indeed, with $\mathcal{I}$ the set of selected components, we have $\widehat{\boldsymbol{\beta}}_{\mathcal{I}} = \left(\mathbf{X}_{\mathcal{I}}^T\mathbf{X}_{\mathcal{I}}\right)^{-1} \mathrm{ST}_\lambda\left(\mathbf{X}_{\mathcal{I}}^T\boldsymbol{Y}\right)$, where $\mathrm{ST}_\lambda(x)$ is the soft-threshold function. In the case of orthogonal design, i.e., when $\mathbf{X}_{\mathcal{I}}^T\mathbf{X}_{\mathcal{I}}$ is the identity matrix, and this reduces to straightforward soft-thresholding in the transformed domain. In the case of non-orthogonal, but still Riesz-stable, design, straightforward thresholding is still a good approximation and a common practice, for instance, in B-spline wavelet thresholding. For the model with exponential response, the objective is to find appropriate values of $\boldsymbol{S}_J$, so that $\widehat{\boldsymbol{\beta}} = \mathbf{X} \cdot \mathrm{ST}_\lambda(\widetilde{\mathbf{X}}\boldsymbol{S}_J)$. can be used as an estimator. For this we need at least that

(C1) the expected value of $\boldsymbol{S}_J$ is close to $\boldsymbol{\theta}$, so that $E(\widetilde{\mathbf{X}}\boldsymbol{S}_J) \approx \widetilde{\mathbf{X}}\boldsymbol{\theta} = \boldsymbol{\beta}$,
(C2) the MLPT decomposition $\boldsymbol{\beta} = \widetilde{\mathbf{X}}\boldsymbol{\theta}$ is sparse,

(C3) the MLPT decomposition of the errors, $\widetilde{\mathbf{X}}(\boldsymbol{S}_J - \boldsymbol{\theta})$ has no outliers, i.e., no heavy tailed distributions.

As $\theta_i = 1/\mu_i = 1/E(Y_i)$, it may be interesting to start the search for appropriate fine scale coefficients $S_{J,i}$ from $S_{J,i}^{[0]} = 1/Y_i$. Unfortunately, $S_{J,i}^{[0]}$ is heavy tailed. Experiments show that the heavy tails cannot be dealt properly by truncation of $1/Y_i$ in $S_{J,i}^{[1]} = \min(1/Y_i, s_{\max})$ without loss of substantial information about the position and nature of the singular points in the density function.

Therefore, a prefilter with a binning effect is proposed; however, keeping track of the original values of $\boldsymbol{Y}$ through the covariate values in the design $\mathbf{X}$. More precisely, let

$$\boldsymbol{S}_J = \boldsymbol{\Pi}\mathbf{D}_{h_{J,0}}\widetilde{\boldsymbol{\Pi}}\boldsymbol{S}_J^{[0]}. \tag{2}$$

The matrices $\widetilde{\boldsymbol{\Pi}}$ and $\boldsymbol{\Pi}$ represent a forward and inverse, one coefficient at-a-time, unbalanced Haar transform defined on the data adaptive knots $t_{J,i} = \sum_{k=0}^{i-1} Y_k$ and $t_{J,0} = 0$. An Unbalanced Haar transform on the vector of knots $\boldsymbol{t}_J$ is defined by

$$s_{j,k} = \frac{\Delta_{j+1,2k}s_{j+1,2k} + \Delta_{j+1,2k+1}s_{j+1,2k+1}}{\Delta_{j,k}} = \frac{\Delta_{j+1,2k}s_{j+1,2k} + \Delta_{j+1,2k+1}s_{j+1,2k+1}}{\Delta_{j+1,2k} + \Delta_{j+1,2k+1}},$$

$$d_{j,k} = s_{j+1,2k+1} - s_{j,k},$$

where $\Delta_{J,k} = t_{J,k} - t_{J,k} = Y_k$ and $\Delta_{j,k} = \Delta_{j+1,2k} + \Delta_{j+1,2k+1}$. In the coefficient at-a-time version, the binning operation $\Delta_{j+1,2k} + \Delta_{j+1,2k+1}$ takes place on a single pair $\Delta_{j+1,k}$ and $\Delta_{j+1,k+1}$, chosen so that $\Delta_{j,k} = \Delta_{j+1,k} + \Delta_{j+1,k+1}$ is as small as possible. Finally, the diagonal matrix $\mathbf{D}_{h_{J,0}}$ in (2), replaces all details $d_{j,k}$ by zero for which the scale $\Delta_{j,k}$ is smaller than a minimum width h_J. The overall effect of (2) is that small values in $\boldsymbol{Y}$ are recursively added to their neighbors until all binned values are larger than $h_{J,0}$. For values of $h_{J,0}$ sufficiently large, it can be analyzed that the coefficients of $\boldsymbol{S}_J$ are close to being normally distributed with expected value asymptotically equal to $\boldsymbol{\theta}$ and variance asymptotically equal to $\boldsymbol{\theta}/h_{J,0}$ [9]. Unfortunately, a large value of $h_{J,0}$ also introduces binning bias. In order to reduce this bias and to let $h_{J,0}$ be sufficiently large, the choice of $h_{J,0}$ is combined with a limit on the number of observations in one bin [9].

5 Fine-Tuning the Selection Threshold

The estimator $\widehat{\boldsymbol{\beta}} = \mathbf{X} \cdot \mathrm{ST}_\lambda(\widetilde{\mathbf{X}}\boldsymbol{S}_J)$. applies a threshold on the MLPT of $\boldsymbol{S}_J$. The input $\boldsymbol{S}_J$ is correlated and heteroscedastic, while the transform is not orthogonal. For all these reasons, the errors on $\widetilde{\mathbf{X}}\boldsymbol{S}_J$ are correlated and heteroscedastic. In an additive normal model where variance and mean are two separate parameters, the threshold would be taken proportional to the standard deviation. In the context of the exponential model with approximate variance function $\mathrm{var}(S_{J,i}) = E(S_{J,i})/h_{J,0}$, coefficients with large variances tend to carry more information, i.e., they have a larger expected value as well. As a result, there is no argument for a threshold linearly depending on the local standard deviation. This paper adopts a single threshold for all coefficients to begin with. Current research also investigates the use of block thresholding methods.

The threshold or any other selection parameter can be finetuned by optimization of the estimated distance between the data generating process and the model under consideration. The estimation of that distance leads to an information criterion. This paper works with an Akaike's Information Criterion for the estimation of the Kullback–Leibler distance. As data generating process, we consider the (asymptotic) independent exponential model for the spacings, and not the asymptotic additive, heteroscedastic normal model for $\boldsymbol{S}_J$. This choice is motivated by the fact that a model for $\boldsymbol{S}_J$ is complicated as it should account for the correlation structure, while the spacings are nearly independent. Moreover, fine-tuning w.r.t. the spacings is not affected by the loss of information in the computation of $\boldsymbol{S}_J$.

The resulting information criterion is given by the sum of two terms, $\text{AIC}(\widehat{\boldsymbol{\theta}}) = \widehat{\ell}(\widehat{\boldsymbol{\theta}}) - \widehat{\nu}(\widehat{\boldsymbol{\theta}})$. The first term is the empirical log-likelihood

$$\widehat{\ell}(\widehat{\boldsymbol{\theta}}) = \sum_{i=1}^{n} \left[\log(\widehat{\theta}_i) - \widehat{\theta}_i Y_i\right],$$

while the second term is an estimator of the degrees of freedom

$$\nu(\widehat{\boldsymbol{\theta}}) = E\left[\widehat{\boldsymbol{\theta}}^T (\boldsymbol{\mu} - \boldsymbol{Y})\right].$$

The degrees of freedom are also the bias of $\widehat{\ell}(\widehat{\boldsymbol{\theta}})$ as estimator of the expected log-likelihood has taken over the unknown data generating process. The expected log-likelihood in its turn is the part of the Kullback–Leibler distance that depends on the estimated parameter vector.

An estimator of the degrees of freedom is developed in [9],
leading to the expression

$$\widehat{\nu}(\widehat{\boldsymbol{\theta}}) = \text{Tr}\left[\mathbf{D}_\lambda \widetilde{\mathbf{X}} \overline{\Upsilon}^{-2} \widetilde{\mathbf{Q}} \Upsilon \widehat{\Theta}^{-1} \mathbf{X}\right],$$

where $\widehat{\Theta}^{-1}$ is a diagonal matrix with slightly shifted versions of the observed values, i.e., $\widehat{\Theta}_{ii}^{-1} = Y_{i-1}$. The matrix Υ is a diagonal matrix with the observations, i.e., $\Upsilon_{ii} = Y_i$. The diagonal matrix $\mathbf{D}_\lambda$ has zeros and ones on the diagonal. The ones correspond to nonzero coefficients in the thresholded MLPT decomposition.

6 Illustration and Concluding Discussion

Ongoing research concentrates on motivating choices for the tuning parameters in the proposed data transformation and processing. In particular, the data transformation depends on the choice of the finest resolution bandwidth h_J, the degree of the local polynomial in the prediction step, the precise design of the updated step. Also, the Unbalanced Haar prefilter is parametrized by a fine scale $h_{J,0}$. Processing parameters include the threshold value, which is selected using the AIC approach of Sect. 5, and the sizes of the blocks in the block threshold procedure.

For the result in Fig. 2, the MLPT adopted a local linear prediction step. In the wavelet literature, the transform is said to have two dual vanishing moments, i.e., $\tilde{p} = 2$, meaning that all detail coefficients of a linear function are zero. The MLPT for the figure also includes an update step designed for two primal vanishing moments,

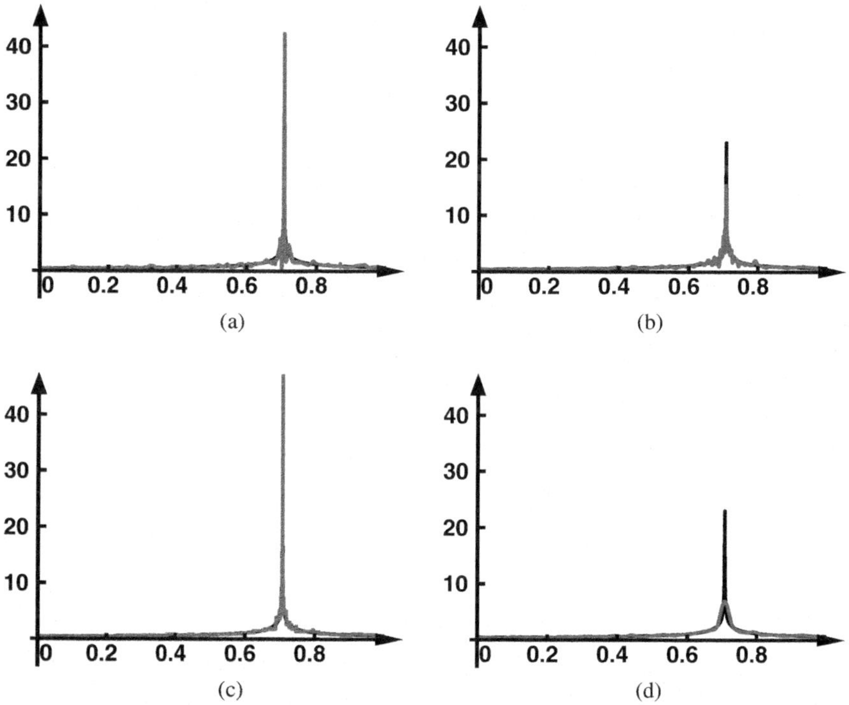

Fig. 2 Panel (a): power law and its estimation from $n = 2000$ observations using the MLPT procedure of this paper. Panel (b): estimation from the same observations using a probit transform centered around the location of the singularity, thus hinges on the knowledge of this location. Panel (c): estimation using the finest possible Haar wavelet transform. This transform involves full processing of many resolution levels. Panel (d): estimation using straightforward uniscale kernel density estimation

meaning that $\int_{-\infty}^{\infty} \Psi_j(x)x^r dx = 0$ for $r = 0$ and $r = 1$. Block sizes were set to one, i.e., classical thresholding was used.

The density function in the simulation study is the power law $f_X(x) = K|x - x_0|^k$ on the finite interval $[0, 1]$, with a singular point $x_0 = 1/2$ in this simulation study and $k = -1/2$. The sample size is $n = 2000$. The MLPT approach, unaware of the presence and location of x_0, is compared with a kernel density estimation applied to a probit transform of the observations, $Y = \Phi^{-1}(X - x_0)$ for $X > x_0$ and $Y = \Phi^{-1}(X - x_0 + 1)$ for $X < x_0$. This transform uses the information on the singularity's location, in order to create a random variable whose density has no end points of a finite interval, nor any singular points inside. In this experiment, the MLPT outperforms the Probit transformed kernel estimation, both in the reconstruction of the singular peak and in the reduction of the oscillations next to the peak. With the current procedure, this is not always the case. Further research concentrates on the design making the MLPT analyses as close as possible to orthogonal projections, using appropriate update steps. With an analysis close to orthogonal projection, the variance propagation throughout the analysis, processing, and reconstruction can be more easily controlled, thereby reducing spurious effects in the reconstruction. Both MLPT and Probit transformation outperform a straightforward uniscale kernel density estimation. This estimation, illustrated the Fig. 2d, oversmooths the sharp peaks of the true density.

References

1. Burt, P.J., Adelson, E.H.: Laplacian pyramid as a compact image code. IEEE Trans. Commun. **31**(4), 532–540 (1983)
2. Deslauriers, G., Dubuc, S.: Symmetric iterative interpolation processes. Constructive Approximation **5**, 49–68 (1989)
3. Donoho, D.L., Johnstone, I.M., Kerkyacharian, G., Picard, D.: Density estimation by wavelet thresholding. The Annals of Statistics **24**(2), 508–539 (1996)
4. D. L. Donoho and T.P.Y. Yu. Deslauriers-Dubuc: ten years after. In S. Dubuc and G. Deslauriers, editors, *Spline Functions and the Theory of Wavelets*, CRM Proceedings and Lecture Notes. American Mathematical Society, 1999
5. Fan, J., Gijbels, I.: Local Polynomial Modelling and its Applications. Chapman and Hall, London (1996)
6. Hall, P., Patil, P.: Formulae for mean integrated squared error of nonlinear wavelet-based density estimators. The Annals of Statistics **23**(3), 905–928 (1995)
7. Jansen, M.: Multiscale local polynomial smoothing in a lifted pyramid for non-equispaced data. IEEE Transactions on Signal Processing **61**(3), 545–555 (2013)
8. M. Jansen. Non-equispaced B-spline wavelets. *International Journal of Wavelets, Multiresolution and Information Processing*, 14(6), 2016
9. Jansen, M.: Density estimation using multiscale local polynomial transforms. Technical report, Department of Mathematics, ULB (2019)
10. Jansen, M., Amghar, M.: Multiscale local polynomial decompositions using bandwidths as scales. Statistics and Computing **27**(5), 1383–1399 (2017)

11. Jansen, M., Nason, G., Silverman, B.: Multiscale methods for data on graphs and irregular multidimensional situations. Journal of the Royal Statistical Society, Series B **71**(1), 97–125 (2009)
12. Sweldens, W.: The lifting scheme: a construction of second generation wavelets. SIAM J. Math. Anal. **29**(2), 511–546 (1998)

On Sensitivity of Metalearning: An Illustrative Study for Robust Regression

Jan Kalina

Abstract Metalearning is becoming an increasingly important methodology for extracting knowledge from a database of available training datasets to a new (independent) dataset. While the concept of metalearning is becoming popular in statistical learning and readily available also for the analysis of economic datasets, not much attention has been paid to its limitations and disadvantages. To the best of our knowledge, the current paper represents the first illustration of metalearning sensitivity to data contamination by noise or outliers. For this purpose, we use various linear regression estimators (including highly robust ones) over a set of 24 datasets with economic background and perform a metalearning study over them as well as over the same datasets after an artificial contamination. The results reveal the whole process to remain rather sensitive to data contamination and some of the standard classifiers turn out to yield unreliable results. Nevertheless, using a robust classification method does not bring a desirable improvement. Thus, we conclude that the task of robustification of the whole metalearning methodology is more complex and deserves a systematic future research.

Keywords Linear regression · Automatic method selection · Contamination · Sensitivity · Robustness

1 Metalearning

Metalearning can be characterized as a methodology for extracting knowledge from a database of training datasets with the ability to apply the knowledge to new independent (validation) datasets. It can be perceived as learning to learn or learning of

J. Kalina (✉)
Institute of Computer Science, The Czech Academy of Sciences, Pod Vodárenskou věží 2, 182 07 Prague 8, Czech Republic
e-mail: kalina@cs.cas.cz

M. La Rocca et al. (eds.), *Nonparametric Statistics*, Springer Proceedings in Mathematics & Statistics 339, https://doi.org/10.1007/978-3-030-57306-5_24

metaknowledge, which is defined as knowledge about whole datasets which serve as a prior knowledge rather than measured values contained in these datasets. Metalearning represents an approach to machine learning (i.e., automated statistical learning) popular in recent computer science and data mining [13, 15]. It also starts to penetrate to economic applications [1], not limited to big data analysis [14].

While the most renowned works on metalearning principles [4, 12] appraise metalearning and list its appealing properties, a truly critical evaluation of metalearning seems to be still missing. It is mainly the fully automatic characteristic of the metalearning process which hinders a profound interpretation of the results, which would standard in the statistical community but not in computer science usually exploiting heuristics and black-box procedures. Other important issues include stability and robustness, while these two concepts do not actually coincide. The instability of metalearning, manifested, e.g., as different recommendations for two rather similar datasets, has been reported with a recommendation for using ensemble methods [4]. However, we are not aware of any discussion of the presence of noise and outlying measurements (outliers) in the data and their influence of the metalearning process, nor we have found any attempts to robustify the metalearning against outliers.

In the current paper, we illustrate the sensitivity of metalearning as its weak point deserving further attention of researchers. The novelty of the current paper is also considering the promising (but not much known) least weighted squares estimator [10, 17] and also a robust version of linear discriminant analysis.

Section 2 of this paper describes principles of our study of metalearning sensitivity, which is performed on 24 real datasets as well as on their artificially contaminated versions by noise or outliers. Section 3 presents results of primary learning as well as metalearning and, finally, Section 4 presents a discussion and conclusions.

2 Description of the Study

We proposed and performed a metalearning study with the aim to compare various linear regression estimators and to find a classification rule allowing to predict the best one for a given (new) dataset. It remains namely unknown (and too difficult to study theoretically) which of the methods should be used for a particular dataset or under particular conditions or which are the most relevant criteria for determining the most suitable method.

The primary learning task is to fit various linear regression estimators for each of the given datasets. The best estimator is found using a specified characteristic of goodness of fit. The subsequent metalearning part has the aim to learn a classification rule allowing to predict the best regression method for a new dataset not present in the training database. Its input data are only selected features of individual datasets together with the result of the primary learning, which typically has the form of the index of the best method for each of the training datasets.

In general, the user of metalearning must specify a list of essential components (parameters), which have been systematically described by [12] and denoted as P,

A, F, Y, and S, where some (P, A, Y) are used in the task of primary learning (base learning) and the remaining ones (F, S) in the subsequent metalearning. Their meaning and our specific choices will be now described.

2.1 Primary Learning

(P) **Datasets**. Metalearning should always use real datasets because any random generation of data is performed in a too specific (i.e., non-representative, biased) way. However, we are not aware of any public repository of metadata (at least for a regression task). Therefore, we use 24 datasets listed in Table 1, which are publicly available datasets investigated primarily with economic motivation. In addition, we also modified the datasets by artificially added contamination as described below.

(A) **Algorithms**. In each of the datasets, we consider the standard linear regression model

$$Y_i = \beta_0 + \beta_1 X_{i1} + \cdots + \beta_p X_{ip} + e_i, \quad i = 1, \ldots, n, \tag{1}$$

where there are p regressors and n observations. We use four different estimators of the parameters:

- Least squares,
- Hampels's M-estimator [6],
- Least trimmed squares (LTS), investigated, e.g., in [16] defined as

$$\arg\min_{b \in \mathbb{R}^{p+1}} \sum_{i=1}^{h} u_{(i)}^2(b), \tag{2}$$

  where $u_i(b)$ is a residual corresponding to the i-th observation for a given $b \in \mathbb{R}^{p+1}$ and $u_{(1)}^2(b) \leq \cdots \leq u_{(n)}^2(b)$ are values arranged in ascending order. We use the probably most common choice $h = \lfloor \frac{3n}{4} \rfloor$, where $\lfloor x \rfloor$ denotes the integer part of x.
- Least weighted squares (LWS) of [17] with linearly decreasing weights $w_i = 1 - (i-1)/n$ for $i = 1, \ldots, n$ of [10] is defined using the same notation as

$$\arg\min_{b \in \mathbb{R}^{p+1}} \sum_{i=1}^{n} w_i u_{(i)}^2(b). \tag{3}$$

(Y) **Prediction measure**. We use the prediction mean square error (PMSE) in the form $\sum_{i=1}^{n}(Y_i - \hat{Y}_i)^2/n$, where $\hat{Y}_i$ denotes the fitted value of the i-th observation (in each of the datasets).

Except for the least squares, the regression estimators presented under (A) were proposed as its robust alternatives [6]. Robust statistics, which gradually becomes important for the analysis of economic data [2, 8], distinguishes between local and global robustness (resistance, insensitivity). From the set of four estimators described above, only Hampel's M-estimator and the LWS are robust in the local sense and only the LTS estimator is highly robust in the global sense; we may refer to [6, 17] for a deeper explanation of the concepts, which are to a large extent contradictory.

We considered two types of data contamination. These can be characterized as a local (i.e., aiming at local sensitivity) and a global (corresponding to global sensitivity) contamination of regressors of individual datasets. For both cases, we will need the following notation. Each measured value will be denoted as X_{ijk}, where i corresponds to a particular dataset, j to an observation within this dataset, and k to a particular variable. The idea is to replace X_{ijk} by $X_{ijk} + \varepsilon_{ijk}$, where ε's are (mutually) independent random variables independent on the given data and ε_{ijk} is generated from normal distribution $\mathsf{N}(0, s\hat{\sigma}^2_{ijk})$, where $\hat{\sigma}^2_{ijk}$ is an estimated variance of the j-th variable within the i-th dataset and s is a chosen constant.

1. Local contamination. Each observation in each dataset is contaminated by a slight noise, i.e., with a small s.
2. Global contamination. A small percentage of observations is contaminated by severe noise, while the remaining ones are retained. Particularly, $c \times 100\ \%$ of the values are randomly chosen for each dataset across all relevant features for a given (and rather large) s.

In the primary learning task, we find the best method for each dataset. This is done using PMSE in a leave-one-out cross-validation, which represents a standard attempt for independent validation. Then, the output of the primary learning is the knowledge (i.e., factor variable, index) of the best method for each of the datasets.

2.2 *Metalearning*

The subsequent metalearning task exploits nine features for each dataset and the factor variable of Table 1 denoting the index of the best method. Parameters of the metalearning will be now described again using the P–A–F–Y–S notation [12].

(F) **Features of datasets**. We select nine features, which can be (avoiding details) for each of the datasets characterized as

1. The number of observations n,
2. The number of regressors p (excluding the intercept),
3. The ratio n/p,
4. Normality of residuals of the least squares evaluated as the p-value of the Shapiro–Wilk test for the least squares,
5. Skewness of residuals of the least squares,

6. Kurtosis of residuals of the least squares,
7. Coefficient of determination R^2 for the least squares,
8. Percentage of outliers estimated by the LTS,
9. Heteroscedasticity of residuals evaluated as the p-value of the Breusch–Pagan test for the least squares.

(S) **Metalearning method**. We exploit the following classification methods:

- Support vector machines (SVM),
- Linear discriminant analysis (LDA),
- MWCD-LDA with linearly decreasing weights, i.e., a robust version of LDA defined in [9], where it was proposed as a linear classification rule based on the minimum weighted covariance determinant (MWCD) estimator of [8, 11],
- k-nearest neighbors for various values of k.

We note that three features, namely n, p, and their ratio, are retained as fixed even if the data are contaminated, while each of the remaining ones is influenced (less or more) by data contamination.

3 Results

3.1 Primary Learning

Table 1 contains together with the list of datasets also the estimated values of σ^2, which were used for obtaining the contaminated datasets as described in Section 2.1. Further, Table 1 shows the best method for raw datasets. Particularly, the best regression method is shown in the table for each of the datasets. Finally, the results are given for datasets modified by each of the two different types of contamination (for different parameters).

Global contamination seems to influence the results of primary learning in a stronger way compared to local contamination. A more detailed analysis, however, reveals that individual features are influenced remarkably in both situations and we can perceive both types of contamination (with selected parameters) to be comparable in terms of severity. We can also say that under global contamination, robust estimators become more commonly the best method with an increasing c, while no clear tendency can be observed for the local contamination.

We also inspected features those are mostly influenced by the contamination. These are features 4 and 9 for the local contamination and features 7 and 9 for the global one. These (and mainly normality and heteroscedasticity of residuals) are, however, crucial ones for the choice of the appropriate regression estimator. Thus, the whole primary learning is influenced strongly by the contamination. While three features remain to be the same under every contamination, these are not so important for the resulting classification rule.

Table 1 Results of primary learning for 24 investigated datasets (raw or contaminated). The best method was found according to the smallest PMSE in a leave-one-study cross-validation study. Columns of the table with the best method serve as factors (responses) for the subsequent classification task of metalearning. Regression methods include (1) least squares, (2) Hampel's M-estimator, (3) LTS with $h = \lfloor 0.75n \rfloor$, and (4) LWS with linearly decreasing weights

Dataset		$\hat{\sigma}^2$	Raw data	Best method					
				Local contam.			Global contam.		
				with $s =$			with $s = 9$ and $c =$		
				0.1	0.2	0.3	0.06	0.12	0.18
1	Aircraft	57.8	3	3	3	3	4	3	3
2	Ammonia	8.9	4	4	3	3	4	4	4
3	Auto MPG	17.9	3	3	3	3	3	4	3
4	Cirrhosis	103	1	2	3	3	1	3	3
5	Coleman	3.2	1	1	1	1	1	1	1
6	Delivery	9.7	2	3	3	2	3	3	3
7	Education	1537	2	2	2	3	2	4	3
8	Electricity	0.85	2	2	2	2	2	2	4
9	Employment	55463	3	4	3	3	3	3	3
10	Furniture 1	0.0019	2	2	2	2	2	3	2
11	Furniture 2	0.056	3	3	3	4	3	3	3
12	GDP growth	9467	2	2	2	2	2	2	3
13	Houseprices	14.6	4	4	3	3	3	3	3
14	Housing	54.3	2	2	2	3	2	4	3
15	Imports	4.2	3	3	3	3	3	3	3
16	Kootenay	22.0	1	1	3	2	1	1	2
17	Livestock	29.4	3	3	3	3	3	4	4
18	Machine	3495	3	3	3	3	3	3	3
19	Murders	17.7	4	4	4	4	4	4	4
20	NOx	0.30	2	3	3	3	4	3	3
21	Octane	0.19	2	2	2	2	2	2	2
22	Pasture	75.4	4	4	4	4	4	4	3
23	Pension	0.24	3	3	3	3	3	3	3
24	Petrol	4022	2	3	3	3	2	3	3

3.2 *Metalearning*

The results of metalearning are overviewed in Table 2, namely as classification performances of the classification rules learned withing the metalearning tasks. There, the performance is evaluated as a classification correctness in a leave-one-out cross-validation study. Comparing both types of contamination, the classification performance remains to be low.

Table 2 Results of metalearning for 24 investigated datasets (raw or contaminated) evaluated as classification correctness in a leave-one-out cross-validation study. The classification rule for the best regression estimator is learned not over the original datasets, but using 9 features of each dataset together with the indicator of the best method obtained from Table 1. The best result in each column is shown in boldface

Classification method	Best method						
	Raw data	Local contam. with			Global contam. ($s = 9$)		
		$s = 0.1$	$s = 0.2$	$s = 0.3$	$c = 0.06$	$c = 0.12$	$c = 0.18$
SVM (linear)	**0.38**	**0.38**	**0.38**	**0.38**	**0.38**	**0.38**	**0.38**
LDA	0.29	0.29	0.29	0.25	0.17	0.29	**0.38**
MWCD-LDA	0.33	0.33	0.33	0.33	0.29	0.33	0.33
k-NN (k=1)	0.29	0.25	0.21	0.25	0.29	0.33	0.29
k-NN (k=3)	0.29	0.29	0.25	0.25	0.33	0.29	0.25
k-NN (k=5)	0.33	0.33	0.33	0.29	**0.38**	0.33	**0.38**

For the local contamination, SVM turns out to be the best method. For most methods, we can observe only small changes (typically a decrease of performance) with an increasing c. A closer analysis exceeding the scope of this paper shows instability again. MWCD-LDA seems robust, but is not very reliable in the classification task, perhaps because MWCD-LDA is reliable in classification tasks to two groups and loses much efficiency with an increasing number of groups. The k-nearest neighbors classifiers suffer from the most dramatic loss of performance, although the method is very common (perhaps the most common) in the metalearning task.

For the global contamination, SVM is again the winner, although its performance is reached also by other methods. With an increasing s, the changes in the best method are quite unpredictable, unstable. SVM seems very robust. It may be a preferable method, although not much used in the metalearning context.

Let us also point out at the increasing performance with an increasing global contamination, e.g., for the standard LDA, which is known as very non-robust (see e.g., [7]). Its performance is improving with an increasing contamination, but this advantage is only illusionary due to excessive effect of outliers.

4 Conclusions

To the best of our knowledge, none of the available metalearning studies has focused on the influence of noise or outliers on the results. Thus, such a unique sensitivity study, which reveals the vulnerability of metalearning, is presented in the current paper. The metalearning task itself, which has the aim to predict the most suitable linear regression estimator for new datasets, is accompanied by a study over datasets with economic background contaminated in two different possible ways.

The SVM is the best method for raw data as well as for any contamination. Its classification performance, although rather low, is not deteriorated by the contaminations under consideration. This cannot be, however, said about most of the remaining classifiers.

The local contamination has the idea to slightly modify each observed value of all training datasets. It is true that the best method shown in Table 1 is retained to a large extent with a similar performance to that obtained for raw (non-contaminated) datasets. However, our further analysis shows individual features to be rather remarkably influenced by the contamination, which is consequently manifested on the metalearning results, e.g., on very different sets of wrongly classified datasets.

The global contamination has another idea to greatly modify a small percentage of selected observed values. Already a smaller percentage of severe outliers has a remarkable influence on the results of metalearning. The results for some of the regression estimators change dramatically in an unpredictable way, which is not monotonous with increasing contamination. The classification rules of non-robust classifiers (such as LDA) are then formally successful, while a more detailed analysis reveals the success to be putting too much influence on outliers, i.e., more information is drawn from errors and randomness than from the signal whose influence on the resulting classification rule is decreased. This idea is supported by the fact that it happens exclusively for a larger percentage of severe outliers that LDA outperforms MWCD-LDA. The classification rule is arbitrary (i.e., useless) determined primarily by outliers.

Thus, the study reveals a weak point of metalearning and motivates a possible future critical evaluation of the metalearning process. Let us now try to list all possible factors which contribute to the sensitivity of metalearning:

- The choice of datasets. We use rather a wide spectrum of datasets with different characteristics from different research tasks, while metalearning is perhaps more suitable only for more homogeneous data (e.g., with analogous dimensionality) or for data from a specific narrow domain.
- Difficult (and unreliable) extrapolation for a very different (outlying) dataset.
- The prediction measure. In our case, PMSE is very vulnerable to outliers.
- The number of algorithms/methods. If their number is larger than very small, we have the experience that learning the classification rule becomes much more complicated and less reliable.
- The classification methods for the metalearning task depend on their own parameters or selected approach, which is another source of uncertainty and thus instability.
- Solving the metalearning method (S) by classification tools increases the vulnerability as well because only the best regression estimator is chosen ignoring information about the performance of other estimators.
- The process of metalearning itself is too automatic so the influence of outliers is propagated throughout the process and the user cannot manually perform an outlier detection or deletion.

Let us also particularly discuss the performance of robust methods withing the metalearning study.

- The LWS estimator turns out to be the best method for some datasets, which is a novel argument in favor of the method. This is interesting because the LWS is using simplistic weights, which could be actually further improved.
- The study presents also a unique comparison of MWCD-LDA with standard LDA. While the robust approach does not improve the performance compared to LDA, its results are not misleading the presence of contamination. MWCD-LDA together with the SVM classifier is the only method with this property, which brings a novel argument in favor of the MWCD-LDA.

Finally, there remain some topics for future research, which can be listed from the simplest to the most difficult (and most important):

- The study can be extended by considering also noise added to the response or additional features, e.g., a robust test of heteroscedasticity.
- A detailed interpretation of the classification rules of metalearning; especially we expect to find arguments that the effect of outliers, although it improves the classification performance, is detrimental.
- Ensemble classification can be used for the metalearning task, which could hopefully improve stability and robustness. In fact, robustness in the task of statistical learning was introduced by Breiman [5], whose ideas have not been exploited in the metalearning context yet.
- We are interested in extending metalearning tasks to extracting association rules from data in the spirit of [3].
- The whole metalearning methodology should be robustified, which remains a more complex task than just a robustification of each of its individual steps.

Acknowledgements The research was supported by the project GA19-05704S of the Czech Science Foundation. The author is thankful to Barbora Peštová for providing the datasets.

References

1. Abbasi, A., Albrecht, C., Vance, A., Hansen, J.: MetaFraud: A meta-learning framework for detecting financial fraud. MIS Quarterly **36**, 1293–1327 (2012)
2. Alfons, A., Templ, M., Filzmoser, P.: Robust estimation of economic indicators from survey samples based on Pareto tail modelling. J. R. Stat. Soc. Series C **62**, 271–286 (2013)
3. Berka, P., Rauch, J.: Meta-learning for post-processing of association rules. Lect. Notes Comput. Sci. **6263**, 251–262 (2010)
4. Brazdil, P., Giradu-Carrier, C., Soares, C., Vilalta, E.: Metalearning: Applications to data mining. Springer, Berlin (2009)
5. Breiman, L., Friedman, J.H., Olshen, R.A., Stone, C.J.: Classification and regression trees. Chapman & Hall/CRC, Boca Raton (1984)
6. Hampel, F.R., Ronchetti, E.M., Rousseeuw, P.J., Stahel, W.A.: Robust statistics: The approach based on influence functions. Wiley, New York (1986)

7. Hubert, M., Rousseeuw, P.J., van Aelst, S.: High-breakdown robust multivariate methods. Statist. Sci. **23**, 92–119 (2008)
8. Kalina, J.: On multivariate methods in robust econometrics. Prague Econ. Pap. **21**, 69–82 (2012)
9. Kalina, J.: Highly robust statistical methods in medical image analysis. Biocyb. Biomed. Eng. **32**, 3–16 (2012)
10. Kalina, J.: Implicitly weighted methods in robust image analysis. J. Math. Imaging Vis. **44**, 449–462 (2012)
11. Roelant, E., van Aelst, S., Willems, G.: The minimum weighted covariance determinant estimator. Metrika **70**, 177–204 (2009)
12. Smith-Miles, K., Baatar, D., Wreford, B., Lewis, R.: Towards objective measures of algorithm performance across instance space. Comp. Oper. Res. **45**, 12–24 (2014)
13. Suh, S.C.: Practical data mining applications. Jones & Bartlett Learning, Sudbury (2012)
14. Varian, H.R.: Big data: New tricks for econometrics. J. Econ. Perspect. **28**, 3–28 (2014)
15. Vilalta, R., Giraud-Carrier, C., Brazdil, P., Soares, C.: Using meta-learning to support data mining. Int. J. Comput. Sci. Appl. **1**, 31–45 (2004)
16. Víšek, J.Á.: The least trimmed squares. Part I: Consistency. Kybernetika **42**, 1–36 (2006)
17. Víšek, J.Á.: Regression with high breakdown point. In: Proceedings of ROBUST 2000, Summer School of JČMF, pp. 324–356. JČMF and Czech Statistical Society, Prague (2001)

Function-Parametric Empirical Processes, Projections and Unitary Operators

Estáte Khmaladze

Abstract We describe another approach to the theory of distribution free testing. The approach uses geometric similarity within various forms of empirical processes: whenever there is an empirical object (like the empirical distribution function) and theoretical parametric model (like a parametric model for distribution function) and a normalised difference of the two, then substitution of estimated values of the parameters leads to projection of this difference. Then one can bring some system in the multitude of these projections. We use unitary operators to describe classes of statistical problems, where one can "rotate" one projection into another, thus creating classes of equivalent problems. As a result, behaviour of various test statistics could be investigated in only one "typical" problem from each class. Thus, the approach promises economy in analytic and numerical work. We also hope to show that the unitary operators involved in "rotations" are of simple and easily implementable form.

Keywords Distribution free testing · Discrete distributions · Uniform empirical process in $[0, 1]^d$ · Linear regression · Equivalence of testing problems

1 Basic Setup

Consider a function parametric empirical process based on a sample $(X_i)_{i=1}^n$ of F-i.i.d. random variables,

$$v_{n,F}(\phi) = \frac{1}{\sqrt{n}} \sum_{i=1}^{n} \left[\phi(X_i) - \int \phi(x)dF(x) \right], \quad \phi \in L_2(F),$$

or its point parametric version, i.e. with $\phi_x(X_i) = \mathbb{I}(X_i \leq x)$,

E. Khmaladze (✉)
Victoria University of Wellington, Wellington, New Zealand
e-mail: Estate.Khmaladze@vuw.ac.nz

M. La Rocca et al. (eds.), *Nonparametric Statistics*, Springer Proceedings in Mathematics & Statistics 339, https://doi.org/10.1007/978-3-030-57306-5_25

$$v_{n,F}(x) = \frac{1}{\sqrt{n}} \sum_{i=1}^{n} [\mathbb{I}(X_i \leq x) - F(x)], \quad x \in \mathbb{R}^d.$$

The following problem of weak convergence: "describe the class $\Psi \subset L_2(F)$, on which $\{v_{n,F}(\phi), \phi \in \Psi\}$ converges to F-Brownian bridge $\{v_F(\phi), \phi \in \Psi\}$" is an extremely important problem, with broad and interesting mathematical theory behind it, see, for example [19]. However, in this short overview, we want to focus on somewhat different direction concerning function parametric empirical processes. The question we ask is "for K a linear operator on $L_2(F)$, find what new processes can be obtained as

$$K^* v_{n,F}(\phi) = v_{n,F}(K\phi), \tag{1}$$

and why can they be interesting". "Linear operator", however, seems too general for specific results, and everywhere below we consider only unitary operators; for the general theory of unitary operators we refer, e.g., to [3].

It is not immediately obvious that the question is sensible. Indeed, the covariance operators of the processes $v_{n,F}$ and $K^* v_{n,F}$ are unitary equivalent, which, with some freedom of speech, is the same as to say they are equal, and therefore the second order properties of the two processes will be the same. Why such thing will be useful?

Nevertheless, we will see that the construction can be applied to various forms of empirical processes in different parts of statistics and lead to a general approach to distribution free testing theory.

Sections 2–4 below contain a review, much shorter, and hopefully clearer, of some of the results already published. Section 5 contains short presentation of new material described in [10]. Farther developments, for example on point processes, also new, will appear in [11].

2 Discrete Distributions

We start with the situation, which will allow us to explain the main point of this approach in a very simple way. Consider a finite-dimensional discrete distribution

$$p = (p(k))_{k=1}^{m}, \; p(k) > 0, \; \sum_{k=1}^{m} p(k) = 1, \quad m < \infty,$$

and let ν_{kn} denote the frequency of the outcome k in n independent trials. Further, consider the vector of "components" of the χ^2-test statistic:

$$Y_n = (Y_{kn})_{k=1}^{m}, \quad Y_{kn} = \frac{\nu_{kn} - np(k)}{\sqrt{np(k)}},$$

so that the χ^2-statistic itself is

$$\langle Y_n, Y_n\rangle = \sum_{k=1}^{m} Y_{kn}^2 \sim \chi_{m-1}^2, \quad n \to \infty.$$

Thus, asymptotic distribution of χ^2-test statistic is free from p.

At the same time, analogues of Kolmogorov-Smirnov statistics, such as

$$\max_{1\le j\le m} \sum_{k=1}^{j} Y_{kn}$$

will have a limit distribution very much dependent on p. As a matter of fact, it is only the χ^2-statistic, if we do not count asymptotically equivalent forms of it, which leads to reasonable goodness of fit test and is, at the same time, asymptotically distribution free. This situation is in contrast with what we have for continuous distributions, where, from the very beginning, we have had a class of distribution free goodness of fit test statistics, see, e.g. [1, 13, 17].

However, the choice of asymptotically distribution free statistics in the discrete case can be broadened to its full extent.

With $\sqrt{p} = (\sqrt{p(k)})_{k=1}^m$, we know that

$$Y_n \xrightarrow{\mathcal{D}} Y, \quad Y = X - \langle X, \sqrt{p}\rangle\sqrt{p}, \tag{2}$$

where $X = (X_k)_{k=1}^m$ is the vector with independent coordinates with standard normal distribution each. The vector $\sqrt{p}$ is vector of unite length. Now let $\sqrt{r}$ be another vector of unit length, and put

$$Z_n = Y_n - \langle Y_n, \sqrt{r}\rangle \frac{1}{1 - \langle\sqrt{p}, \sqrt{r}\rangle}(\sqrt{r} - \sqrt{p}). \tag{3}$$

The transformation of Y_n into Z_n is one-to-one: it is unitary transformation which maps $\sqrt{p}$ to $\sqrt{r}$ and vice versa. That is why, being applied to the projection Y, it maps it into projection Z.

Theorem 1 *Khmaladze [7] If $Y_n \xrightarrow{\mathcal{D}} Y$, then*

$$Z_n \xrightarrow{\mathcal{D}} Z, \quad Z = X - \langle X, \sqrt{r}\rangle\sqrt{r}. \tag{4}$$

Although the proof of the theorem is immediate, it lies at the heart of very wide possibilities of extension, some of which we demonstrate below.

For discrete distribution it implies the following: the transformation (2) *of X to Y is a projection; it projects X parallel to the vector $\sqrt{p}$. If we had a different discrete distribution, say, $q = (q(k))_{k=1}^m$, then the vector, parallel to which we project, will be* *different. Theorem 1 above allows us to choose any vector of unit length, and switch from projection Y to projection Z.*

In more detail, consider the class of discrete distributions of the same dimension m. The vector of components of χ^2-statistic corresponding to each of them, say, to distribution p, can be thus mapped to the vector with the same limit distribution as the vector of components of χ^2-statistic for any other distribution, say, for distribution q. Or any of them can be mapped into vector Z_n corresponding to some, fixed, distribution, say, to uniform distribution on m disjoint events. Therefore, statistics for testing p which are based on the transformed vector Z_n will have limit distribution completely free from this p. At the same time, since the correspondence between Y_n and Z_n is one-to-one, the "statistical information", whichever way we measure it, in both vectors is the same.

It is shown in [7] that the approach can be used in testing hypothesis about parametric families of discrete distributions. It seems to work for quite high dimensions of the parameter. In [15] it was shown how to apply this method to test hypothesis in contingency tables, when parameters can be of dimension 20–25, and the sample size not too large, about $n = 400 - 500$.

3 Uniform Empirical Process on $[0, 1]^d$

Let us use notation v_F for F-Brownian bridge, and w_F for F-Brownian motion, and consider F, which lives on $[0, 1]^d$ and has positive density. Then, see, e.g., [4, 16],

$$v_F(x) = w_F(x) - F(x)w_F(\mathbf{1}),$$

where **1** denotes the vector with all d coordinates equal 1. This process can not be normalized to something standard:

$$\frac{dv_F(x)}{\sqrt{f(x)}} = \frac{dw_F(x)}{\sqrt{f(x)}} - \frac{dF(x)}{\sqrt{f(x)}} w_F(\mathbf{1}),$$

and although $dw_F(x)/\sqrt{f(x)}$ behaves in distribution as differential of the standard Brownian motion, this is not enough to standardize $dv_F(x)$ as there is another differential on the right hand side, dependent on F. However, with use of one extra Winer stochastic integral, the normalisation becomes possible.

Theorem 2 *Khmaladze [9] (i) The process with differential*

$$u(dx) = \frac{v_F(dx)}{\sqrt{f(x)}} - \int_{[0,1]^d} \frac{v_F(dy)}{\sqrt{f(y)}} \frac{(1 - \sqrt{f(x)})}{1 - \int_{[0,1]^d} \sqrt{f(y)}dy} dx$$

is the standard Brownian bridge on $[0, 1]^d$.

(ii) If $G \ll F$ and $l(x) = \sqrt{dG(x)/dF(x)}$, then the process with differential

$$v_G(dx) = l(x)v_F(dx) - \int_{[0,1]^d} l(y)v_F(dy) \frac{l^2(x) - l(x)}{1 - \int_{[0,1]^d} l(y)dF(y)} dx$$

is G-Brownian bridge.

Being applied to empirical process $v_{n,F}$ based on a sample form distribution F the transformation in part (i) will transform it into a process, u_n, with the same limit distribution as the uniform empirical process, that is the one, based on uniform random variables on $[0, 1]^d$, although in the transformation there are no other random variables, but those with distribution F. Transformation of $v_{n,F}$ as in (ii) will map it into a process $v_{n,G}$, with the same limit distribution as the empirical process based on sample from the distribution G.

In order to show how this theorem follows from the general construction of (1) *consider the subspace of functions $\mathcal{L}(G) = \{\alpha \in L_2(G) : \langle\alpha, 1\rangle_G = 0\}$, where 1 stands for function identically equal to number 1. This is the subspace on which the process v_G "lives": Brownian bridge v_G is Brownian motion w_G restricted to $\mathcal{L}(G)$, see, e.g., [8]; for Gaussian measures on Hilbert spaces see also [14]. Similarly, the process v_F lives on the subspace $\mathcal{L}(F) = \{\alpha \in L_2(F) : \langle\alpha, 1\rangle_F = 0\}$. If $G \sim F$, the operator of multiplication by l, i.e. $l\alpha(x) = l(x)\alpha(x)$ will map $L_2(G)$ into $L_2(F)$ isometrically, so that the function 1 (from $L_2(G)$) is mapped into function l, while the subspace $\mathcal{L}(G)$ is mapped into the subspace of functions, orthogonal to l. What remains is to rotate this latter subspace into $\mathcal{L}(F)$, the subspace of functions, orthogonal to 1 (from $L_2(F)$). For this we use appropriate unitary operator in $L_2(F)$:*

$$K\beta = \beta - (l - 1)\frac{1}{\langle l, l - 1\rangle_F}\langle\beta, l - 1\rangle_F.$$

If $\beta \perp l, 1$, then $K\beta = \beta$, while $Kl = 1$, and $K1 = l$. Therefore, as a result,

$$v_G(\alpha) = v_F(Kl\alpha),$$

and this is equivalent to statement (ii).

Shifting orthogonality constrain. *We know that a Brownian bridge v_F is a Brownian motion, subjected to orthogonality condition $v_F(1) = 0$, or, equivalently, restricted to the subspace $\mathcal{L}(F)$. We can, however, use unitary operator to "move" this orthogonality condition "further away", which may lead to unexpected results. We illustrate the fact in Theorem 3 although we did not investigate statistical implications of these possibilities enough.*

Choose η_A to be a density on "small" set $A \subset [0, 1]^d$. For $\psi \in \mathcal{L}(U)$, choose

$$K_A\psi = \frac{\psi}{\sqrt{f}} - (\sqrt{\frac{\eta_A}{f}} - 1)\frac{1}{1 - \langle\sqrt{\eta_A}, \sqrt{f}\rangle}\langle\sqrt{\eta_A} - \sqrt{f}, \psi\rangle \in \mathcal{L}(F).$$

Consider

$$b(\psi) = v_F(K_A\psi).$$

In the statement below, we can use ψ *equal indicator function of* $[0, x]$ *and thus speak about differential of* $b(\psi)$.

Theorem 3 *The process with differential*

$$b(dx) = \frac{v_F(dx)}{\sqrt{f(x)}} - \int_{y\in A}\sqrt{\frac{\eta_A(y)}{f(y)}}v_F(dy)\,\frac{(\sqrt{\eta_A(x)} - \sqrt{f(x)})}{1 - \int_{y\in A}\sqrt{\eta_A(y)f(y)}dy}dx$$

is a standard Brownian motion on $[0, 1]^d \setminus A$, *while*

$$\int_{y\in A}\eta_A(y)b(dy) = 0.$$

On the interval $[0, 1]$, $A = [0, \Delta]$ *and uniform* F *it takes the form*

$$b(dt) = u(dt) - \frac{u(\Delta)}{\Delta}dt, \qquad t \le \Delta,$$
$$b(dt) = u(dt) + \frac{u(\Delta)}{\sqrt{\Delta} - \Delta}dt, \; t > \Delta,$$

and represents $b(x)$ *as Brownian bridge on* $[0, \Delta]$ *and Brownian motion on* $[\Delta, 1]$. *This is very different from the usual form*

$$w(dt) = u(dt) + \frac{u(t)}{1-t}dt,$$

with w *also a Brownian motion.*

4 Parametric Hypotheses in $\mathbb{R}^d$

Let $\mathbb{F} = \{F_\theta(x), x \in \mathbb{R}^d, \theta \in \Theta \subseteq \mathbb{R}^m\}$ be a parametric family of distributions in $\mathbb{R}^d$, which depend on m-dimensional parameter θ. Suppose we need to test hypothesis that an unknown distribution F belongs to this family. Let $v_{n,F}(x, \theta) = \sqrt{n}[F_n(x) - F_\theta(x)]$ be empirical process where we have to substitute an estimation for θ based on the sample, and let $\hat{\theta}$ be the MLE for θ.

The first order Taylor expansion in θ of the parametric empirical process $v_{n,F}(x, \hat{\theta})$ produces

$$v_{n,F}(x, \hat{\theta}) = v_n(x, \theta) - \frac{\partial}{\partial\theta}F_\theta(x)\,\sqrt{n}(\hat{\theta} - \theta) + R_n(x),$$

where, for a regular family $\mathbb{F}$, the residual R_n is asymptotically negligible [2, 4, 16]. With notation Γ_θ for Fisher information matrix, and notation

$$a(x) = \Gamma_\theta^{-1/2} \frac{\dot{f}_\theta}{f_\theta}(x)$$

for the orthonormal version of the score function, the function parametric version of $v_{n,F}$ can be written as

$$\widehat{v}_{n,F}(\phi) = v_{n,F}(\phi, \hat{\theta}) = v_{n,F}(\phi) - \langle \phi, a \rangle^T v_{n,F}(a) + R_n(\phi).$$

As a result, one can see that the limit in distribution for $\widehat{v}_{n,F}$ is the process

$$\widehat{v}_F(\phi) = v_F(\phi) - \langle \phi, a \rangle^T v_F(a) = v_F(\phi - \langle \phi, a \rangle^T a). \tag{5}$$

It is easy to verify that $\widehat{v}_F$ is projection of v_F. This fact was shown in more general context in [5], and was subsequently, to some surprise, often overlooked. It explains, however, interesting phenomena in asymptotic behaviour of empirical processes with estimated parameters—for example, that even if one knows the value of true parameter, it is usually better to substitute an estimator, because the power of a test based on $\widehat{v}_{n,F}$ will be higher than of the same tests based on $v_{n,F}$.

The projection structure of the right hand side in (5) can therefore be established for any regular parametric family, and, generally, at any particular value of the parameter within the same family. Therefore, we end up with lots of projections; distribution of a tests statistic based on one of them will differ from that based on another one. There seems to be endless need for numerical approximations of these distributions. However, there is a way to glue wide classes of them all in one single problem, as explained below.

From one parametric hypotheses to another. With $F = F_\theta$, $a = a_\theta$ and similarly $G = G_\vartheta$, $b = b_\vartheta$, consider

$$\mathcal{L}(F, a) = \{\phi \in L_2(F) : \langle \phi, 1 \rangle_F = \langle \phi, a \rangle_F = 0\},$$

$$\mathcal{L}(G, b) = \{\phi \in L_2(G) : \langle \phi, 1 \rangle_G = \langle \phi, b \rangle_G = 0\}$$

Similarly to what we said about v_F and v_G in Sect. 3, the limiting processes $\widehat{v}_F$ and $\widehat{v}_G$ live on these subspaces, respectively. On these subspaces, their distribution is the same as corresponding Brownian motions. Therefore, in order to transform $\widehat{v}_F$ into $\widehat{v}_G$ one needs unitary operator, which maps one subspace into another:

$$U_{b,a} : \mathcal{L}_b(G) \longrightarrow \mathcal{L}_a(F),$$

and, consequently,

$$\widehat{v}_G(\psi) = \widehat{v}_F(U_{b,a}\psi).$$

Consider again Hellinger function

$$l(x) = \sqrt{\frac{dG_\vartheta(x)}{dF_\theta(x)}},$$

and denote $a_0(x) = 1$ and $b_0(x) = 1$, the same function but considered as elements of different spaces. Then the operator K of the previous section, which we now denote K_{a_0,lb_0} will rotate the function lb_0 into a_0, but it will not necessarily rotate lb_1 into a_1, but only into some $\widetilde{lb}_1$. Here a_1 and b_1 denote first coordinates of normalized score-functions a and b. In general, a_k and b_k, $k = 1, \ldots, m$, will denote their respective k-th coordinates.

Since $K_{a_0,\,lb_0}$ is a unitary operator, it preserves norms and angles, and therefore $\widetilde{lb}_1 \perp a_0$. Now we can rotate $\widetilde{lb}_1$ further into a_1 using operator $K_{a_1,\,\widetilde{lb}_1}$, and then consider the product

$$U_{a,\,b,\,1}(\phi) = K_{a_0,\,\widetilde{lb}_1} K_{a_0,\,lb_0}(l\phi).$$

As a product of unitary operators, it is a unitary operator itself. It maps b_0 into a_0 and b_1 into a_1, and it will leave all functions orthogonal to a_1 and $\widetilde{lb}_1$ unchanged, see [9], Sect. 4, or [11], Sect. 3.

For parametric families with m-dimensional parameter, we use induction. Given $j \in \{0, 1, \ldots, m\}$, suppose we have a unitary operator $U_{a,\,b,\,j}$ that maps lb_i to a_i for $0 \le i \le j$. For example, we have constructed above $U_{a,\,b,\,0} = K_{a_0,\,lb_0}$ and $U_{a,\,b,\,1} = K_{a_1,\,\widetilde{lb}_1} K_{a_0,\,lb_0}$. Now define the function

$$\widetilde{lb}_{j+1} := U_{a,\,lb,j} lb_{j+1},$$

and introduce

$$U_{a,\,lb,\,j+1} = K_{a_{j+1},\,\widetilde{lb}_{j+1}} U_{a,\,lb.\,j}.$$

Then $U_{a,\,lb,\,j+1}$ is a unitary operator that maps lb_i to a_i for $0 \le i \le j+1$. Continuing in this fashion, we see that $U_{a,\,lb,\,m}$ is a unitary operator that maps lb_i to a_i for all $i = 0, \ldots, m$. Therefore by an analogous argument as in the case of Brownian bridge we have the following theorem:

Theorem 4 *Khmaladze [9] Suppose $\widehat{v}_F$ is projected Brownian bridge, parallel to an (orthonormal) m-dimensional vector-function a, and, similarly, suppose $\widehat{v}_G$ is projected Brownian bridge, parallel to an (orthonormal) m-dimensional vector-function b. If measures F and G are equivalent (mutually absolutely continuous), then $\widehat{v}_F$ can be unitarily mapped into $\widehat{v}_G$ as follows:*

$$\widehat{v}_F(U_{a,\,lb,\,m}\, l\psi) \stackrel{d}{=} \widehat{v}_G(\psi). \tag{6}$$

5 Distribution Free Testing for Linear Regression

A wider version of this section, although still in a draft form, can be found in [10]. It includes general parametric regression models and multi-dimensional covariates.

Consider a sequence of pairs $(X_i, Y_i)_{i=1}^n$, where Y_i is a "response variable" and X_i is the corresponding "covariate". We, basically, will not assume anything about probabilistic nature of $(X_i)_{i=1}^n$, except that their empirical distribution function weakly converges to some distribution function:

$$F_n(x) = \frac{1}{n}\sum_{i=1}^{n} \mathbb{I}_{(X_i \le x)}, \quad \text{and} \quad F_n \overset{w}{\to} F.$$

About $(Y_i)_{i=1}^n$ we assume, that given $(X_i)_{i=1}^n$, they are independent, and moreover, that there exists a function $m(x)$, such that the differences, or "errors" $(\varepsilon_i)_{i=1}^n$, with $\varepsilon_i = Y_i - m(X_i)$, are G- i.i.d. random variables.

How shall we test a simple linear regression, which states, that $m(x) = x^T\theta$ or

$$Y_i = X_i^T\theta + \varepsilon_i, \ i = 1, \ldots, n,$$

with some constant θ? Here not any test will do. Tests we want should have two properties: they should be able to detect all contiguous alternatives to the linearity, i.e. the local deviations form linearity of order $1/\sqrt{n}$, and they should have limit distribution independent from the vector of covariates $(X_i)_{i=1}^n$ and the distribution G of G-i.i.d. errors ε_i. One method to create class of such tests was described in [12] and in [18]. Both of these papers have been based on the approach suggested in [6], although in several respects they are technically different from each other. In this section we outline another method, which is much simpler. Its implementation is straightforward.

In vector form, one can write the regression above as

$$Y = X^T\theta + \varepsilon, \quad \theta \in \mathbb{R}^d,$$

where X^T is a matrix, with i-th row X_i^T. The residuals can be written as

$$\hat{\varepsilon} = Y - X^T\hat{\theta} \ \text{ with } \ \hat{\theta} = (XX^T)^{-1}XY,$$

or, using normalised vector of residuals $z = (XX^T)^{-1/2}X$,

$$\hat{\varepsilon} = Y - z^T zY = \varepsilon - z^T z\,\varepsilon. \tag{7}$$

The empirical regression process (partial sum process)

$$\hat{R}_{n,\varepsilon}(x) = \frac{1}{\sqrt{n}} \sum_{i=1}^{n} \hat{\varepsilon}_i \mathbb{I}_{(X_i \le x)}$$

is the natural object to base test statistics upon, cf. [12] and in [18]. Therefore, we should be interested in asymptotic behaviour of this process.

Let

$$w_{\varepsilon,n}(x) = \frac{1}{\sqrt{n}} \sum_{i=1}^{n} \varepsilon_i \mathbb{I}_{(X_i \le x)}.$$

It is clear that if the errors ε_i have finite variance, the process $w_{\varepsilon,n}$ converges weakly to a Brownian motion w_F in time F. Then it is possible to describe the process $\hat{R}_{n,\epsilon}$ as asymptotically one-dimensional projection of $w_{\varepsilon,n}$, cf. [10]. However, $\hat{R}_{n,\epsilon}$ is not a Brownian bridge. Indeed, its variance is

$$E\hat{R}^2_{n,\epsilon}(x) = F_n(\min(x, y)) - \frac{1}{n} \sum_{i=1}^{n} z_i^T z_i \, \mathbb{I}_{(X_i \le x)},$$

which, clearly, is not of the form $F_n(x) - F_n^2(x)$. Thus, the limit distribution of $\hat{R}_{n,\epsilon}$ depends on values of the covariates, and that in unfamiliar fashion. The covariance matrix of $\hat{\varepsilon}$ also depends on covariates:

$$E\hat{\varepsilon}\,\hat{\varepsilon}^T = I - zz^T.$$

Therefore, limit distribution of tests statistics based on $\hat{R}_{n,\epsilon}$ needs to be calculated anew for new values of the covariates.

To present the main step below, we do not need $X_i \in \mathbb{R}^d$, $d > 1$, it is enough that $d = 1$. Consider the operator in $\mathbb{R}^n$,

$$U_{a,b} = I - \frac{\langle a - b, \cdot \rangle_{F_n}}{1 - \langle a, b \rangle_{F_n}} (a - b) \text{ with } \|a\| = \|b\| = 1.$$

This operator is unitary, and (cf., e.g., [7])

$$U_{a,b}a = b, \; U_{a,b}b = a, \; U_{a,b}c = c, \text{ if } c \perp a, b.$$

Now choose $a = z$ and choose b equal $r = (1, \ldots, 1)^T/\sqrt{n}$, the vector not depending on covariates at all. Since $\hat{\varepsilon} \perp z$ we obtain:

$$\hat{e} = U_{z,r}\hat{\varepsilon} = \hat{\varepsilon} - \frac{\langle \hat{\varepsilon}, r \rangle}{1 - \langle z, r \rangle}(r - z),$$

or

$$\hat{e}_i = \hat{\varepsilon}_i - \frac{\sum_{j=1}^n \hat{\varepsilon}_j/\sqrt{n}}{1 - \frac{1}{\sqrt{n}}\sum_{j=1}^n z_j}(\frac{1}{\sqrt{n}} - z_i).$$

The new residuals have covariance matrix

$$E\hat{e}\hat{e}^T = I - rr^T.$$

This would be the covariance matrix of the residuals in the problem of testing

$$Y_i = \theta + e_i, \quad i = 1, 2, \ldots, n, \tag{8}$$

which is completely free from covariates.

For the partial sum process based on the new residuals,

$$\hat{R}_{n,e}(x) = \frac{1}{\sqrt{n}}\sum_{i=1}^n \hat{e}_i \mathbb{I}_{(X_i \le x)},$$

we have

$$E\hat{R}^2_{n,e}(x) = F_n(x) - F_n^2(x).$$

Thus, this process is asymptotically Brownian bridge in time F and the class of distribution free test statistics based on $\hat{R}_{n,e}$ is broad and well known.

Linear regression with constant term. This extension can be made with no extra difficulty. Let now

$$Y = \theta_0 1 + (X - \bar{X}1)\theta_1 + \varepsilon,$$

(here 1 stands for the n-dimensional vector with all coordinates equal number 1). Substituting the usual least square estimators for θ_0 and θ_1 and using again notation r and notation

$$\tilde{z} = \frac{1}{\sqrt{\sum_{j=1}^n (X_j - \bar{X})^2}}(X - \bar{X}),$$

one can write the residuals in succinct form

$$\hat{\varepsilon} = Y - \langle Y, r\rangle r - \langle Y, \tilde{z}\rangle\tilde{z} = \varepsilon - \langle \varepsilon, r\rangle r - \langle \varepsilon, \tilde{z}\rangle\tilde{z}.$$

From this it follows that the covariance matrix of $\hat{\epsilon}$ is

$$E\hat{\varepsilon}\hat{\varepsilon}^T = I - rr^T - \tilde{z}\tilde{z}^T,$$

and it still depends on the values of the covariates. The regression process with these residuals,

$$\hat{R}_{n,\varepsilon}(x) = \frac{1}{\sqrt{n}} \sum_{i=1}^{n} \hat{\varepsilon}_i \mathbb{I}_{(X_i \le x)},$$

will, therefore, have asymptotic distribution which depends on $\tilde{z}$.

Vector r is, obviously, free from covariates and there is no reason to replace it, but it will be useful to replace the vector $\tilde{z}$. Introduce another vector $\tilde{r}$, different from $\tilde{z}$, which also has unit norm and is orthogonal to r. It is simpler to arrange the coordinates of both vectors $\tilde{z}$ and $\tilde{r}$ in increasing order. As an example of $\tilde{r}$ consider the function

$$\tilde{r}\left(\frac{i}{n}\right) = \sqrt{12}\left[\frac{i}{n} - \frac{n+1}{2n}\right] \text{ and let } Q_n(t) = \sum_{i=1}^{nt} \tilde{r}\left(\frac{i}{n}\right)/n, \tag{9}$$

where now i equals the rank of X_i. What we will do now is to rotate $\tilde{z}$ into $\tilde{r}$, leaving vectors orthogonal to them unchanged. Define

$$\hat{e} = U_{\tilde{z},\tilde{r}}\hat{\varepsilon} = \hat{\varepsilon} - \frac{\langle \hat{\varepsilon}, \tilde{r} - \tilde{z} \rangle}{1 - \langle z, r \rangle}(\tilde{r} - \tilde{z}) = \hat{\varepsilon} - \frac{\langle \hat{\varepsilon}, \tilde{r} \rangle}{1 - \langle \tilde{z}, \tilde{r} \rangle}(\tilde{r} - \tilde{z}). \tag{10}$$

Thus calculation of new residuals in this case is as simple as in the previous one.

Theorem 5 *Khmaladze [10] (i) Covariance matrix of residuals $\hat{e}$ in* (10) *is*

$$E\hat{e}\hat{e}^T = I - rr^T - \tilde{r}\tilde{r}^T.$$

(ii) The regression process based on $\hat{e}$,

$$\hat{R}_{n,e}(x) = \frac{1}{\sqrt{n}} \sum_{i=1}^{n} \hat{e}_i \mathbb{I}_{(X_i \le x)},$$

has the covariance function

$$E\hat{R}_{n,e}(x)\hat{R}_{n,e}(y) = F_n(\min(x, y)) - F_n(x)F_n(y) - Q_n(F_n(x))Q_n(F_n(y)) + O(1/n),$$

As a corollary of (ii), the process $\hat{R}_{n,e}$, with change of time $t = F(x)$, converges in distribution to projection of standard Brownian motion on [0,1] parallel to functions t and Q:

$$\hat{R}_e(x) = w(x) - tw(1) - Q(t) \int \tilde{r}(s)dw(s),$$

and statistics based on $\hat{R}_{n,e}$, invariant under the time transformation above, will be asymptotically distribution free.

References

1. Anderson, T.W., Darling, D.A.: Asymptotic theory of certain "Goodness of Fit" criteria based on stochastic processes. Ann. Math. Stat. **23**, 193–212 (1952)
2. Cramér, H.: Mathematical Methods of Statistics. Princeton University Press (1942)
3. Glazman, I.M., Ljubich, JuI.: Finite-Dimensional Linear Analysis. MIT Press, Cambridge, Mass (1974)
4. Janssen, A.: Asymptotic properties of Neyman-Pearson tests for infinite Kullback-Leibler information. Ann. Stat. **14**, 1068–1079 (1986)
5. Khmaladze, E.: The use of ω^2-tests for testing parametric hypotheses. Theory Probab. Appl. **24**, 280–297 (1979)
6. Khmaladze, E.: Goodness of fit problem and scanning innovation martingales. Ann. Stat. **21**, 798–830 (1993)
7. Khmaladze, E.: Note on distribution free testing for discrete distributions. Ann. Stat. **41**, 2979–2993 (2013)
8. Khmaladze, E.: Some new connections between the Brownian bridges and the Brownian motions. Commun. Stoch. Anal. **9**, 401–412 (2015)
9. Khmaladze, E.: Unitary transformations, empirical processes and distribution free testing. Bernoulli **22**, 563–599 (2016)
10. Khmaladze, E.: Distribution free approach to testing linear regression using unitary transformations. Extension to general parametric regression, Research Report, SMS VUW (2019). http://sms.victoria.ac.nz/Main/ResearchReportSeries
11. Khmaladze, E.:Towards asymptotically distribution-free testing for point processes, SMS VUW (2019). http://homepages.ecs.vuw.ac.nz/Users/Estate/WebHome
12. Khmaladze, E., Koul, H.L.: Martingale transform goodness of fit tests in regression models. Ann. Stat. **32**, 955–1034 (2004)
13. Kolmogorov, A.: Sulla determinazione empirica di una legge di distribuzione. Ital. Attuari. Giorn. **4**, 1–11 (1933)
14. Kuo, H.-H.: Gaussian measures in Banach spaces, 463, Lecture Notes in Mathematics, Springer (2006)
15. Nguien, T.M.: A new approach to distribution-free goodness of fit test in in contingency tables. Metrika **80**, 153–170 (2017)
16. Shorack, G.R., Wellner, J.A.: Empirical Processes with Applications to Statistics. Wiley, New York (1986)
17. Smirnov, N.V.: On the distribution of von Mises ω^2 test. Mat. Sbornik **2**, 973–993 (1937)
18. Stute, W., Thies, S., Zhu, L.-X.: Model checks for regression: an innovation process approach. Ann. Stat. **26**, 1916–1934 (1998)
19. van der Vaart, A., Wellner, J.A.: Weak Convergence of Empirical Processes. Springer (1996)

Rank-Based Analysis of Multivariate Data in Factorial Designs and Its Implementation in R

Maximilian Kiefel and Arne C. Bathke

Abstract Recently, a completely nonparametric rank-based approach for inference regarding multivariate data from factorial designs has been introduced, with theoretical results for two different asymptotic settings. Namely, for the situation of few factor levels with large sample sizes at each level, and for the situation of a large number of factor levels with small sample sizes in each group. In this article, we examine in detail how this theory can be translated into practical application. A challenge in this regard has been feasibly implementing consistent covariance matrix estimation in the setting of small sample sizes. The finite sampling distributions are approximated using moment estimators. In order to make the results widely available, we introduce the R package **nparMD** which performs nonparametric analysis of multivariate data in a two-way layout. Multivariate data in a one-way layout have already been addressed by the **npmv** package. Similar to the latter, within the **nparMD** package, there are no assumptions met about the underlying distribution of the multivariate data. The components of the response vector do not necessarily have to be measured on the same scale, but they have to be at least binary or ordinal. Due to the factorial design, hypotheses to be tested include the main effects of both factors, as well as their interaction. The new R package is equipped with two versions of the testing procedure, corresponding to the two asymptotic situations mentioned above.

Keywords Nonparametric model · Multivariate test · Rank statistic · MANOVA · Factorial design · Non-normalitt

1 Introduction

This paper demonstrates how to perform nonparametric inference on multivariate responses in factorial designs. In order to allow for immediate application to real

M. Kiefel (✉) · A. C. Bathke
Fachbereich Mathematik, Universität Salzburg, 5020 Salzburg, Austria
e-mail: maximilian.kiefel@sbg.ac.at

A. C. Bathke
e-mail: Arne.Bathke@sbg.ac.at

M. La Rocca et al. (eds.), *Nonparametric Statistics*, Springer Proceedings in Mathematics & Statistics 339, https://doi.org/10.1007/978-3-030-57306-5_26

data, we introduce the R [15] package **nparMD** [10], which performs fully nonparametric, rank-based analysis of multivariate data samples with two fully crossed design factors. The package is available at the Comprehensive R Archive Network (CRAN) under https://CRAN.R-project.org/package=nparMD. While the results presented here pertain to a design with two factors, a generalization to higher-way layouts is methodologically straightforward but rather technical.

The underlying asymptotic theory which is related to semiparametric heteroscedastic two-factor MANOVA (see [8]) has largely been described in [1, 2, 5, 9, 12]. However, translating this theory into applicable procedures requires three major steps. The first one is to find a feasible way to estimate the covariance matrix in the setting of several samples with small sample sizes each. The second one is to devise reasonable finite-sample approximations to the sampling distributions of the test statistics considered. And the final step is to provide an effective way for researchers to actually apply these methods to their data—that is, developing an adequate statistical software package.

As the methods considered here present a generalization of the multivariate nonparametric inference procedures described by [2–4, 7, 9, 11], the implementation in R is also partially related to the methods used in the corresponding **npmv** package [6] which is designed for comparing multivariate data samples in a one-way layout. Similarities appear, for example, in terms of the classes of test statistics known as Wilk's Lambda (LR), the ANOVA-type or Dempster's (D), the Lawley–Hotelling (LH), and the Bartlett–Nanda–Pillai (BNP) criteria which are used frequently in this context. The nonparametric versions and finite approximations of these test statistics have been investigated and discussed in the publications cited above.

However, differences between the setting considered in the present paper and the one-way layout discussed in the above publications appear in many aspects regarding hypotheses to be tested and their interpretation, estimation of covariance matrices, asymptotics, and further details including computational challenges. An essential part of the nonparametric model is their reliance on the nonparametric relative effect as a statistical functional and the subsequent construction of relative effect estimators which are based on midranks. In contrast to methods where longitudinal data are examined for simple factor effects [13], the current method is based on variablewise ranks. This means that separate rankings are performed for each component within the p-dimensional observation vector $\mathbf{X}_{ijr} = (X_{ijr}^{(1)}, \ldots, X_{ijr}^{(p)})'$, where $i = 1, \ldots, a$ and $j = 1, \ldots, b$ denote the factor levels, while $r = 1, \ldots, n_{ij}$ denotes the experimental units (subjects) or replications within a certain factor level combination. The underlying model states that for each value of i and j, all n_{ij} observations within the group (i, j) follow the same p-variate distribution $\mathbf{X}_{ijr} \sim F_{ij}$. All observation vectors $\mathbf{X}_{ijr}$ are also stated to be independent while the components of the response vector are allowed to be dependent with arbitrary dependence structure. Let $\mathbf{I}_d$ be the $d \times d$ identity matrix, $\mathbf{J}_d$ be the $d \times d$ matrix of ones, and $\mathbf{P}_d = \mathbf{I}_d - d^{-1}\mathbf{J}_d$. Then, the hypotheses of interest can be formulated via the vector of cumulative distribution functions $\mathbf{F} = (F_{11}, \ldots, F_{1b}, F_{21}, \ldots, F_{ab})$ and a suitable contrast matrix as $\mathbf{CF} \equiv \mathbf{0}$, with the following choices of $\mathbf{C}$.

$$\mathbf{C}_A = \mathbf{P}_a \otimes \frac{1}{b}\mathbf{J}_b \qquad \text{(no main effect of factor A)}$$

$$\mathbf{C}_B = \frac{1}{a}\mathbf{J}_a \otimes \mathbf{P}_b \qquad \text{(no main effect of factor B)}$$

$$\mathbf{C}_{AB} = \mathbf{P}_a \otimes \mathbf{P}_b \qquad \text{(no interaction effect between A and B).}$$

Note that $\mathbf{A} \otimes \mathbf{B}$ denotes the Kronecker product and $\mathbf{A} \bigoplus \mathbf{B}$ will denote a block diagonal matrix with blocks $\mathbf{A}$ and $\mathbf{B}$. Let $c(x) = 0, \frac{1}{2}, 1$ if $x <, =, > 0$. For $l = 1, \ldots, p$ let $R_{ijr}^{(l)} = \frac{1}{2} + \sum_{i'=1}^{a} \sum_{j'=1}^{b} \sum_{r'=1}^{n_{i'j'}} c(X_{ijr}^{(l)} - X_{i'j'r'}^{(l)})$ denote the midrank among all $N = \sum_{i=1}^{a} \sum_{j=1}^{b} n_{ij}$ observations $X_{111}^{(l)}, \ldots, X_{abn_{ab}}^{(l)}$ which is equal to row-wise ranking when all observations are arranged into a $p \times N$-matrix. Define

$$\mathbf{H}^{(A)} = \frac{b}{(a-1)N^2} \sum_{i=1}^{a} (\tilde{\mathbf{R}}_{i..} - \tilde{\mathbf{R}}_{...})(\tilde{\mathbf{R}}_{i..} - \tilde{\mathbf{R}}_{...})'$$

$$\mathbf{H}^{(B)} = \frac{a}{(b-1)N^2} \sum_{j=1}^{b} (\tilde{\mathbf{R}}_{.j.} - \tilde{\mathbf{R}}_{...})(\tilde{\mathbf{R}}_{.j.} - \tilde{\mathbf{R}}_{...})'$$

$$\mathbf{H}^{(AB)} = \frac{1}{(a-1)(b-1)N^2} \sum_{i=1}^{a} \sum_{j=1}^{b} (\bar{\mathbf{R}}_{ij.} - \tilde{\mathbf{R}}_{i..} - \tilde{\mathbf{R}}_{.j.} + \tilde{\mathbf{R}}_{...})(\bar{\mathbf{R}}_{ij.} - \tilde{\mathbf{R}}_{i..} - \tilde{\mathbf{R}}_{.j.} + \tilde{\mathbf{R}}_{...})'$$

$$\widehat{\mathbf{S}}_{ij} = \frac{1}{(n_{ij}-1)N^2} \sum_{k=1}^{n_{ij}} (\mathbf{R}_{ijk} - \bar{\mathbf{R}}_{ij})(\mathbf{R}_{ijk} - \bar{\mathbf{R}}_{ij})'$$

$$\mathbf{G} = \frac{1}{ab} \sum_{i=1}^{a} \sum_{j=1}^{b} \frac{1}{n_{ij}} \widehat{\mathbf{S}}_{ij},$$

where $\mathbf{R}_{ijr} = (R_{ijr}^{(1)}, \ldots, R_{ijr}^{(p)})'$, $\bar{\mathbf{R}}_{ij.} = \frac{1}{n_{ij}} \sum_{k=1}^{n_{ij}} \mathbf{R}_{ijk}$, $\tilde{\mathbf{R}}_{i..} = \frac{1}{b} \sum_{j=1}^{b} \bar{\mathbf{R}}_{ij}$, $\tilde{\mathbf{R}}_{.j.} = \frac{1}{a} \sum_{i=1}^{a} \bar{\mathbf{R}}_{ij}$, $\tilde{\mathbf{R}}_{...} = \frac{1}{ab} \sum_{i=1}^{a} \sum_{j=1}^{b} \bar{\mathbf{R}}_{ij}$. In the definition of $\mathbf{H}^{(\cdot)}$ and $\mathbf{S}_{ij}$, the sums of squares and cross-product matrices are divided by N^2 indicating the use of *rank transforms (RT)* $\hat{Y}_{ijr}^{(l)} = N^{-1}(R_{ijr}^{(l)} - \frac{1}{2})$ which themselves are related to *asymptotic rank transforms (ART)* $Y_{ijr}^{(l)} = H^{(l)}(X_{ijr}^{(l)})$. Here, $H^{(l)}(x) = \frac{1}{N} \sum_{i=1}^{a} \sum_{j=1}^{b} n_{ij} F_{ij}^{(l)}(x)$ is defined as average *cdf* for variable *(l)* of the response vector. It would also be possible to directly formulate the test statistics using RT instead of original ranks ($\hat{Y}_{ijr}^{(l)}$ instead of $R_{ijr}^{(l)}$ and without division by N^2). However, ranks are more intuitive, due to their straightforward interpretation.

Now let $\psi = (A, B, AB)$. The **nparMD** package uses the following core test statistics which are based in construction on the homonymous classical parametric test statistics that had been proposed for the analysis of multivariate data in the normal model.

$$\text{ANOVA Type (Dempster's) criterion}: T_D = \text{tr}(\mathbf{H}^{(\psi)})/\text{tr}(\mathbf{G}) \tag{1}$$

$$\text{Wilk's Lambda (Likelihood Ratio) criterion}: T_{LR} = \log|\mathbf{I} + \mathbf{H}^{(\psi)}\mathbf{G}^-| \tag{2}$$

$$\text{The Lawley–Hotelling criterion}: T_{LH} = \text{tr}(\mathbf{H}^{(\psi)}\mathbf{G}^-) \tag{3}$$

$$\text{The Bartlett–Nanda–Pillai criterion}: T_{BNP} = \text{tr}(\mathbf{H}^{(\psi)}\mathbf{G}^-(\mathbf{I} + \mathbf{H}^{(\psi)}\mathbf{G}^-)^-). \tag{4}$$

As non-singularity of $\mathbf{G}$ or $\mathbf{I} + \mathbf{H}^{(\psi)}\mathbf{G}^-$ can not be assumed in general we use a so-called pseudoinverse, the Moore–Penrose generalized inverse which is defined as matrix satisfying the Penrose conditions [14]. For each of these types of tests, **nparMD** provides two testing procedures tailored to the two different asymptotic settings mentioned above.

2 Large Sample Sizes n_{ij}

The **nparMD** package provides a function for the large sample case (at least seven observations per factor-level combination are recommended), where the hypotheses are tested by nonparametric analogs to Dempster's ANOVA and the Lawley–Hotelling criterion. Following the recommendations of [5, 12], the distribution of the ANOVA-type statistic T_D is approximated by a central $F_{(\widehat{f}_\psi, \widehat{f}_0)}$ distribution with estimated degrees of freedom, as follows.

$$\widehat{f}_A = \frac{(a-1)^2 N^2 tr(\widehat{\mathbf{V}}_N)^2}{(abN)^2 tr(\mathbf{T}_A \widehat{\mathbf{V}}_N \mathbf{T}_A \widehat{\mathbf{V}}_N)} \tag{5}$$

$$\widehat{f}_B = \frac{(b-1)^2 N^2 tr(\widehat{\mathbf{V}}_N)^2}{(abN)^2 tr(\mathbf{T}_B \widehat{\mathbf{V}}_N \mathbf{T}_B \widehat{\mathbf{V}}_N)} \tag{6}$$

$$\widehat{f}_{AB} = \frac{(a-1)^2(b-1)^2 N^2 tr(\hat{\mathbf{V}}_N)^2}{(abN)^2 tr(\mathbf{T}_{AB} \widehat{\mathbf{V}}_N \mathbf{T}_{AB} \widehat{\mathbf{V}}_N)} \tag{7}$$

$$\widehat{f}_0 = \frac{N^2 tr(\widehat{\mathbf{V}}_N)^2}{N^2 \sum_{i=1}^{a} \sum_{j=1}^{b} \frac{1}{n_{ij}-1} tr\left(\frac{\widehat{\mathbf{S}}_{ij}}{n_{ij}}\right)}, \tag{8}$$

where

$$\mathbf{T}_A = \mathbf{P}_a \otimes \frac{1}{b}\mathbf{J}_b \otimes \mathbf{I}_p \tag{9}$$

$$\mathbf{T}_B = \frac{1}{a}\mathbf{J}_a \otimes \mathbf{P}_b \otimes \mathbf{I}_p \tag{10}$$

$$\mathbf{T}_{AB} = \mathbf{P}_a \otimes \mathbf{P}_b \otimes \mathbf{I}_p \tag{11}$$

$$\widehat{\mathbf{V}}_N = N \cdot \bigoplus_{i=1}^{a} \bigoplus_{j=1}^{b} \widehat{\mathbf{S}}_{ij} \cdot \frac{1}{n_{ij}}. \tag{12}$$

As $\widehat{f}_0$ tends to be very large in this setting, an $F_{(\widehat{f}_\psi, \infty)}$ approximation is actually used in the implementation. The distribution of the Lawley–Hotelling type criterion T_{LH} as defined before is approximated by a central χ^2-distribution since $gT_{LH}^{(\psi)}$ is approximately χ_f^2 distributed [8], where $g = (a-1, b-1, (a-1)(b-1))$ and $f = g \cdot p$.

Table 1 shows the results of a simulation study that has been carried out in order to demonstrate the actual performance of the test under several conditions, that is, different underlying distributions and different sample size settings. The simulated power of the test is shown in Fig. 1. Alternatives were formulated as location shifts for the first level of factor A (see also figure captions for details). Underlying distributions were homo- and heteroscedastic multivariate normal, as well as multinomial. Heteroscedasticity and response dependency were modeled by symmetric $p \times p$ covariance matrices Σ_{ij} with off-diagonal elements $\rho_{ij} = \sqrt{ij}/(1 + ij)$ and diagonal elements $1 - \rho_{ij}$.

In order to simulate the power for underlying ordinal response data, including dependency of the components, we drew samples from a multinomial distribution with parameters $n = 5$ and $p = (0.2, 0.3, 0.5)$ (if $i \geq 2$) and $p = (0.2 - \delta p, 0.3, 0.5 + \delta p)$ (if $i = 1$). δp denotes the probability shift inducing a main effect

Table 1 Simulated α (nominal $\alpha = 5\%$); $a = 3$; $b = 2$; $p = 3$

Underlying distribution	ψ	$T_D^{(\psi)}$	$T_{LH}^{(\psi)}$	$T_D^{(\psi)}$	$T_{LH}^{(\psi)}$	$T_D^{(\psi)}$	$T_{LH}^{(\psi)}$
		$7 \leq n_{ij} \leq 12$		$15 \leq n_{ij} \leq 20$		$25 \leq n_{ij} \leq 30$	
mvrnorm	A	0.043	0.078	0.050	0.065	0.049	0.060
	AB	0.047	0.084	0.045	0.062	0.047	0.060
	B	0.050	0.077	0.050	0.059	0.047	0.054
ordinal	A	0.049	0.082	0.052	0.068	0.050	0.062
	AB	0.046	0.077	0.050	0.066	0.046	0.056
	B	0.048	0.072	0.049	0.058	0.049	0.054
lognormal	A	0.047	0.078	0.051	0.068	0.053	0.063
	AB	0.051	0.080	0.050	0.068	0.051	0.059
	B	0.050	0.071	0.054	0.066	0.050	0.054
mvrnorm (high correlation)*	A	0.057	0.081	0.054	0.069	0.060	0.057
	AB	0.058	0.077	0.058	0.066	0.057	0.055
	B	0.053	0.068	0.056	0.063	0.051	0.055

*covariance matrix (σ_{mn}) with diagonal entries 1 and off-diagonal entries $1–0.1 \cdot |m - n|$

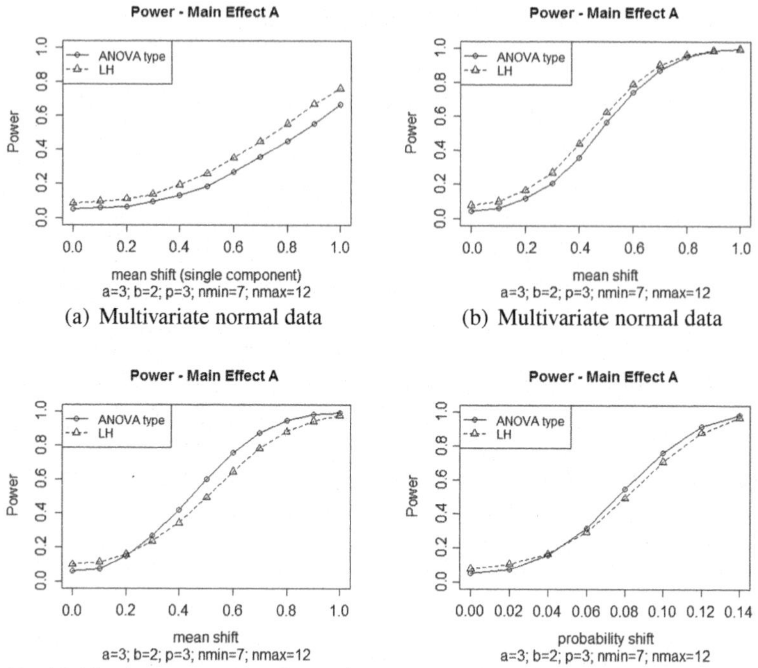

(a) Multivariate normal data

(b) Multivariate normal data

(c) Multivariate normal data; heteroscedasticity and response dependency modeled by Σ_{ij}

(d) Multinomial (ordinal) data

Fig. 1 Simulated power

of A. As shown in Fig. 1 δp ranges from 0 to 0.14. Sample sizes were chosen randomly between 7 and 12 (discrete uniform distribution).

3 Small Sample Sizes n_{ij}, Large Number a of Samples

The setting described in this section applies to small samples, but with a minimum requirement of four observations per factor-level combination. Asymptotics in this situation rely on the number of samples, that is, the number of levels of one factor (here, without loss of generality, a being large). A semiparametric approach to the small sample case has been described in [8]. Under suitable centering and scaling, all four test statistics are shown to have an asymptotic normal distribution for increasing number of factor levels a. The covariance matrix estimation, which has to be done for each group individually, is one of the main challenges within the theoretical part but also in terms of implementation. Bathke and Harrar [1] proposed a consistent variance and covariance matrix estimator based on the theory of U-statistics. For practical reasons it is formulated in terms of RT $\hat{\mathbf{Y}}_{ijr} = (\hat{Y}_{ijr}^{(1)}, \ldots, \hat{Y}_{ijr}^{(p)})$—recall that ART are not observable. Define

$$\boldsymbol{\Psi}_{ij}(\boldsymbol{\Omega}) = \frac{1}{4c_{ij}} \sum_{(k_1,k_2,k_3,k_4)\in\mathcal{K}}^{n_{ij}} \boldsymbol{\Omega}(\hat{\mathbf{Y}}_{ijk_1} - \hat{\mathbf{Y}}_{ijk_2})(\hat{\mathbf{Y}}_{ijk_1} - \hat{\mathbf{Y}}_{ijk_2})' \times \boldsymbol{\Omega}(\hat{\mathbf{Y}}_{ijk_3} - \hat{\mathbf{Y}}_{ijk_4})(\hat{\mathbf{Y}}_{ijk_3} - \hat{\mathbf{Y}}_{ijk_4})', \quad (13)$$

where $\mathcal{K}$ is the set of all quadruples $\kappa = (k_1, k_2, k_3, k_4)$ without replication, $c_{ij} = n_{ij}(n_{ij} - 1)(n_{ij} - 2)(n_{ij} - 3)$ and $\boldsymbol{\Omega}$ a matrix of constants with dimension $p \times p$. Obviously, this construction requires $n_{ij} \geq 4$ while $|\mathcal{K}|$ is growing very fast for increasing n_{ij} which might lead to high computational cost in practice. Therefore, the nparMD package performs a *randomized covariance matrix estimation* if the groupsize n_{ij} exceeds a default limit n_{max} for a certain factor-level combination. If necessary, $\mathcal{K}$ is replaced by a random subset $\mathcal{K}' \subset \mathcal{K}$, where $|\mathcal{K}'| = c_{max} = n_{max}(n_{max} - 1)(n_{max} - 2)(n_{max} - 3)$. Within the simulation study, a default limit of $n_{max} = 6$ has proved as an appropriate tradeoff between computational cost and estimation accuracy. To avoid misunderstanding, this procedure is not equivalent to drawing 6 observations for the covariance matrix estimation of larger groups since that would lead to high loss of information. An example to demonstrate the runtime difference and results in a difference between using full $\mathcal{K}$ and using $\mathcal{K}'$ instead is shown in Table 2. Without explicit functional modeling of the actual runtime with regard to the sample size, it appeared to improve from approximately exponential to linear within the simulation study.

The underlying asymptotic theorem of the inference procedure requires centering and scaling of the four test statistics such that a unified null distribution can be obtained: $\sqrt{a}(\ell T_{\mathcal{G}}^{(\psi)} + h) = \sqrt{a}\,\mathrm{tr}(\mathbf{H}^{(\psi)} - \mathbf{G})\boldsymbol{\Omega} + o_P(1)$, where $\ell = 1, 2, 1, 4$, $h = 1, 2p\log 2, p, 2p$ and $\boldsymbol{\Omega} = (\frac{1}{tr\mathbf{G}})\mathbf{I}_p, \mathbf{G}^-, \mathbf{G}^-, \mathbf{G}^-$ for $\mathcal{G} = D, LR, LH, BNP$. Then the null distribution is given by the following theorem.

Theorem 1 ([1]) *Let $\psi = A, AB$. Under the null hypothesis (no ψ effect) and for any fixed matrix of constants $\boldsymbol{\Omega}$*

$\sqrt{a}\, tr(\mathbf{H}^{(\psi)} - \mathbf{G})\boldsymbol{\Omega}\, \tau_{\psi}^{-1}(\boldsymbol{\Omega}) \xrightarrow{\mathcal{L}} N(0, 1)$ *as $a \to \infty$ and n_{ij} and b bounded,*

where $\tau_{\psi}^2 = \begin{cases} \frac{2}{b}\{\nu_1(\boldsymbol{\Omega}) + \frac{\nu_2(\boldsymbol{\Omega})}{(b-1)^2}\} \text{ when } \psi = AB \\ \frac{2}{b}\{\nu_1(\boldsymbol{\Omega}) + \nu_2(\boldsymbol{\Omega})\} \text{ when } \psi = A. \end{cases}$

Here, $\nu_1(\boldsymbol{\Omega}) = \frac{1}{ab}\sum_{i=1}^{a}\sum_{j=1}^{b} \frac{1}{n_{ij}(n_{ij}-1)} tr(\boldsymbol{\Psi}_{ij}(\boldsymbol{\Omega}))$,

and $\nu_2(\boldsymbol{\Omega}) = \frac{1}{ab}\sum_{i=1}^{a}\sum_{j\neq j'}^{b} \frac{tr(\boldsymbol{\Omega}\mathbf{S}_{ij}\boldsymbol{\Omega}\mathbf{S}_{ij'})}{n_{ij}n_{ij'}}$.

Figure 2 shows the actual size of α (desired size $\alpha = 0.05$) of the test for different values of the number a of factor levels when Theorem 1 is applied.

Figure 3 shows the simulated power of the test under true alternatives. **1** denotes the p-dimensional vector of ones. The true main effect of factor A is modeled by multivariate normal distributions of the form $N(\mathbf{1} \cdot \delta, \Sigma_{ij})$ (if $i < 10$) and $N(-\mathbf{1} \cdot \delta, \Sigma_{ij})$

Table 2 Example: elapsed time and results difference (error)

n_{ij}	t_1^*	t_2^*	Error T_D	Error T_{LR}	Error T_{LH}	Error T_{BNP}
≤ 6	0.32	0.28	0	0	0	0
≤ 10	2.64	0.42	0.00012	0.00137	0.00022	0.00056
≤ 15	14.12	0.64	0.00062	0.00044	0.00026	0.00014
≤ 20	45.66	1.03	0.00075	0.00021	$<10^{-5}$	0.00023

* t_1 versus t_2 show the elapsed time (stated in seconds) of the testing procedure using full $\mathcal{K}$ (t_1) versus $\mathcal{K}'$ (t_2) with $n_{max} = 6$

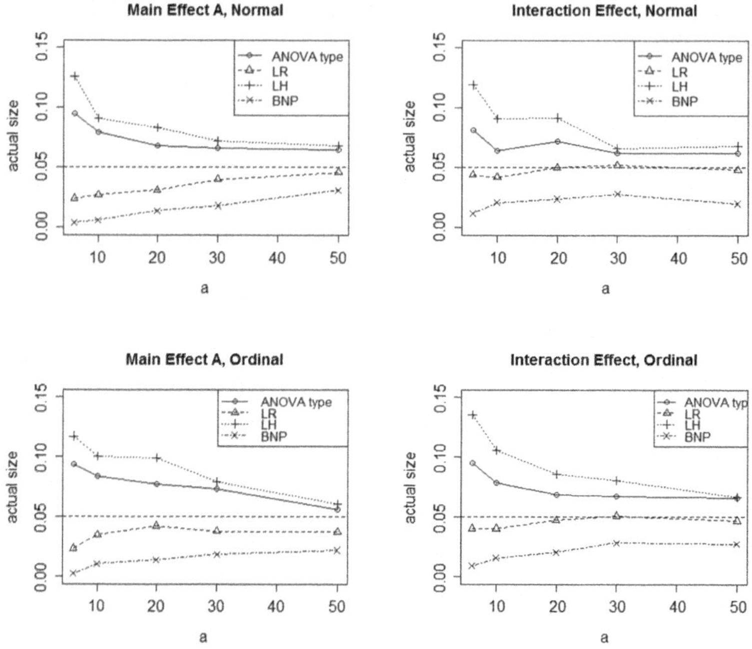

Fig. 2 Simulated α under null hypothesis ($b = 3$, $p = 3$, $4 \leq n_{ij} \leq 10$)

(else) with location shift δ, where Σ_{ij}, as defined above, induces heteroscedasticity and response dependency. A true interaction is simulated in a similar way, that is $N(\mathbf{1} \cdot \delta, \Sigma_{ij})$ (if $i < 10$ and if $j < 2$) and $N(-\mathbf{1} \cdot \delta, \Sigma_{ij})$ (else). As shown in Fig. 3, δ ranges from 0 to 0.5 It appeared that Σ_{ij} did not affect the actual size of the test when there was no true effect in terms of location shift. See Fig. 2 at $a = 20$. Again, to simulate the power for underlying ordinal response data, including dependency of the components, we drew samples from a multinomial distribution with parameters $n = 5$ and $p = (0.2 - \delta p, 0.3, 0.5 + \delta p)$ (if $i \leq 10$) and $p = (0.2 + \delta p, 0.3, 0.5 - \delta p)$ (else) with δp denoting a shift in probability again. In order to induce an interaction effect the setting is changed to $p = (0.2 - \delta p, 0.3, 0.5 + \delta p)$ (if $i \leq 10$ and $j \leq 2$) and $p = (0.2 + \delta p, 0.3, 0.5 - \delta p)$ (else).

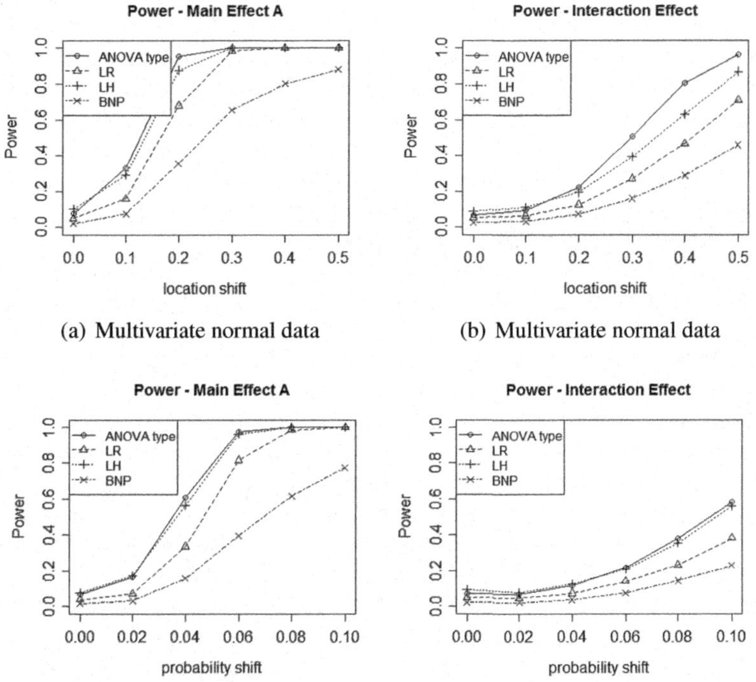

(a) Multivariate normal data (b) Multivariate normal data

(c) Multinomial (ordinal) data (d) Multinomial (ordinal) data

Fig. 3 Simulated Power of all four test statistics ($a = 20, b = 3, p = 3, 4 \leq n_{ij} \leq 10$)

Within this setting of small sample sizes, but many samples (large a setting), the null hypothesis "no main effect of factor B" is not considered explicitly, as this situation would actually correspond to having several observations for each level of factor B. Thus, by relabeling the factors, it fits into the previously discussed asymptotic framework.

4 Conclusion

Nonparametric rank-based inference procedures for multivariate data in two-way factorial designs have been developed by adapting theoretical results from [1]. This includes the development of an R package **nparMD**. Note that the response variables are not required to be metric—in fact, a mix of metric, ordinal, and binary responses is just fine.

At a glance, the R package **nparMD** consists of two major functions (`nparml` and `nparms`), and it is designed to cover a large number of situations in which multivariate data occur, as, for example, in many biological, biomedical, behavioral, and clinical studies. The function `nparml` should be used for larger samples, that is, at least seven observations per factor-level combination according to recommendations and simulation results. In case of smaller samples, the `nparms` function can be used—provided that, with regard to one of the explanatory factors, there are many samples available.

Future versions of the package will include the Wilk's Lambda and the Bartlett–Nanda–Pillai criteria also for the large sample case as well as for explicitly testing the main effect of factor B in the small sample case.

Acknowledgments The research was supported by Austrian Science Fund (FWF) I 2697-N31.

References

1. Bathke, A., Harrar, S.: Rank-based inference for multivariate data in factorial designs. In: Springer Proceedings in Mathematics and Statistics, vol. 168, pp. 121–139 (2016)
2. Bathke, A.C., Harrar, S.W.: Nonparametric methods in multivariate factorial designs for large number of factor levels. J. Stat. Plan. Inference **138**(3), 588–610 (2008)
3. Bathke, A.C., Harrar, S.W., Madden, L.V.: How to compare small multivariate samples using nonparametric tests. Comput. Stat. Data Anal. **52**(11), 4951–4965 (2008)
4. Bathke, A.C., Harrar, S.W., Rauf Ahmad, M.: Some contributions to the analysis of multivariate data. Biom. J. **51**(2), 285–303 (2009)
5. Brunner, E., Dette, H., Munk, A.: Box-type approximations in nonparametric factorial designs. J. Am. Stat. Assoc. **92**(440), 1494–1502 (1997)
6. Burchett, W.W., Ellis, A.R., Harrar, S.W., Bathke, A.C.: Nonparametric inference for multivariate data: the r package NPMV. J. Stat. Softw. **76**(1), 1–18 (2017). https://doaj.org/article/24ea0c31eae94c9d9951b512db2c69b4
7. Harrar, S., Bathke, A.: A nonparametric version of the Bartlett-Nanda-Pillai multivariate test. Asymptotics, approximations, and applications. Am. J. Math. Manag. Sci. **28**(3–4), 309–335 (2008)
8. Harrar, S., Bathke, A.: A modified two-factor multivariate analysis of variance: asymptotics and small sample approximations. Ann. Inst. Stat. Math. **64**(1), 135–165 (2012)
9. Harrar, S.W., Bathke, A.C.: Nonparametric methods for unbalanced multivariate data and many factor levels. J. Multivar. Anal. **99**(8), 1635–1664 (2008)
10. Kiefel, M., Bathke, A.C.: nparMD: nonparametric analysis of multivariate data in factorial designs. R package version 0.1.0 (2018). https://CRAN.R-project.org/package=nparMD
11. Liu, C., Bathke, A.C., Harrar, S.W.: A nonparametric version of wilks lambda: asymptotic results and small sample approximations. Stat. Probab. Lett. **81**(10), 1502–1506 (2011)
12. Munzel, U., Brunner, E.: Nonparametric methods in multivariate factorial designs. J. Stat. Plan. Inference **88**(1), 117–132 (2000)
13. Noguchi, K., Gel, Y.R., Brunner, E., Konietschke, F.: NPARLD: an r software package for the nonparametric analysis of longitudinal data in factorial experiments. J. Stat. Softw. **50**(12) (2012). https://doaj.org/article/2d24c3d127ad4b258a4c7e05b562f6f7
14. Penrose, R.: A generalized inverse for matrices. Math. Proc. Camb. Philos. Soc. **51**(3), 406–413 (1955)
15. R Core Team: R: a language and environment for statistical computing. R Foundation for Statistical Computing, Vienna, Austria (2016). https://www.R-project.org/

Tests for Independence Involving Spherical Data

Pierre Lafaye De Micheaux, Simos Meintanis, and Thomas Verdebout

Abstract We propose consistent procedures for testing the independence of circular variables based on the empirical characteristic function. The new methods are first specified for observations lying on a torus, i.e., for bivariate circular data, but it is shown that these methods can readily be extended to arbitrary dimension. The large-sample behavior of the test statistic is investigated under fixed alternatives. Finite-sample results are also presented.

Keywords Empirical distribution function · Directional statistics · L2-type test

1 Introduction

Circular distributions naturally arise in many areas of applied research such as biology, meteorology, animal behavior, geology, etc. Realizations of such random vectors are interpreted as *directions*, and analogously to the problem with conventional multivariate random vectors there exist circumstances where two or more directions may or may not be independent. An obvious way to go about testing independence is to use methods for classical (non-circular) random variables, such as the correlation coefficient. Due to the periodicity through classical methods do not automatically carry over from the linear domain, and need to be properly modified for circular obser-

P. Lafaye De Micheaux (✉)
School of Mathematics and Statistics, UNSW Sydney, Sydney, Australia
e-mail: lafaye@unsw.edu.au

S. Meintanis
Department of Economics, National and Kapodistrian University of Athens, Athens, Greece
e-mail: simosmei@econ.uoa.gr

Unit for Business Mathematics and Informatics, North-West University, Potchefstroom, South Africa

T. Verdebout
Mathematics Department and ECARES, Université libre de Bruxelles (ULB), Bruxelles, Belgium
e-mail: tverdebo@ulb.ac.be

M. La Rocca et al. (eds.), *Nonparametric Statistics*, Springer Proceedings in Mathematics & Statistics 339, https://doi.org/10.1007/978-3-030-57306-5_27

vations. In this connection, Watson and Beran [20] propose a correlation coefficient for circular time series data and calibrate the test via the permutation distribution of this coefficient, while Epp et al. [5] provide a large-sample normal approximation of the distribution of this test criterion. Likewise, the standard Cramér–von Mises has been adapted to the circular case by Rothman [18], whereas Shieh et al. [19] employ a version of Kendall's tau statistic. There are also a number of (semi)parametric approaches such as testing independence under axial symmetry or testing independence with von Mises marginals; see Rao and Puri [16] and Mardia and Puri [13], respectively. For testing independence as well as for general treatment of statistical problems associated with the circular domain, the readers are referred to the monographs of Jammalamadaka and SenGupta [9] and Ley and Verdebout [12]. Recently, there is also some interest in testing independence between a spherical random variable in general dimension and a corresponding univariate linear random variable as in the case of the statistic suggested by García–Portugués et al. [8] and calibrated in García–Portugués et al. [7].

While the aforementioned procedures for independence are either tailored to the bivariate case and/or often test for lack of linear association rather than independence and in the spherical/linear case employ smoothing techniques, our approach is meant for general dimension and is consistent against arbitrary deviations from independence (not just correlation), and makes no use of smoothing techniques with the familiar problems associated to bandwidth selection and slow convergence. Specifically, we suggest a test for independence for a pair of random variables Z_1 and Z_2 with arbitrary distributions. The test statistic utilizes the familiar factorization property of the joint characteristic function (CF) into the product of the corresponding marginals. In this connection note that while for general random variables uniqueness requires that the CF be computed over all possible arguments, as it will be argued further down the paper the circular domain is exceptional in this respect, and thus we adopt as a population measure of discrepancy from independence

$$\mathscr{T}_w = \sum_{r_1}\sum_{r_2} |\varphi(r_1, r_2) - \varphi_1(r_1)\varphi_2(r_2)|^2 \, w(r_1, r_2), \tag{1}$$

where $\varphi(\cdot, \cdot)$ denotes the joint CF of Z_1 and Z_2, and $\varphi_i(\cdot)$, $i = 1, 2$, the corresponding marginal CFs. The probability measure $w(\cdot, \cdot)$, as well as the range of summation, will be specified later.

While the test statistics considered herein may be referenced to the general formulation (1), we emphasize particular convenient instances which regardless of dimension are shown to be free of the usual computational difficulties arising from the inherent multivariate nature of the problem.

The remainder of this work is outlined as follows. In Sect. 2, we state the null hypothesis and the suggested new criteria in the bivariate case. Corresponding computations are presented in Sect. 3 where we also carry out an extension to more general situations and show the consistency of our test. Finally, the finite-sample properties of the methods are investigated by means of a simulation study in Sect. 4.

2 Testing Independence in the Bivariate Case

Let $\Theta := (\Theta^{(1)}, \Theta^{(2)})^\top$, denote an arbitrary pair of circular random variables with a joint distribution function (DF) $F(\vartheta^{(1)}, \vartheta^{(2)}) = \mathbb{P}(\Theta^{(1)} \le \vartheta^{(1)}, \Theta^{(2)} \le \vartheta^{(2)})$. We wish to test the null hypothesis of independence

$$\mathscr{H}_0 : F(\vartheta^{(1)}, \vartheta^{(2)}) = F_1(\vartheta^{(1)}) F_2(\vartheta^{(2)}), \forall (\vartheta^{(1)}, \vartheta^{(2)}) \in (0, 2\pi) \times (0, 2\pi), \quad (2)$$

where $F_i(\cdot)$, $i = 1, 2$, denote the corresponding marginal DFs.

By the well-known factorization property of CFs, it follows that the null hypothesis $\mathscr{H}_0$ in (2) may equivalently be stated as

$$\varphi(r_1, r_2) = \varphi_1(r_1)\varphi_2(r_2), \forall (r_1, r_2) \in \mathbb{R} \times \mathbb{R}, \quad (3)$$

where $\varphi(r_1, r_2) := \mathbb{E}(e^{\mathrm{i}(r_1\Theta^{(1)} + r_2\Theta^{(2)})})$, $\mathrm{i} = \sqrt{-1}$ defines the (joint) CF of Θ and $\varphi_i(r) := \mathbb{E}(e^{\mathrm{i}r\Theta^{(i)}})$, $r \in \mathbb{R}$ stands for the marginal CF of $\Theta^{(i)}$, $i = 1, 2$. While for conventional random variables on the real line, the CF needs to be defined for all real r, due to periodicity, in the case of circular random variables, it is sufficient to consider the marginal CF only for integer values of the corresponding argument; see Jammalamadaka and SenGupta [9, Sect. 2.1]. Hence, the null hypothesis is equivalent to (3) being true for $(r_1, r_2) \in \mathbb{Z} \times \mathbb{Z}$, where $\mathbb{Z} = \{0, \pm 1, \pm 2, \ldots\}$. Suppose $\vartheta_j := (\vartheta_j^{(1)}, \vartheta_j^{(2)})$, $j = 1, \ldots, n$ are independent copies of the spherical random variable Θ. Then the joint CF figuring in the left-hand side of (3) may be estimated by means of the empirical CF

$$\widehat{\varphi}(r_1, r_2) = \frac{1}{n} \sum_{j=1}^{n} e^{\mathrm{i}(r_1\vartheta_j^{(1)} + r_2\vartheta_j^{(2)})}, \quad (4)$$

while the marginal empirical CF $\widehat{\varphi}_1(r_1)$ (resp. $\widehat{\varphi}_2(r_2)$) can be obtained by setting $r_2 = 0$ (resp. $r_1 = 0$) in (4).

In view of (1), we employ the quantity

$$D_n(r_1, r_2) = \widehat{\varphi}(r_1, r_2) - \widehat{\varphi}_1(r_1)\widehat{\varphi}_2(r_2), \quad (5)$$

which if (3) holds true should be close to zero as $n \to \infty$, for all $r_1, r_2 = 0, \pm 1, \pm 2, \ldots$. This leads us to suggest the test statistic

$$T_{n,w} = n \sum_{r_1=-\infty}^{\infty} \sum_{r_2=-\infty}^{\infty} |D_n(r_1, r_2)|^2 \, w(r_1, r_2), \quad (6)$$

which is an estimated version of the population counterpart figuring in (1) specified to the bivariate case with integration taking place on point masses over $\mathbb{Z} \times \mathbb{Z}$, with respect to the measure $w(\cdot,\cdot)$ (to be further particularized below). Clearly, rejection of the null hypothesis $\mathscr{H}_0$ of independence stated in (2) is for large values of $T_{n,w}$.

We close this section by noting that the use of the CF in order to test independence is not new in the literature. In fact, there are several works dealing with CF-based methods for testing independence with conventional (non-spherical) random variables; see for instance Csörgő [4], Kankainen and Ushakov [10], Bilodeau and Lafaye de Micheaux [1], Székely et al. [17], Meintanis and Iliopoulos [14], and Fan et al. [6]. These methods are very convenient from the computational point of view, a feature which is particularly important in the multivariate context where the corresponding methods based on the DF, apart from being computationally demanding, suffer from the lack of definite order in $\mathbb{R}^p$, $p > 1$. On top of this, CF-based procedures for independence have proved to compete well and often outperform DF-based methods even when the latter methods are available.

3 Computations and Extensions

3.1 Test on the Torus

We first study the computational aspects of the test statistic figuring in (6). Specifically, following some algebra we have from (4) to (5)

$$
\begin{aligned}
|D_n(r_1, r_2)|^2 &= \frac{1}{n^2}\sum_{j,k=1}^{n} \cos\left(r_1\vartheta_{jk}^{(1)} + r_2\vartheta_{jk}^{(2)}\right) \\
&+ \frac{1}{n^4}\sum_{j,k,\ell,m=1}^{n} \cos\left(r_1\vartheta_{jk}^{(1)} + r_2\vartheta_{\ell m}^{(2)}\right) \\
&- \frac{2}{n^3}\sum_{j,k,\ell=1}^{n} \cos\left(r_1\vartheta_{jk}^{(1)} + r_2\vartheta_{j\ell}^{(2)}\right)
\end{aligned} \tag{7}
$$

with $\vartheta_{jk}^{(i)} = \vartheta_j^{(i)} - \vartheta_k^{(i)}$, $j, k = 1, \ldots, n$, $i = 1, 2$. Clearly, then the test statistic is invariant with respect to origin. Expression (7) also shows that $T_{n,w}$ involves periodic components, and hence clarifies the need for introducing a probability measure $w(\cdot,\cdot)$, such that $\sum_{r_1}\sum_{r_2} w(r_1, r_2) < \infty$, in order to temper these periodic components of the test statistic figuring in (6).

Moreover, by straightforward algebra and, by using (7) in (6), we readily obtain the test statistic in the form

$$T_{n,w} = \frac{1}{n}\sum_{j,k=1}^{n} C_w(\vartheta_{jk}^{(1)}, \vartheta_{jk}^{(2)}) + \frac{1}{n^3}\sum_{j,k,\ell,m=1}^{n} C_w(\vartheta_{jk}^{(1)}, \vartheta_{\ell m}^{(2)}) - \frac{2}{n^2}\sum_{j,k,\ell=1}^{n} C_w(\vartheta_{jk}^{(1)}, \vartheta_{j\ell}^{(2)}) \tag{8}$$

where

$$C_w(x, y) = \sum_{r_1=-\infty}^{\infty}\sum_{r_2=-\infty}^{\infty} \cos(r_1 x + r_2 y) w(r_1, r_2). \tag{9}$$

A little reflection on (9) shows that if $w(r_1, r_2)$ is a symmetric around zero probability mass function (PMF) in the domain $(r_1, r_2) \in \mathbb{Z} \times \mathbb{Z}$, then the infinite sum figuring in the right-hand side of this equation may be interpreted as the CF corresponding to this PMF computed at the argument (x, y). In order to construct such a PMF, we choose any univariate PMF $f(r)$ defined on the non-negative integers, and obtain a new symmetrized PMF, say $v(\cdot)$, on $\mathbb{Z}$ by setting

$$\begin{aligned} v(\pm r) &= (1/2) f(r), \ r = 1, 2, \ldots, \\ v(0) &= f(0). \end{aligned} \tag{10}$$

Clearly, the new PMF $v(\cdot)$ so constructed is symmetric around zero and, hence, the imaginary part of its CF vanishes identically; see Meintanis and Verdebout [15]. In fact, it may easily be shown that $v(\cdot)$ has as CF the real part of the CF corresponding to the PMF $f(r)$, which will be henceforth denoted by $c_f(\cdot)$.

From this observation and adopting the decomposition $w(r_1, r_2) = v(r_1)v(r_2)$ for the probability measure $w(\cdot, \cdot)$, we deduce from (9) by means of the trigonometric identity $\cos(x + y) = \cos(x)\cos(y) - \sin(x)\sin(y)$, that $C_w(x, y) = c_f(x)c_f(y)$, which renders the test statistic in the following convenient form

$$T_{n,w} = \frac{1}{n}\sum_{j,k=1}^{n} c_f(\vartheta_{jk}^{(1)})c_f(\vartheta_{jk}^{(2)}) + \frac{1}{n^3}\left(\sum_{j,k=1}^{n} c_f(\vartheta_{jk}^{(1)})\right)\left(\sum_{j,k=1}^{n} c_f(\vartheta_{jk}^{(2)})\right) - \frac{2}{n^2}\sum_{j,k,\ell=1}^{n} c_f(\vartheta_{jk}^{(1)})c_f(\vartheta_{j\ell}^{(2)}). \tag{11}$$

A standard example is to choose the Poisson distribution as the 'core' PMF $f(r)$, in which case the test statistic results from (11) with $c_f(\theta) = \cos(\lambda \sin\theta)e^{\lambda(\cos\theta - 1)}$, where $\lambda > 0$ denotes the Poisson parameter.

3.2 Extension to Arbitrary Dimension

Suppose now that $\Theta := (\Theta^{(1)\top}, \Theta^{(2)\top})^\top$ is composed of two random vectors, $\Theta^{(1)}$ and $\Theta^{(2)}$ of dimensions p_1 and p_2, respectively. By analogous reasoning, it follows that the null hypothesis of independence reduces to

$$\varphi(r_1, r_2) = \varphi_1(r_1)\varphi_2(r_2), \forall (r_1, r_2) \in \mathbb{Z}^{p_1} \times \mathbb{Z}^{p_2}, \tag{12}$$

where $\mathbb{Z}^p$ denotes the Cartesian product space resulting from the set of integers $\mathbb{Z}$.

Along the lines of Sect. 3.1, we suggest the test criterion

$$T_{n,w} = n \sum_{r_1} \sum_{r_2} |D_n(r_1, r_2)|^2 \, w(r_1, r_2), \tag{13}$$

with summation taking place on point masses over $\mathbb{Z}^{p_1} \times \mathbb{Z}^{p_2}$, with respect to the measure $w(\cdot, \cdot)$.

The line of reasoning of the previous subsection follows through and the test statistic referring to the population distance defined by (1) reduces to (8) with

$$C_W(x, y) = \sum_{r_1 \in \mathbb{Z}^{p_1}} \sum_{r_2 \in \mathbb{Z}^{p_2}} \cos(r_1^\top x + r_2^\top y) W(r_1, r_2), \tag{14}$$

where $W(\cdot, \cdot)$ is a point mass measure over $\mathbb{Z}^{p_1} \times \mathbb{Z}^{p_2}$. Let again $f(r)$ be a univariate core PMF and adopt the decomposition $W(r_1, r_2) = V_1(r_1)V_2(r_2)$, and the independent-marginal factorizations $V_i(r_i) = \prod_{s=1}^{p_i} v(r_{si})$, where r_{si}, $s = 1, \ldots, p_i$, stand for the components of the vector r_i, and where $v(r)$ is constructed as in (10).

Then proceeding from (14), we have

$$\begin{aligned} C_W(x, y) &= \sum_{r_1 \in \mathbb{Z}^{p_1}} \cos(r_1^\top x) V_1(r_1) \sum_{r_2 \in \mathbb{Z}^{p_2}} \cos(r_2^\top y) V_2(r_2) \\ &\quad - \sum_{r_1 \in \mathbb{Z}^{p_1}} \sin(r_1^\top x) V_1(r_1) \sum_{r_2 \in \mathbb{Z}^{p_2}} \sin(r_2^\top y) V_2(r_2) \\ &= \sum_{r_1 \in \mathbb{Z}^{p_1}} \cos(r_1^\top x) V_1(r_1) \sum_{r_2 \in \mathbb{Z}^{p_2}} \cos(r_2^\top y) V_2(r_2), \end{aligned} \tag{15}$$

with the last equation justified because the PMFs V_1 and V_2 so constructed are symmetric around their origins.

To proceed further, we invoke the trigonometric identity

$$\cos\left(\sum_{\ell=1}^{p} \theta_\ell\right) = \sum_{q=0}^{\lfloor \frac{p}{2} \rfloor} (-1)^q \sum_{1 \le j_1 < j_2 \ldots < j_{2q} \le p} \prod_{k=1}^{2q} \sin \theta_{j_k} \prod_{\ell \ne j_1, j_2, \ldots, j_{2q}}^{p} \cos \theta_\ell. \tag{16}$$

A little reflection on this identity shows that the only terms of (16) that 'survive' when plugged in (15) are those that contain only products of cosines, while the products in (16) that contain even a single sine factor vanish when plugged in (15) since $s(x) := \sum_{r=-\infty}^{\infty} \sin(rx)v(r) = 0$, due to the fact that $s(x)$ is by definition the imaginary part of the CF corresponding to the symmetric around zero PMF $v(\cdot)$, and hence $s(x)$ vanishes at each argument $x \in \mathbb{R}$. Consequently, the sum in the right-hand side of (15) reduces to

$$C_W(x, y) = \prod_{s=1}^{p_1} c_f(x_s) \prod_{s=1}^{p_2} c_f(y_s), \tag{17}$$

where $x_s,\ s = 1, \ldots, p_1$ and $y_s,\ s = 1, \ldots, p_2$, stand for the components of the vectors x and y, respectively.

Clearly, then we have arrived at a particularly user-friendly expression for the test statistic which we report below for definiteness:

$$\begin{aligned} n^{-1}T_w = &\frac{1}{n^2} \sum_{j,k=1}^{n} C_f(\vartheta_{jk}^{(1)}) C_f(\vartheta_{jk}^{(2)}) + \frac{1}{n^4} \left(\sum_{j,k=1}^{n} C_f(\vartheta_{jk}^{(1)}) \right) \left(\sum_{j,k=1}^{n} C_f(\vartheta_{jk}^{(2)}) \right) \\ &- \frac{2}{n^3} \sum_{j,k,\ell=1}^{n} C_f(\vartheta_{jk}^{(1)}) C_f(\vartheta_{j\ell}^{(2)}), \end{aligned} \tag{18}$$

where

$$C_f(\vartheta_{jk}^{(i)}) = \prod_{s=1}^{p_i} c_f(\vartheta_{s,jk}^{(i)}), \tag{19}$$

with $\vartheta_{s,jk}^{(i)},\ s = 1, \ldots, p_i$, being the components of the vector $\vartheta_{jk}^{(i)},\ i = 1, 2$.

3.3 Consistency

In the following proposition, we consider the limit behavior of the test statistic against arbitrary deviations from independence.

Proposition 1 *Let $T_{n,w}$ be the the test statistic in (13) with $w(\cdot, \cdot)$ being a given PMF. Then*

$$\frac{T_{n,w}}{n} \longrightarrow \mathcal{T}_w, \text{ a.s. as } n \to \infty, \tag{20}$$

with $\mathcal{T}_w$ defined in (1).

Proof From the strong law of large numbers (see also Csörgő [3]) for more refined limit results on the empirical CF process), we have for each $(r_1, r_2) \in \mathbb{Z}^{p_1} \times \mathbb{Z}^{p_2}$,

$$\widehat{\varphi}(r_1, r_2) \longrightarrow \varphi(r_1, r_2), \text{ a.s. as } n \to \infty,$$

and likewise for the marginal empirical CFs. Therefore, (20) follows from the Lebesque's dominated convergence theorem since $|D_n(r_1, r_2)|^2 \leq 4$. Moreover, in view of (12), the almost sure limit $\mathcal{T}_w$ in the right-hand side of (20) is positive unless $\mathcal{H}_0$ holds true, which in turn implies that under alternatives

$$T_{n,w} \longrightarrow \infty, \text{ a.s. as } n \to \infty,$$

and, consequently, (20) is equivalent to the strong consistency of the test which rejects the null hypothesis $\mathcal{H}_0$ of independence for large values of $T_{n,w}$ ■

4 Simulations

The bivariate von Mises cosine distribution $vM_2^c(\mu_1, \mu_2, \kappa_1, \kappa_2, \kappa_3)$ with mean parameters μ_1, μ_2, concentration parameters κ_1, κ_2, and association parameter κ_3 has the following density function

$$f_{vM_2^c}(\theta_1, \theta_2|\mu_1, \mu_2, \kappa_1, \kappa_2, \kappa_3) \propto \exp[\kappa_1 \cos(\theta_1 - \mu_1) + \kappa_2 \cos(\theta_2 - \mu_2) + \kappa_3 \cos(\theta_1 - \mu_1 - \theta_2 + \mu_2)].$$

It is thus easily seen that a value of $\kappa_3 = 0$ means independence while increasing κ_3 creates more dependence between two random variable $\Theta^{(1)}$ and $\Theta^{(2)}$ from this distribution.

In order to confirm that our test is able to capture dependence for spherical data, we generated observations from the above density using the `rvmcos()` function from the `BAMBI` R package; see [2]. We kept the default values $\kappa_1 = 1, \kappa_2 = 1, \mu_1 = 0$ and $\mu_2 = 0$.

First, we generated $M = 10{,}000$ samples of size $n = 100$ under the null hypothesis of independence (i.e., with $\kappa_3 = 0$). The 5% critical value for the test (0.511) was computed as the 95% empirical quantile of the corresponding 10,000 test statistic values of (18). As a first approximation, we assumed that this critical value is valid for any value of κ_3 (i.e., that it is independent of the marginal distributions).

Next, we computed the proportion of times (over $M = 10{,}000$) that our test statistic $T_{n,w}$ (with a Poisson weight with parameter $\lambda = 1$) exceeds this critical value, namely the empirical power of the test, for 10 increasing equispaced values of κ_3, ranging from 0 to 1. Results, displayed in Table 1, confirm that our test behaves as expected. In this limited simulation study, we used the knowledge of the underlying distribution of $(\Theta^{(1)}, \Theta^{(2)})$ which in practice is rarely available. In a future version of this work, we will explore numerically the potential of our test to detect dependence in a purely non-parametric way; we will resort to permutations.

Table 1 Empirical power of our new test for increasing values of κ_3. Number of Monte Carlo replications: $M = 10{,}000$. Sample size: $n = 100$. Total computation time: 92 mn

κ_3	0/9	1/9	2/9	3/9	4/9	5/9	6/9	7/9	8/9	9/9
Power	0.047	0.058	0.100	0.187	0.307	0.452	0.612	0.740	0.847	0.911

All simulations were done on a DELL XPS13 laptop, equipped with a 64-bit Linux Debian 10 operating system. We used R version 3.6.1 and set the seed to the value 1. Our test of independence was programmed in C/C++ following guidelines from [11].

References

1. Bilodeau, M., Lafaye de Micheaux, P.: A multivariate empirical characteristic function test of independence with normal marginals. J. Multivar. Anal. **95**, 345–369 (2005)
2. Chakraborty, S., Wong, S.W.K.: BAMBI: an R package for fitting bivariate angular mixture model (2017). arXiv:1708.07804
3. Csörgő, S.: Multivariate empirical characteristic functions. Z. Wahr. Verw. Geb. **55**, 203–229 (1981)
4. Csörgő, S.: Testing for independence by the empirical characteristic function. J. Multivar. Anal. **95**, 290–299 (1985)
5. Epp, R.J., Tukey, J.W., Watson, G.S.: Testing unit vectors for correlation. J. Geophys. Res. **76**, 8480–8483 (1971)
6. Fan, Y., Lafaye de Micheaux, P., Penev, S., Salopek, D.: Multivariate nonparametric test of independence. J. Multivar. Anal. **153**, 180–210 (2017)
7. García-Portugués, E., Crujeiras, R.M., González Manteiga, W.: A test for linear-directional independence, with applications to wildfire orientation and size. Stoch. Environ. Res. Risk Assess. **28**, 1261–1275 (2014)
8. García-Portugués, E., Crujeiras, R.M., González Manteiga, W.: Central limit theorems for directional and linear variables with applications. Stat. Sin. **25**, 1207–1229 (2015)
9. Jammalamadaka, S.R., SenGupta, A.: Topics in Circular Statistics. World Scientific, Singapore (2001)
10. Kankainen, A., Ushakov, N.G.: A consistent modification of a test for independence based on the empirical characteristic function. J. Math. Sci. **89**, 1486–1493 (1999)
11. Lafaye De Micheaux, P., Tran, V.A.: PoweR: a reproducible research tool to ease Monte Carlo power simulation studies for goodness-of-fit tests in R. J. Stat. Softw. **69**(1), 1–44 (2016)
12. Ley, C., Verdebout, T.: Modern Directional Statistics. CRC Press, New York (2017)
13. Mardia, K.V., Puri, M.L.: A spherical correlation coefficient robust against scale. Biometrika **65**, 391–395 (1978)
14. Meintanis, S.G., Iliopoulos, G.: Fourier test for multivariate independence. Comput. Stat. Data Anal. **52**, 1884–1895 (2008)
15. Meintanis, S.G., Verdebout, T.: Le Cam maximin tests for symmetry of circular data based on the characteristic function. Stat. Sin. (2018). To appear
16. Rao, S.J., Puri, M.L.: Testing independence of bivariate circular data and weighted U-statistics. In: Krishnaiah, P.R. (ed.) Multivariate Analysis, vol. 4, pp. 513–522. North Holland, Amsterdam (1977)
17. Rizzo, M.L., Székely G.J., Bakirov, N.K.: Measuring and testing dependence by correlation distances. Ann. Stat. **35**, 2769–2794 (2007)

18. Rothman, E.D.: Test for coordinate independence for a bivariate sample on torus. Ann. Math. Stat. **42**, 1962–1969 (1971)
19. Shieh, G.S., Johnson, R.A., Frees, E.W.: Testing independence of bivariate circular data and weighted U-statistics. Stat. Sin. **4**, 729–747 (1994)
20. Watson, G.S., Beran, R.: Testing a sequence of unit vectors for serial correlation. J. Geophys. Res. **72**, 5655–5659 (1967)

Interval-Wise Testing of Functional Data Defined on Two-dimensional Domains

Patrick B. Langthaler, Alessia Pini, and Arne C. Bathke

Abstract Functional Data Analysis is the statistical analysis of data sets composed of functions of a continuous variable on a given domain. Previous work in this area focuses on one-dimensional domains. In this work, we extend a method developed for the one-dimensional case, the interval-wise testing procedure (IWT), to the case of a two-dimensional domain. We first briefly explain the theory of the IWT for the one-dimensional case, followed by a proposed extension to the two-dimensional case. We also discuss challenges that appear in the two-dimensional case but do not exist in the one-dimensional case. Finally, we provide results of a simulation study to explore the properties of the new procedure in more detail.

Keywords Functional Data Analysis · Type I Error Control · Rectangular Domain

1 Introduction

Functional Data Analysis is the statistical analysis of data sets composed of functions of a continuous variable (time, space, ...), observed in a given domain (see Ramsay and Silverman [8]). In this work, we focus on the inference for functional data,

P. B. Langthaler (✉) · A. C. Bathke
Department of Mathematics, Paris-Lodron University Salzburg,
Hellbrunnerstraße 34, 5020 Salzburg, Austria
e-mail: PatrickBenjamin.Langthaler@stud.sbg.ac.at

A. C. Bathke
e-mail: arne.bathke@sbg.ac.at

A. Pini
Department of Statistical Sciences, Università Cattolica del Sacro Cuore,
Largo A. Gemelli, 1-20123 Milan, Italy
e-mail: alessia.pini@unicatt.it

M. La Rocca et al. (eds.), *Nonparametric Statistics*, Springer Proceedings
in Mathematics & Statistics 339, https://doi.org/10.1007/978-3-030-57306-5_28

a particularly challenging problem since functional data are objects embedded in infinite-dimensional spaces, and the traditional inferential tools cannot be used in this case. The challenge of deriving inference for functional data is currently tackled by literature from two different perspectives: global inference involves testing a (typically simple) null hypothesis against an alternative extending over the whole (remaining) domain of the parameter space (see e.g. Benko et al. [1], Cuevas et al. [2], Hall and Keilegom [4], Horváth et al. [5], Horváth and Kokoszka [5]); local inference instead addresses the problem of selecting the areas of the domain responsible for the rejection of the null hypothesis, assigning a p-value to each point of the domain (see e.g. Pini and Vantini [6, 7]).

Here, we take the second line of research as a starting point. More precisely, Pini and Vantini [6] suggest performing inference on the coefficients of a B-spline basis expansion, while in extension of the previous work, the same authors propose the interval-wise testing procedure (IWT) which performs inference directly on the functional data (without requiring a basis expansion) [7]. Both methods propose to adjust local p-values in order to control the interval-wise error rate, that is, the probability of wrongly rejecting the null hypothesis in any interval.

In this paper, we extend the IWT to functional data defined in two-dimensional domains. Indeed, all current works addressing local inference deal with one-dimensional domains. Their extension to two (or more) dimensions is not trivial since it would require to define a proper notion of 'interval' in two dimensions. We start from a brief overview of the IWT and its properties (Sect. 2, then we discuss how to extend this approach to two-dimensional domains (Sect. 3). Finally, we report in Sect. 4 on a simulation study investigating the properties of this new method, and draw some conclusions (Sect. 5).

2 Previous Works: The IWT for Functional Data Defined on One-Dimensional Domains

We give here a brief overview of the IWT. For a thorough treatment of the method, see Pini and Vantini [6, 7]. The setting in which the IWT can be used is this: assume that for each unit of analysis (a subject, mechanical object, etc.) we have observed a function $x_i(t)$, $i = 1, \ldots, N$, $t \in D = [a, b]$, with $a, b \in \mathbb{R}$. For ease of notation, we assume here that functional data are continuous. However, this assumption can be relaxed (see Pini and Vantini [7]).

Assume that we aim at locally testing, for each point of the domain, a null hypothesis H_0^t against an alternative H_1^t. For example, assume that the sample is divided into a different groups. We indicate our functional data as $x_{ij}(t)$, where $j = 1, \ldots, a$ denotes the group, and $i = 1, \ldots, n_j$ denotes the units in group j. We could be interested in testing mean differences between the groups:

$$H_0^t : \mu_1(t) = \mu_2(t) = \cdots = \mu_a(t), \quad H_1^t = H_0^C, \tag{1}$$

where $\mu_j(t) = \mathbb{E}[x_{ij}(t)]$. Testing each hypothesis (1) is straightforward since it involves univariate data. The challenge is to adjust the results in order to take the multiplicity of tests into account. The IWT executes this as follows: First, we test each hypothesis (1) separately, and denote the corresponding p-value as $p(t)$. This is an unadjusted p-value function defined for all $t \in D$. Next, we test the null and alternative hypotheses over each interval $\mathscr{I} = [t_1, t_2] \subseteq D$ and the complementary set of each interval $\mathscr{I}^C = D \setminus \mathscr{I}$:

$$H_0^{(\mathscr{I})} : \bigcap_{t \in \mathscr{I}} H_0^{(t)} : \mu_1(t) = \mu_2(t) = ... = \mu_a(t) \; \forall t \in \mathscr{I}; \; H_1^{(\mathscr{I})} = H_0^{(\mathscr{I})^C}; \quad (2)$$

$$H_0^{(\mathscr{I}^C)} : \bigcap_{t \in \mathscr{I}^C} H_0^{(t)} : \mu_1(t) = \mu_2(t) = ... = \mu_a(t) \; \forall t \in \mathscr{I}^C; \; H_1^{(\mathscr{I}^C)} = H_0^{(\mathscr{I}^C)^C}. \quad (3)$$

Denote the p-values of test (2) and (3) as $p^{\mathscr{I}}$ and $p^{\mathscr{I}^C}$, respectively.

For each point $t \in D$, we now define an adjusted p-value for interval-wise testing as the maximum p-value of all tests including the point t:

$$\widetilde{p}_{IWT}(t) := \max \left\{ \max_{\mathscr{I}: t \in \mathscr{I}} p^{\mathscr{I}}; \max_{\mathscr{I}^C: t \in \mathscr{I}^C} p^{\mathscr{I}^C} \right\}. \quad (4)$$

These p-values provide interval-wise control of the type 1 error rate, that is,

$$\forall \alpha \in (0, 1) : \forall \mathscr{I} : P_{H_0^{\mathscr{I}}} \left(\bigcup_{t \in \mathscr{I}} \{\widetilde{p}_{IWT}(t) \leq \alpha\} \right) \leq \alpha.$$

In practice, it is obviously not possible to perform a statistical test in every point of the domain, and in every interval and complementary interval. So, the implementation of the IWT requires discretizing functional data on a dense grid of p equally sized subintervals. Functional data are approximated with a constant in each subinterval. Then, the unadjusted p-value is computed on each subinterval, and the p-value of tests (2) and (3) is computed for every interval and the complementary interval that can be created as a union of the p subintervals. Finally, the adjusted p-value $\widetilde{p}_{IWT}(t)$ is computed applying formula (4) on the performed tests. For details, see Pini and Vantini [7].

3 Methods

The primary task in extending the IWT to functional data defined on two-dimensional domains is to find a suitable neighbourhood over which the tests are performed (corresponding to the intervals in the one-dimensional case). If one has decided on a neighbourhood, the very same principle of p-value adjustment as in the one-dimensional case applies. Instead of interval-wise control, we then get control for every neighbourhood of the specified type. What constitutes a *good* neighbourhood can depend on the specific data one wants to analyse. If there is reason to believe that the two-dimensional subset of the domain in which there is a significant difference between groups takes on a certain shape, then it is reasonable to take this kind of shape as the neighbourhood. The choice of neighbourhood might also depend on the shape of the domain of the functional data. In this contribution, we will try to stay as general as possible and make no assumptions about the area in which significant differences may be found. We do however have to make an assumption about the shape of the two-dimensional domain. We will assume that the data have been recorded on a rectangular grid. This shape makes the use of rectangles or squares as neighbourhoods plausible.

Before we continue, we would like to make one important remark: When using intervals as neighbourhoods in the one-dimensional case, the complement of an interval can be seen as an interval that *wraps around.* When using rectangles or squares in the two-dimensional scenario, this is not the case. A rectangle that leaves the domain on the right or top and comes in on the left or bottom can not necessarily be described as the complementary set of a rectangle fully contained in the domain. For ease of computation, we decided to test for all possible squares and the complements thereof and to not test squares that *wrap around.*

Here, we describe the extension of IWT to the two-dimensional domain D, starting from a general definition of neighbourhood. In the following subsection, we discuss instead different possible choices of neighbourhoods.

3.1 The Extension of the IWT to Functional Data Defined on Two-Dimensional Domains

Assume to observe functional data $x_i(t)$, with $t \in D = [a_1, b_1] \times [a_2, b_2]$, and $i = 1, \dots, n$. Assume that the functions $x_i(t)$ are continuous on D. Also, in this case, we aim at locally testing a null hypothesis against an alternative, and selecting the points of the domain where the null hypothesis is rejected. For instance, assume again that units belong to a groups, and that we aim at testing mean equality between the groups (1).

The two-dimensional extension of the IWT requires defining a notion of 'interval' in two dimensions, or neighbourhood. Let us assume that a proper family of neighbourhoods has been defined (e.g. all rectangles and rectangles' complements included in the domain), and denote as $\mathcal{N}$ a generic neighbourhood. Then, the two-dimensional IWT requires testing the null and alternative hypotheses on every possible neighbourhood, and on every complement of it, i.e. performing the tests

$$H_0^{(\mathscr{N})} : \bigcap_{t \in \mathscr{N}} H_0^{(t)} : \mu_1(t) = \mu_2(t) = \cdots = \mu_a(t) \ \forall t \in \mathscr{N}; \ H_1^{(\mathscr{N})} = H_0^{(\mathscr{N})^C}; \tag{5}$$

$$H_0^{(\mathscr{N}^C)} : \bigcap_{t \in \mathscr{N}^C} H_0^{(t)} : \mu_1(t) = \mu_2(t) = \ldots = \mu_a(t) \ \forall t \in \mathscr{N}^C; \ H_1^{(\mathscr{N}^C)} = H_0^{(\mathscr{N}^C)^C} \tag{6}$$

for all $\mathscr{N} \in \mathscr{F}$. Let us denote with $p^{\mathscr{N}}$ the p-value of such test. Then, the adjusted p-value at point $t \in D$ can be computed as

$$\widetilde{p}_{IWT}(t) := \max\left\{ \max_{\mathscr{N}:t\in\mathscr{N}} p^{\mathscr{N}}; \max_{\mathscr{N}^C:t\in\mathscr{N}} p^{\mathscr{N}} \right\}. \tag{7}$$

It is then straightforward to prove, extending the result of IWT for one-dimensional data, that such p-values provide interval-wise control of the type 1 error rate over all neighbourhoods, that is,

$$\forall \alpha \in (0, 1) : \forall \mathscr{N} : P_{H_0^{\mathscr{N}}}\left(\bigcup_{t\in\mathscr{N}} \{\widetilde{p}_{IWT}(t) \leq \alpha\} \right) \leq \alpha.$$

3.2 The Problem of Dimensionality in the Choice of the Neighbourhood

In the one-dimensional case, we have two parameters that fully characterise an interval: the starting point of the interval and its length. There are p different starting points and p different lengths. There is, however, only one interval of length p, giving us a total of $p^2 - p + 1$ possible intervals for which p-values need to be computed. Thus, the computational cost is of order p^2. If we want to use rectangles as neighbourhoods in the two-dimensional case, we can first freely chose the lower left corner of the rectangle, giving us $p_1 p_2$ possibilities, where p_1 is the number of grid points on the x-axis and p_2 is the number of grid points on the y-axis. Once the lower left corner is chosen, the rectangle can then be fully characterised by its length in the x-direction and its length in the y-direction. These can however not be chosen freely since the rectangle has to remain inside the domain. Overall, this puts us as $\frac{p_1(p_1-1)p_2(p_2-1)}{2}$ possible neighbourhoods, setting the computational cost to the order of $p_1^2 p_2^2$.

If we are content with only using squares, assuming for now that the domain D is observed on a square grid discretised in $p \times p$ points, we only need to test for $\frac{p(p-1)(2p-1)}{6}$ neighbourhoods. The computational cost is thus of the order p^3, an order lower than the one of the rectangle case.

What if we want to have the benefit of lower computational cost but our domain is a rectangular grid? In this case, we can simply rescale the grid: assume we start from a $p_1 \times p_2$ grid and assume $p_1 \neq p_2$. We fix $p := \min\{p_1, p_2\}$. Let, w.l.o.g, $p_1 < p_2$. We then rescale the axis with the p_2 observations by using new points with coordinates $a_2 + \frac{i(b_2-a_2)}{p-1}$, $i = 0, ..., p-1$. If $\frac{p_2-1}{p_1-1}$ is not an integer, then the functions were not observed at the resulting new grid points. We can however use a simple nearest neighbour imputation, using only the nearest neighbour. Note that when we speak of squares and rectangles, we mean in terms of the grid, not in terms of the physical units of the observations. Accordingly, by the nearest neighbour, we mean the nearest observation, assuming that the distance between two adjacent grid points is the same for the two dimensions. Our *squares* thus can be thought of as rectangles whose sides have the same ratio as the sides of the original domain.

4 Simulation Study

We chose to conduct a simulation study looking at the following scenarios:

- (S0) The grid is quadratic and the null hypothesis is true everywhere. In this case, we should see that we have weak control of the error rate.
- (S1) The grid is quadratic and the null hypothesis is violated on a square region. In this case, we should have our square-wise control of the error rate.
- (S2) The grid is quadratic and the null hypothesis is violated on a region that is not a square but a rectangle with unequal sides. Thus, we have no control of the FWER in this scenario.
- (S3) The grid is rectangular and the null hypothesis is violated on a rectangular region, the ratio of the sides of which is the same as the ratio of the sides of the grid. If our rescaling works as is should, we should see the same results as in (S1).

4.1 Simulation Settings

For all our simulation scenarios, we followed the following scheme: First, we created two mean functions on some domain. We discretised the mean functions by evaluating them on a grid. We created observations by adding a realisation of a random variable

$$y = (y_{1,1}, y_{1,2}, \ldots, y_{1,p_2}, y_{2,1}, \ldots, y_{p_1,1}, \ldots, y_{p_1,p_2}) \tag{8}$$

to each mean grid. This realisation was drawn from a multivariate Gaussian distribution with mean zero and covariance function

$$Cov(Y_{i,j}, Y_{i',j'}) = 0.1 \cdot e^{-10((i-i')^2+(j-j')^2)}. \tag{9}$$

This simulates data that come from an infinitely differentiable stochastic process (see Ramussen and Williams [9] pp. 83). Between subjects and groups, the errors were uncorrelated. We did this 10 times for the first mean function and 10 times for the second, giving us a sample of 20 observations divided into two groups. The hypothesis of interest was the equality of distribution between the two groups, and the specific test used was a permutation test by Hall and Tajvidi [3]. In scenarios (S0), (S1) and (S2), the domain was the square $[0, 1] \times [0, 1]$. In (S3), the domain was $[0, 2] \times [0, 1]$. The first mean function was always a constant zero. The second mean function was as follows:

- (S0) The second mean function was also a constant zero.
- (S1) The second mean function was defined as

$$f(x, y) = \begin{cases} 0 & \text{if } x \leq 0.25 \text{ or } x \geq 0.75 \text{ or } y \leq 0.25 \text{ or } y \geq 0.75 \\ \frac{x-0.25}{0.25} & \text{if } 0.25 < x \leq 0.5 \text{ and } x \leq y \leq 1-x \\ \frac{0.75-x}{0.25} & \text{if } 0.5 < x < 0.75 \text{ and } x \leq y \leq 1-x \\ \frac{y-0.25}{0.25} & \text{if } 0.25 < y \leq 0.5 \text{ and } y \leq x \leq 1-y \\ \frac{0.75-y}{0.25} & \text{if } 0.5 < y < 0.75 \text{ and } \leq x \leq 1-y \end{cases}$$

  This is a quadratic pyramid of height 1 and with base $[0.25, 0.75] \times [0.25, 0.75]$.
- (S2) The second mean function was defined as

$$f(x, y) = \begin{cases} 0 & \text{if } y \leq 0.25 \text{ or } y \geq 0.75 \\ \frac{y-0.25}{0.25} & \text{if } 0.25 < y \leq 0.5 \\ \frac{0.75-y}{0.25} & \text{if } 0.5 < y < 0.75 \end{cases}$$

  This is a triangular prism of height 1 and with base $[0, 1] \times [0.25, 0.75]$.
- (S3) The second mean function was defined as

$$f(x, y) = \begin{cases} 0 & \text{if } x \leq 0.5 \text{ or } x \geq 1.5 \text{ or } y \leq 0.25 \text{ or } y \geq 0.75 \\ 2x-1 & \text{if } 0.5 < x \leq 1 \text{ and } 0.5x \leq y \leq 1-0.5x \\ 3-2x & \text{if } 1 < x < 1.5 \text{ and } 0.5x \leq y \leq 1-0.5x \\ \frac{y-0.25}{0.25} & \text{if } 0.25 < y \leq 0.5 \text{ and } 2y \leq x \leq 2-2y \\ \frac{0.75-y}{0.25} & \text{if } 0.5 < y < 0.75 \text{ and } 2y \leq x \leq 2-2y \end{cases}$$

  This is a pyramid of height 1 with base $[0.5, 1.5] \times [0.25, 0.75]$.

As to the number of grid points, we used 21×21 grid points in scenarios (S0), (S1) and (S2), and 41×21 grid points in scenario (S3). The mean functions for the second group for the scenarios (S1), (S2) and (S3) are illustrated in Figure 1

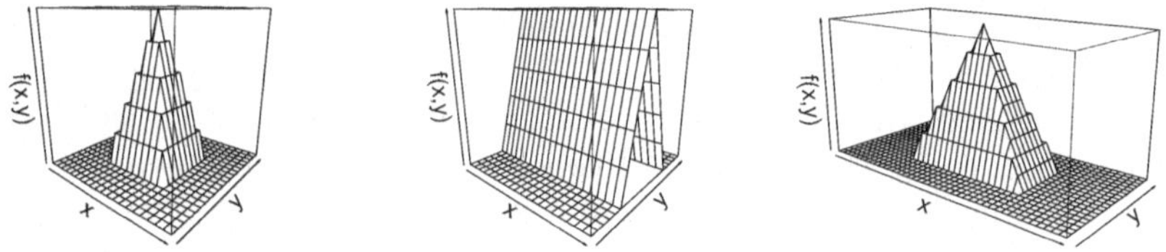

Fig. 1 Perspective plots of the mean functions used in scenarios (S1), (S2) and (S3) (from left to right)

Table 1 FWER estimated over 1000 runs in scenarios (S0), (S1), (S2) and (S3)

	Scenario 0	Scenario 1	Scenario 2	Scenario 3
FWER	0.014	0.021	0.21	0.032

4.2 *Results of Simulation Study*

For each scenario, we estimated the FWER of the IWT and the pointwise probability of rejection over 1000 simulation runs. The nominal FWER level was set to $\alpha = 0.05$. The estimated FWER is reported in Table 1, and the probability of rejection in Fig. 2. Since the estimated probability of rejection was zero in all points in (S0), we decided to show in the figure only the results of (S1), (S2) and (S3). Looking at the FWER, the simulations confirmed what was expected from the theory. When the null hypothesis was true over the whole domain (S0) when it was violated on a square (S1), and when it was violated on a rectangle with the same aspect ratio as the domain (S3), the FWER was controlled, and in fact, the procedure was conservative (the actual FWER was significantly lower than its nominal value in all three cases). However, when the null hypothesis was violated on a region that was different from a square (S2), the FWER was not controlled. Indeed, in this scenario, it was slightly higher than its nominal level.

Regarding the power, we can see that the two-dimensional IWT was able to correctly detect the portions of the domain where the null hypothesis was false with a reasonably good power (see Figure 2). As expected, the power was relatively low at the boundary between the region where the null hypothesis was true and the region where it was false, but it reached the value 1 inside the latter region.

5 Discussion

In this paper, we extended the IWT by Pini and Vantini [7] to two-dimensional functional data defined on a rectangular domain. We performed a simulation study to assess the performance of the method when using squares and/or rectangles with the same ratio of sides as the domain and the complement of such shapes as neighbour-

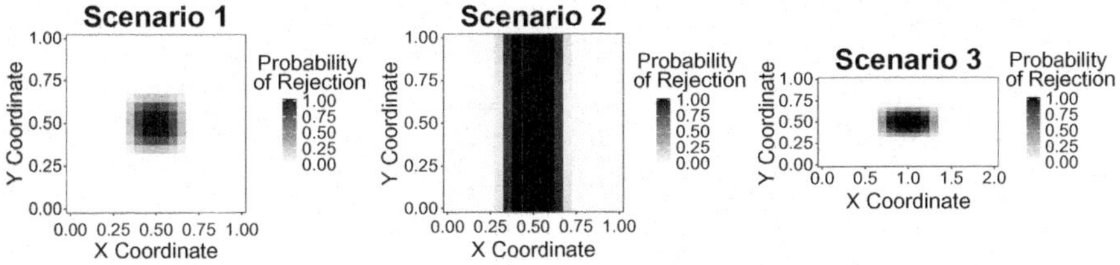

Fig. 2 Probability of rejection of each grid point estimated over 1000 runs in scenarios (S1), (S2) and (S3) (from left to right)

hoods. The results of the simulation study show that the FWER is controlled when the null hypothesis is true in such neighbourhoods, but not necessarily when it is true on neighbourhoods of a different shape. The simulation also shows that the method can be quite conservative in some instances. Future work will target further improving the respective performance of the method in these situations while keeping the computational complexity manageable.

Acknowledgements The presented research was funded by the Austrian Science Fund (FWF): KLI657-B31 and I 2697-N31 and by PMU-FFF: A-18/01/029-HÖL.

References

1. Benko, M., Härdle, W., Kneip, A., et al.: Common functional principal components. Annal. Stat. **37**(1), 1–34 (2009)
2. Cuevas, A., Febrero, M., Fraiman, R.: An anova test for functional data. Comput. Stat. Data Anal. **47**(1), 111–122 (2004)
3. Hall, P., Tajvidi, N.: Permutation tests for equality of distributions in high-dimensional settings. Biometrika **89**(2), 359–374 (2002)
4. Hall, P., Van Keilegom, I.: Two-sample tests in functional data analysis starting from discrete data. Statistica Sinica, pp. 1511–1531 (2007)
5. Horváth, L., Kokoszka, P., Reeder, R.: Estimation of the mean of functional time series and a two-sample problem. J. R. Stat. Soc.: Ser. B (Stat. Methodol.) **75**(1), 103–122 (2013)
6. Pini, A., Vantini, S.: The interval testing procedure: a general framework for inference in functional data analysis. Biometrics **73**(3), 835–845 (2016)
7. Pini, A., Vantini, S.: Interval-wise testing for functional data. J. Nonparametric Stat. **29**(2), 407–424 (2017)
8. Ramsay, J.O., Silverman, B.W.: Functional Data Analysis, second edn. Springer (2005)
9. Edward Rasmussen, C., Williams, C.K.I.: Gaussian Processes in Machine Learning. MIT Press (2006)

Assessing Data Support for the Simplifying Assumption in Bivariate Conditional Copulas

Evgeny Levi and Radu V. Craiu

Abstract The paper considers the problem of establishing data support for the simplifying assumption (SA) in a bivariate conditional copula model. It is known that SA greatly simplifies the inference for a conditional copula model, but standard tools and methods for testing SA in a Bayesian setting tend to not provide reliable results. After splitting the observed data into training and test sets, the method proposed will use a flexible Bayesian model fit to the training data to define tests based on randomization and standard asymptotic theory. Its performance is studied using simulated data. The paper's supplementary material also discusses theoretical justification for the method and implementations in alternative models of interest, e.g. Gaussian, Logistic and Quantile regressions.

Keywords Calibration function · Conditional copula · Permutation · Simplifying assumption

1 Introduction

A copula is a mathematical concept often used to model the joint distribution of several random variables. The applications of copula models permeate a number of fields where of interest is the simultaneous study of dependent variables, e.g. [8, 10, 13, 17]. The propagation of copula-related ideas in probability and statistics started with [19] which proved that for a random vector $(Y_1, \dots, Y_k)$ with cumulative distribution function (CDF) $H(y_1, \dots, y_k)$ and marginal continuous CDFs $F_i(y_i)$, $i = 1, \dots, k$, there exists a unique *copula* $C : [0, 1]^k \rightarrow [0, 1]$ such that $H(y_1, \dots, y_k) = C(F_1(y_1), \dots, F_k(y_k))$. For statistical modelling, it is also useful to note that a k-dimensional copula C and marginal continuous CDFs $F_i(y_i), i = 1, \dots, k$ are building blocks for a valid k-dimensional CDF, $C(F_1(y_1), \dots, F_k(y_k))$

E. Levi · R. V. Craiu (✉)
Department of Statistical Sciences, University of Toronto, Toronto, ON, Canada
e-mail: craiu@utstat.toronto.edu

E. Levi
e-mail: evgeny@utstat.utoronto.ca

M. La Rocca et al. (eds.), *Nonparametric Statistics*, Springer Proceedings in Mathematics & Statistics 339, https://doi.org/10.1007/978-3-030-57306-5_29

with ith marginal CDF equal to $F_i(y_i)$, thus providing much-needed flexibility in modelling multivariate distributions. The previous construction can be extended when conditioning on a covariate vector $X \in \mathbf{R}^q$ [14, 17] so that

$$H(y_1, \ldots, y_k|X) = C_X(F_1(y_1|X), \ldots, F_k(y_k|X)), \tag{1}$$

where all CDFs and the copula are conditional on X. For the rest of this paper, we follow [16] and assume: (1) that the copula in (1) belongs to a parametric family that remains the same across the whole range of X, and (2) its one-dimensional parameter depends on X through some unknown function $\theta(X) : \mathbf{R}^q \to \Theta \subset \mathbf{R}$. The range of $\theta(X)$ is usually restricted, so we introduce a known one-to-one link function $g : \Theta \to \mathbf{R}$ such that the *calibration function*, $\eta : \mathbf{R}^q \to \mathbf{R}$, defined as $\eta(X) = g(\theta(X))$ has unrestricted range. Sometimes, it is convenient to parametrize a copula family in terms of Kendall's tau, $\tau(X) : \mathbf{R}^q \to [-1, 1]$, which, for any given value of X, is in one-to-one correspondence with $\theta(X)$ when the copula parameter is one-dimensional. Thus, there is also a known one-to-one function $g'(\cdot)$ such that $\eta(X) = g'(\tau(X))$.

The simplifying assumption (SA) [5] states that copula C_X in (1) is independent of X, or that $\eta(X)$ is constant. Clearly, SA greatly simplifies the estimation in conditional copula models, including their use in vines (see, for instance, [1]). Acar et al. [2] showed that assuming SA when the data generative process has non-constant calibration may bias the inference, while Levi and Craiu [16] showed that SA is violated when important covariates are not included in the model (1). In light of these results, there is a genuine demand for strategies that effectively test whether the SA is appropriate or not. A number of research contributions address this issue for frequentist analyses, e.g. [3, 6, 9, 11].

This contribution belongs within the Bayesian paradigm, following the general philosophy expounded also in Klein and Kneib [12]. In this setting, it was observed in Craiu and Sabeti [4] that generic model selection criteria tend to choose a more complex model even when SA holds. In the next section, we present the problem in mathematical terms and review some of the Bayesian model selection procedures used for SA. A new approach, based on permutations, is described in Sect. 3. The Appendix contains a theoretical justification of the proposed algorithm and a discussion of extensions to other regression problems. A merit of the proposal is that it is quite general in its applicability, but this comes, unsurprisingly, at the expense of power. In order to investigate whether the trade-off is reasonable, we design a simulation study and present its conclusions in Sect. 4. The paper ends with a summary and discussion of future work.

2 Problem Setup

We consider observed data that consist of n independent triplets $\mathscr{D} = \{(x_i, y_{1i}, y_{2i}), i = 1, \ldots, n\}$ where $y_{ji} \in \mathbf{R}$, $j = 1, 2$, and $x_i \in \mathbf{R}^q$. Denote $\mathbf{y}_1 = (y_{11}, \ldots, y_{1n})$, $\mathbf{y}_2 = (y_{21}, \ldots, y_{2n})$ and $\mathbf{X} \in \mathbf{R}^{n \times q}$ the matrix with i^{th} row equal to x_i^T. We rely on (1) to express the *full conditional model* density for Y_1 and Y_2 given $\mathbf{X}$

$$p(\mathbf{y}_1, \mathbf{y}_2|\mathbf{X}, \omega) = \prod_{i=1}^{n} f_1(y_{1i}|\omega, x_i) f_2(y_{2i}|\omega, x_i) c_{\theta(x_i)} \left(F_1(y_{1i}|\omega, x_i), F_2(y_{2i}|\omega, x_i)\right), \tag{2}$$

where f_j, F_j are the density and, respectively, the CDF for Y_j, and ω denotes all the parameters and latent variables in the joint and marginals models. The copula density function is denoted by c, and it depends on X through unknown function $\theta(X) = g^{-1}(\eta(X))$. The copula family can be selected using several model selection criteria (e.g. [16, 18]). Once the copula family is selected, the objective is to check whether the SA is valid, in other words whether (2) becomes the *reduced model*

$$P(\mathbf{y}_1, \mathbf{y}_2|\mathbf{X}, \omega) = \prod_{i=1}^{n} f_1(y_{1i}|\omega, x_i) f_2(y_{2i}|\omega, x_i) c_{\theta} \left(F_1(y_{1i}|\omega, x_i), F_2(y_{2i}|\omega, x_i)\right), \tag{3}$$

in which the copula depends only on one parameter, θ. Flexible Bayesian models usually yield posteriors that are analytically intractable, so their characteristics will be estimated using draws $\{\omega^{(t)}\}_{t=1}^{M}$ obtained via a Markov chain Monte Carlo (MCMC) algorithm (e.g. [16, 18]). Data support for the full and reduced models, (2) and (3), may be established using predictive power as a criterion.

2.1 The Cross-Validated Pseudo Marginal Likelihood and Its Conditional Variant

The cross-validated pseudo marginal likelihood (CVML) [7] calculates the average (over parameter values) prediction power for model $\mathscr{M}$ via

$$\text{CVML}(\mathscr{M}) = \sum_{i=1}^{n} \log\left(P(y_{1i}, y_{2i}|\mathscr{D}_{-i}, \mathscr{M})\right), \tag{4}$$

where $\mathscr{D}_{-i}$ is the data set from which the ith observation has been removed. An estimate of (4) for a given model is estimated using posterior draws $\omega^{(t)}$ given the whole data set $\mathscr{D}$, (detailed derivations can be found in Levi and Craiu [16]) via

$$\text{CVML}_{est}(\mathcal{M}) = -\sum_{i=1}^{n} \log\left(\frac{1}{M}\sum_{t=1}^{M} P(y_{1i}, y_{2i}|\omega^{(t)}, \mathcal{M})^{-1}\right). \tag{5}$$

The model with the largest CVML is preferred.

The conditional CVML (CCVML), introduced by Levi and Craiu [16] specifically for copula models, exploits conditional rather than joint predictions

$$\text{CCVML}(\mathcal{M}) = \frac{1}{2}\left\{\sum_{i=1}^{n} \log\left[P(y_{1i}|y_{2i}, \mathcal{D}_{-i}, \mathcal{M})\right] + \sum_{i=1}^{n} \log\left[P(y_{2i}|y_{1i}, \mathcal{D}_{-i}, \mathcal{M})\right]\right\}.$$

Again this criterion can be estimated from posterior samples using

$$\begin{aligned}\text{CCVML}_{est}(\mathcal{M}) = -\tfrac{1}{2}\textstyle\sum_{i=1}^{n} &\left\{\log\left[\tfrac{1}{M}\textstyle\sum_{t=1}^{M} \frac{P(y_{2i}|\omega^{(t)},\mathcal{M})}{P(y_{1i},y_{2i}|\omega^{(t)},\mathcal{M})}\right]\right. \\ &\left. + \log\left[\tfrac{1}{M}\textstyle\sum_{t=1}^{M} \frac{P(y_{1i}|\omega^{(t)},\mathcal{M})}{P(y_{1i},y_{2i}|\omega^{(t)},\mathcal{M})}\right]\right\}. \end{aligned} \tag{6}$$

Similar to CVML, the model with the largest CCVML is selected.

2.2 *Watanabe–Akaike Information Criterion*

The Watanabe–Akaike Information Criterion [21] is an information-based criterion that is closely related to CVML, as discussed in [20]. The WAIC is defined as

$$\text{WAIC}(\mathcal{M}) = -2\text{fit}(\mathcal{M}) + 2\text{p}(\mathcal{M}), \tag{7}$$

where the model fitness is

$$\text{fit}(\mathcal{M}) = \sum_{i=1}^{n} \log E\left[P(y_{1i}, y_{2i}|\omega, \mathcal{M})\right], \tag{8}$$

and the penalty

$$\text{p}(\mathcal{M}) = \sum_{i=1}^{n} \text{Var}[\log P(y_{1i}, y_{2i}|\omega, \mathcal{M})]. \tag{9}$$

The expectation in (8) and the variance in (9) are with respect to the conditional distribution of ω given the data and can easily be estimated using the $\omega^{(t)}$ draws. The model with the smallest WAIC measure is preferred.

3 Detecting Data Support for SA

As will be shown in Sect. 4, the criteria described above exhibit unsatisfactory performances when the reduced model is the generative one. While it is expected that the flexibility of the full model will yield good predictions even when SA holds, it was surprising to see that the penalty term in (9) is not large enough to downgrade the full model under the SA null. Therefore, we base our diagnostics on some of the properties that are invariant to the group of permutations when SA holds.

In the first stage, we randomly divide the data $\mathscr{D}$ into training and test sets, $\mathscr{D}_1$ and $\mathscr{D}_2$, with n_1 and n_2 sample sizes, respectively. The full model defined by (2) is fitted on $\mathscr{D}_1$, and we denote $\omega^{(t)}$ the tth draw sampled from the posterior. For the ith item in $\mathscr{D}_2$, compute point estimates $\hat{\eta}_i$ and $\hat{U}_i = (\hat{U}_{1i}, \hat{U}_{2i})$, where $\hat{U}_{ji} = F_j(y_{ji}|\hat{\omega}_j, x_i)$, $j = 1, 2$, $i = 1, \ldots, n_2$, and ω_j denotes the vector of all the parameters and latent variables related to the jth marginal distribution. The marginal parameter estimates, $\hat{\omega}_j$, are obtained from the training data posterior draws. For instance, if the marginal models are $Y_{1i} \sim \mathscr{N}(f_1(x_i), \sigma_1^2)$ and $Y_{2i} \sim \mathscr{N}(f_2(x_i), \sigma_2^2)$, then each of the MCMC sample $\omega^{(t)}$ leads to an estimate $\hat{f}_1^t(x_i)$, $\hat{f}_2^t(x_i)$, $\hat{\sigma}_1^t$, $\hat{\sigma}_2^t$, $\hat{\eta}^t(x_i)$. Then $\hat{U}_i = (\hat{U}_{1i}, \hat{U}_{2i})$ are obtained using

$$(\hat{U}_{1i}, \hat{U}_{2i}) = (\Phi((y_{1i} - \overline{\hat{f}_1(x_i)})/\overline{\hat{\sigma}_1}), \Phi((y_{2i} - \overline{\hat{f}_2(x_i)})/\overline{\hat{\sigma}_2})),$$

where the overline $\overline{a}$ signifies the averages of Monte Carlo draws a^t.

Given the vector of calibration function evaluations at the test points, $\hat{\eta} = (\hat{\eta}_1, \ldots, \hat{\eta}_{n_2})$, and a partition $\min(\hat{\eta}) = a_1 < \ldots < a_{K+1} = \max(\hat{\eta})$ of the range of η into K disjoint intervals, define the set of observations in $\mathscr{D}_2$ that yield calibration function values between a_k and a_{k+1}, $B_k = \{i : a_k \le \hat{\eta}_i < a_{k+1}\}$ $k = 1, \ldots, K$. We choose the partition such that each "bin" B_k has approximately the same number of elements, n_2/K.

Under SA, the bin-specific estimates for various measures of dependence, e.g. Kendall's τ or Spearman's ρ, computed from the samples $\hat{U}_i$, are invariant to permutations, or swaps across bins. Based on this observation, we consider the procedure described in Table 1 for identifying data support for SA. The distribution of the resulting test statistics obtained in Method 1 is determined empirically, via permutations. Alternatively, one can rely on the asymptotic properties of the bin-specific dependence parameter estimates and construct a Chi-square test. Specifically, suppose the bin-specific Pearson correlations $\hat{\rho}_k$ are computed from samples $\{\hat{U}_i : i \in B_k\}$), for all $k = 1, \ldots, K$, and let $\hat{\rho} = (\hat{\rho}_1, \ldots, \hat{\rho}_K)^T$ and $\tilde{n} = n_2/K$ be the number of points in each bin. It is known that $\hat{\rho}_k$ is asymptotically normal distributed for each k so that

$$\sqrt{\tilde{n}}(\hat{\rho}_k - \rho_k) \xrightarrow{d} \mathscr{N}(0, (1 - \rho_k^2)^2),$$

where ρ_k is the true correlation in bin k. If we assume that $\{\hat{\rho}_k\colon k = 1, \ldots, K\}$ are independent, and set $\rho = (\rho_1, \ldots, \rho_K)^T$ and $\Sigma = diag((1 - \rho_1^2)^2, \ldots, (1 - \rho_K^2)^2)$, then we have

Table 1 Method 1: A permutation-based procedure for assessing data support in favour of SA

A1 Compute the kth bin-specific Kendall's tau $\hat{\tau}_k$ from $\{\hat{U}_i : i \in B_k\}$) $k = 1, \ldots, K$.

A2 Compute the observed statistic $T^{obs} = SD_k(\hat{\tau}_k)$ (where $SD_k(a_k)$ is a standard deviation of a_k over index k). Note that if SA holds, we expect the observed statistic to be close to zero.

A3 Consider J permutations $\lambda_j : \{1, \ldots, n_2\} \to \{1, \ldots, n_2\}$. For each permutation λ_j:

- A3.1 Compute $\hat{\tau}_{jk} = \tau(\{\hat{U}_i : \lambda_j(i) \in B_k\})$ $k = 1, \ldots, K$.
- A3.2 Compute test statistic $T_j = SD_k(\hat{\tau}_{jk})$. Note if SA holds, then we expect T_j to be close to T^{obs}.

A4 We consider that there is support in favour of SA at significance level α if T^{obs} is smaller than the $(1-\alpha)$-th empirical quantile calculated from the sample $\{T_j : 1 \le j \le J\}$.

$$\sqrt{\tilde{n}}(\hat{\rho} - \rho) \xrightarrow{d} \mathcal{N}(0, \Sigma).$$

In order to combine evidence across bins, we define the matrix $A \in \mathbf{R}^{(K-1)\times K}$ as

$$A = \begin{bmatrix} 1 & -1 & 0 & \cdots & 0 \\ 0 & 1 & -1 & \cdots & 0 \\ \vdots & \vdots & \vdots & \vdots & \vdots \\ 0 & 0 & \cdots & 1 & -1 \end{bmatrix}. \tag{10}$$

Since under the null hypothesis SA holds, one gets $\rho_1 = \ldots = \rho_K$, implying

$$\tilde{n}(A\hat{\rho})^T (A \Sigma A^t)^{-1} (A\hat{\rho}) \xrightarrow{d} \chi^2_{K-1}.$$

Method 2, with its steps detailed in Table 2, relies on the ideas above to test SA.

Method 1 evaluates the p-value using a randomization procedure [15], while the second is based on the asymptotic normal theory of Pearson correlations. To get reliable results, it is essential to assign test observations to "correct" bins which is true when calibration predictions are as close as possible to the true unknown values, i.e. $\hat{\eta}(x_i) \approx \eta(x_i)$. The latter heavily depends on the estimation procedure and sample size of the training set. Therefore, it is advisable to apply very flexible models for the calibration function estimation and have enough data points in the training set. The trade-off we notice is that as more observations are assigned to $\mathcal{D}$, the calibration test predictions improve, even as power decreases due to a smaller sample size in $\mathcal{D}_2$. For our simulations, we have used $n_1 \approx 0.5n$ and $n_2 \approx 0.5n$, and $K \in \{2, 3\}$.

Remarks: The equivalence between SA and equality of $\eta(x)$ across bins is central to both methods and requires some discussion. Below, we assume that only two bins are used and that the estimation is based on very large data so that finite-sample variability is ignored.

1. *Necessity* If SA is true then indeed η must be constant across bins *as long as the copula family is the same* across the whole range of X.

Table 2 Method 2: A Chi-square test for assessing data support in favour of SA

B1 Compute the bin-specific Pearson correlation $\hat{\rho}_k$ from samples $\{\hat{U}_i : i \in B_k\}$), for all $k = 1,\ldots,K$. Let $\hat{\rho} = (\hat{\rho}_1,\ldots,\hat{\rho}_K)^T$, and $\tilde{n} = n_2/K$, the number of points in each bin.

B2 Define $\rho = (\rho_1,\ldots,\rho_K)^T$, $\Sigma = diag((1-\rho_1^2)^2,\ldots,(1-\rho_K^2)^2)$ and $A \in \mathbf{R}^{(K-1)\times K}$ as in (10) then under SA we have that $\rho_1 = \ldots = \rho_K$ and

$$\tilde{n}(A\hat{\rho})^T(A\Sigma A^t)^{-1}(A\hat{\rho}) \xrightarrow{d} \chi^2_{K-1}.$$

Compute $T^{obs} = \tilde{n}(A\hat{\rho})^T(A\hat{\Sigma}A^t)^{-1}(A\hat{\rho})$.

B3 Compute p-value = $P(\chi^2_{K-1} > T^{obs})$ and reject SA if p-value$< \alpha$.

2. *Sufficiency* If SA does not hold, assume that the calibration function takes two values. Assuming consistency of the calibration's estimator, it is expected that bin 1 and bin 2 will contain pairs (u_1, u_2) following distributions $\pi_1(u_1, u_2)$ and $\pi_2(u_1, u_2)$ with corresponding correlations $\rho_1 < \rho_2$, respectively. After a random permutation, pairs in each bin will follow a mixture distribution $\lambda\pi_1(u_1, u_2) + (1-\lambda)\pi_2(u_1, u_2)$ and $(1-\lambda)\pi_1(u_1, u_2) + \lambda\pi_2(u_1, u_2)$ in bins 1 and 2, respectively, with $\lambda \in (0, 1)$. Thus, the post-permutation correlations in bins 1 and 2 are $\lambda\rho_1 + (1-\lambda)\rho_2$ and $(1-\lambda)\rho_1 + \lambda\rho_2$. Observe that each correlation is between ρ_1 and ρ_2 which implies that the absolute difference between them will be less than $\rho_2 - \rho_1$, so we expect to reject the null. This argument offers heuristic support for the method, but obviously cannot be extended to cases where η is non-constant in each bin and finite sample variability must be accounted for.

A theoretical justification for Method 2 and extensions of this idea to other models are available in Appendix.

4 Simulations

In this section, we present the performance of the proposed methods and comparisons with generic CVML and WAIC criteria on simulated data sets. Different functional forms of calibration function, sample sizes and magnitude of deviation from SA will be explored.

4.1 Simulation Details

We generate samples of sizes $n = 500$ and $n = 1000$ from 3 scenarios described below. For all scenarios, the Clayton copula will be used to model dependence between responses, while covariates are independently sampled from $\mathcal{U}[0, 1]$. For all scenarios, the covariate dimension $q = 2$. Marginal conditional distributions $Y_1|X$

and $Y_2|X$ are modelled as Gaussian with constant variances σ_1^2, σ_2^2 and conditional means $f_1(X), f_2(X)$, respectively. The model parameters must be estimated jointly with the calibration function $\eta(X)$. For convenience, we parametrize calibration on Kendall's tau $\tau(X)$ scale.

Sc1 $f_1(X) = 0.6\sin(5x_1) - 0.9\sin(2x_2)$,
$f_2(X) = 0.6\sin(3x_1 + 5x_2)$,
$\tau(X) = 0.5, \sigma_1 = \sigma_2 = 0.2$.
Sc2 $f_1(X) = 0.6\sin(5x_1) - 0.9\sin(2x_2)$,
$f_2(X) = 0.6\sin(3x_1 + 5x_2)$,
$\tau(X) = \delta + \gamma \times \sin(10X^T\beta)$
$\beta = (1, 3)^T/\sqrt{10}, \sigma_1 = \sigma_2 = 0.2$.
Sc3 $f_1(X) = 0.6\sin(5x_1) - 0.9\sin(2x_2)$,
$f_2(X) = 0.6\sin(3x_1 + 5x_2)$,
$\tau(X) = \delta + \gamma \times 2(x_1 + \cos(6x_2) - 0.45)/3$
$\sigma_1 = \sigma_2 = 0.2$.

Sc1 corresponds to SA since Kendall's τ is independent of covariate level. The calibration function in **Sc2** has single index form for the calibration function, while in **Sc3** it has an additive structure on τ scale (generally not additive on η scale); these simulations are useful to evaluate performance under model misspecification. We note that τ in **Sc2** and **Sc3** depends on parameters δ (average correlation strength) and γ (deviation from SA), which in this study take values $\delta \in \{0.25, 0.75\}$ and $\gamma \in \{0.1, 0.2\}$, respectively.

4.2 Simulation Results

For each sample size and scenario, we have repeated the analysis using 250 independently replicated data sets. For each data, the GP-SIM model suggested by Levi and Craiu [16] is fitted. This method implements sparse Gaussian Process (GP) priors for marginal conditional means and sparse GP-Single Index for calibration function. These non-parametric models are more flexible than parametric ones and can effectively capture various patterns. The inference is based on 5000 MCMC samples for all scenarios, as the chains were run for 10,000 iterations with 5000 samples discarded as burn-in. The number of inducing inputs was set to 30 for all GP. For generic SA testing, GP-SIM fitting is done for the whole data sets, and posterior draws are used to estimate CVML and WAIC. Since the proposed methods requires data splitting, we first randomly divide the data equally into training and testing sets. We fit GP-

Table 3 Simulation results: generic, proportion of rejection of SA for each scenario, sample size and model selection criterion

Scenario	$n = 500$			$n = 1000$		
	CVML (%)	CCVML (%)	WAIC (%)	CVML (%)	CCVML (%)	WAIC (%)
Sc1	33.3	31.1	34.7	38.2	37.3	37.8
Sc2 ($\delta = 0.75, \gamma = 0.1$)	99.1	98.7	99.1	100	100	100
Sc2 ($\delta = 0.75, \gamma = 0.2$)	100	100	100	100	100	100
Sc2 ($\delta = 0.25, \gamma = 0.1$)	80.1	84.4	80.1	99.1	100	99.1
Sc2 ($\delta = 0.25, \gamma = 0.2$)	100	100	100	100	100	100
Sc3 ($\delta = 0.75, \gamma = 0.1$)	76.9	73.3	77.8	85.7	82.2	85.8
Sc3 ($\delta = 0.75, \gamma = 0.2$)	99.1	97.3	99.1	99.1	97.8	99.1
Sc3 ($\delta = 0.25, \gamma = 0.1$)	54.7	56.4	55.6	65.3	68.4	64.9
Sc3 ($\delta = 0.25, \gamma = 0.2$)	89.8	92.0	91.1	99.6	100	99.6

SIM on the training set and then use the obtained posterior draws to construct point estimates of $F_1(y_{1i}|x_i)$, $F_2(y_{2i}|x_i)$ and $\eta(x_i)$ for every observation in the test set. In Method 1, we used 500 permutations. Table 3 shows the percentage of SA rejections for generic Bayesian selection criteria. The presented results clearly illustrate that generic methods have difficulties identifying SA. This leads to a loss of statistical efficiency since a complex model is selected over a much simpler one. Moreover, CVML or CCVML fails to identify SA as both measures do not penalize directly for the complexity of the model. The simulations summarized in Table 4 show that the proposed methods (setting $\alpha = 0.05$) have much smaller probability of Type I error which vary around the threshold of 0.05. It must be pointed, however, that under SA the performance of χ^2 test worsens with the number of bins K, which is not surprising since as K increases, the number of observations in each bin goes down, and normal approximation for the distribution of Pearson correlation becomes tenuous, while the permutation-based test is more robust to small samples. The performance of both methods improves with sample size. We also notice a loss of power between Scenarios 2 and 3, which is due to model misspecification, since in the latter case the generative model is different from the postulated one. All methods break down when the departure from SA is not large, e.g. $\gamma = 0.1$. Although not desirable, this has limited impact in practice since, in our experience, in this case the predictions produced by either model are very similar.

Table 4 Simulation results: proposed method, proportion of rejection of SA for each scenario, sample size, number of bins (K) and method

Scenario	Permutation test				χ^2 test			
	$n = 500$		$n = 1000$		$n = 500$		$n = 1000$	
	$K = 2$ (%)	$K = 3$ (%)	$K = 2$ (%)	$K = 3$ (%)	$K = 2$ (%)	$K = 3$ (%)	$K = 2$ (%)	$K = 3$ (%)
Sc1	4.9	6.2	3.5	5.3	9.7	11.1	10.7	13.7
Sc2($\delta = 0.75, \gamma = 0.1$)	90.2	80.4	99.6	99.1	94.7	94.2	99.6	99.1
Sc2($\delta = 0.75, \gamma = 0.2$)	100	100	100	100	100	100	100	100
Sc2($\delta = 0.25, \gamma = 0.1$)	25.8	18.7	55.1	47.1	30.2	21.8	58.7	53.8
Sc2($\delta = 0.25, \gamma = 0.2$)	91.6	84.9	99.6	99.6	92.4	91.1	99.6	99.6
Sc3($\delta = 0.75, \gamma = 0.1$)	28.0	24.0	57.3	52.9	41.3	45.8	72.4	72.9
Sc3($\delta = 0.75, \gamma = 0.2$)	88.4	85.8	98.7	98.7	94.2	92.0	100	99.1
Sc3($\delta = 0.25, \gamma = 0.1$)	8.0	7.5	11.1	10.7	9.8	10.7	15.1	12.9
Sc3($\delta = 0.25, \gamma = 0.2$)	19.6	18.2	63.6	60.9	24.9	23.6	70.2	69.3

5 Conclusion

We propose two methods to check data support for the simplifying assumption in conditional bivariate copula problems. Both are based on data splitting into training and test sets, partitioning the test set into bins using calibration values obtained in training and using randomization or χ^2 tests to determine if the dependence is constant across bins. Empirically, it was shown that the probability of Type I error is controlled when SA holds. When the generative process does not satisfy SA, these two methods also perform well, showing larger power than generic model selection criteria. Future work will address questions related to the proportion of data that should be assigned to training and test sets as well as bin sizes.

Acknowledgments We thank Stanislav Volgushev and an anonymous referee for suggestions that have greatly improved the paper. Financial support for this work was provided by the Canadian Statistical Sciences Institute (CANSSI) and by NSERC of Canada.

Appendix

Theoretical Discussion

In this section, we prove that under canonical assumptions, the probability of Type I error for Method 2 in Sect. 3 converges to α when SA is true.

Suppose we have independent samples from K populations (groups), $(u_{1i}^1, u_{2i}^1)_{i=1}^{n_1} \sim (U_1^1, U_2^1)$,...,$(u_{1i}^K, u_{2i}^K)_{i=1}^{n_K} \sim (U_1^K, U_2^K)$, the goal is to test $\rho_1 = \cdots = \rho_K$ (here ρ is Pearson correlation).

To simplify notation, we assume $n_1 = \ldots, n_K = n$. Let $\hat{\rho} = (\hat{\rho}_1, \ldots, \hat{\rho}_K)$ be the vector of sample correlations , $\Sigma = diag((1-\rho_1^2)^2, \ldots, (1-\rho_K^2)^2)$ and $(K-1) \times K$ matrix A as defined in Sect. 3, then canonical asymptotic results imply that if $\rho_1 = \cdots = \rho_K$ and as $n \to \infty$,

$$T = n(A\hat{\rho})^T (A\Sigma A^T)^{-1}(A\hat{\rho}) \xrightarrow{d} \chi^2_{K-1}. \tag{11}$$

Based on the model fitted on $\mathscr{D}_1$, we define estimates of $F_1(y_{1i}|x_i)$ and $F_2(y_{2i}|x_i)$ by $\hat{U} = \{\hat{U}_i = (\hat{F}_1(y_{1i}|x_i), \hat{F}_2(y_{2i}|x_i))\}_{i=1}^{n_2}$. Note that $\hat{U}$ depends on $\mathscr{D}_1$ and X. Given a fixed number of bins K and assuming, without loss of generality, equal sample sizes in each bin $\tilde{n} = n_2/K$, we define a test statistic $T(\hat{U})$ as in (11) with $\hat{\rho}_j$ estimated from $\{\hat{U}_{(j-1)\tilde{n}+1}, \ldots, \hat{U}_{j\tilde{n}}\}$, for $1 \le j \le K$.

Note that in Method 2, test cases are assigned to "bins" based on the value of predicted calibration function $\hat{\eta}(x_i)$ which is not taken into account in the generic definition of test statistic $T(\hat{U})$ above. To close this gap, we introduce a permutation $\lambda^* : \{1, \ldots, n_2\} \to \{1, \ldots, n_2\}$ that "sorts" $\hat{U}$ from smallest $\hat{\eta}(x)$ value to largest, i.e. $\hat{U}_{\lambda^*} = \{\hat{U}_{\lambda^*(i)}\}_{i=1}^{n_2}$ with $\hat{\eta}(x_{\lambda^*(1)}) < \hat{\eta}(x_{\lambda^*(2)}) < \cdots < \hat{\eta}(x_{\lambda^*(n_2)})$. Hence, the test statistic in Method 2 has the form $T(\hat{U}_{\lambda^*})$ as in (11) but in this case test cases with smallest predicted calibrations are assigned to the first group, or bin, and with largest calibrations to the Kth group/bin. Finally, define a test function ϕ with specified significance level α to test SA:

$$\phi(\hat{U}|\lambda^*) = \begin{cases} 1 & \text{if} \quad T(\hat{U}_{\lambda^*}) > \chi^2_{K-1}(1-\alpha), \\ 0 & \text{if} \quad T(\hat{U}_{\lambda^*}) \le \chi^2_{K-1}(1-\alpha). \end{cases} \tag{12}$$

Intuitively, if SA is false then we would expect $T(\hat{U}_{\lambda^*})$ to be larger then the critical value $\chi^2_{K-1}(1-\alpha)$.

The goal is to show that this procedure have probability of type I error equal to α, which is equivalent to the expectation of the test function:

$$\text{P(Type I error)} = \int \phi(\hat{U}|\lambda^*)P(\lambda^*|\mathscr{D}_1, X)P(\hat{U}|\mathscr{D}_1, X)P(\mathscr{D}_1)P(X)d\hat{U}d\mathscr{D}_1 dX d\lambda^*. \tag{13}$$

Note that λ^* does not depend on $\hat{U}$ because of the data splitting to training and test sets. Also usually $P(\lambda^*|\mathscr{D}_1, X)$ is just a point mass at some particular permutation. In general the above integral cannot be evaluated, however if we assume that for all test cases:

$$\begin{aligned} \hat{F}_1(y_{1i}|x_i) &\xrightarrow{p} F_1(y_{1i}|x_i) \quad \text{as} \quad n \to \infty \quad \forall i, \\ \hat{F}_2(y_{2i}|x_i) &\xrightarrow{p} F_2(y_{2i}|x_i) \quad \text{as} \quad n \to \infty \quad \forall i, \end{aligned} \tag{14}$$

then under SA and as $n \to \infty$, $P(\hat{U}|\mathscr{D}_1, X) = P(\hat{U}) \approx \prod_{i=1}^{n_2} c(\hat{u}_{1i}, \hat{u}_{2i})$ where c is a copula density and the expectation becomes

$$\text{P(Type I error)} = \int \phi(\hat{U}|\lambda^*)P(\lambda^*|\mathscr{D}_1, X)P(\hat{U})P(\mathscr{D}_1)P(X)d\hat{U}d\mathscr{D}_1 dXd\lambda^* =$$
$$= \int \left(\int \phi(\hat{U}|\lambda^*)P(\hat{U})d\hat{U} \right) P(\lambda^*|\mathscr{D}_1, X)P(\mathscr{D}_1)P(X)d\mathscr{D}_1 dXd\lambda^* = \alpha. \tag{15}$$

Since if SA is true, $\int \phi(\hat{U}|\lambda^*)P(\hat{U})d\hat{U} = \alpha$ for any λ^*. Therefore, if marginal CDF predictions for test cases are consistent then this procedure has the required probability of Type I error for sufficiently large sample size.

Extensions to Other Models

The proposed idea of dividing the data into training and test subsets, splitting the observations on the test set to bins defined in first stage and then using a test to check distributional difference between bins can be extended to other models. For example, one can use a similar construction in a regression problem with conditional mean $f(X)$ and constant variance. First assign test cases to bins by the values of $\hat{f}(X)$, and then compare means in each bin either by randomization procedures or using χ^2 test. This approach can be especially useful when $f(X)$ is assumed to have a complex form, such as generalized additive models, including additive tree structures, splines or based on Bayesian non-parametric methods. Simulations we conducted (not reported here) suggest that for large covariate dimensions, standard F-tests for Gaussian error regression yield large Type I error probabilities, a problem that is attenuated using the permutation-based ideas described in the paper.

References

1. Aas, K., Czado, C., Frigessi, A., Bakken, H.: Pair-copula constructions of multiple dependence. Insur. Math. Econ. **44**, 182–198 (2009)
2. Acar, E., Genest, C., Nešlehová, J.: Beyond simplified pair-copula constructions. J. Multivar. Anal. **110**, 74–90 (2012)
3. Acar, E.F., Craiu, R.V., Yao, F., et al.: Statistical testing of covariate effects in conditional copula models. Electron. J. Stat. **7**, 2822–2850 (2013)
4. Craiu, R.V., Sabeti, A.: In mixed company: Bayesian inference for bivariate conditional copula models with discrete and continuous outcomes. J. Multivar. Anal. **110**, 106–120 (2012)
5. Czado, C.: Pair-copula constructions of multivariate copulas. Copula Theory and its Applications, pp. 93–109. Springer, Berlin (2010)
6. Derumigny, A., Fermanian, J.-D.: About tests of the "simplifying" assumption for conditional copulas (2016). arXiv:1612.07349
7. Geisser, S., Eddy, W.F.: A predictive approach to model selection. J. Am. Stat. Assoc. **74**, 153–160 (1979)

8. Genest, C., Favre, A.-C.: Everything you always wanted to know about copula modeling but were afraid to ask. J. Hydrol. Eng. **12**, 347–368 (2007)
9. Gijbels, I., Omelka, M., Veraverbeke, N.: Estimation of a copula when a covariate affects only marginal distributions. Scand. J. Stat. **42**, 1109–1126 (2015)
10. Hougaard, P.: Analysis of Multivariate Survival Data. Statistics for Biology and Health. Springer, New York (2000)
11. Killiches, M., Kraus, D., Czado, C.: Examination and visualisation of the simplifying assumption for vine copulas in three dimensions. Aust. N. Z. J. Stat. **59**, 95–117 (2017)
12. Klein, N., Kneib, T.: Simultaneous inference in structured additive conditional copula regression models: a unifying Bayesian approach. Stat. Comput. 1–20 (2015)
13. Lakhal, L., Rivest, L.-P., Abdous, B.: Estimating survival and association in a semicompeting risks model. Biometrics **64**, 180–188 (2008)
14. Lambert, P., Vandenhende, F.: A copula-based model for multivariate non-normal longitudinal data: analysis of a dose titration safety study on a new antidepressant. Stat. Med. **21**, 3197–3217 (2002)
15. Lehmann, E.L., Romano., J.P.: Testing Statistical Hypotheses. Springer Science & Business Media (2006)
16. Levi, E., Craiu, R.V.: Bayesian inference for conditional copulas using gaussian process single index models. Comput. Stat. Data Anal. **122**, 115–134 (2018)
17. Patton, A.J.: Modelling asymmetric exchange rate dependence*. Int. Econ. Rev. **47**, 527–556 (2006)
18. Sabeti, A., Wei, M., Craiu, R.V.: Additive models for conditional copulas. Statistics **3**, 300–312 (2014)
19. Sklar, A.: Fonctions de répartition à *n* dimensions et leurs marges. Publications de l'Institut de Statistique de l'Université de Paris **8**, 229–231 (1959)
20. Vehtari, A., Gelman, A., Gabry, J.: Practical Bayesian model evaluation using leave-one-out cross-validation and WAIC. Stat. Comput. **27**, 1413–1432 (2017)
21. Watanabe, S.: Asymptotic equivalence of Bayes cross validation and widely applicable information criterion in singular learning theory. J. Mach. Learn. Res. **11**, 3571–3594 (2010)

Semiparametric Weighting Estimations of a Zero-Inflated Poisson Regression with Missing in Covariates

M. T. Lukusa and F. K. H. Phoa

Abstract We scrutinize the problem of missing covariates in the zero-inflated Poisson regression model. Under the assumption that some covariates for modeling the probability of *the* zero and the nonzero states are missing at random, the complete-case estimator is known to be biased and inefficient. Although the inverse probability weighting estimator is unbiased, it remains inefficient. We propose four types of semiparametric weighting estimations where *the conditional probabilities and the conditional expected score functions are estimated* either by using the generalized additive models (GAMs) and the Nadaraya kernel smoother method. In addition, we allow the *conditional probabilities* and the conditional expectations to be *either of the same types or of different types.* Moreover, a Monte Carlo experiment is used to investigate the merit of the proposed method.

Keywords Excess zeroes · Nonparametric selection probability · Generalized additive models · Augmentation part · Missing at random · Estimating equation

1 Preliminaries

Zero-inflated data are quite common in various sectors in the real world such as ecology, manufacturing, medicine, agriculture, and transportation, [18]. In the presence of the zero-inflated feature, the traditional count regression models, i.e., Poisson regression model and the negative binomial regression model may fail to provide an adequate fit of the data set. *Differently,* the zero-inflated (ZI) models naturally become the best tools in this situation. ZI models can accommodate the excess zeros feature and other features that lead to the overdispersion. Among many ZI models, zero-inflated Poisson (ZIP) model is the most popular one [6]. In practice, the

M. T. Lukusa (✉) · F. K. H. Phoa
Institute of Statistical Science, Academia Sinica, Taipei, Republic of China
e-mail: lukusa@stat.sinica.edu.tw

F. K. H. Phoa
e-mail: fredphoa@webmail.stat.sinica.edu.tw

M. La Rocca et al. (eds.), *Nonparametric Statistics*, Springer Proceedings in Mathematics & Statistics 339, https://doi.org/10.1007/978-3-030-57306-5_30

parameters of a ZIP regression model are functions of covariates, but some covariates may be missing at random (MAR) [12]. The complete case estimation is reliable only when missings are completely at random (MCAR) [12]. Missing data problem in ZI models has not received much attention in the literature. Following [15], we extend [7] that proposed a semiparametric inverse probability method (IPW). To gain efficiency, [11] proposed the augmentation inverse probability method estimation. There are many works about semiparametric estimations in the presence of the missing data, for instance, [2, 9, 14, 16, 17]. These works are mostly kernel-based approaches, and a few are the generalized additive models based for missing data. *Note that the generalized additive models (GAMs) proposed by* [5] *appear to be flexible, easier to implement, and help to reduce the curse of dimensionality.* We propose four estimators where the selection probability and the augmentation term to be plugged in the estimating equations (EE) are both estimated by the Nadaraya kernel function [8, 19] or the GAMs. In two EEs, the selection probability and the augmentation term are of the same nature, and in the other two, they are of different nature. Numerical results revealed that our proposed estimators are unbiased, consistent, and perform all better than the IPW estimator proposed by [7]. The rest of this work is as follows. A brief review of the ZIP distribution and the missing data problem is given in Sect. 2. In Sect. 3, we develop the augmentation semiparametric IPW methods for the ZIP regression model. In Sect. 4, we conduct some numerical experiments. Finally, Sect. 5 provides a summary.

2 Review of Zero-Inflated Model and Naive Estimation

Let

$$Y = 0\omega + (1-\omega)U$$

with $\omega = Bin(p), p = H(\boldsymbol{\gamma}_0^T \mathscr{X}_1), H(u) = [1+\exp(-u)]^{-1}, U = Pois(\lambda)$, and $\lambda = Pois(\boldsymbol{\beta}_0^T \mathscr{X}_2)$. Denote with $\boldsymbol{\theta}_0 = \left(\boldsymbol{\gamma}_0^T, \boldsymbol{\beta}_0^T\right)^T$ the vector parameters to estimate. Moreover, $\mathscr{X}_1$ and $\mathscr{X}_2$ are subsets of $\mathscr{X}$ which is the design matrix. Here, $\mathscr{X} = (X, Z)$, where X is the covariate subject to missingness whereas Z is another covariate that is always observed. Let $V = (Z, W)$, with W a surrogate variable of X. *A ZIP random variable (Y) can be seen as a mixture of two populations which is given by the following expression:*

$$Y = \begin{cases} y = 0, & \text{the subject is in the group not at risk with probability } \omega \\ y > 0, & \text{the subject is in the group at risk with probability } 1-\omega. \end{cases} \tag{1}$$

Then, the parametric ZIP regression model developed by [6] *is*

$$P(Y = y|\mathscr{X}) = H(\boldsymbol{\gamma}^T \mathscr{X}_{1i}) I_{(y=0)} + I_{(Y>0)}[1 - H(\boldsymbol{\gamma}^T \mathscr{X}_{1i})]U. \tag{2}$$

Let $\{(Y_i, \mathscr{X}_i) : i = 1, \ldots, n\}$ *be a sample of independent observations of* $(Y, \mathscr{X})$. *Then, the likelihood function of a ZIP model* [7] *is expressed by*

$$\begin{aligned}\mathscr{L}(\boldsymbol{\theta}) &= \prod_{i=1}^{n} \left[H\left(\boldsymbol{\gamma}^T \mathscr{X}_{1i}\right) H^{-1}\left(\boldsymbol{\gamma}^T \mathscr{X}_{1i} + \exp(\boldsymbol{\beta}^T \mathscr{X}_{2i})\right)\right]^{I(Y_i=0)} \\ &\prod_{i=1}^{n} \left\{\left[1 - H(\boldsymbol{\gamma}^T \mathscr{X}_{1i})\right] \frac{\exp[-\exp(\boldsymbol{\beta}^T \mathscr{X}_{2i})][\exp(\boldsymbol{\beta}^T \mathscr{X}_{2i})]^{Y_i}}{Y_i!}\right\}^{I(Y_i>0)}.\end{aligned}$$

By applying the natural logarithm to $\mathscr{L}(\boldsymbol{\theta})$, we obtained the log-likelihood function of ZIP distribution which is $\log[\mathscr{L}(\boldsymbol{\theta})] = \ell(\boldsymbol{\theta}) = \sum_{i=1}^{n} \ell_i(\boldsymbol{\theta})$,

$$\begin{aligned}\ell_i(\boldsymbol{\theta}) = I(Y_i = 0) &\left\{\log\left[H\left(\boldsymbol{\gamma}^T \mathscr{X}_{1i}\right)\right] - \log\left[H\left(\boldsymbol{\gamma}^T \mathscr{X}_{1i} + \exp(\boldsymbol{\beta}^T \mathscr{X}_{2i})\right)\right]\right\} \\ I(Y_i > 0) &\left\{\log\left[1 - H(\boldsymbol{\gamma}^T \mathscr{X}_{1i})\right] + Y_i \boldsymbol{\beta}^T \mathscr{X}_{2i} - \exp(\mathscr{X}_{2i})\right\}. \end{aligned} \tag{3}$$

In the absence of missing data and by optimizing $\ell(\boldsymbol{\theta})$, we can obtain $\widehat{\boldsymbol{\theta}}$, the ZIP unbiased estimator of $\boldsymbol{\theta}$. Following [7], we assume that some covariates that model ω and λ are missing at random (MAR) [12]. *Let* δ *be the missing indicator such that* $\delta = 1$ *if* X *is observed, and* 0 *otherwise. We* aim at obtaining $\widehat{\boldsymbol{\theta}}$, which is a robust and unbiased estimator of $\boldsymbol{\theta} = \left(\boldsymbol{\gamma}^T, \boldsymbol{\beta}^T\right)^T$. A naive estimation method refers to the complete case (CC) estimating equation (EE) given by

$$U_{CC,n}(\boldsymbol{\theta}) = \frac{1}{\sqrt{n}} \sum_{i=1}^{n} \delta_i S_i(\boldsymbol{\theta}), \tag{4}$$

where $S_i(\boldsymbol{\theta}) = \partial \ell_i(\boldsymbol{\theta})/\partial \boldsymbol{\theta}$, and where $S_{i1}(\boldsymbol{\theta}) = \partial \ell_i(\boldsymbol{\theta})/\partial \boldsymbol{\gamma}$ and $S_{i2}(\boldsymbol{\theta}) = \partial \ell_i(\boldsymbol{\theta})/\partial \boldsymbol{\beta}$ are the components of the score function ($S_i(\boldsymbol{\theta})$), respectively, expressed by

$$\begin{aligned}S_{i1}(\boldsymbol{\theta}) = &\tfrac{\partial}{\partial \boldsymbol{\gamma}} I(Y_i = 0) \left\{\log\left[H(\boldsymbol{\gamma}^T \mathscr{X}_{1i})\right] - \log\left[H\left(\boldsymbol{\gamma}^T \mathscr{X}_{1i} + \exp(\boldsymbol{\beta}^T \mathscr{X}_{2i})\right)\right]\right\} \\ &+ \tfrac{\partial}{\partial \boldsymbol{\gamma}} I(Y_i > 0) \left\{\log\left[1 - H(\boldsymbol{\gamma}^T \mathscr{X}_{1i})\right] + Y_i \boldsymbol{\beta}^T \mathscr{X}_{2i} - \exp(\boldsymbol{\beta}^T \mathscr{X}_{2i})\right\}\end{aligned}$$

and

$$\begin{aligned}S_{i2}(\boldsymbol{\theta}) = &\tfrac{\partial}{\partial \boldsymbol{\beta}} I(Y_i = 0) \left\{\log\left[H(\boldsymbol{\gamma}^T \mathscr{X}_{1i})\right] - \log\left[H\left(\boldsymbol{\gamma}^T \mathscr{X}_{1i} + \exp(\boldsymbol{\beta}^T \mathscr{X}_{1i})\right)\right]\right\} \\ &+ \tfrac{\partial}{\partial \boldsymbol{\beta}} I(Y_i > 0) \left\{\log\left[1 - H(\boldsymbol{\gamma}^T \mathscr{X}_{1i})\right] + Y_i \boldsymbol{\beta}^T \mathscr{X}_{1i} - \exp(\boldsymbol{\beta}^T \mathscr{X}_{2i})\right\}.\end{aligned}$$

For simplicity, we choose $\mathscr{X}_1 = \mathscr{X}_2$. By solving $U_{CC,n}\left(\boldsymbol{\theta}\right) = \mathbf{0}$, we obtain $\widehat{\boldsymbol{\theta}}_{CC,n}$ (naive solution), which is a biased and not efficient estimator of $\boldsymbol{\theta}$. Under MAR assumption, $E\left[U_{CC,n}\left(\boldsymbol{\theta}\right)\right] \neq 0$. Besides the CC method, [7] proposed a semiparametric IPW of ZIP model with missings[1] whose estimating equation is

---

[1] IPW denotes the inverse probability weighting. When π is known in preliminary stage, Eq. (5) refers to true IPW estimating equation (see, [7]), and its solution is $\widehat{\boldsymbol{\theta}}_{Wt}$.

$$U_{Ws,n}\left(\boldsymbol{\theta},\hat{\pi}\right)=\frac{1}{\sqrt{n}}\sum_{i=1}^{n}\frac{\delta_i}{\hat{\pi}\left(Y_i,V_i\right)}S_i(\boldsymbol{\theta}). \tag{5}$$

Here $\hat{\pi}$ is the Nadaraya kernel estimator [8] of π given by

$$\widehat{\pi}_k=\widehat{\pi}(y,v)=\frac{\sum_{k=1}^{n}\delta_k K_h(Y_k=y,V_k-v)}{\sum_{i=1}^{n}K_h(Y_i=y,V_i-v)}, \tag{6}$$

where $K_h(u)$ is a specific kernel function and h is the bandwidth parameter. When V is categorical, (6) $\hat{\pi}(y,v)=\sum_{k=1}^{n}\delta_k I(Y=y_k,V_k=v)\sum_{i=1}^{n}I(Y=y_i,V_i=v)$ as in [7]. By solving $U_{Ws,n}(\boldsymbol{\theta},\hat{\pi})=\mathbf{0}$, we obtain $\hat{\boldsymbol{\theta}}_{Ws}$ as a semiparametric estimator of $\boldsymbol{\theta}$. Since $U_{Ws,n}(\boldsymbol{\theta},\hat{\pi})$ does not use all data, $\hat{\boldsymbol{\theta}}_{Ws}$ is less efficient in general. Hence, (5) needs to incorporate all available data to become more efficient.

3 Proposed Methods

Robins et al. [11] proposed the augmentation inverse probability weighting (AIPW) estimator where the selection probability (π) and the augmentation term ($\mathscr{A}$) were estimated parametrically. Often, $\mathscr{A}$ refers to the conditional expectation of the score function given *the observed data.* It is expressed as $\mathscr{A}=E[S_1(\boldsymbol{\theta})|Y,V]$. When π and $\mathscr{A}$ are known (true), the general form of AIPW estimating equation is

$$U_{Wa,n}\left(\theta,\pi,\mathscr{A}\right)=\frac{1}{\sqrt{n}}\sum_{i=1}^{n}\Big[\frac{\delta_i}{\pi\left(Y_i,V_i\right)}S_i(\boldsymbol{\theta})+\big(1-\frac{\delta_i}{\pi\left(Y_i,V_i\right)}\big)\mathscr{A}(Y_i,V_i)\Big]. \tag{7}$$

In general (7) is not available because the knowledge of π and $\mathscr{A}$ is often hypothetical. When the distribution assumptions are not well understood, the parametric estimators of π and $\mathscr{A}$ may not be a good idea. Thus, we propose four semiparametric double robust AIPW estimating equations.[2]

3.1 Fully Kernel-Assisted Estimation

The fully kernel-assisted estimating equation of a ZIP regression model is

[2]Double robustness feature [10, 13]. These properties prevent the covariance matrix from suffering from misspecification provided that at least one of the preliminary estimates including π_g or $\mathscr{A}_g$ plugged in Eq. (7) is correct. Only in this way, all ASEs and 95% CP of the proposed estimators are valid for inference.

$$U_{Wa1,n}(\boldsymbol{\theta}, \hat{\pi}_k, \hat{\mathscr{A}_k}) = \frac{1}{\sqrt{n}} \sum_{i=1}^{n} \left[\frac{\delta_i}{\hat{\pi}_k(Y_i, V_i)} S_i(\boldsymbol{\theta}) + (1 - \frac{\delta_i}{\hat{\pi}_k(Y_i, V_i)}) \hat{\mathscr{A}_k}(Y_i, V_i) \right], \quad (8)$$

where $\widehat{\pi}_k$ is given in (6) and $\hat{\mathscr{A}_k}$ is expressed in (9) by

$$\hat{\mathscr{A}_k} = \hat{\mathscr{A}_k}(y, v) = \frac{\sum_{i=1}^{n} \delta_i S_i(\boldsymbol{\theta}) K_h(Y_i = y, V_i - v)}{\sum_{i=1}^{n} \delta_i K_h(Y_i = y, V_i - v)}, \quad (9)$$

where K_h is the kernel function and h is the bandwidth parameter. Here $\hat{\pi}$ and $\hat{\mathscr{A}}$ are both the Nadaraya estimators [8, 19]. When V is categorical, $\hat{\mathscr{A}_k} = \sum_{i=1}^{n} \delta_i S_i(\boldsymbol{\theta}) I(Y_i = y, V_i = v) | \sum_{i=1}^{n} \delta_i I(Y_i = y, V_i = v)$, where I (.) is an indicator function as it is in [14].

3.2 Fully GAMs-Assisted Estimation

The generalized additive models (GAMs) method is one of the popular nonparametric techniques [5]. The GAMs regression in the sense of [5] is

$$\eta = a_0 + \sum_{k=1}^{m} \beta_k X_k + \sum_{j=m+1}^{q} s_j(X_j), \quad (10)$$

where $\eta = g[E(Y^*|\mathbf{X}))]$ is the mean function, $\mathbf{X} = (X_1, \ldots, X_{m+1}, \ldots, X_q)^T$ are covariates without missings, g is a link function, $s_j(X_j)$ are arbitrary unknown smooth functions, a_0 is the intercept term, and β_k are regression coefficients. We assume that (10) is identifiable by imposing $E[Y^*] = a_0$ and $E[s_j(X_j)] = 0$ hold for all j. We obtain $\hat{\pi}(Y, Z, W)$ by fitting a binary GAMs regression given by

$$logit(\pi) = \log\left(\frac{\pi}{1-\pi}\right) = a_0 + s_1(Y) + s_2(Z) + \beta_1(W), \quad (11)$$

where $\pi = P(\delta = 1|Y, Z, W)$ and δ is the MAR indicator variable.[3] To obtain $\hat{\mathscr{A}_g}$, we implement model (10) where the response variable is $Y^* = E\left[S(\boldsymbol{\theta})|Y, Z, W\right]$. Both $\hat{\pi}_g$ and $\hat{\mathscr{A}_g}$ are the GAMs estimators of π and $\mathscr{A}$, respectively. Then the AIPW fully GAMs estimating equation of a ZIP model is expressed by

$$U_{Wa2,n}(\boldsymbol{\theta}, \hat{\pi}_g, \hat{\mathscr{A}_g}) = \frac{1}{\sqrt{n}} \sum_{i=1}^{n} \left[\frac{\delta_i}{\hat{\pi}_g(Y_i, V_i)} S_i(\boldsymbol{\theta}) + (1 - \frac{\delta_i}{\hat{\pi}_g(Y_i, V_i)}) \hat{\mathscr{A}_g}(Y_i, V_i) \right]. \quad (12)$$

[3]With regard to Eq. (11), δ is used as the outcome variable.

The natural *spline* functions are used to *estimate* s_1 and s_2. Since $\mathbf{X}$ is the observed data, the R package mcgv is used to obtain $\hat{\pi}_g$ and $\hat{\mathscr{A}}_g$. In (8), $\hat{\pi}_k$ and $\hat{\mathscr{A}}_k$ are nuisance components, whereas in (12), $\hat{\pi}_g$ and $\hat{\mathscr{A}}_g$ are nuisance components.

3.3 Mixed Nuisance Functions-Assisted Estimation

Similar to (8) and (12), we construct two mixed estimating equations. The first one is obtained by plugging in $\hat{\pi}_g$ and $\hat{\mathscr{A}}_k$ in (7). We have

$$U_{Wa3,n}\big(\boldsymbol{\theta}, \hat{\pi}_g, \hat{\mathscr{A}}_k\big). \tag{13}$$

The second one is obtained by plugging in $\hat{\pi}_k$ and $\hat{\mathscr{A}}_g$ in (7). We have

$$U_{Wa4,n}\big(\boldsymbol{\theta}, \hat{\pi}_k, \hat{\mathscr{A}}_g\big). \tag{14}$$

Here, $U_{Wa1,n}\big(\boldsymbol{\theta}, \hat{\pi}_k, \hat{\mathscr{A}}_k\big)$, $U_{Wa2,n}\big(\boldsymbol{\theta}, \hat{\pi}_g, \hat{\mathscr{A}}_g\big)$, $U_{Wa3,n}\big(\boldsymbol{\theta}, \hat{\pi}_g, \hat{\mathscr{A}}_k\big)$, and $U_{Wa4,n}\big(\boldsymbol{\theta}, \hat{\pi}_k, \hat{\mathscr{A}}_g\big)$ differ only in terms of their plugged-in estimators. By solving $U_{Wa1,n}\big(\boldsymbol{\theta}, \hat{\pi}_k, \hat{\mathscr{A}}_k\big) = \mathbf{0}$, $U_{Wa2,n}\big(\boldsymbol{\theta}, \hat{\pi}_g, \hat{\mathscr{A}}_g\big) = \mathbf{0}$, $U_{Wa3,n}\big(\boldsymbol{\theta}, \hat{\pi}_k, \hat{\mathscr{A}}_g\big) = \mathbf{0}$, and $U_{Wa4,n}\big(\boldsymbol{\theta}, \hat{\pi}_g, \hat{\mathscr{A}}_k\big) = \mathbf{0}$, we obtain, respectively, $\hat{\boldsymbol{\theta}}_{W_{\pi_k A_k}}$, $\hat{\boldsymbol{\theta}}_{W_{\pi_g A_g}}$, $\hat{\boldsymbol{\theta}}_{W_{\pi_g A_k}}$, and $\hat{\boldsymbol{\theta}}_{W_{\pi_k A_g}}$. These solutions *are the semiparametric and double robust estimators* of $\boldsymbol{\theta}$ in the sense of [10, 13]. Compared to other proposed estimators, the fully kernel-assisted estimator [14] $\hat{\boldsymbol{\theta}}_{W_{\pi_k A_k}}$ is well known in the literature of missing data. *Details about the fully GAMs asymptotic variance are derived from* [1].[4]

4 Large Sample

Main results

Theorem 1 [5] *Under some specific regularity conditions and provided that* $nh^{2r} \to 0$ *and* $nh^{2d} \to \infty$ *as* $n \to \infty$, $\widehat{\boldsymbol{\theta}}_{W_{\pi_k A_k}} \to \boldsymbol{\theta}$, $\widehat{\boldsymbol{\theta}}_{W_{\pi_g A_g}} \to \boldsymbol{\theta}$, $\widehat{\boldsymbol{\theta}}_{W_{\pi_g A_k}} \to \boldsymbol{\theta}$, *and* $\hat{\boldsymbol{\theta}}_{W_{\pi_k A_g}} \to \boldsymbol{\theta}$ *in probability. Then* $\sqrt{n}(\hat{\boldsymbol{\theta}}_{W_{\pi_k A_k}} - \boldsymbol{\theta})$, $\sqrt{n}(\hat{\boldsymbol{\theta}}_{W_{\pi_g A_g}} - \boldsymbol{\theta})$, $\sqrt{n}(\hat{\boldsymbol{\theta}}_{W_{\pi_g A_k}} - \boldsymbol{\theta})$, *and* $\sqrt{n}$ $(\hat{\boldsymbol{\theta}}_{W_{\pi_k A_g}} - \boldsymbol{\theta})$ *are all asymptotic* $\mathscr{N}(0, \Delta_{wa})$, *where*

$$\Delta_{Wa} = G_F^{-1}(\boldsymbol{\theta})\left\{Var\left[U_{Wa,n}\big(\boldsymbol{\theta}, \pi, \mathscr{A}\big)\right]\right\}[G_F^{-1}(\boldsymbol{\theta})]^T,$$

[4] The theoretical substance of GAMs framework follows from the general discussion of 2-step semiparametric estimations from [1].

[5] The proof of Theorem 1 follows along the same lines related to Theorem 2 from [15]. Also, the proof uses some previous results from [7].

and where $Var\left[U_{Wa}(\boldsymbol{\theta}, \pi, \mathscr{A})\right] = E\left[\left\{U_{Wa}(\boldsymbol{\theta}, \pi, \mathscr{A})\right\}^{\otimes 2}\right]$ and $G_F(\boldsymbol{\theta}) = E\left[-\frac{\partial S(\boldsymbol{\theta})}{\partial \boldsymbol{\theta}^T}\right]$. This is the background of our asymptotic. Following [7, 9, 15], a consistent asymptotic variance of Δ_{Wa} is given by

$$\hat{\Delta}_{Wa} = G_{F,n}^{-1}(\hat{\boldsymbol{\theta}}, \hat{\pi}, \hat{\mathscr{A}}) \left\{Var\left[U_{Wa,n}(\hat{\boldsymbol{\theta}}, \hat{\pi}, \hat{\mathscr{A}})\right]\right\} [G_{F,n}^{-1}(\hat{\boldsymbol{\theta}}, \hat{\pi}, \hat{\mathscr{A}})]^T, \quad (15)$$

where $G_{F,n}(\hat{\boldsymbol{\theta}}, \hat{\pi}, \hat{\mathscr{A}}) = \frac{\partial}{\partial \boldsymbol{\theta}^T} U_{Waj,n}(.)$ with $j = 1, 2, 3, 4$. Here, $Var\left[U_{Wa}(\boldsymbol{\theta}, \pi, \mathscr{A})\right]$ is explicitly estimated by $\{\hat{J}(.) - (\hat{J}^*(.) - \hat{J}^{**}(.))\}$, where $\hat{J}_n(\boldsymbol{\theta}, \pi) = \frac{1}{n}\sum_{i=1}^n \frac{\delta_i}{\pi^2(Y_i, V_i)} [S_i(\boldsymbol{\theta})]^{\otimes 2}$, $\hat{J}_n^*(\boldsymbol{\theta}, \pi, \mathscr{A}) = \frac{1}{n}\sum_{i=1}^n \frac{\delta_i}{\pi^2(Y_i, V_i)} \left[\hat{S}_i^*(\boldsymbol{\theta})\right]^{\otimes 2}$, and $\hat{J}_n^{**}(\boldsymbol{\theta}, \pi, \mathscr{A}) = \frac{1}{n}\sum_{i=1}^n \left[\hat{S}_i^*(\boldsymbol{\theta})\right]^{\otimes 2}$. Here, $\hat{S}_i^*(\boldsymbol{\theta}) = \frac{\sum_{j=1}^n \delta_j S_j(\boldsymbol{\theta}) K_h(Y_j = Y_i, V_j - V_i)}{\sum_{k=1}^n \delta_k K_h(Y_j = Y_i, V_k - V_i)}$ with $i = 1, \ldots, n$ for fully-kernel model [15]. As for the AIPW GAMs idea, its asymptotic properties are based on the idea from [1]. Thus, we notice that

1. The asymptotic variances (ASV) Δ_{Wa} of $\sqrt{n}(\hat{\boldsymbol{\theta}}_{W_{\pi_k A_k}} - \boldsymbol{\theta})$, $\sqrt{n}(\hat{\boldsymbol{\theta}}_{W_{\pi_g A_g}} - \boldsymbol{\theta})$, $\sqrt{n}(\hat{\boldsymbol{\theta}}_{W_{\pi_g A_k}} - \boldsymbol{\theta})$, and $\sqrt{n}(\hat{\boldsymbol{\theta}}_{W_{\pi_k A_g}} - \boldsymbol{\theta})$, respectively, are all similar in structure.[6]
2. The consistent ASV estimators of Δ_{wa} $(\hat{\Delta}_{Wa})$ are both sandwich types.
3. The ASV Δ_{Wa} is known to be relatively more efficient than Δ_{Ws} $(\hat{\Delta}_{Ws})$ which refers to the ASV of semiparametric IPWs provided in [7].
4. $\hat{\boldsymbol{\theta}}_{W_{\pi_k A_k}}, \hat{\boldsymbol{\theta}}_{W_{\pi_g A_g}}, \hat{\boldsymbol{\theta}}_{W_{\pi_g A_k}}$, and $\hat{\boldsymbol{\theta}}_{W_{\pi_k A_g}}$ are all asymptotically equivalent and double robust, whereas $\hat{\boldsymbol{\theta}}_{Ws}$ is robust only.

5 Simulation Study

We investigate the performance of the proposed semiparametric estimators using two different cases. In Case 1, X is $U(-2, 2)$, Z is $U(-2, 2)$, and t is $N(0, 1)$, W is 1 if $(X - \sigma * t) < 0$ or 0 elsewhere with $\sigma = 0.75$. Since Z is continuous, we use uniform kernel defined by $K_h(u) = 0.5$ if $u \in (-1, 1)$ or 0 elsewhere. The bandwidth (h) is $h = 1/3\, \hat{\sigma}_{Zyw} n^{-1/3}$ as in [20]. In Case 2, X is $U(-2, 1)$, Z is a multinomial $(0, 1, 2, 3)$ with probability $(0.1, 0.4, 0.3, 0.2)$ respectively, and W is 1 if $X \leq 0$, and 0 otherwise. Unlike in Case 1, in Case 2, $V = (Z, W)^T$ is categorical. Moreover, Y is obtained as $Y = 0 * \omega + (1 - \omega)P(\lambda)$, where $\omega = H(\boldsymbol{\gamma}^T \mathscr{X})$ and $\lambda = \exp(\boldsymbol{\beta}^T \mathscr{X})$, where $\mathscr{X} = (1, X, Z)^T$. We define δ such that $\delta = 1$ if X is observed and $\delta = 0$ otherwise. Under MAR assumption, the selection probability $P(\delta|Y, V, W)$ is $\pi(\boldsymbol{\alpha}) = \left(1 - \exp(\alpha_0 + \boldsymbol{\alpha}_1 Y + \boldsymbol{\alpha}_2 V)\right)^{-1}$, where $\boldsymbol{\alpha}$ the nuisance parameter. We compare the performance of $\widehat{\boldsymbol{\theta}}_{W_{\pi_k A_k}}, \widehat{\boldsymbol{\theta}}_{W_{\pi_g A_g}}, \widehat{\boldsymbol{\theta}}_{W_{\pi_g A_k}}, \widehat{\boldsymbol{\theta}}_{W_{\pi_k A_g}}, \widehat{\boldsymbol{\theta}}_{Ws}, \widehat{\boldsymbol{\theta}}_{W}$, and $\widehat{\boldsymbol{\theta}}_{CC}$ computationally

[6] Contrarily to the approach used in [3, 4, 10, 13], etc., the asymptotic variance we propose follows from [1, 7, 15]. In the proposed framework, the preliminary nonparametric functions contribute implicitly to sandwich-type covariance matrix via the augmentation term $A(Y, V)$ and $S_i^*(\boldsymbol{\theta})$.

via their Bias of the estimator (Bias), asymptotic standard errors (ASE), standard deviations (SD), and coverage probabilities (CP) of their 95% confidence intervals. *The true values are* $\boldsymbol{\theta}_0 = (\gamma_0, \gamma_1, \gamma_2, \beta_0, \beta_1, \beta_2)^T$ and $\boldsymbol{\alpha} = (\alpha_0, \alpha_1, \alpha_2, \alpha_3, \alpha_4)^T$. With regard to Case 1, $\boldsymbol{\theta}_0 = (\boldsymbol{\gamma}_0^T, \boldsymbol{\beta}_0^T) = (1.0, 0.5, 0.5, 1.0, 0.7, 1.0)^T$ and $\boldsymbol{\alpha} = (-.5, 0.3, 10, -0.5, 1)^T$. *The true values,* $\boldsymbol{\theta}_0$ and $\boldsymbol{\alpha}$, in Case 2 are set as follows: $\boldsymbol{\theta}_0 = (-1, -1, 0.5, 1, 0.7, 1)^T$ and $\boldsymbol{\alpha} = (0.5, 0.5, 0.5, 1.0, 0.7, 1.0)^T$. In both cases, the sample size is 750 and the number of repetitions is 500. Regarding the GAMs implementation on the observed data, *the natural spline functions* are used to estimate $s_1(.)$ and $s_2(.)$, and the R package mcgv is used to estimate $\pi(Y, V)$ and $\mathscr{A}(Y, V)$.

Table 1 Simulation result Case 1 with bandwidth $h = 4\hat{\sigma}_{Zyw}n^{-1/3}$

Parameter		$n =$ **750** and mr$=$**0.47**						
		$\hat{\boldsymbol{\theta}}_{CC}$	$\hat{\boldsymbol{\theta}}_{Wt}$	$\hat{\boldsymbol{\theta}}_{Ws}$	$\hat{\boldsymbol{\theta}}_{W_{\pi_k A_k}}$	$\hat{\boldsymbol{\theta}}_{W_{\pi_g A_g}}$	$\hat{\boldsymbol{\theta}}_{W_{\pi_k A_g}}$	$\hat{\boldsymbol{\theta}}_{W_{\pi_g A_k}}$
Logistic model								
γ_1	Bias	−0.4570	0.0027	0.0395	0.0071	0.0037	0.0171	0.0013
	SD	0.1335	0.1434	0.1226	0.1151	0.1189	0.1174	0.1153
	ASE	0.1373	0.1489	**0.1357**	0.1334	0.1342	0.1337	0.1339
	CP	0.0740	0.9640	0.9620	0.9860	0.9740	0.9820	0.9840
γ_2	Bias	−0.2790	0.0034	−0.0073	−0.0082	0.0062	0.0017	0.0054
	SD	0.1202	0.1297	0.1188	0.1154	0.1159	0.1173	0.1144
	ASE	0.1211	0.1286	**0.1202**	0.1174	0.1192	0.1186	0.1187
	CP	0.3480	0.9500	0.9520	0.9540	0.9560	0.9560	0.9520
γ_3	Bias	−0.3982	0.0074	0.0021	0.0016	0.0086	0.0017	0.0013
	SD	0.1337	0.1436	0.1214	0.1164	0.1173	0.1164	0.1153
	ASE	0.1312	0.1407	**0.1324**	0.1290	0.1308	0.1300	0.1339
	CP	0.1760	0.9520	0.9620	0.9680	0.9740	0.9700	0.9840
Poisson model								
β_1	Bias	0.0835	−0.0045	0.0067	0.0046	−0.0047	0.0069	−0.0050
	SD	0.0644	0.0691	0.0648	0.0639	0.0660	0.0641	0.0652
	ASE	0.0690	0.0715	**0.0669**	0.0651	0.0659	0.0651	0.0658
	CP	0.7940	0.9500	0.9540	0.9520	0.9340	0.9520	0.9440
β_2	Bias	−0.0261	0.0008	−0.0056	−0.0046	0.0014	−0.0057	0.0014
	SD	0.0384	0.0404	0.0390	0.0387	0.0393	0.0387	0.0391
	ASE	0.0413	0.0413	**0.0391**	0.0375	0.0384	0.0382	0.0383
	CP	0.9140	0.9520	0.9400	0.9360	0.9480	0.9420	0.9480
β_3	Bias	−0.0374	0.0036	0.0039	0.0042	0.0033	0.0026	0.0034
	SD	0.0476	0.0505	0.0480	0.0475	0.0497	0.0478	0.0485
	ASE	0.0500	0.0504	**0.0478**	0.0465	0.0475	0.0469	0.0474
	CP	0.8840	0.9440	0.9480	0.9460	0.9360	0.9420	0.9420

1. On average 47% of X were missing in 500 replications
2. The average rate of $Y = 0$ was 81% in 500 fully simulated data sets
3. The average rate of $Y = 0$ was 72% in 500 validated simulated samples

6 Discussion

The simulation result of Case 1 revealed that in 500 repetitions, the missing rate of X was 0.47, and the rates of $Y = 0$ were 0.81 and 0.72 in simulated and validated samples, respectively. Details about ZIP regression model fit are provided in Table 1. Likewise, the simulation result of Case 2 revealed that in 500 repetitions, the missing rate of X was 0.49, and the rates of $Y = 0$ were 0.71 and 0.53 in simulated and validated samples, respectively. More details are given in Table 2.

We presented the numerical results as evidence for the main results. We found that the complete case estimator ($\widehat{\boldsymbol{\theta}}_{CC}$) is biased and its empirical 95% CP is significantly far from nominal 95% CI. The true weight IPW estimator ($\widehat{\boldsymbol{\theta}}_{Wt}$) and semiparametric

Table 2 Simulation results Case 2 with categorical V

Parameter		$n = \mathbf{750}$ and mr$=\mathbf{0.49}$						
		$\widehat{\boldsymbol{\theta}}_{CC}$	$\widehat{\boldsymbol{\theta}}_{Wt}$	$\widehat{\boldsymbol{\theta}}_{Ws}$	$\widehat{\boldsymbol{\theta}}_{W_{\pi_k A_k}}$	$\widehat{\boldsymbol{\theta}}_{W_{\pi_g A_g}}$	$\widehat{\boldsymbol{\theta}}_{W_{\pi_k A_g}}$	$\widehat{\boldsymbol{\theta}}_{W_{\pi_g A_k}}$
Logistic model								
γ_1	Bias	−0.3274	0.0055	0.0287	0.0122	−0.0005	0.0174	0.0053
	SD	0.2582	0.2670	0.2273	0.2248	0.2351	0.2210	0.2226
	ASE	0.2465	0.2524	**0.2071**	0.2043	0.2031	0.2027	0.2029
	CP	0.7280	0.9360	0.9180	0.9260	0.9080	0.9300	0.9240
γ_2	Bias	−0.2712	0.0025	−0.0045	−0.0035	−0.0008	0.0067	0.0037
	SD	0.1246	0.1280	0.1150	0.1146	0.1084	0.1054	0.1140
	ASE	0.1260	0.1296	**0.1135**	0.1125	0.1115	0.1114	0.1118
	CP	0.4100	0.9540	0.9560	0.9580	0.9560	0.9580	0.9460
γ_3	Bias	−0.4493	−0.0033	−0.0113	−0.0086	0.0008	−0.0037	−0.0010
	SD	0.1278	0.1334	0.1099	0.1079	0.1182	0.1110	0.1079
	ASE	0.1293	0.1329	**0.1061**	0.1044	0.1052	0.1051	0.1052
	CP	0.0620	0.9480	0.9480	0.9460	0.9220	0.9360	0.9480
Poisson model								
β_1	Bias	0.0751	0.0026	0.0090	0.0095	0.0035	0.0056	0.0019
	SD	0.0689	0.0738	0.0707	0.0703	0.0746	0.0721	0.0725
	ASE	0.0706	0.0730	**0.0685**	0.0674	0.0672	0.0678	0.0672
	CP	0.8040	0.9460	0.9320	0.9260	0.9240	0.9300	0.9260
β_2	Bias	−0.0204	−0.0010	−0.0035	−0.0035	−0.0010	−0.0030	−0.0007
	SD	0.0273	0.0285	0.0280	0.0279	0.0275	0.0276	0.0284
	ASE	0.0265	0.0266	**0.0248**	0.0244	0.0244	0.0246	0.0244
	CP	0.8660	0.9220	0.8900	0.8840	0.9080	0.9000	0.8960
β_3	Bias	−0.0309	−0.0020	−0.0041	−0.0043	−0.0023	−0.0031	−0.0017
	SD	0.0292	0.0305	0.0297	0.0296	0.0305	0.0301	0.0301
	ASE	0.0304	0.0309	**0.0293**	0.0287	0.0289	0.0290	0.0290
	CP	0.8380	0.9440	0.9300	0.9300	0.9400	0.9360	0.9360

1. On average 49% of X were missing in 500 replications
2. The average rate of $Y = 0$ was 71% in 500 full simulated data sets
3. The average rate of $Y = 0$ was 53% in 500 validated simulated samples

IPW estimator ($\widehat{\boldsymbol{\theta}}_{Ws}$) yield good estimates, but they were still less efficient due to listwise deletion. The $\widehat{\boldsymbol{\theta}}_{W_{\pi_k A_k}}$, $\widehat{\boldsymbol{\theta}}_{W_{\pi_g A_g}}$, $\widehat{\boldsymbol{\theta}}_{W_{\pi_g A_k}}$, and $\widehat{\boldsymbol{\theta}}_{W_{\pi_k A_g}}$ performed much better than $\widehat{\boldsymbol{\theta}}_{Wt}$ and slightly better than $\widehat{\boldsymbol{\theta}}_{Ws}$. In addition, $\widehat{\boldsymbol{\theta}}_{W_{\pi_k A_k}}$, $\widehat{\boldsymbol{\theta}}_{W_{\pi_g A_g}}$, $\widehat{\boldsymbol{\theta}}_{W_{\pi_g A_k}}$, and $\widehat{\boldsymbol{\theta}}_{W_{\pi_k A_g}}$ were all double robust estimators [11]. Moreover, $\widehat{\boldsymbol{\theta}}_{W_{\pi_k A_k}}$, $\widehat{\boldsymbol{\theta}}_{W_{\pi_g A_g}}$, $\widehat{\boldsymbol{\theta}}_{W_{\pi_g A_k}}$, and $\widehat{\boldsymbol{\theta}}_{W_{\pi_k A_g}}$ are root-n consistent. Its related proof is beyond the scope of this work. Moreover, the fully kernel- and fully GAMs- assisted estimations are well known in the semiparametric method for missing data literature, contrarily to the semiparametric mixed Kernel-GAMs and GAMs-Kernel estimators which are novel ideas and should be the object of further investigation. *Although* the focus was mostly methodological than theoretical, our work shares many similarities with other studies, i.e., [3, 9] in nature.

Acknowledgments The authors gratefully thank Miss Ula Tzu-Ning Kung for improving the English in this paper. *We are also thankful to the referee for the constructive comments and suggestions that have improved considerably the quality of this manuscript.* This work was supported by the Career Development Award of Academia Sinica (Taiwan) grant number 103-CDA-M04 and the Ministry of Science and Technology (Taiwan) grant numbers 105-2118-M-001-007-MY2, 107-2118-M-001-011-MY3 and 107-2321-B-001-038.

References

1. Ackerberg, D., Chen, X.H., Hahn, J.Y.: A practical asymptotic variance for two-steps semiparametric estimators. Rev. Econ. Stat. **94**, 481–498 (2012)
2. Creemers, A., Aerts, M., Hens, N., Molenberghs, G.: A nonparametric approach to weighted estimating equations for regression analysis with missing covariates. Comput. Stat. Data Anal. **56**, 100–113 (2012)
3. Firpo, S., Rothe, C.: Properties of doubly robust estimators when nuisance functions are estimated nonparametrically. Econ. Theory (2019). To appear
4. Hahn, J., Ridder, G.: Asymptotic variance of semiparametric estimators with generated regressors. Econometrica **81**, 315–340 (2012)
5. Hastie, T.J., Tibshirani, R.J.: Generalized additive models. Stat. Sci. **1**, 297–318 (1986)
6. Lambert, D.: Zero-inflated Poisson regression, with an application to defects in manufacturing. Technometrics **34**, 1–14
7. Lukusa, T.M., Lee, S.M., Li, C.S.: Semiparametric estimation of a zero-inflated Poisson regression model with missing covariates. Metrika **79**, 457–483 (2016)
8. Nadaraya, E.A.: On estimating regression. Theory Probab. Appl **9**, 141–142 (1964)
9. Newey, W.K.: Kernel estimation of partial means and a general variance estimator. Econ. Theory **10**, 233–253 (1994)
10. Robins, J.M., Rotnitzky, A.: Comment on "Inference for semiparametric models: some questions and an answer" by P. Bickel and J. Kwon. Stat. Sin. **11**, 920–936 (2001)
11. Robins, J.M., Rotnitzky, A., Zhao, L.P.: Estimation of regression coefficients when some regressors are not always observed. J. Am. Stat. Assoc. **89**, 846–866 (1994)
12. Rubin, D.B.: Inference and missing data. Biometrika **63**, 581–592 (1976)
13. Scharfstein, D., Rotnitzky, A., Robins, J.: Adjusting for nonignorable drop-out using semiparametric nonresponse models. J. Am. Stat. Assoc. **94**, 1096–1120 (1999)
14. Wang, C.Y., Chen, H.Y.: Augmented inverse probability weighted estimator for Cox missing covariate regression. Biometrics **57**, 414–419 (2001)

15. Wang, C.Y., Wang, S.: A note on kernel assisted estimators in missing covariate regression. Stat. Probab. Lett. **55**, 439–449 (2001)
16. Wang, Q., Linton, O., Härdle, W.: Semiparametric regression analysis with missing response at random. J. Am. Stat. Assoc. **99**, 334–345 (2004)
17. Tsatis, A.A.: Semiparametric Theory and Missing Data. Springer, USA (2006)
18. Tu, W., Liu, H.: Zero-Inflated Data. John Wiley & Sons, Ltd. (2016)
19. Watson, G.S.: Smooth regression analysis. Sankhya Ser. A **26**, 359–372 (1964)
20. Wang, C.Y., Wang, S., Zhao, L.P., Ou, S.T.: Weighted semiparametric estimation in regression with missing covariates data. J. Am. Stat. Assoc. **92**, 512–525 (1997)

The Discrepancy Method for Extremal Index Estimation

Natalia Markovich

Abstract We consider the nonparametric estimation of the extremal index of stochastic processes. The discrepancy method that was proposed by the author as a data-driven smoothing tool for probability density function estimation is extended to find a threshold parameter u for an extremal index estimator in case of heavy-tailed distributions. To this end, the discrepancy statistics are based on the von Mises–Smirnov statistic and the k largest order statistics instead of an entire sample. The asymptotic chi-squared distribution of the discrepancy measure is derived. Its quantiles may be used as discrepancy values. An algorithm to select u for an estimator of the extremal index is proposed. The accuracy of the discrepancy method is checked by a simulation study.

Keywords Nonparametric estimation · Discrepancy method · von Mises–Smirnov statistic · Extremal index · Heavy-tailed distribution

1 Introduction

Let $X^n = \{X_1, \ldots, X_n\}$ be a sample of random variables (rvs) with cumulative distribution function (cdf) $F(x)$. We consider the nonparametric estimation of the extremal index (EI) of stochastic processes. There are nonparametric methods like the well-known blocks and runs estimators of the EI which require the selection of two parameters, where an appropriate threshold u is among them [3]. Modifications of the blocks estimator [6, 20] and sliding blocks estimators [17, 19] require only the block size without u. The intervals estimator depends only on u [9]. Less attention is devoted to the estimation of parameters required for these estimators.

The discrepancy method was proposed and studied in [13, 23] as a data-driven smoothing tool for light-tailed probability density function (pdf) estimation by independent identically distributed (iid) data. The idea is to find a smoothing parameter

N. Markovich (✉)
V.A. Trapeznikov Institute of Control Sciences of Russian Academy of Sciences, Moscow 117997, Russia
e-mail: nat.markovich@gmail.com

M. La Rocca et al. (eds.), *Nonparametric Statistics*, Springer Proceedings in Mathematics & Statistics 339, https://doi.org/10.1007/978-3-030-57306-5_31

(i.e., a bandwidth) h as a solution of the discrepancy equation:

$$\rho(\widehat{F}_h, F_n) = \delta.$$

Here, $\widehat{F}_h(x) = \int_{-\infty}^{x} \widehat{f}_h(t)dt$, $\hat{f}_h(t)$ is some pdf estimate, and δ is a known discrepancy value of the estimation of $F(x)$ by the empirical distribution function $F_n(t)$, i.e., $\delta = \rho(F, F_n)$, $\rho(\cdot, \cdot)$ is a metric in the space of cdf's. Since δ is usually unknown, some quantiles of limit distributions of von Mises–Smirnov (M-S)

$$\omega_n^2 = n \int_{-\infty}^{\infty} (F_n(x) - F(x))^2 dF(x),$$

and Kolmogorov–Smirnov statistics were proposed as δ. Distributions of these statistics are invariant regarding $F(x)$, [4]. In practice the bandwidth h may be found as a solution of the equation [13]

$$\hat{\omega}_n^2(h) = 0.05, \qquad (1)$$

where

$$\hat{\omega}_n^2(h) = \sum_{i=1}^{n} \left(\widehat{F}_h(X_{i,n}) - \frac{i - 0.5}{n} \right)^2 + \frac{1}{12n}$$

is based on the order statistics $X_{1,n} \le \cdots \le X_{n,n}$ corresponding to the sample X^n. The value 0.05 corresponding to the maximum (mode) of the pdf of the statistic ω_n^2 was found by tables of statistic ω_n^2 [4] as the discrepancy value δ.

It is noted in [14, 16] that for heavy-tailed distributions, the statistic ω_n^2 may not reach the value 0.05 and hence, the discrepancy equation (1) may have no solutions, or the solutions provide too small values of h that are unsatisfactory for pdf estimation. In order to estimate heavy-tailed pdf's, the modification of the discrepancy method based on the k largest order statistics instead of the entire sample was considered in [16]. Namely, the statistic

$$\hat{\omega}_n^2(h) = \sum_{i=n-k+1}^{n} \left(\widehat{F}_h(X_{i,n}) - \frac{i - 0.5}{n} \right)^2 + \frac{1}{12n}$$

was proposed to be used in (1). A similar idea was explored in [15] to estimate the EI.

Definition 1 ([12, p. 67]) The stationary sequence $\{X_n\}_{n\ge1}$ is said to have EI $\theta \in [0, 1]$ if for each $0 < \tau < \infty$ there is a sequence of real numbers $u_n = u_n(\tau)$ such that it holds

$$\lim_{n\to\infty} n(1 - F(u_n)) = \tau, \qquad \lim_{n\to\infty} P\{M_n \le u_n\} = e^{-\tau\theta}, \qquad (2)$$

where $M_n = \max\{X_1, \ldots, X_n\}$.

The EI reflects a cluster structure of a sequence. If $X_1, \ldots, X_n$ are independent, $\theta = 1$ holds. One can determine the cluster as the number of consecutive observations exceeding a threshold u between two consecutive non-exceedances, [9]. Then the values of inter-cluster times $T_1(u)$ for a given threshold u are stated as the numbers of consecutive non-exceedances between two consecutive clusters, [9]. We have

$$T_1(u) = \min\{j \geq 1 : M_{1,j} \leq u, X_{j+1} > u | X_1 > u\},$$

$M_{1,j} = \max\{X_2, \ldots, X_j\}$, $M_{1,1} = -\infty$. Observations of $T_1(u_n)$ normalized by the tail function $\{Y_i = \overline{F}(u_n) T_1(u_n)_i\}$, $i = 1, \ldots, L$, $L = L(u_n)$, $L < n$,[1] are derived to be asymptotically exponentially distributed with weight θ, i.e.,

$$P\{\overline{F}(u_n) T_1(u_n) > t\} \to \theta \exp(-\theta t) \qquad \text{for } t > 0$$

as $n \to \infty$ under a specific mixing condition and u_n satisfying (2), [9].

The discrepancy equation may be based on the k, $1 \leq k \leq L-1$, largest order statistics of a sample $\{Y_i = (N_u/n) T_1(u)_i\}$ as follows:

$$\hat{\omega}_L^2(u) = \sum_{i=L-k+1}^{L} \left(\widehat{G}(Y_{i,L}) - \frac{i-0.5}{L} \right)^2 + \frac{1}{12L} = \delta. \tag{3}$$

Here, $N_u = \sum_{i=1}^{n} \mathbf{1}\{X_i > u\}$ is the number of observations which exceed a predetermined high threshold u. $\widehat{G}(Y_{i,L})$ is determined by $G(t) = 1 - \theta \exp(-\theta t)$ with the replacement of θ by some estimate $\widehat{\theta}(u)$ and t by the order statistic $Y_{i,L}$, [15]. An appropriate value of the threshold u can be found as a solution of the discrepancy equation (3) with a predetermined value δ with regard to any consistent nonparametric estimator of EI. The calculation (3) by an entire sample may lead to the lack of a solution of the discrepancy equation regarding u the same way as for the heavy-tailed pdf estimation or to too large values u which may not be appropriate for the estimation of θ. The selection of k and δ remains a problem. We aim to obtain a limit distribution of the discrepancy statistic related to (3) depending on k and to use its quantiles as δ.

The paper is organized as follows. In Sect. 2, related work is recalled. In Sect. 3, a limit distribution of the normalized statistic $\hat{\omega}_L^2(u)$ is obtained, and an algorithm of the discrepancy method based on the M-S statistic is given. Simulation study is shown in Sect. 4. Conclusions are presented in Sect. 5.

[1] $L = 1$ holds when $\theta = 0$.

2 Related Work

Our achievements are based on the following Lemmas 3.4.1, 2.2.3 by [7] concerning limit distributions of the order statistics. They are recalled here.

Lemma 1 ([7, p. 89]) *Let $X, X_1, X_2, \ldots, X_n$ be iid rvs with common cdf F, and let $X_{1,n} \le X_{2,n} \le \cdots \le X_{n,n}$ be the n order statistics. The joint distribution of $\{X_{i,n}\}_{i=n-k+1}^{n}$ given $X_{n-k,n} = t$, for some $k \in \{1, \ldots, n-1\}$, equals the joint distribution of the set of the order statistics $\{X^*_{i,k}\}_{i=1}^{k}$ of iid rvs $\{X^*_i\}_{i=1}^{k}$ with cdf*

$$F_t(x) = P\{X \le x | X > t\} = (F(x) - F(t))/(1 - F(t)), \quad x > t. \tag{4}$$

Lemma 2 ([7, p. 41]) *Let $U_{1,n} \le U_{2,n} \le \cdots \le U_{n,n}$ be the n order statistics from a standard uniform distribution. Then, as $n \to \infty$, $k \to \infty$, $n - k \to \infty$,*

$$(U_{k,n} - b_n)/(a_n)$$

is asymptotically standard normal with

$$b_n = (k-1)/(n-1), \qquad a_n = \sqrt{b_n(1-b_n)/(n-1)}.$$

Definition 2 ([9]) For real u and integers $1 \le k \le l$, let $\mathscr{F}_{k,l}(u)$ be the σ-field generated by the events $\{X_i > u\}$, $k \le i \le l$. Define the mixing coefficients $\alpha_{n,q}(u)$,

$$\alpha_{n,q}(u) = \max_{1 \le k \le n-q} \sup |P(B|A) - P(B)|,$$

where the supremum is taken over all $A \in \mathscr{F}_{1,k}(u)$ with $P(A) > 0$ and $B \in \mathscr{F}_{k+q,n}(u)$ and k, q are positive integers.

Theorem 1 ([9, p. 547]) *Let $\{X_n\}_{n \ge 1}$ be a stationary process of rvs with tail function $\overline{F}(x) = 1 - F(x)$. Let the positive integers $\{r_n\}$ and the thresholds $\{u_n\}$, $n \ge 1$ be such that $r_n \to \infty$, $r_n \overline{F}(u_n) \to \tau$, and $P\{M_{r_n} \le u_n\} \to exp(-\theta\tau)$ hold as $n \to \infty$ for some $\tau \in (0, \infty)$ and $\theta \in [0, 1]$. If there are positive integers $q_n = o(r_n)$ such that $\alpha_{cr_n, q_n}(u_n) = o(1)$ for any $c > 0$, then we get for $t > 0$*

$$P\{\overline{F}(u_n)T_1(u_n) > t\} \to \theta \exp(-\theta t), \qquad n \to \infty.$$

For declustering the sample into approximately independent inter-cluster times $\{(T_1(u))_i\}$, one can take $k - 1 = \lfloor \theta \sum_{i=1}^{n} \mathbf{1}(X_i > u) \rfloor$ of the largest inter-exceedance times, [9]. The larger u corresponds to larger inter-cluster times whose number $L(u)$ may be small. This leads to a larger variance of the estimates based on $\{(T_1(u))_i\}$.

The intervals estimator of the EI follows from Theorem 1 and depends only on u. It is defined as [3, p. 391],

$$\hat{\theta}_n(u) = \begin{cases} \min(1, \hat{\theta}_n^1(u)), & \text{if } \max\{(T_1(u))_i : 1 \le i \le L-1\} \le 2, \\ \min(1, \hat{\theta}_n^2(u)), & \text{if } \max\{(T_1(u))_i : 1 \le i \le L-1\} > 2, \end{cases} \tag{5}$$

$$\hat{\theta}_n^1(u) = \frac{2(\sum_{i=1}^{L-1}(T_1(u))_i)^2}{(L-1)\sum_{i=1}^{L-1}(T_1(u))_i^2},$$

$$\hat{\theta}_n^2(u) = \frac{2(\sum_{i=1}^{L-1}((T_1(u))_i - 1))^2}{(L-1)\sum_{i=1}^{L-1}((T_1(u))_i - 1)((T_1(u))_i - 2)}.$$

The intervals estimator is derived to be consistent for m-dependent processes, [9]. Asymptotic normality of $\sqrt{k_n}(\hat{\theta}_n(u) - \theta)$, where $k_n = \lfloor n/r_n \rfloor$ and $r_n \to \infty$, $r_n = o(n)$ as $n \to \infty$ is proved for the intervals estimator in [18].

3 Main Results

3.1 Theory

Let us rewrite the left-hand side of (3) in the following form:

$$\hat{\omega}_L^2(u) = \sum_{i=L-k+1}^{L} \left(1 - \theta \exp(-Y_{i,L}\theta) - \frac{i-0.5}{L}\right)^2 + \frac{1}{12L}.$$

We disregard the marginal distribution of the random number of inter-cluster times $L = L(u)$. This approach is referred to as a conditional one. This is possible since the limit distribution of $L(u_n)$ does not depend on u_n as $n \to \infty$. Following [3], Sect. 10.3.1, the probability $P\{L(u_n) = i\}$ may be approximated by a binomial distribution with probability $p_n^* = P\{M_{r_n} \le u_n\}$ that tends to $e^{-\tau\theta}$ by (2). Here, r_n denotes the length of a data block. The cluster is defined as a block of data with at least one exceedance over u_n. The same is true for the cluster defined as in [9]. We have

$$1 - \theta \exp(-Y_{i,L}\theta) =^d U_{i,L}, \tag{6}$$

where $\{U_{i,L}\}$ are order statistics derived from an iid uniform [0, 1] sample $\{U_i\}_{i=1}^L$.

By Lemma 1, the joint distribution of upper order statistics $(U_{L-k+1,L}, \ldots, U_{L,L})$ given $U_{L-k,L} = t, t \in [0, 1)$, for some $k = 1, \ldots, L-1$, equals the joint distribution of order statistics $(U_{1,k}^*, \ldots, U_{k,k}^*)$ of associated iid rvs $\{U_i^*\}_{i=1}^k$ with cdf

$$F_t(x) = (x-t)/(1-t), \qquad 0 \le t < x \le 1. \tag{7}$$

Lemma 3 derives the normal distribution of $U^*_{i,k}$ given $U_{L-k,L} = t$ after normalization.

Lemma 3 *Let $U^*_{1,k} \le U^*_{2,k} \le \cdots \le U^*_{k,k}$ be the k order statistics of iid rvs $\{U^*_i\}_{i=1}^k$ with cdf (7). Then, as $k \to \infty$, $i \to \infty$, $k-i \to \infty$,*

$$(U^*_{i,k} - b_{i,k})/a_{i,k}$$

is asymptotically standard normal with

$$b^*_{i,k} = \frac{i-1}{k-1} = \frac{b_{i,k}-t}{1-t}, \qquad a_{i,k} = (1-t)\sqrt{\frac{b^*_{i,k}(1-b^*_{i,k})}{k-1}}. \tag{8}$$

Proof The pdf of $(U^*_{i,k} - b_{i,k})/a_{i,k}$ is given by [2]

$$\begin{aligned}
&\frac{k!}{(i-1)!(k-i)!} \cdot f(a_{i,k}x + b_{i,k}) F_t(a_{i,k}x + b_{i,k})^{i-1}(1 - F_t(a_{i,k}x + b_{i,k}))^{k-i} \\
&= \frac{k!}{(i-1)!(k-i)!} \cdot \frac{a_{i,k}}{1-t} \cdot \left(\frac{xa_{i,k} + b_{i,k} - t}{1-t}\right)^{i-1} \cdot \left(1 - \frac{xa_{i,k} + b_{i,k} - t}{1-t}\right)^{k-i} \qquad (9) \\
&= \left(\frac{k!}{(i-1)!(k-i)!}(b^*_{i,k})^{i-1}(1-b^*_{i,k})^{k-i}\frac{a_{i,k}}{1-t}\right) \\
&\cdot \left(1 + \frac{xa_{i,k}}{b^*_{i,k}(1-t)}\right)^{i-1}\left(1 - \frac{xa_{i,k}}{(1-b^*_{i,k})(1-t)}\right)^{k-i}.
\end{aligned}$$

In the same way as in the proof of Lemma 2.2.3 [7], we obtain

$$\begin{aligned}
&(i-1)\log\left(1 + \frac{xa_{i,k}}{b^*_{i,k}(1-t)}\right) + (k-i)\log\left(1 - \frac{xa_{i,k}}{(1-b^*_{i,k})(1-t)}\right) \\
&= (i-1)\left(x\frac{a_{i,k}}{b^*_{i,k}(1-t)} - \frac{x^2}{2}\left(\frac{a_{i,k}}{b^*_{i,k}(1-t)}\right)^2 + \cdots\right) \\
&+ (k-i)\left(-x\frac{a_{i,k}}{(1-b^*_{i,k})(1-t)} - \frac{x^2}{2}\left(\frac{a_{i,k}}{(1-b^*_{i,k})(1-t)}\right)^2 - \cdots\right).
\end{aligned}$$

From (8) it follows $(i-1)\left(\frac{a_{i,k}}{b^*_{i,k}(1-t)}\right)^2 + (k-i)\left(\frac{a_{i,k}}{(1-b^*_{i,k})(1-t)}\right)^2 = 1$. The other terms are of smaller order. Using Stirlings's formula for $k!$ we find that the factor in the third string of (9) tends to $(2\pi)^{-1/2}$. Thus, the statement follows.

Lemma 4 *Let the conditions of Lemma 3 be fulfilled. Then the statistic*

$$\chi^2 = \sum_{i=1}^{k^*} \left((U^*_{i,k} - b_{i,k})/a_{i,k}\right)^2 \tag{10}$$

is asymptotically χ^2 *distributed with* $k^* = [k/2]$ *degrees of freedom.*

Proof Let us denote $Y^*_{i,k} = (U^*_{i,k} - b_{i,k})/a_{i,k}$ and obtain the distribution $\Phi_\zeta(y)$ of $\zeta = \chi/\sqrt{k}$. From Lemma 3 $Y^*_{i,k} \sim N(0,1)$ holds asymptotically. Due to the symmetry, we get $Y^*_{1,k} \le \cdots \le Y^*_{k^*,k}$ and $Y^*_{k,k} \le \cdots \le Y^*_{k^*+1,k}$ for odd k ($Y^*_{k,k} \le \cdots \le Y^*_{k^*,k}$ for even k). By Lemma 1, the joint pdf of the k^* order statistics $Y^*_{1,k}, \ldots, Y^*_{k^*,k}$ is determined by [2]

$$f(x_1, \ldots, x_{k^*}) = (k^*)! \prod_{i=1}^{k^*} f(x_i) = \frac{(k^*)!}{(\sqrt{2\pi})^{k^*}} \exp\left(-\frac{\sum_{i=1}^{k^*} x_i^2}{2}\right),$$
$$t < x_1 < x_2 < \cdots < x_{k^*} < +\infty.$$

For positive y, the cdf $\Phi_\zeta(y)$ is equal to the probability to fall inside the k^*-dimensional sphere $\sum_{i=1}^{k^*} \left(Y^*_{i,k}\right)^2 = y^2\sqrt{k^*}$. For negative y, we have $\Phi_\zeta(y) = 0$. Hence, we obtain

$$\Phi_\zeta(y) = \frac{(k^*)!}{(\sqrt{2\pi})^{k^*}} \cdot$$
$$\int \cdots \int_{\sum_{i=1}^{k^*} x_i^2 < y^2 k^*} \exp\left(-\frac{\sum_{i=1}^{k^*} x_i^2}{2}\right) \mathbf{1}(t < x_1 < x_2 < \cdots < x_{k^*} < +\infty) dx_1 dx_2 \ldots dx_{k^*}.$$

Using spherical coordinates and replacing $x_1 = \rho\cos\theta_1\cos\theta_2\ldots\cos\theta_{k^*-1}$, $x_2 = \rho\cos\theta_1\cos\theta_2\ldots\sin\theta_{k^*-1}$, $\ldots$, $x_{k^*-1} = \rho\cos\theta_1\sin\theta_2$, $x_{k^*} = \rho\sin\theta_1$, we find the intervals of each variable ρ and θ_i, $i = 1, \ldots, k^* - 1$. Since $t < x_1 < x_2 < \cdots < x_{k^*} < +\infty$ holds, we get $x_2/x_1 = \tan(\theta_{k^*-1}) > 1$ and hence, $\pi/4 < \theta_{k^*-1} < \pi/2$, $\sqrt{2}/2 < \sin(\theta_{k^*-1}) < 1$. From $x_3/x_2 = \tan(\theta_{k^*-2})/\sin(\theta_{k^*-1}) > 1$ we then have $\tan(\theta_{k^*-2}) > \sin(\theta_{k^*-1})$, and since the largest value of $\sin(\theta_{k^*-1})$ is 1, we get $\tan(\theta_{k^*-2}) > 1$ and $\pi/4 < \theta_{k^*-2} < \pi/2$. Finally, we get $\pi/4 < \theta_i < \pi/2, i = 1, \ldots, k^* - 1$. Now, we may take the following integral:

$$\Phi_\zeta(y) = \frac{(k^*)!}{(\sqrt{2\pi})^{k^*}} \cdot$$
$$\int_{\pi/4}^{\pi/2} \cdots \int_{\pi/4}^{\pi/2} \int_0^{y\sqrt{k^*}} \exp\left(-\frac{\rho^2}{2}\right) \rho^{k^*-1} D(\theta_1, \ldots, \theta_{k^*-1}) d\rho d\theta_{k^*-1} \ldots d\theta_1$$
$$= C_{k^*} \int_0^{y\sqrt{k^*}} \exp\left(-\frac{\rho^2}{2}\right) \rho^{k^*-1} d\rho,$$

where

$$C_{k^*} = \frac{(k^*)!}{(\sqrt{2\pi})^{k^*}} \int_{\pi/4}^{\pi/2} \cdots \int_{\pi/4}^{\pi/2} D(\theta_1, \ldots, \theta_{k^*-1}) d\theta_{k^*-1} \ldots d\theta_1.$$

Then the constant C_{k^*} can be obtained from the equation:

$$\Phi_\zeta(+\infty) = 1 = C_{k^*} \int_0^\infty \exp\left(-\frac{\rho^2}{2}\right) \rho^{k^*-1} d\rho = C_{k^*} \Gamma\left(\frac{k^*}{2}\right) 2^{k^*/2-1}.$$

Hence, it follows

$$\Phi_\zeta(y) = \frac{1}{\Gamma(k^*/2)\, 2^{k^*/2-1}} \int_0^{y\sqrt{k^*}} \exp\left(-\frac{\rho^2}{2}\right) \rho^{k^*-1} d\rho.$$

The pdf of ζ is given by

$$\varphi_\zeta(y) = \frac{\sqrt{2k^*}}{\Gamma(k^*/2)} \left(\frac{y\sqrt{k^*}}{\sqrt{2}}\right)^{k^*-1} \exp\left(-\frac{k^* y^2}{2}\right).$$

Hence, one can get the chi-squared pdf of χ^2:

$$p(x) = \frac{x^{k^*/2-1} \exp(-x/2)}{2^{k^*/2} \Gamma(k^*/2)}.$$

3.2 Discrepancy Equation Based on the Chi-Squared Statistic

Regarding a consistent estimate $\widehat{\theta}(u)$ by (6) and denoting $i^* = i - L + k^*$, $i = L - k^* + 1, \ldots, L$, then u can be selected as a solution of the discrepancy equation

$$\sum_{i^*=2}^{k^*-1} \left((1 - \widehat{\theta}(u) \exp(-Y_{i^*+L-k^*,L} \widehat{\theta}(u)) - b_{i^*,k^*})/a_{i^*,k^*}\right)^2 = \delta \quad (11)$$

for a given k such that $k^* - 1 \geq 2$ and $k^* = [(k-2)/2]$. Here δ is a mode $\max\{k^* - 2, 0\}$ of the $\chi^2(k^*)$ distribution and $t = 1 - \widehat{\theta}(u) \exp(-Y_{L-k,L} \widehat{\theta}(u))$. $b_{i,k}$, $a_{i,k}$, and k^* are calculated as in Lemmas 3 and 4. This could be an alternative to the method (3). In Fig. 1a, c, one can see that the empirical cdf of the left-hand side statistic in (11), where $\widehat{\theta}(u)$ is based on (5), and a chi-squared cdf are rather close. Figure 1b, d shows that the discrepancy equation (11) may have solutions since the left-hand side statistic in (11) crosses the mode of the χ^2-distribution.

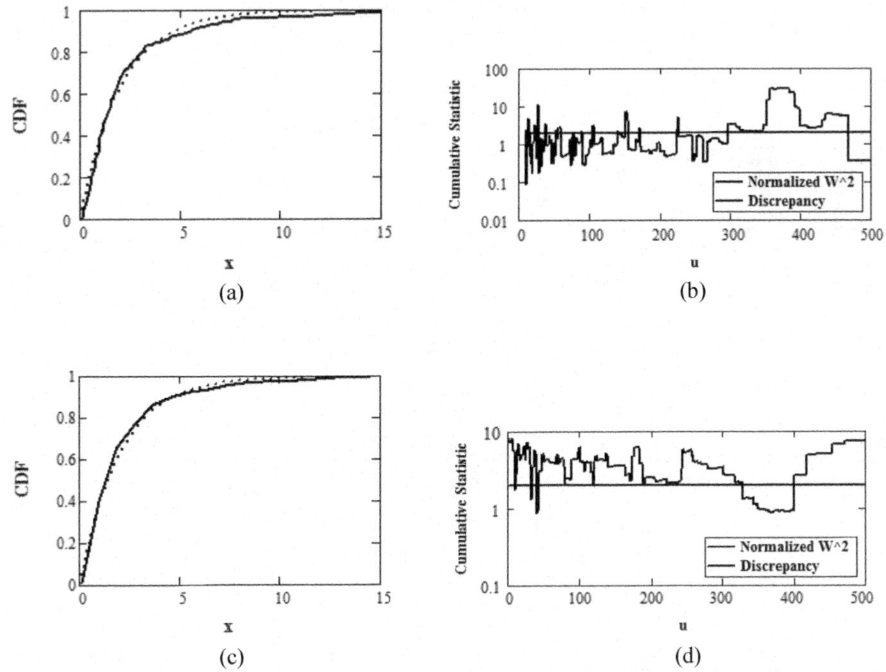

Fig. 1 Empirical cdf of the left-hand side of (11) built by 300 re-samples by a Moving Maxima (MM) process with sample size $n = 4 \cdot 10^4$, the EI $\theta = 0.8$, $k^* = 5$, $t = 0.867$, and $u = 10$ (solid line) and chi-squared cdf (points), Fig. 1a; Left-hand side statistic in (11) for $k^* = 5$ against threshold u and the χ^2 mode as the discrepancy, Fig. 1(b). The same statistics for an ARMAX process with $\theta = 0.25$, $k^* = 5$, $t = 0.953$, and $u = 50$, Fig. 1c and for $k^* = 5$ Fig. 1d

Remark 1 The discrepancy methods (3) and (11) are universal and can be used for any nonparametric estimator $\widehat{\theta}(u)$.

Algorithm 3.1 *1. Using* $X^n = \{X_1, X_2, \ldots, X_n\}$ *and taking thresholds* u *corresponding to quantile levels* $q \in \{0.90, 0.905, \ldots, 0.995\}$, *generate samples of inter-cluster times* $\{T_1(u)_i\}$ *and the normalized rvs*

$$\{Y_i\} = \{\overline{F}(u)T_1(u)_i\} = \{(N/n)T_1(u)_i\}, \qquad i = \overline{1, L}, \qquad L = L(u),$$

where N *is the number of exceedances over threshold* u.

2. *For each* u, *select* $k = sL(u)$, $0 < s < 1$, *e.g.,* $s = 0.0001$.
3. *Use a sorted sample* $Y_{L-k+1,L} \leq \cdots \leq Y_{L,L}$ *and find all solutions* $u_1, \ldots, u_l$ *(here,* l *is a random number) of the discrepancy equation (11).*
4. *For each* u_j, $j \in \{1, \ldots, l\}$, *calculate* $\widehat{\theta}(u_j)$ *and find*

$$\widehat{\theta}_1(u) = \frac{1}{l}\sum_{i=1}^{l}\widehat{\theta}(u_i), \quad \widehat{\theta}_2(u) = \widehat{\theta}(u_{min}), \qquad \widehat{\theta}_3(u) = \widehat{\theta}(u_{max})$$

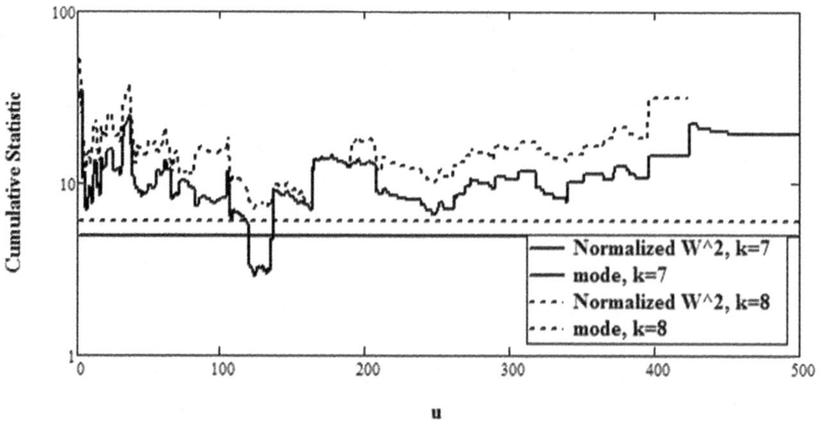

Fig. 2 Left-hand side statistic in (11) for $k^* = 7, 8$ against threshold u and the χ^2 mode $\max\{k^* - 2, 0\}$ as the discrepancy for an ARMAX process with $\theta = 0.75$

as resulting estimates, where u_{min} and u_{max} are the minimal and maximal values among $\{u_1, \ldots, u_l\}$.

3.3 Estimation of k

It remains to select k. For declustering purposes, i.e., to have approximately independent clusters of exceedances over u, it is recommended in [9] to take the largest value k such that $(T_1(u))_{L-k,L}$ is strictly larger than $(T_1(u))_{L-k-1,L}$.

We propose another approach. For each predetermined threshold u and for a corresponding $L(u)$, one may decrease the k-value until the discrepancy equations have solutions and select the largest one among such k's. Figure 2 shows that the solution of (11) exists for $k^* = 7$ and it does not for $k^* = 8$. Due to several possible solutions u, the average may be taken over all estimates $\widehat{\theta}(u)$ with such u's.

4 Simulation Study

Our simulation study, enhancing the behavior of Algorithm 3.1 is based on 1000 replicas of samples $\{X_1, \ldots, X_n\}$ with size $n = 10^5$ generated from a set of models. These models are Moving Maxima (MM), Autoregressive Maximum (ARMAX), AR(1), AR(2), MA(2), and GARCH. The AR(1) process is considered with uniform noise (ARu) and with Cauchy distributed noise (ARc). Using Algorithm 3.1, we check the accuracy of the intervals estimator (5), where u is selected based on (11).

The root mean squared error (RMSE) and the absolute bias are given in Tables 1 and 2. The best results are shown in bold numbers.

4.1 Models

Let us shortly recall the processes under study. The mth order MM process is $X_t = \max_{i=0,\ldots,m}\{\alpha_i Z_{t-i}\}$, $t \in \mathbb{Z}$, where $\{\alpha_i\}$ are constants with $\alpha_i \geq 0$, $\sum_{i=0}^{m}\alpha_i = 1$, and Z_t are iid standard Fréchet distributed rvs with the cdf $F(x) = \exp(-1/x)$, for $x > 0$. Its EI is equal to $\theta = \max_i\{\alpha_i\}$, [1]. Values $m = 3$ and $\theta \in \{0.5, 0.8\}$ corresponding to $\alpha = (0.5, 0.3, 0.15, 0.05)$ and $\alpha = (0.8, 0.1, 0.008, 0.02)$ are taken.

The ARMAX process is determined as $X_t = \max\{\alpha X_{t-1}, (1-\alpha)Z_t\}$, $t \in \mathbb{Z}$, where $0 \leq \alpha < 1$, $\{Z_t\}$ are i.i.d standard Fréchet distributed rvs and $P\{X_t \leq x\} = \exp(-1/x)$ holds assuming $X_0 = Z_0$. Its EI is given by $\theta = 1 - \alpha$, [3]. We consider $\theta \in \{0.25, 0.75\}$.

The ARu process is defined by $X_j = (1/r)X_{j-1} + \epsilon_j$, $j \geq 1$ and $X_0 \sim U(0, 1)$ with X_0 independent of ϵ_j. For a fixed integer $r \geq 2$, let ϵ_n, $n \geq 1$ be iid rvs with $P\{\epsilon_1 = k/r\} = 1/r$, $k = 0, 1, \ldots, r-1$. The EI of AR(1) is $\theta = 1 - 1/r$ [5]. $\theta \in \{0.5, 0.8\}$ are taken.

The MA(2) process is determined by $X_i = pZ_{i-2} + qZ_{i-1} + Z_i$, $i \geq 1$, with $p > 0, q < 1$, and iid Pareto rvs $Z_{-1}, Z_0, Z_1, \ldots$ with $P\{Z_0 > x\} = 1$ if $x < 1$, and $P\{Z_0 > x\} = x^{-\alpha}$ if $x \geq 1$ for some $\alpha > 0$ [20]. Its EI is $\theta = (1 + p^\alpha + q^\alpha)^{-1}$. We consider $\alpha = 2$, $(p, q) = (1/\sqrt{2}, 1/\sqrt{2}), (1/\sqrt{3}, 1/\sqrt{6})$ with corresponding $\theta = 1/2, 2/3$.

We consider also processes studied in [8, 17, 22]. The ARc process is $X_j = 0.7X_{j-1} + \epsilon_j$, where ϵ_j is standard Cauchy distributed and $\theta = 0.3$. The AR(2) process is $X_j = 0.95X_{j-1} - 0.89X_{j-2} + \epsilon_j$, where ϵ_j is Pareto distributed with tail index 2 and $\theta = 0.25$. The GARCH(1,1) is $X_j = \sigma_j\epsilon_j$, with $\sigma_j^2 = \alpha + \lambda X_{j-1}^2 + \beta\sigma_{j-1}^2$, $\alpha = 10^{-6}$, $\beta = 0.7$, $\lambda = 0.25$, with an iid sequence of standard Gaussian rvs $\{\epsilon_j\}_{j\geq 1}$ and $\theta = 0.447$ [11].

4.2 Estimators and Their Comparison

In Tables 1 and 2, we insert apart from our estimates $\widehat{\theta}_1(u) - \widehat{\theta}_3(u)$ the available results of the simulation study by [8, 17, 20, 21]. The estimators are notated as follows. $\widehat{\theta}^{db}$ denotes the disjoint blocks and $\widehat{\theta}^{sb}$ the sliding blocks estimators [17, 18]; $\widehat{\theta}^{r}$ the runs estimator [24]; $\widehat{\theta}^{ml}$ the multilevel estimator [20]; $\widehat{\theta}^{bcml}$ and $\widehat{\theta}^{mlsb}$ the bias-corrected multilevel and the multilevel sliding blocks estimators [20, 21]; $\widehat{\theta}^{C}$ and $\widehat{\theta}^{Cms}$ the cycles and the max-stable cycles estimators [8]. We can compare only results related to processes overlapping with our experiment. We calculate the

Table 1 The root mean squared error

$RMSE \cdot 10^4/\theta$	MM		ARMAX		ARu		MA(2)		ARc	AR(2)	GARCH
	0.5	0.8	0.25	0.75	0.5	0.8	0.5	2/3	0.3	0.25	0.328
$s = 0.001$											
$\widehat{\theta}_1$	146	188	136	173	2719	1217	227	545	66	**426**	237
$\widehat{\theta}_2$	141	164	122	156	2163	946	295	813	104	498	288
$\widehat{\theta}_3$	360	440	291	430	3330	1527	336	470	67	432	467
$s = 0.0005$											
$\widehat{\theta}_1$	105	155	100	148	2519	1127	268	696	50	498	**231**
$\widehat{\theta}_2$	139	144	97	151	1906	854	434	1107	161	692	860
$\widehat{\theta}_3$	355	451	296	463	3325	1527	355	449	67	439	484
$s = 0.0001$											
$\widehat{\theta}_1$	**96**	**120**	88	**115**	2224	975	331	846	151	620	416
$\widehat{\theta}_2$	149	135	117	143	**1704**	768	503	1197	358	953	1127
$\widehat{\theta}_3$	350	453	285	434	3331	1518	336	441	67	431	474
$\widehat{\theta}^{Kimt}$	217	569	**69**	498	1883	**199**	309	466	**33**	3630	4028
$\widehat{\theta}^{db}$	630		550	3640					320		1000
$\widehat{\theta}^{sb}$	550		450	950					320		840
$\widehat{\theta}^{r}$	550								140		1550
$\widehat{\theta}^{ml}$			550	3640			**220**	485			
$\widehat{\theta}^{bcml}$			400	1070			375	**210**			
$\widehat{\theta}^{mlsb}$			420	853							
$\widehat{\theta}^{C}$	660								950		1580
$\widehat{\theta}^{Cms}$	320								6080		3520

$K-$gaps estimates by [22] with IMT-selected pairs (u, K) (for details regarding the IMT test, see [10]) which are denoted as $\widehat{\theta}^{Kimt}$.

We may conclude the following. The intervals estimator coupling with the discrepancy method demonstrates a good performance in comparison with other investigated estimators. It is not appropriate for light-tailed distributed processes (by its definition) as one can see by the example of the ARu process. The K-gaps estimator is indicated as one of the most promising methods in [8], [17]. Our estimators, especially $\widehat{\theta}_1(u)$ for smaller s (that reflect the smaller number of the largest order statistics k), may perform better.

Comparing Tables 1 and 2 with Fig. 1 by [21], where the multilevel and the bias-corrected multilevel estimators were compared by data simulated from an ARMAX model only, one can see that the latter estimates demonstrate much larger accuracy values. Particularly, our estimate gives the best RMSE equal to 0.0088 and 0.0115 as far as the best among these estimates show about 0.04 and a bit less than 0.15 for $\theta = 0.25$ and $\theta = 0.75$, respectively.

Table 2 The absolute bias

$\lvert Bias \rvert$ $10^4/\theta$	MM		ARMAX		ARu		MA(2)		ARc	AR(2)	GARCH
	0.5	0.8	0.25	0.75	0.5	0.8	0.5	2/3	0.3	0.25	0.328
$s = 0.001$											
$\widehat{\theta}_1$	7.4016	**6.2708**	2.9461	**3.6254**	2709	1204	182	516	66	399	**70**
$\widehat{\theta}_2$	41	50	22	49	2139	921	267	791	104	477	139
$\widehat{\theta}_3$	38	40	36	62	3302	1474	72	246	67	314	174
$s = 0.0005$											
$\widehat{\theta}_1$	37	12	15	13	2513	1118	251	684	50	485	137
$\widehat{\theta}_2$	104	59	52	65	1893	841	424	1017	161	679	809
$\widehat{\theta}_3$	11	61	48	45	3296	1474	47	227	67	319	210
$s = 0.0001$											
$\widehat{\theta}_1$	56	35	48	43	2221	970	322	842	151	613	391
$\widehat{\theta}_2$	126	81	97	97	**1701**	760	497	1194	358	949	1119
$\widehat{\theta}_3$	43	59	31	43	3303	1464	51	246	67	**304**	186
$\widehat{\theta}^{Kimt}$	**0.14862**	567	54	496	1878	**196**	306	462	**33**	3627	4027
$\widehat{\theta}^{db}$	160		450	3530					80		690
$\widehat{\theta}^{sb}$	100		180	340					80		630
$\widehat{\theta}^{r}$	50								160		630
$\widehat{\theta}^{ml}$			450	4170			87.5	270			
$\widehat{\theta}^{bcml}$			**0**	1070			**40**	**25**			
$\widehat{\theta}^{mlsb}$			179	320							
$\widehat{\theta}^{C}$	130								200		230
$\widehat{\theta}^{Cms}$	20								6000		3230

In [8] the cycles, the max-stable cycles, the runs, the K-gaps, the disjoint blocks, and sliding blocks estimators were compared. For the first three estimators, the misspecification IMT test was applied as a choice method of the threshold-run parameter. As an alternative, quantiles $q \in \{0.95, 0.975, 0.90\}$ were used for these estimators as thresholds with the run parameter estimated by the latter test. We can compare only results related to MM with $\theta = 0.5$, ARMAX with $\theta = 0.75$, and AR(1) with $\theta = 0.5$ processes. The best bias equal to 0.002 and the RMSE equal to 0.032 for an MM process were achieved by the max-stable cycles estimator $\widehat{\theta}^{Cms}$. For our estimator, the best absolute bias is 0.00074 and the RMSE is 0.0096 for an MM process. For an ARMAX process, the best were the cycles estimated with the bias equal to 0.003 and the max-stable cycles estimated with the RMSE equal to 0.032. Our estimator provides the best absolute bias 0.00036 and the RMSE 0.0115.

The MA(2) process has been studied in [20] regarding the multilevel and the bias-corrected multilevel blocks estimators with two specific weighted functions. For MA(2) with $\theta = 0.5$, the obtained best absolute bias is inside the interval $(2.5, 3.0) \cdot 10^{-3}$, and the MSE is in $(0.75, 1.0) \cdot 10^{-3}$. Our estimator provides the best absolute bias $4.7 \cdot 10^{-3}$ and the MSE $5.15259 \cdot 10^{-4}$. For MA(2) with $\theta = 2/3$, we find in

[20] the bias about 0.0025 and the MSE $0.7 \cdot 10^{-3}$. Our estimator shows 0.0227 and $2.025 \cdot 10^{-3}$ for the bias and the MSE, respectively.

5 Conclusions

The discrepancy method proposed for smoothing of pdf estimates is modified to select the threshold parameter u for the EI estimation. We derive the χ^2 asymptotic distribution of the statistic relating to the M-S statistic. This allows us to use its mode as an unknown discrepancy value δ. Since the discrepancy method may be applied for different estimators of the EI, one can find other parameters such as the block size for the blocks estimator of the EI instead of the threshold in the same way. The accuracy of the intervals estimator (5) with u selected by the new discrepancy method (11) is provided by a simulation study. The comparison with several EI estimators shows its good performance regarding heavy-tailed distributed processes.

Acknowledgments The author appreciates the partial financial support by the Russian Foundation for Basic Research, grant 19-01-00090.

References

1. Ancona-Navarrete, M.A., Tawn, J.A.: A comparison of methods for estimating the extremal index. Extremes **3**(1), 5–38 (2000)
2. Balakrishnan, N., Rao, C.R. (eds.): Handbook of Statistics, vol. 16. Elsevier Science B.V, Amsterdam (1998)
3. Beirlant, J., Goegebeur, Y., Teugels, J., Segers, J.: Statistics of Extremes: Theory and Applications. Wiley, Chichester (2004)
4. Bolshev, L.N., Smirnov, N.V.: Tables of Mathematical Statistics. Nauka, Moscow (1965). (in Russian)
5. Chernick, M.R., Hsing, T., McCormick, W.P.: Calculating the extremal index for a class of stationary. Adv. Appl. Prob. **23**, 835–850 (1991)
6. Drees, H.: Bias correction for estimators of the extremal index (2011). arXiv:1107.0935
7. de Haan, L., Ferreira, A.: Extreme Value Theory: An Introduction. Springer (2006)
8. Ferreira, M.: Analysis of estimation methods for the extremal index. Electron. J. Appl. Stat. Anal. **11**(1), 296–306 (2018)
9. Ferro, C.A.T., Segers, J.: Inference for clusters of extreme values. J. R. Stat. Soc. B. **65**, 545–556 (2003)
10. Fukutome, S., Liniger, M.A., Süveges, M.: Automatic threshold and run parameter selection: a climatology for extreme hourly precipitation in Switzerland. Theor. Appl. Climatol. **120**, 403–416 (2015)
11. Laurini, F., Tawn, J.A.: The extremal index for GARCH(1,1) processes. Extremes **15**, 511–529 (2012)
12. Leadbetter, M.R., Lingren, G., Rootzén, H.: Extremes and Related Properties of Random Sequence and Processes. Springer, New York (1983)
13. Markovich, N.M.: Experimental analysis of nonparametric probability density estimates and of methods for smoothing them. Autom. Remote. Control. **50**, 941–948 (1989)

14. Markovich, N.M.: Nonparametric Analysis of Univariate Heavy-Tailed data: Research and Practice. Wiley (2007)
15. Markovich, N.M.: Nonparametric estimation of extremal index using discrepancy method. In: Proceedings of the X International conference System identification and control problems" SICPRO-2015 Moscow. V.A. Trapeznikov Institute of Control Sciences, 26–29 January 2015, pp. 160–168 (2015). ISBN 978-5-91450-162-1
16. Markovich, N.M.: Nonparametric estimation of heavy-tailed density by the discrepancy method. In: Cao, R. et al. (eds.) Nonparametric Statistics. Springer Proceedings in Mathematics & Statistics, vol. 175, pp. 103–116. Springer International Publishing, Switzerland (2016)
17. Northrop, P.J.: An efficient semiparametric maxima estimator of the extremal index. Extremes **18**(4), 585–603 (2015)
18. Robert, C.Y.: Asymptotic distributions for the intervals estimators of the extremal index and the cluster-size probabilities. J. Stat. Plan. Inference **139**, 3288–3309 (2009)
19. Robert, C.Y., Segers, J., Ferro, C.A.T.: A sliding blocks estimator for the extremal index. Electron. J. Stat. **3**, 993–1020 (2009)
20. Sun, J., Samorodnitsky, G.: Estimating the extremal index, or, can one avoid the threshold-selection difficulty in extremal inference? Technical report, Cornell University (2010)
21. Sun, J., Samorodnitsky, G.: Multiple thresholds in extremal parameter estimation. Extremes (2018). https://doi.org/10.1007/s10687-018-0337-5
22. Süveges, M., Davison, A.C.: Model misspecification in peaks over threshold analysis. Ann. Appl. Stat. **4**(1), 203–221 (2010)
23. Vapnik, V.N., Markovich, N.M., Stefanyuk, A.R.: Rate of convergence in L_2 of the projection estimator of the distribution density. Autom. Remote. Control. **53**, 677–686 (1992)
24. Weissman, I., Novak, SYu.: On blocks and runs estimators of the extremal index. J. Stat. Plan. Inference **66**, 281–288 (1978)

Correction for Optimisation Bias in Structured Sparse High-Dimensional Variable Selection

Bastien Marquis and Maarten Jansen

Abstract In sparse high-dimensional data, the selection of a model can lead to an overestimation of the number of nonzero variables. Indeed, the use of an ℓ_1 norm constraint while minimising the sum of squared residuals tempers the effects of false positives, thus they are more likely to be included in the model. On the other hand, an ℓ_0 regularisation is a non-convex problem and finding its solution is a combinatorial challenge which becomes unfeasible for more than 50 variables. To overcome this situation, one can perform selection via an ℓ_1 penalisation but estimate the selected components without shrinkage. This leads to an additional bias in the optimisation of an information criterion over the model size. Used as a stopping rule, this IC must be modified to take into account the deviation of the estimation with and without shrinkage. By looking into the difference between the prediction error and the expected Mallows's Cp, previous work has analysed a correction for the optimisation bias and an expression can be found for a signal-plus-noise model given some assumptions. A focus on structured models, in particular, grouped variables, shows similar results, though the bias is noticeably reduced.

1 Introduction

This paper falls under the scope of variable selection in high-dimensional data. The number of variables might then be larger than the number of observations, thus leading to difficulties in the computation of the models. Indeed, the classical tools for low-dimensional data can no longer be used. In order to find a solution, the assumption of sparsity is often made, meaning that the number of nonzero variables is supposed to be smaller than the number of observations. Besides an improvement in calculations, another benefit from sparsity results from the fact that the obtained

B. Marquis (✉) · M. Jansen
Universit libre de Bruxelles, Brussels, Belgium
e-mail: bastien.marquis@ulb.ac.be

M. Jansen
e-mail: maarten.jansen@ulb.ac.be

M. La Rocca et al. (eds.), *Nonparametric Statistics*, Springer Proceedings
in Mathematics & Statistics 339, https://doi.org/10.1007/978-3-030-57306-5_32

models are more interpretable. Finding the nonzeros becomes the main objective. The obvious method would be to test all subsets; however, this proves to be unfeasible for more than a few dozen variables. More greedy algorithms exist, such as forward selection or backward elimination, but they are likely to miss the true model in addition to remaining computationally heavy. An other way to tackle the problem is to use regularisation on the sum of squared residuals. For a linear model $\mathbf{Y} = \mathbf{X}\boldsymbol{\beta} + \boldsymbol{\varepsilon}$ with $\mathbf{Y}, \boldsymbol{\varepsilon} \in \mathbb{R}^n$ and $\boldsymbol{\beta} \in \mathbb{R}^m$, this consists in finding the estimator $\widehat{\boldsymbol{\beta}}$ that solves an equation usually of the form:

$$\min_{\boldsymbol{\beta}} \frac{1}{n} \|\mathbf{Y} - \mathbf{X}\boldsymbol{\beta}\|_2^2 + \lambda \|\boldsymbol{\beta}\|_p^p,$$

where the regularisation parameter λ is a kind of 'budget': acting as a weight, it limits the norm of $\widehat{\boldsymbol{\beta}}$. Ridge regression and Lasso are special cases where $p = 2$ and $p = 1$, respectively, and their solutions can be calculated even when $\mathbf{X}$ is not of full rank.

Ideally, one would perform ℓ_0 regularisation, i.e. penalising the sum of squared residuals by the number of nonzero variables, but this is a combinatorial problem, and therefore intractable from the computational point of view. On the other hand, Lasso [8] and other ℓ_1-type regularisations (penalisation by the sum of the absolute values of the selected variables) offer a quadratic programming alternative whose solution is still a proper variable selection, as it contains many zeros.

Lasso has many advantages. First, it applies shrinkage on the variables which can lead to better predictions than simple least squares due to Stein's phenomenon [7]. Second, the convexity of its penalty means Lasso can be solved numerically. Third, ℓ_1 regularisation is variable selection consistent under certain conditions, provided that the coefficients of the true model are large enough compared to the regularisation parameter [6, 10, 14]. Fourth, Lasso can take into account structures; simple modifications of its penalisation term result in structured variable selection. Such variations among others are the fused lasso [9], the graphical lasso [3] and the composite absolute penalties [13] including the group-lasso [12]. Fifth, for a fixed regularisation parameter, ℓ_1 regularisation has nearly the same sparsity as ℓ_0 regularisation [2]. However, this does not hold anymore when the regularisation parameter is optimised in a data-dependent way, using an information criterion such as AIC [1] or Mallows's Cp [5].

Mallows's Cp, like many other information criteria, takes the form of a penalised—likelihood or—sum of squared residuals whose penalty depends on the number of selected variables. Therefore, among all models of equal size, the selection is based on the sum of squared residuals. Because of sparsity, it is easy to find a well-chosen combination of falsely significant variables that reduces the sum of squared residuals, by fitting the observational errors. The effects of these false positives can be tempered by applying shrinkage. The optimisation of the information criterion then overestimates the number of variables needed, including too many false positives in the model. In order to avoid this scenario, one idea could be to combine both ℓ_1 and ℓ_0 regularisation: the former to select the nonzeros and the latter to estimate their value. The optimal balance between the sum of squared residuals and the ℓ_0 regularisation should shift towards smaller models. Of course, this change must be taken into account and the expression of the information criterion consequently adapted. The correction for the difference between ℓ_0 and ℓ_1 regularisation has been described as a 'mirror' effect [4].

In Sect. 2, we explain more the mirror effect. The main contribution of this paper follows and concerns the impact of a structure among the variables and how it affects the selection. More precisely, the behaviour of the mirror effect for unstructured and structured signal-plus-noise models is investigated in Sect. 3. A simulation is presented in Sect. 4 to support the previous sections.

2 The Mirror Effect

Consider the linear model

$$\mathbf{Y} = \mathbf{X}\boldsymbol{\beta} + \boldsymbol{\varepsilon}$$

where $\boldsymbol{\varepsilon}$ is a n-vector of independent and identically distributed errors with $E(\varepsilon_i) = 0$ and $\mathrm{var}(\varepsilon_i) = \sigma^2$ for $i = 1, \ldots, n$. The design matrix $\mathbf{X}$ has size $n \times m$ with n possibly smaller than m. We assume $\boldsymbol{\beta} \in \mathbb{R}^m$ is sparse in the sense that the unknown number of nonzeros n_1 in $\boldsymbol{\beta}$ is smaller than n. For a given k, let $\mathbf{S}^k$ be a binary m-vector with $m - k$ zeros, provided by a procedure $\mathcal{S}(\mathbf{Y}, k)$ which can be unstructured best k selection or any structured procedure. An example of such a procedure could be an implementation of Lasso with the regularisation parameter fine-tuned to obtain the appropriate model size. Also, define $\mathbf{O}^k$ as the selection found by an oracle knowing $\mathbf{X}\boldsymbol{\beta}$ without noise, using the same procedure as for $\mathbf{S}^k$, i.e. $\mathbf{O}^k = \mathcal{S}(\mathbf{X}\boldsymbol{\beta}, k)$. The notations $\mathbf{X}_{\mathbf{S}^k}$ and $\mathbf{X}_{\mathbf{O}^k}$ are used for the $n \times k$ submatrices of $\mathbf{X}$ containing the k columns corresponding to the 1s in $\mathbf{S}^k$ and $\mathbf{O}^k$, respectively.

One way to look at the mirror effect comes from investigating the difference between the expected average squared prediction error and Mallows's Cp criterion[1]. The quality of the oracle least squares projection, $\widehat{\boldsymbol{\beta}}_{\mathbf{O}^k} = (\mathbf{X}_{\mathbf{O}^k}^T \mathbf{X}_{\mathbf{O}^k})^{-1} \mathbf{X}_{\mathbf{O}^k}^T \mathbf{Y}$, is measured by the prediction error $\mathrm{PE}(\widehat{\boldsymbol{\beta}}_{\mathbf{O}^k})$ which can be, in turn, estimated unbiasedly by $\Delta_p(\widehat{\boldsymbol{\beta}}_{\mathbf{O}^k})$, where, for a generic selection $\mathbf{S}$,

$$\mathrm{PE}(\widehat{\boldsymbol{\beta}}_{\mathbf{S}}) = \frac{1}{n} E\left(\|\mathbf{X}\boldsymbol{\beta} - \mathbf{X}_{\mathbf{S}}\widehat{\boldsymbol{\beta}}_{\mathbf{S}}\|_2^2\right) \tag{1}$$

and $\Delta_p(\widehat{\boldsymbol{\beta}}_{\mathbf{S}})$ is a non-studentised version of Mallows's Cp

$$\Delta_p(\widehat{\boldsymbol{\beta}}_{\mathbf{S}}) = \frac{1}{n}\|\mathbf{Y} - \mathbf{X}_{\mathbf{S}}\widehat{\boldsymbol{\beta}}_{\mathbf{S}}\|_2^2 + \frac{2|\mathbf{S}|}{n}\sigma^2 - \sigma^2. \tag{2}$$

---

[1] A similar discussion would hold for any distance between selected and true model.

The selection $\mathbf{S}^k = \mathcal{S}(\mathbf{Y}, k)$, however, depends on $\mathbf{Y}$. Hence, for the corresponding least squares projection $\widehat{\boldsymbol{\beta}}_{\mathbf{S}^k} = (\mathbf{X}_{\mathbf{S}^k}^T \mathbf{X}_{\mathbf{S}^k})^{-1} \mathbf{X}_{\mathbf{S}^k}^T \mathbf{Y}$, the expectation of $\Delta_p(\widehat{\boldsymbol{\beta}}_{\mathbf{S}^k})$ will not be equal to $\text{PE}(\widehat{\boldsymbol{\beta}}_{\mathbf{S}^k})$.

Among all selections of size k for k large enough so that the important variables are in the model, the procedure consisting of minimising (2), i.e. $\mathcal{S}(\mathbf{Y}, k) = \arg\min_{|\mathbf{S}|=k} \Delta_p(\widehat{\boldsymbol{\beta}}_{\mathbf{S}})$, adds seemingly nonzero variables—that should, in fact, be zeros—in order to fit the observational errors by minimising further the distance between $\mathbf{X}_{\mathbf{S}^k} \widehat{\boldsymbol{\beta}}_{\mathbf{S}^k}$ and $\mathbf{Y}$. The consequence is a better-than-average appearance of $E\Delta_p(\widehat{\boldsymbol{\beta}}_{\mathbf{S}^k})$ contrasting with the worse-than-average true prediction error: indeed the false positives perform worse in staying close to the signal without the errors than variables selected in a purely arbitrary way. This two-sided effect of appearance versus reality is described as a mirror effect [4].

Whereas information criteria have been designed to evaluate the quality of one specific model, the optimisation over the randomised $\Delta_p(\widehat{\boldsymbol{\beta}}_{\mathbf{S}^k})$ affects the statistics of the selected variables. Because the selection $\mathbf{O}^k$ does not depend on $\mathbf{Y}$, leaving the statistics of the selected variables unchanged, the oracle curve $\text{PE}(\widehat{\boldsymbol{\beta}}_{\mathbf{O}^k})$ acts as a mirror reflecting $\text{PE}(\widehat{\boldsymbol{\beta}}_{\mathbf{S}^k})$ onto $\Delta_p(\widehat{\boldsymbol{\beta}}_{\mathbf{S}^k})$:

$$\text{PE}(\widehat{\boldsymbol{\beta}}_{\mathbf{S}^k}) - \text{PE}(\widehat{\boldsymbol{\beta}}_{\mathbf{O}^k}) \approx m_k \approx \text{PE}(\widehat{\boldsymbol{\beta}}_{\mathbf{O}^k}) - E\Delta_p(\widehat{\boldsymbol{\beta}}_{\mathbf{S}^k}). \tag{3}$$

Figure 1 plots $\text{PE}(\widehat{\boldsymbol{\beta}}_{\mathbf{S}^k})$ and $\Delta_p(\widehat{\boldsymbol{\beta}}_{\mathbf{S}^k})$ in dashed and dotted lines, respectively, as functions of the model size k. Also, the mirror curve $\text{PE}(\widehat{\boldsymbol{\beta}}_{\mathbf{O}^k}) \approx \Delta_p(\widehat{\boldsymbol{\beta}}_{\mathbf{S}^k}) + m_k$ is represented as the solid curve. Details of the calculations can be found in Sect. 4.

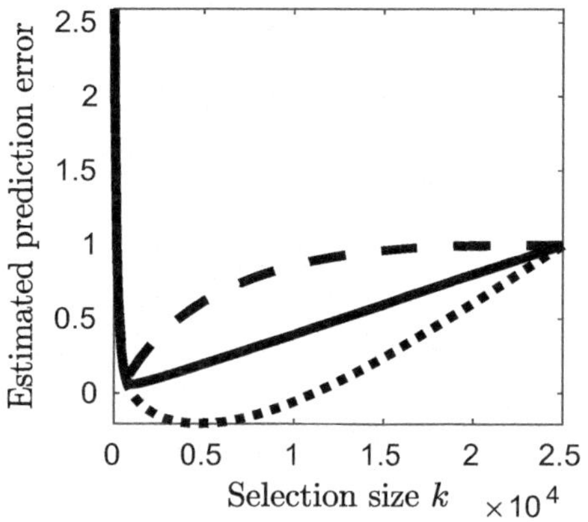

Fig. 1 Illustration of the mirror effect. The mirror curve is plotted in solid line as a function of the model size and reflects the prediction error and Mallows's Cp, represented as dashed and dotted curves, respectively

The Mirror and Degrees of Freedom

The mirror effect is closely related to the concept of generalised degrees of freedom [11]. Defining the residual vector $\mathbf{e}_k = \mathbf{Y} - \mathbf{X}_{\mathbf{S}^k}\widehat{\boldsymbol{\beta}}_{\mathbf{S}^k}$, the generalised degrees of freedom are given by

$$\nu_k = \frac{1}{\sigma^2}E[\boldsymbol{\varepsilon}^T(\boldsymbol{\varepsilon} - \mathbf{e}_k)] = \frac{1}{n}E(\boldsymbol{\varepsilon}^T\mathbf{X}_{\mathbf{S}^k}\widehat{\boldsymbol{\beta}}_{\mathbf{S}^k}) \tag{4}$$

and $\Lambda_p(\widehat{\boldsymbol{\beta}}_{\mathbf{S}^k}) = n^{-1}\|\mathbf{Y} - \mathbf{X}_{\mathbf{S}^k}\widehat{\boldsymbol{\beta}}_{\mathbf{S}^k}\|_2^2 + 2\nu_k n^{-1}\sigma^2 - \sigma^2$ is then an unbiased estimator of $\mathrm{PE}(\widehat{\boldsymbol{\beta}}_{\mathbf{S}^k})$ for any choice of the selection $\mathbf{S}^k$. Under sparsity assumptions [4], we have

$$\nu_k = E[\|\mathbf{P}_{\mathbf{S}_k}\boldsymbol{\varepsilon}\|_2^2]\sigma^{-2} + o[\mathrm{PE}(\widehat{\boldsymbol{\beta}}_{\mathbf{S}^k})] \text{ as } n \to \infty \tag{5}$$

with the projection matrix $\mathbf{P}_{\mathbf{S}^k} = \mathbf{X}_{\mathbf{S}^k}(\mathbf{X}_{\mathbf{S}^k}^T\mathbf{X}_{\mathbf{S}^k})^{-1}\mathbf{X}_{\mathbf{S}^k}^T$. Given such a projection, the mirror correction m_k is found to be

$$m_k = \frac{1}{n}E[\|\mathbf{P}_{\mathbf{S}^k}\boldsymbol{\varepsilon}\|_2^2; \boldsymbol{\beta}] - \frac{k}{n}\sigma^2, \tag{6}$$

meaning that $m_k = (\nu_k - k)\sigma^2/n + o[\mathrm{PE}(\widehat{\boldsymbol{\beta}}_{\mathbf{S}^k})]$. This expression explicitly writes the parametric dependence on $\boldsymbol{\beta}$. An unbiased estimator is given by

$$\widehat{m}_k = \frac{1}{n}E[\|\mathbf{P}_{\mathbf{S}^k}\boldsymbol{\varepsilon}\|_2^2|\mathbf{S}^k; \boldsymbol{\beta}] - \frac{k}{n}\sigma^2. \tag{7}$$

3 Qualitative Description of the Mirror

Using (7) within the signal-plus-noise model $\mathbf{Y} = \boldsymbol{\beta} + \boldsymbol{\varepsilon}$, an estimator of the mirror effect m_k can be written as

$$\widehat{m}_k = \frac{1}{n}E(\|\boldsymbol{\varepsilon}_{\mathbf{S}^k}\|_2^2|\mathbf{S}^k; \boldsymbol{\beta}) - \frac{k}{n}\sigma^2 \tag{8}$$

for any selection $\mathbf{S}^k$. The behaviour of this estimator is described in the two following subsections.

3.1 Unstructured Signal-Plus-Noise Models

The mirror effect can be decomposed into three stages depending on the model size k.

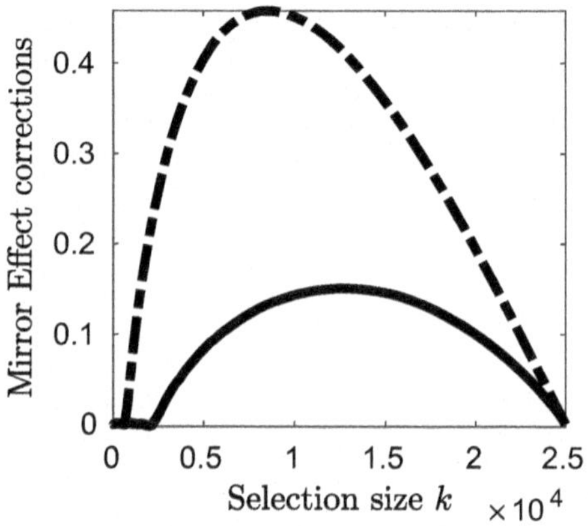

Fig. 2 Illustration of the mirror correction behaviour for unstructured (dot-dashed line) and group selections (solid) as functions of the model size

Small selection sizes: especially if $k \leq n_1$ (the true number of nonzeros in $\boldsymbol{\beta}$), only nonzero variables should be selected. The errors associated with these variables have an expected variance equal to σ^2 since they are randomly distributed among these variables. Hence, m_k is close to zero.

Intermediate selection sizes: the large errors are selected accordingly to their absolute value. Their expected variance is obviously greater than σ^2, meaning that m_k is increasing. Its growth is high at first and decreases to zero as smaller errors are added to the selection, leading to the last stage.

Large selection sizes: finally, only the remaining small errors are selected. This has the effect of diminishing the (previously abnormally high) variance to its original expected value; m_k drops to zero which is achieved for the full model.

The illustration of the mirror correction is depicted in Fig. 2 (dot-dashed line). The three stages appear approximately in the intervals [0, 1000], [1000, 8000] and [8000, 25000]. Details of the calculations can be found in Sect. 4.

3.2 Structured Models: Grouped Variables

We now focus on group selection where l groups are selected and variables that belong to the same group are added to the model altogether. Group-lasso may be the selection procedure used for the construction of the estimator $\widehat{\boldsymbol{\beta}}_{S^l}$. See Sect. 4 for a more complete definition of $\widehat{\boldsymbol{\beta}}_{S^l}$.

Small selection sizes: groups of nonzero variables are selected first as they have major effects on the linear regression. As before, the associated errors are randomly distributed, so their expected variance equals σ^2 and m_l is roughly zero.

Intermediate selection sizes: the groups containing only errors are selected accordingly to their norms. These groups typically contain high value errors and some low value errors. Hence, their expected variance is greater than σ^2 but smaller than in the case of unstructured selection as the latter just selects the largest errors. The consequence on m_l is that it increases but its growth, although high at first, is smaller than for the unstructured case.

Large selection size: groups of small errors are selected (although they can still contain some large errors), meaning that their expected variance decreases to σ^2 which is achieved for the full model.

The explanations above hold for any type of structure, so we can deduce that the unstructured mirror has the largest amplitude in signal-plus-noise models. Indeed, as there is no constraint on the variables, once the nonzeros are selected, the errors are included in the model given their absolute value and more correction is needed in order to temper their effects.

This description is represented in Fig. 2 where the mirror corrections for group and unstructured selections can be visually compared as they are plotted in dot-dashed and solid lines, respectively. The three stages of group selection can be seen in the intervals [0, 2500], [2500, 125000] and [12500, 25000]. Details of the calculations can be found in Sect. 4.

4 Simulation

In this simulation, $r = 2500$ groups containing $w = 10$ coefficients $\boldsymbol{\beta}_j$ are generated so that $\boldsymbol{\beta} = (\boldsymbol{\beta}_j)_{j=1,\ldots,2500}$ is a n-dimensional vector with $n = 25000$. Within group j, the $\boldsymbol{\beta}_j$ coefficients have the same probability p_j of being nonzero and, for each group j, a different value p_j is randomly drawn from the set $\mathbf{P} = (0.95, 0.80, 0.50, 0.05, 0)$ with the respective probability $\mathbf{Q} = (0.02, 0.02, 0.01, 0.20, 0.75)$. The expected proportion of nonzeros is $\langle \mathbf{P}, \mathbf{Q} \rangle = 1/20$ for the whole vector $\boldsymbol{\beta}$. The nonzeros from $\boldsymbol{\beta}$ are distributed according to the zero inflated Laplace model $f_{\beta|\beta \neq 0}(\beta) = (a/2)\exp(-a|\beta|)$ where $a = 1/5$. The observations then are computed as the vector of groups $(\mathbf{Y}_j)_{j=1,\ldots,r} = (\boldsymbol{\beta}_j)_{j=1,\ldots,r} + (\boldsymbol{\varepsilon}_j)_{j=1,\ldots,r}$, where $\boldsymbol{\varepsilon}$ is an n-vector of independent, standard normal errors.

Estimates $\widehat{\boldsymbol{\beta}}$ are calculated considering two configurations: groups of size $w = 10$ (initial setting) and groups of size 1 (unstructured selection). In the latter scenario, $\widehat{\beta}_i = Y_i S_i^k$ where $\mathbf{S}^k$ is the binary n-vector selecting the k largest absolute values. The 10-group estimator is $\widehat{\boldsymbol{\beta}}_j = \mathbf{Y}_j S_j^l$ where $\mathbf{S}^l$ is the binary r-vector selecting the l groups whose ℓ_2 norms are the largest. Using Lasso and group-lasso, respectively, would provide us with the same selections $\mathbf{S}^k$ and $\mathbf{S}^l$, because of the signal-plus-noise model. Mirror corrections for both configurations are found using (8).

A comparison of the false discovery rates (FDR) of nonzero variables for unstructured and group selections is presented in Fig. 3: because we allow groups to contain zeros and nonzeros in this simulation, at first the recovery rate of the nonzeros in the group setting is below the recovery rate of the unstructured nonzeros. However, the two rates quickly cross and we find the recovery of the nonzeros in the group setting

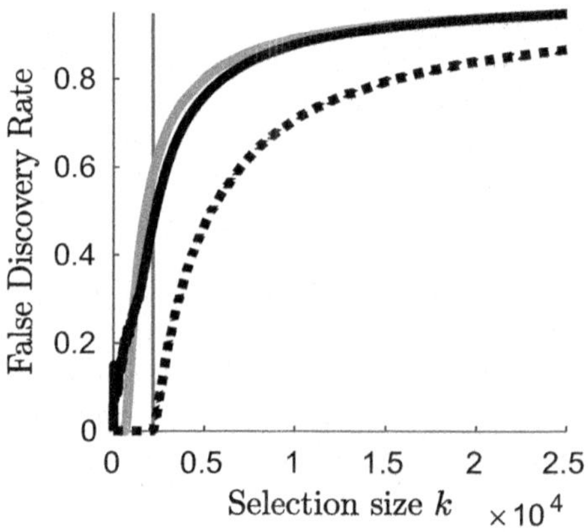

Fig. 3 Illustration of the false discovery rate (FDR) of nonzero variables for unstructured (grey solid curve) and group selections (black solid curve) as functions of the model size. The vertical line represents the selection size of the best model found with the group-mirror-corrected Cp and the black dotted curve is the FDR of nonzero groups

to be better afterwards: particularly, this is the case for the selection that minimises the group-mirror-corrected Cp (marked with the vertical line in Fig. 3).

The FDR of the nonzero groups (groups containing at least one nonzero variable) suggests that the recovery of nonzeros in the group setting will be consistently superior to the recovery of unstructured nonzeros if groups are either fully nonzero or fully zero. In that case, the group-mirror-corrected Cp should perform really well as information criterion to obtain the optimal selection size.

5 Conclusion

During the optimisation of an information criterion over the model size, using both ℓ_1 and ℓ_0-regularisation for the selection and estimation of variables allows us to take advantage of quadratic programming for the former and least squares projection for the latter. This technique avoids an overestimation of the number of selected variables; however, it requires a corrected expression for the information criterion: the difference between ℓ_0 and ℓ_1 regularisation is compensated using the mirror effect.

In this paper, we described the behaviour of the mirror effect in signal-plus-noise models, observing three stages depending on the model size. This way we can distinguish the selection of nonzero variables, of large false positives and of small false positives for which the mirror is, respectively, close to zero, then increasing and finally decreasing to zero again. In the special case of structured selection, we note a similar behaviour for the mirror although its amplitude is smaller, meaning that the information criterion needs less correction.

References

1. Akaike, H.: Information theory and an extension of the maximum likelihood principle. In: B.N. Petrov, F. Csáki (eds) Proceedings of the Second International Symposium on Information Theory, pp. 267–281. Akadémiai Kiadó, Budapest (1973)
2. Donoho, D.L.: For most large underdetermined systems of equations, the minimal ℓ_1-norm near-solution approximates the sparsest near-solution. Commun. Pure Appl. Math. **59**, 907–934 (2006)
3. Friedman, J., Hastie, T., Tibshirani, R.: Sparse inverse covariance estimation with the graphical lasso. Biostatistics **9**(3), 432–441 (2008)
4. Jansen, M.: Information criteria for variable selection under sparsity. Biometrika **101**, 37–55 (2014)
5. Mallows, C.L.: Some comments on cp. Technometrics **15**, 661–675 (1973)
6. Meinshausen, N., Bühlmann, P.: High-dimensional graphs and variable selection with the lasso. Annal. Stat. **34**, 1436–1462 (2006)
7. Stein, C.: Inadmissibility of the usual estimator for the mean of a multivariate normal distribution. In: J. Neyman (ed), Proceedings of the Third Berkeley Symposium on Mathematical Statistics and Probability, pp. 197–206. University of California Press (1956)
8. Tibshirani, R.: Regression shrinkage and selection via the lasso. J. R. Stat. Soc. Ser. B **58**, 267–288 (1996)
9. Tibshirani, R., Saunders, M., Rosset, S., Zhu, J., Knight, K.: Sparsity and smoothness via the fused lasso. J. R. Stat. Soc. Ser. B **67**(1), 91–108 (2005)
10. Wainwright, M.J.: Sharp thresholds for high-dimensional and noisy sparsity recovery using ℓ_1-constrained quadratic programming (lasso). IEEE Trans. Inform. Theory **55**(5), 2183–2202 (2009)
11. Ye, J.: On measuring and correcting the effects of data mining and model selection. J. Am. Stat. Assoc. **93**, 120–131 (1998)
12. Yuan, M., Lin, Y.: Model selection and estimation in regression with grouped variables. J. R. Stat. Soc. Ser. B **68**(1), 49–67 (2007)
13. Zhao, P., Rocha, G., Yu, B.: The composite absolute penalties family for grouped and hierarchical variable selection. Annal. Stat. **37**, 3468–3497 (2009)
14. Zhao, P., Yu, B.: On model selection consistency of lasso. J. Mach. Learn. Res. **7**, 2541–2563 (2006)

United Statistical Algorithms and Data Science: An Introduction to the Principles

Subhadeep Mukhopadhyay

Abstract Developing algorithmic solutions to tackle the rapidly increasing variety of data types, by now, is recognized as an outstanding open problem of modern statistics and data science. But why does this issue remain difficult to solve programmatically? Is it merely a passing trend, or does it have the potential to radically change the way we build learning algorithms? Discussing these questions without falling victim to the big data hype is not an easy task. Nonetheless, an attempt will be made to better understand the core statistical issues, in a manner to which *every* data scientist can relate.

Keywords United data science · Modern data representation · Algorithmic portability

1 The Gorilla in the Room

Developing algorithmic solutions to the "variety problem" is now recognized (by experts from academia to industry—Sam Madden, Michael Stonebraker, Thomas Davenport, and Clive Longbottom) as an *unsolved*[1] problem of modern data science. According to the experts, solving this problem will allow us to unlock the enormous potential of analytics. But why does this issue remain difficult to solve programmatically? Does it have any impact on day-to-day statistical practice? Is it merely a passing trend, or does it have the potential to radically change the way we build learning algorithms? To start with, it is not even clear how to formulate the "variety problem" precisely, let alone find an effective and useful solution to it.

---

[1]Michael Stonebraker calls data variety a "Research problem! Killing most CIO's" and "If there is any achilles heel it's going to be this."

S. Mukhopadhyay (✉)
Department of Statistics, Fox Business School, Temple University, Philadelphia, USA
e-mail: deep@unitedstatalgo.com

M. La Rocca et al. (eds.), *Nonparametric Statistics*, Springer Proceedings in Mathematics & Statistics 339, https://doi.org/10.1007/978-3-030-57306-5_33

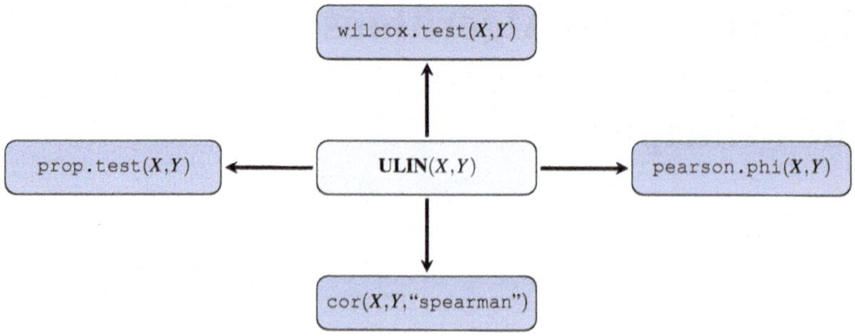

Fig. 1 The vision of "Algorithm of Algorithms." We have displayed the computing code for implementing four fundamental statistical methods for learning from (X, Y) data

To get to the heart of the matter quickly, we shall focus on a concrete problem that arises in everyday data-modeling tasks. Figure 1 displays four fundamental statistical methods, widely considered blockbusters of twentieth-century statistics [10, 12]– for the ubiquitous (X, Y) learning problem, each of which is specially crafted to perform a *specific* modeling task. For example, Wilcoxon statistics is applicable for two-sample problem (X binary 0/1 and Y continuous); Spearman's correlation applies to both X and Y continuous; and Pearson's ϕ-coefficient and two-proportion Z-statistics can be applied when we have samples from X and Y both binary.
Therefore, to compute appropriate statistical measures, practitioners need to pick the algorithms after performing a painstaking manual inspection of each data type. The situation becomes scary as the number of variables increases, creating serious roadblocks for automation and ease of programming. The question naturally arises: How can we develop a unified computing formula that yields appropriate statistical measures *without* having the data-type information from the user?[2] By doing so, we would dramatically reduce the complexity and cost of analytics. Accordingly, there is a growing demand for an easy and systematic data analysis pipeline by designing algorithms with a better data-adaptation property.
Instead of coding each one of these specialized algorithms separately in a "compartmentalized" manner, the goal is to develop a more integrated and methodical approach by designing a *master algorithm* (denoted as `ULIN` in Fig. 1) capable of avoiding the case-specific data operations for efficient programming. This leads us to the concept of "United Statistical Algorithms"—algorithms that can simultaneously model heterogeneous data types. By doing so, such algorithms substantially reduce the complexity of programming and help us to see the connection between different isolated pieces.

---

[2]It is appalling to note that the simple problem depicted in Fig 1 becomes a highly non-trivial if we try to attack it using old run-of-the-mill thinking and analysis. A somewhat amusing exercise would be to see how many can pass this "litmus test."

2 The Design Principle

What we need is an easy-to-implement unified computing formula. The two fundamental ingredients for constructing such learning algorithms are

- First, we need to represent data through some intelligently designed transformation, else we will not be able to abstract out the complexity that arises due to different data types.
- Second, no solution lies in parametric-statistics land. The algorithms have to be data-driven (also known as nonparametric). As Richard [6] said "Without proper regard for the *individuality* of the problem the task of computation will become hopeless."

The above two principles make it clear that a marriage between applied harmonic data analysis and modern nonparametric statistics will be a common theme in this pursuit of designing smart computational algorithms. However, the real challenge is to figure out the appropriate choice of such nonparametric data-transformation technique.

> However, the difficulty that challenges the inventive skill of the applied mathematician is to find suitable coordinate functions.—Richard Courant [6].

2.1 *From Principles to Construction*

Given $X_1, \ldots, X_n$ a random sample from an unknown distribution F, denote the empirical mid-distribution function by $\widetilde{F}_X^{\text{mid}}(x) = \widetilde{F}_X(x) - \frac{1}{2}\widetilde{p}_X(x)$, where $\widetilde{F}_X(x) = n^{-1}\sum_{i=1}^n \mathbb{I}\{X_i \le x\}$, and the sample probability mass function by $\widetilde{p}_X(x) = n^{-1}\sum_{i=1}^n \mathbb{I}\{X_i = x\}$. Then we have the following data-driven linear transformation:

$$\Psi(x; \widetilde{F}_X) = \frac{\sqrt{12}\{\widetilde{F}_X^{\text{mid}}(x) - 1/2\}}{\sqrt{1 - \sum_x \widetilde{p}_X^3(x)}} \tag{1}$$

which, by design, is orthonormal with respect to the Hilbert space $\mathscr{L}^2(\widetilde{F})$: $\int \Psi(x; \widetilde{F}_X)\, \mathrm{d}\widetilde{F}_X = 0$ and $\int \Psi^2(x; \widetilde{F}_X)\, \mathrm{d}\widetilde{F}_X = 1$. Figure 2 displays the shape of this empirical linear transform (`eLT`) for three different kinds of random variables. Next, we introduce the concept of Universal Linear Statistics (abbreviated as `ULIN`) as an inner product in the new transformed domain:

$$\text{ULIN}(Y, X) = \mathbb{E}[\Psi(X; \widetilde{F}_X)\Psi(Y; \widetilde{F}_Y); \widetilde{F}_{X,Y}]. \tag{2}$$

The statistical interpretation of these orthonormal transforms will be discussed next.

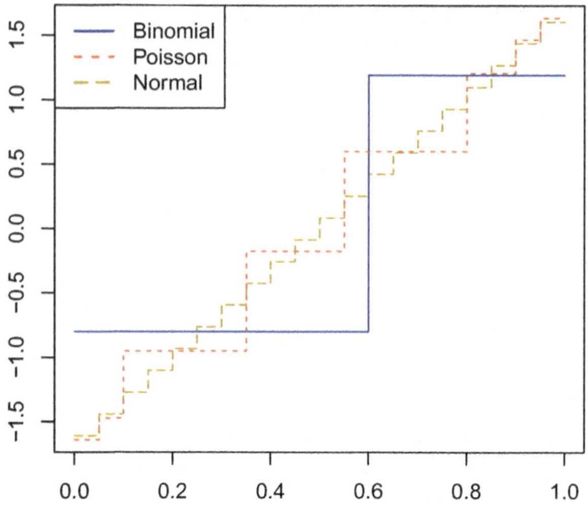

Fig. 2 (color online) The shapes of empirical transform based on $n = 20$ random samples generated from Poisson ($\lambda = 5$) (in red), Binomial(1, .4) (in blue), and standard normal (in goldenrod) distributions. They are plotted over unit interval by joining $\{\widetilde{F}_X(x_i), \Psi(x_i; \widetilde{F}_X)\}$ for distinct values. Note the changing shapes of $\Psi(x; \widetilde{F}_X)$ for different varieties of data—key characteristic of a nonparametric transform

3 Universality Properties

The `ULIN` statistic enjoys some remarkable universality properties. Different special cases of this general formulation will be discussed to gain more intuition as to how it brings surprising unity to the algorithm design.

X, Y both Continuous Case

More than a century back, Charles Spearman [22] introduced the radical idea of measuring the statistical association between two sets of continuous measurements $\{X_1, \ldots, X_n\}$ and $\{Y_1, \ldots, Y_n\}$ *via their ranks*. The following result connects the ULIN and Spearman's correlation coefficient (proof is given in Appendix A1).

Theorem 1 *Given n-pairs of continuous measurements* (X_i, Y_i), *we have the following equivalence with the Spearman's rank correlation coefficient:*

$$\text{ULIN}(Y, X) = 1 - \frac{6\sum_{i=1}^{n} d_i^2}{n(n^2-1)}, \tag{3}$$

where $d_i = \mathscr{R}(X_i) - \mathscr{R}(Y_i)$, *the difference in ranks.*

X, Y both Binary Case

Consider the following problem where X denotes the incidence of a heart attack (yes/no) and Y denotes the consumption of a daily aspirin tablet or sugar pill (placebo). A 5-year-long randomized study, carried out by Physicians' Health Study Research Group at Harvard Medical School [1, 23], yielded the following data: out of $n_1 = 11{,}034$ male physicians taking a placebo, 189 suffered heart attack during the study, a proportion of $\widetilde{p}_1 = 189/11,034 = 0.0171$. On the other hand, the proportion of heart attack in the aspirin group was $\widetilde{p}_2 = 104/11,037 = 0.0094$. Looking at the data, can we conclude that regular intake of aspirin reduces the risk of heart attack? Let p_1 be the true proportion of heart attack who were given placebo and p_2 is the true proportion of heart attack among the aspirin group. Then this is a classical problem of a two-sample Z-test. Interestingly, as shown in the following theorem, for X and Y both binary, our `ULIN` statistic *automatically* produces (4) for comparing two population proportions (for proof see Appendix A2).

Theorem 2 *Given n-pairs of binary* (X_i, Y_i)*, our* `ULIN` *computing formula reproduces the two-proportion Z-test statistic:*

$$\sqrt{n}\,\text{ULIN}(Y, X) = \frac{\widetilde{p}_1 - \widetilde{p}_2}{\sqrt{\widetilde{p}(1-\widetilde{p})(\frac{1}{n_1} + \frac{1}{n_2})}}, \tag{4}$$

where $\widetilde{p} = \frac{n_1\widetilde{p}_1 + n_2\widetilde{p}_2}{n_1+n_2}$ *is the pooled sample proportion.*

The next result shows another delightful equivalence between the Pearson correlation and `ULIN` statistics for X and Y both binary.

Corollary 1 *For* X, Y *binary, we have the following equality between Pearson correlation (also known as* ϕ *coefficient for* 2×2 *contingency table setup) and* ULIN *statistics:*

$$\text{Pearson}(Y, X) = \text{ULIN}(Y, X).$$

A one-line proof is based on the crucial observation that $\mathscr{Z}(X) = \Psi(X; F_X)$ for binary X, where $\mathscr{Z}(X) = \frac{X - \mathbb{E}(X)}{\sigma(X)}$ is the standardized X. For details, see Appendix.

The Devil Is In The Details

What if a data scientist ignores the binary nature of the data and applies the Spearman's correlation anyway? As we will see, neglecting the data-type information can lead to disastrous result. To appreciate this better, consider (X, Y) Bernoulli random variables with marginal and joint distribution $\Pr(X = 0) = \Pr(Y = 0) = \Pr(X = 0, Y = 0) = p \in (0, 1)$, which implies $Y = X$ almost surely. However, the traditional $\text{Spearman}(X, Y) = p(1 - p) < 1$, *not attaining the value* 1! The problem lies in ignoring the discreteness (i.e., the data-type information) of X and Y [15], which implies that data scientists have no choice but to manually probe the data type for selecting the appropriate method. This is where `ULIN` comes to the rescue due to its auto-adaptation property. In this example, it is straightforward to show that $\text{ULIN}(X, Y) = 1$.

X binary, Y Continuous

Finally, consider the case where X is binary and Y is continuous. The surprising connection between two-sample Wilcoxon test [24] and `ULIN` statistic is formalized in the following result (proof is given in Appendix A3).

Theorem 3 *For a two-sample problem with* $\{(X_i, Y_i), i = 1, \ldots, n = n_1 + n_2\}$ *where* $X \in \{0, 1\}$ *and* Y *is continuous with the pooled rank denoted by* $\mathscr{R}(Y_i)$*, we have the following equivalent representation of the Wilcoxon rank-sum statistic:*

$$\text{ULIN}(Y, X) = \sqrt{\frac{12}{n^2\, n_1 n_2}} \left[\sum_{i=n_2+1}^{n} \mathscr{R}(Y_i) - \frac{n_1(n+1)}{2} \right]. \tag{5}$$

Few Remarks

- In short, we have shown that ULIN(Y, X) acts as a "4-in-1" algorithm that is smart enough to automatically adjust itself (self-correcting) for various types of data, without *any* input from the user.
- Another practical benefit of our unification lies in its easy "correlation" interpretation (2) by bringing different statistics into a *common* range $[-1, 1]$. Accordingly, one can use this ULIN statistic for learning as well as interpreting the strength of relationships between mixed varieties of data types using a single computing code (see Appendix B).

4 The Age of Unified Algorithms Is Here

The *science* of finding good representation systems is at the heart of many applications, including approximation, compression, and sparsity, which revolutionized computer vision, signal and image processing [7, 8, 13]. This field was born out of Joseph Fourier's [11] revolutionary insight , which he presented on December 21, 1807 before the French Academy.[3] Since then, efficient data representation "trick" has played an important role for developing faster numerical algorithms.

In contrast, our development was driven by a fundamentally different motivation—to use carefully designed (in a data-adaptive or nonparametric manner) representation system as a theoretical tool to unify diverse statistical algorithms rather than making an existing algorithm faster. It is inspired by the statistical considerations instead of applied harmonic analysis concepts.

In fact, one can build the whole basis system for the Hilbert space $\mathscr{L}^2(\widetilde{F})$ starting from $\Psi(x; \widetilde{F})$ (Eq. 1): perform Gram-Schmidt orthonormalization on $\{\Psi, \Psi^2, \ldots\}$. This empirically constructed basis functions act as a universal coordinate system for data analysis, which has been applied to a range of problems, including time series analysis [20], multiple testing [16], distributed learning [3], bump-hunting [17], generalized empirical-Bayes [19], and many others. What is important is that in all of the above cases, our custom-designed orthonormal system unifies a wide class of parametric

[3]Fourier was unable to publish his results until 1822 because his 1807 presentation was not well received by Joseph Lagrange and accordingly the publication in the Memoirs of the Academy was refused. The idea that signals can be built from the sum of harmonic functions was too radical to consider.

and nonparametric, linear and non-linear, discrete and continuous statistical methods, thereby marking a big stride forward in the development of "United Data Science." This modeling philosophy is crucial for three reasons:

- Theoretical: There is a dire need to put some order into the current inventory of algorithms that are mushrooming at a staggering rate, in order to better understand the statistical core. Our modeling philosophy can provide that "organizing principle."
- Practical: It provides construction of "distribution-free" [4] automated algorithms. Wisely designed transformation acts as an abstraction to overcome the "variety" problem that significantly reduces the complexity of programming for analyzing heterogeneous data types.

> Extracting new insights from the data sets currently being generated will require not only faster computers, but also smarter algorithms [14, p. 627].

- Pedagogical: There is a growing need to develop a comprehensive training curriculum covering the fundamental statistical learning methods for building a twenty-first-century data-capable workforce. The first action plan was proposed by [5] where he argued that only 20% of the total curriculum should be allotted for teaching theoretical foundation of data science, rest being computing, collaboration, and software tool development. It immediately raises the following pedagogical challenge: how can we cover this wide range of topics (which [4] calls "Greater statistics: learning from data") [5] within the allotted time—the dilemma of *"too many topics, too little time"* [18]. Our united statistical learning viewpoint can offer some concrete solutions by providing a "concise, comprehensive, and connected" (I call it the **three C's** of teaching) view of the subject, thereby accelerating students' learning.

5 The Three Pillars

How can we minimize the manual heavy lifting that every practicing data scientist has to go through simply to get their statistical analysis right? What we need is a modern language of data analysis that can abstract away all the bells and whistles of the underlying complexity into a simple-to-compute autonomous formula—moving from *compartmentalized* algorithm design to a unified computing that can radically change the way we practice and teach "programming with data." There is little doubt

---

[4]If the algorithmic logic and computing formulas are only valid for any specific distribution, then it is not "distribution-free" by design.

[5]Also see, for example, David Donoho's (2017) "Greater Data Science" curriculum structure that is composed of six categories of activities: Data Exploration and Preparation, Data Representation and Transformation, Computing with Data, Data Modeling, Data Visualization and Presentation, and Science about Data Science.

that this field will keep evolving and expanding rapidly in the coming years due to its growing popularity and pervasive necessity across many disciplines.
Three pillars of modern data science are statistical efficiency [21], computational efficiency [2], and, third, the emerging paradigm, design efficiency *via* unified algorithms. The critical task is to assemble different "mini-algorithms" into a coherent master algorithm for increased simplification of theory and computation. This whole field is still very nascent and desperately needs new ideas. But the key lies in *modern data representation* techniques whose intuition comes from understanding the shared statistical structure of a class of working algorithms.

Acknowledgements An earlier draft of this paper was presented at a seminar organized by Ecole polytechnique, Applied Mathematics Department, France on June 19, 2018, and at the International Society of Nonparametric Statistics (ISNPS) conference, Italy, June 14, 2018. I owe special thanks to the participants of the conference and the seminar for the stimulating and congenial discussions.

Appendix

Here we outline the details for proving the main results along with the numerical R-code.

Appendix A1. Proof of Theorem 1
When all X_i $(i = 1, \ldots, n)$ are distinct, following (1), we have

$$\Psi(x_i; \widetilde{F}_X) = \sqrt{\frac{12}{n^2-1}}\left(\mathscr{R}(X_i) - \frac{n+1}{2}\right), \tag{6}$$

since $\widetilde{F}^{\text{mid}}(x_i) = (\mathscr{R}(X_i) - .5)/n$ and $1 - \sum_i \widetilde{p}^3(x_i) = 1 - n^{-2}$. Similar expression holds for Y. Substituting these expressions into our universal inner product formula

$$\text{ULIN}(X, Y) = n^{-1} \sum_{i=1}^{n} \Psi(x_i; \widetilde{F}_X)\Psi(y_i; \widetilde{F}_Y),$$

immediately yields

$$\begin{aligned}\text{ULIN}(X, Y) &= \frac{12}{n(n^2-1)}\left[\sum_{i=1}^{n}\left(\mathscr{R}(X_i) - \frac{n+1}{2}\right)\left(\mathscr{R}(Y_i) - \frac{n+1}{2}\right)\right] \\ &= \frac{12}{n(n^2-1)}\left[\sum_{i=1}^{n}\mathscr{R}(X_i)\mathscr{R}(Y_i) - n\left(\frac{n+1}{2}\right)^2\right]. \end{aligned} \tag{7}$$

Complete the proof by verifying that (7) can be rewritten as

$$\mathrm{ULIN}(X, Y) = \frac{12}{n(n^2-1)}\left[\frac{n(n^2-1)}{12} - \sum_{i=1}^{n}\frac{d_i^2}{2}\right] \equiv 1 - \frac{6\sum_{i=1}^{n} d_i^2}{n(n^2-1)},$$

where $d_i = \mathscr{R}(X_i) - \mathscr{R}(Y_i)$. □

Appendix A2. Proof of Theorem 2

For binary X and Y, it is not difficult to verify that for $i = 1, \ldots, n_1, n_1+1, \ldots, n_1+n_2$

$$\Psi(x_i;\widetilde{F}_X) = \begin{cases}\sqrt{\frac{n_2}{n_1}} & \text{for } x_i = 1\\ -\sqrt{\frac{n_1}{n_2}} & \text{for } x_i = 0,\end{cases} \qquad \Psi(y_i;\widetilde{F}_Y) = \begin{cases}\sqrt{\frac{1-\widetilde{p}}{\widetilde{p}}} & \text{for } y_i = 1\\ -\sqrt{\frac{\widetilde{p}}{1-\widetilde{p}}} & \text{for } y_i = 0,\end{cases} \tag{8}$$

where $\widetilde{p} = (n_1\widetilde{p}_1 + n_2\widetilde{p}_2)/(n_1+n_2)$ denotes the pooled sample proportion for Y. With this in hand, we start by explicitly writing down the expression for $\mathrm{ULIN}(X, Y)$:

$$\frac{1}{n}\Big\{n_1\widetilde{p}_1\Psi(1;\widetilde{F}_X)\Psi(1;\widetilde{F}_Y) + n_1(1-\widetilde{p}_1)\Psi(1;\widetilde{F}_X)\Psi(0;\widetilde{F}_Y)$$
$$+ n_2\widetilde{p}_2\Psi(0;\widetilde{F}_X)\Psi(1;\widetilde{F}_Y) + n_2(1-\widetilde{p}_2)\Psi(0;\widetilde{F}_X)\Psi(0;\widetilde{F}_Y)\Big\}.$$

Let us now simplify the first two terms of the above expression:

$$n_1\widetilde{p}_1\Psi(1;\widetilde{F}_X)\big[\Psi(1;\widetilde{F}_Y) - \Psi(0;\widetilde{F}_Y)\big] + n_1\Psi(1;\widetilde{F}_X)\Psi(0;\widetilde{F}_Y) = -\sqrt{n_1 n_2}\frac{\widetilde{p}_2}{\sqrt{\widetilde{p}(1-\widetilde{p})}}. \tag{9}$$

Following the same steps, we also have

$$n_2\widetilde{p}_2\Psi(0;\widetilde{F}_X)\big[\Psi(1;\widetilde{F}_Y) - \Psi(0;\widetilde{F}_Y)\big] + n_2\Psi(0;\widetilde{F}_X)\Psi(0;\widetilde{F}_Y) = \sqrt{n_1 n_2}\frac{\widetilde{p}_1}{\sqrt{\widetilde{p}(1-\widetilde{p})}}. \tag{10}$$

Combining (9) and (10), we immediately have

$$\sqrt{n}\,\mathrm{ULIN}(Y, X) = \frac{\widetilde{p}_1 - \widetilde{p}_2}{\sqrt{\widetilde{p}(1-\widetilde{p})(\frac{1}{n_1}+\frac{1}{n_2})}}.$$

This completes the proof. □

Appendix A3. Proof of Theorem 3

Here we show that our $\mathrm{ULIN}(X, Y)$ automatically reproduces the Wilcoxon rank-sum statistics for X binary and Y continuous. Substituting the expressions of the linear transforms from (6) and (8), we have

$$\mathrm{ULIN}(X,Y) = n^{-1}\sum_{i=1}^{n}\Psi(x_i;\widetilde{F}_X)\Psi(y_i;\widetilde{F}_Y)$$
$$= \frac{\sqrt{12}}{n^2}\left[\sqrt{\frac{n_2}{n_1}}\sum_{i=1}^{n_1}\left(\mathscr{R}(Y_i)-\frac{n+1}{2}\right) - \sqrt{\frac{n_1}{n_2}}\sum_{i=n_1+1}^{n}\left(\mathscr{R}(Y_i)-\frac{n+1}{2}\right)\right].$$

Straightforward algebraic manipulation yields the following:

$$\mathrm{ULIN}(X,Y) = \sqrt{\frac{12}{n^2\,n_1 n_2}}\left[\sum_{i=n_2+1}^{n_1+n_2}\mathscr{R}(Y_i) - \frac{n_1(n+1)}{2}\right]. \tag{11}$$

Thus, the theorem holds. □

Appendix B. R-computing Code

Our Universal Linear Statistics (`ULIN`) is easy-to-compute. In the following, we provide the R-code. Recall that the practitioners have to write only "one" master function which is, as shown in the paper, intelligent enough to adapt itself to different varieties of X and Y.

```
ULIN <- function(X,Y){

n <- length(Y)

u.x <- (rank(X,ties.method = c("average")) - .5)/n

phi.x <- scale(poly(u.x ,1))

u.y <- (rank(Y,ties.method = c("average")) - .5)/n

phi.y <- scale(poly(u.y ,1))

ULIN <- as.vector(cov(phi.x,phi.y))

return(ULIN)

}
```

References

1. Agresti, A.: An Introduction to Categorical Data Analysis, vol. 135. Wiley, New York (1996)
2. Aho, A.V., Hopcroft, J.E.: The Design and Analysis of Computer Algorithms. Pearson Education India (1974)
3. Bruce, S., Li, Z., Yang, H., Mukhopadhyay, S.: Nonparametric distributed learning architecture for big data: Algorithm and applications. IEEE Transactions on Big Data (in press) (2018)
4. Chambers, J.M.: Greater or lesser statistics: a choice for future research. Stat. Comput. **3**(4), 182–184 (1993)

5. Cleveland, W.S.: Data science: an action plan for expanding the technical areas of the field of statistics. Int. Stat. Rev. **69**(1), 21–26 (2001)
6. Courant, R.: Variational methods for the solution of problems of equilibrium and vibrations. Bull. Am. Math. Soc. **49**(1), 1–23 (1943)
7. Daubechies, I.: Ten Lectures on Wavelets, vol. 61. Siam (1992)
8. Donoho, D.L.: Compressed sensing. IEEE Trans. Inf. Theory **52**(4), 1289–1306 (2006)
9. Donoho, D.L.: 50 Years of Data Science. J. Comput. Graph. Stat. **26**, 745–766 (2017)
10. Efron, B.: The statistical century. Roy. Stat. Soc. News **22**(5), 1–2 (1995)
11. Fourier, J.: Theorie Analytique de la Chaleur, par M. Chez Firmin Didot, père et fils, Fourier (1822)
12. Hacking, I.: Trial by number. Science **5**(9), 69–70 (1984)
13. LeCun, Y., Bengio, Y., Hinton, G.: Deep learning. Nature **521**(7553), 436 (2015)
14. Loh, P.-R., Baym, M., Berger, B.: Compressive genomics. Nat. Biotechnol. **30**(7), 627 (2012)
15. Marshall, A.W.: Copulas, marginals, and joint distributions. Lect. Notes-Monogr. Ser. 213–222 (1996)
16. Mukhopadhyay, S.: Large scale signal detection: A unifying view. Biometrics **72**(2), 325–334 (2016)
17. Mukhopadhyay, S.: Large-scale mode identification and data-driven sciences. Elect. J. Stat. **11**(1), 215–240 (2017a)
18. Mukhopadhyay, S.: Statistics educational challenge in the 21st century. Biostat. Biometrics J. **2**(2), 1–2 (2017b)
19. Mukhopadhyay, S., Fletcher, D.: Generalized empirical Bayes modeling via frequentist goodness of fit. Natu. Sci. Rep. *8 (9983)*(9983), 1–15 (2018)
20. Mukhopadhyay, S., Parzen, E.: Nonlinear time series modeling: A unified perspective, algorithm, and application. J. Risk Finan. Manag., Spec. Issue "Appl. Econ." *8*(37), 1–18 (2018)
21. Rao, C.R.: Information and the accuracy attainable in the estimation of statistical parameters. In: Breakthroughs in statistics, pp. 235–247. Springer (1992)
22. Spearman, C.: The proof and measurement of association between two things. Am. J. Psychol. **15**(1), 72–101 (1904)
23. Steering Committee of the Physicians' Health Study Research Group: Preliminary report: Findings from the aspirin component of the ongoing physicians' health study. N. Engl. J. Med. **318**(4), 262–264 (1988)
24. Wilcoxon, F.: Individual comparisons by ranking methods. Biometrics Bull. **1**(6), 80–83 (1945)

The Halfspace Depth Characterization Problem

Stanislav Nagy

Abstract The halfspace depth characterization conjecture states that for any two distinct (probability) measures P and Q in the d-dimensional Euclidean space, there exists a point at which the halfspace depths of P and Q differ. Until recently, it was widely believed that this conjecture holds true for all integers $d \geq 1$. In several research papers dealing with this problem, partial positive results towards the complete characterization of measures by their depths can be found. We provide a comprehensive review of this literature, point out to certain difficulties with some of these earlier results and construct examples of distinct (probability or finite) measures whose halfspace depths coincide at all points of the sample space, for all integers $d > 1$.

Keywords Characterization · Depth · Floating body · Halfspace depth · Tukey depth

1 Introduction: Depth and Characterization of Measures

In nonparametric statistics of multivariate data, the concept of data depth has attracted a lot of attention in the past decades. Denote by $\mathscr{P}$ the collection of all Borel probability measures on the Euclidean space $\mathbb{R}^d$, and by $\mathscr{M}$ all finite Borel measures on $\mathbb{R}^d$. For $P \in \mathscr{M}$ given, the depth is a function $D\colon \mathbb{R}^d \to \left[0, P\left(\mathbb{R}^d\right)\right] : x \mapsto D(x; P)$ whose aim is to evaluate a 'centrality index' of points $x \in \mathbb{R}^d$ with respect to the main bulk of mass of P. Loci of points whose depth is large enough define central parts of the distribution and low depth value $D(x; P)$ indicates that x aligns poorly with P. Using depth, one is able to recover analogues of rank tests, order statistics and other important nonparametric and robust procedures also for multivariate data.

S. Nagy (✉)
Department of Probability and Mathematical Statistics, Faculty of Mathematics and Physics, Charles University, Sokolovská 83, Prague, Czech Republic
e-mail: nagy@karlin.mff.cuni.cz

M. La Rocca et al. (eds.), *Nonparametric Statistics*, Springer Proceedings in Mathematics & Statistics 339, https://doi.org/10.1007/978-3-030-57306-5_34

Arguably, the most important depth is due to Tukey [16], who defined the *halfspace depth* of $x \in \mathbb{R}^d$ with respect to $P \in \mathscr{P}$ (or $P \in \mathscr{M}$) as

$$hD(x; P) = \inf \left\{ P(H^+) \colon H^+ \text{ is a (closed) halfspace in } \mathbb{R}^d \text{ with } x \in \partial H^+ \right\}, \quad (1)$$

where by ∂K we mean the topological boundary of a set K. In the sequel, we write H^+ and H^- for the two closed halfspaces associated with their boundary hyperplane H in $\mathbb{R}^d$. Halfspace depth is equivalent with a multivariate trimming procedure. Denote by $P_\delta = \left\{ x \in \mathbb{R}^d \colon hD(x; P) \geq \delta \right\}$ the upper level set of the depth. Then P_δ can be expressed as the intersection of closed halfspaces H^+ with the property $P(H^+) > P\left(\mathbb{R}^d\right) - \delta$ [14, Proposition 6]. Thus, the level sets P_δ are, in fact, the convex sets obtained by clipping off all the tail regions that correspond to projections $\langle X, u \rangle$ of the the random vector $X \sim P$, with $u \in \mathbb{R}^d \setminus \{0\}$. The depth hD thus presents a plausible multivariate alternative to quantiles used in $\mathbb{R}$.

An essential property of quantiles in $\mathbb{R}$ is that their complete set fully determines any measure $P \in \mathscr{M}$ on $\mathbb{R}$. Therefore, no information is lost when solely quantile-based inference about distributions is conducted. A natural question whether such a property holds true also for the multivariate quantiles based on hD is called the halfspace depth characterization conjecture.

Characterization conjecture
For any two distinct (probability, or finite) Borel measures P and Q on $\mathbb{R}^d$ there exists a point $x \in \mathbb{R}^d$ such that $hD\,(x; P) \neq hD\,(x; Q)$.

This problem has a long history—a series of partial positive results to the conjecture can be found in Table 1. We know that, for instance, if $P \in \mathscr{P}$ is uniform on a finite number of points (that is, P is an empirical distribution), the position of the supporting points of P can be recovered from its depth only.

It was widely believed that the general characterization conjecture holds true ([2, p. 2306], [6, p. 1598]). Surprisingly, recently it turned out that this conjecture is false. In the present note, we review some known results that relate to this problem. In Sect. 2, we provide two examples of measures that are quite different, yet their depths are the same. In Sect. 3, we scrutinize some proofs listed in Table 1 and point to several issues with the known results. Concluding remarks and two major open problems are given in Sect. 4.

2 Negative Results

The general characterization conjecture is not valid. To illustrate this, in Sect. 2.1, we review a counter-example of a set of probability measures with the same depth from

Table 1 Review of positive results on the depth characterization problem

	Year	Ref.	Authors	Characterization for	Comment
Discrete	1999	[15]	Struyf and Rousseeuw	Empirical distributions	
	2002	[9]	Koshevoy	Finitely supported distributions	
	2007	[4]	Hassairi and Regaieg	Finitely supported distributions	
	2008	[2]	Cuesta-Albertos and Nieto-Reyes	Discrete distributions	Section 3.1.1
Continuous	2003	[10]	Koshevoy	Continuous integrable distributions	Section 3.2.1
	2008	[5]	Hassairi and Regaieg	Distributions with smooth densities	Section 3.2.2
	2010	[7]	Kong and Zuo	Distributions with smooth depth	Section 3.2.3
	2018	[13]	Nagy, Schütt, and Werner	Distributions with floating bodies	Section 3.2.3

[12]. In Sect. 2.2, we provide a new example of two mutually singular finite measures whose densities can be written explicitly, yet their halfspace depths coincide.

2.1 *Counter-Example for Probability Measures*

In a construction provided in [12], it is demonstrated that for any integer $d > 1$ there exist uncountable sets of probability measures in $\mathbb{R}^d$ whose depths are identical. The proof involves advances from [11] on the depth of α-symmetric random vectors. For $0 < \alpha \leq \infty$ given, denote for $x = (x_1, \dots, x_d)^\mathsf{T} \in \mathbb{R}^d$

$$\|x\|_\alpha = \begin{cases} \left(\sum_{i=1}^d |x_i|^\alpha\right)^{1/\alpha} & \text{for } \alpha \in (0, \infty), \\ \max_{i=1,\dots,d} |x_i| & \text{for } \alpha = \infty. \end{cases}$$

We say that a random vector $X = (X_1, \dots, X_d)^\mathsf{T} \sim P \in \mathscr{P}$ has an α-symmetric distribution, or is α-symmetric, if

$$\text{for any } u \in \mathbb{R}^d,\ \langle X, u\rangle \text{ has the same distribution as } \|u\|_\alpha X_1. \tag{2}$$

The collection of all 2-symmetric distributions is exactly the set of all spherically symmetric probability measures. The multivariate distribution with independent Cauchy marginals is an example of a 1-symmetric distribution. Generally, α-symmetric distributions generalize the stable distributions, extensively studied in probability, to multivariate sample spaces.

The distinctive property (2) of projections of α-symmetric distributions makes them well suited for depth-based analysis. For any $X \sim P$ that is α-symmetric, it is immediate that

$$\begin{aligned} hD(x; P) &= \inf_{u \in \mathbb{R}^d \setminus \{0\}} \mathsf{P}(\langle X, u \rangle \leq \langle x, u \rangle) = \inf_{u \in \mathbb{R}^d \setminus \{0\}} \mathsf{P}(\|u\|_\alpha X_1 \leq \langle x, u \rangle) \\ &= \mathsf{P}\left(X_1 \leq \inf_{u \in \mathbb{R}^d \setminus \{0\}} \langle x, u \rangle / \|u\|_\alpha \right) = \mathsf{P}(X_1 \leq - \|x\|_{\alpha^*}) \end{aligned} \tag{3}$$

for α^* the conjugate index of α. For $\alpha > 1$, the index conjugate to α satisfies $\alpha^{-1} + (\alpha^*)^{-1} = 1$; for $\alpha \leq 1$ we define $\alpha^* = \infty$. The last equality in (3) follows from a version of Hölder's inequality that can be found, for instance, in [1, Lemma A.1].

From (3), we see that the depths of all α-symmetric distributions with $0 < \alpha \leq 1$ share the same sets of contours—concentric (hyper)-cubes, and their depth decreases with $\|x\|_\infty \to \infty$ at a rate given by the cumulative distribution function of the univariate marginal distribution X_1 of X. Therefore, to construct distributions with the same depth it is enough to find different α-symmetric distributions with $\alpha \leq 1$ that have the same laws of their univariate marginals X_1. In that case, the depth of all such distributions takes the form $hD(x; P) = F(- \|x\|_\infty)$ for all $x \in \mathbb{R}^d$, where F is the cumulative distribution function of X_1.

As shown in [12], there exist large collections of different α-symmetric random vectors X with identical univariate marginals X_1. This can be concluded, for instance, from some classical advances in functional analysis, where it was demonstrated already in 1930s that certain special functions are positive definite, and as such they correspond to well-defined characteristic functions of d-dimensional α-symmetric random vectors. For technical details, we refer to [12] and references therein.

2.2 *Counter-Example for Finite Measures*

The only α-symmetric distribution discussed in Sect. 2.1 with an analytically expressible density is the 1-symmetric distribution $P^1 \in \mathscr{P}$ with independent Cauchy marginals. Its density takes the form

$$f(x) = \prod_{i=1}^{d} \frac{1}{\pi \left(1 + x_i^2\right)} \quad \text{for } x \in \mathbb{R}^d,$$

and its depth can be expressed as

$$hD\left(x; P^1\right) = \frac{1}{2} - \frac{\arctan\left(\|x\|_\infty\right)}{\pi} \quad \text{for } x \in \mathbb{R}^d. \tag{4}$$

We start from this distribution and find a couple of easily expressible finite measures that share the same depth for all $x \in \mathbb{R}^d$. In particular, we show that for some heavy-tailed symmetric measures supported on the coordinate axes

$$A_i = \left\{x \in \mathbb{R}^d : x_j = 0 \text{ for all } j \neq i\right\} \quad \text{for } i = 1, \ldots, d,$$

it follows that their depth takes the form $F\left(-\|x\|_\infty\right)$ for a fixed function F for all $x \in \mathbb{R}^d \setminus \{0\}$.

Let $P \in \mathscr{P}$ be a measure that is absolutely continuous with respect to the one-dimensional Hausdorff measure λ_S supported on the set $S = \bigcup_{i=1}^d A_i \setminus \{0\}$. The Radon-Nikodym derivative (the density) of P with respect to λ_S is given by

$$g(x) = \frac{1}{d} \sum_{i=1}^{d} \frac{\mathbb{I}\left[x \in A_i \setminus \{0\}\right]}{\pi\left(1 + x_i^2\right)} \quad \text{for } x \in \mathbb{R}^d.$$

From the symmetry of g we get $hD(0; P) = 1/2$, as any closed halfspace whose boundary passes through the origin $0 \in \mathbb{R}^d$ halves the mass of P. It is also easy to see that because P is supported only on the coordinate axes, for any $x \in A_i \setminus \{0\}$ the halfspace H^+ whose boundary H passes through x that minimizes the P-mass is

$$H^+ = \left\{y \in \mathbb{R}^d : \operatorname{sgn}(x_i)\, y_i \geq |x_i|\right\} \tag{5}$$

for sgn the signum function, i.e. H^+ with H parallel to all coordinate axes A_j for $j \neq i$, and $0 \notin H^+$. For the depth of $X \sim P$ this implies

$$hD\,(x; P) = \mathsf{P}\left(X_i \geq \|x\|_\infty\right) = \frac{1}{d}\left(\frac{1}{2} - \frac{\arctan\left(|x_i|\right)}{\pi}\right) \quad \text{for } x \in A_i \setminus \{0\}.$$

A straightforward computation shows[1] that, due to the heavy-tailedness of P, also for a general point $x \in \mathbb{R}^d \setminus \{0\}$ an analogue of the previous formula holds true, and if $j \in \{1, \ldots, d\}$ is an index such that $\left|x_j\right| = \|x\|_\infty$, we can also write

$$hD\,(x; P) = \mathsf{P}\left(X_j \geq \|x\|_\infty\right) = \begin{cases} \frac{1}{d}\left(\frac{1}{2} - \frac{\arctan(\|x\|_\infty)}{\pi}\right) & \text{for } x \in \mathbb{R}^d \setminus \{0\}, \\ 1/2 & \text{for } x = 0. \end{cases} \tag{6}$$

In particular, the depth $hD\,(\cdot; P)$ is a constant multiple of $hD\left(\cdot; P^1\right)$ from (4) at all points $x \neq 0$. To obtain two finite measures with exactly the same depth in $\mathbb{R}^d$, we modify the Cauchy distribution P^1 by taking a measure $Q \in \mathscr{M}$ to be the sum of $1/d$ times P^1, and $1/2 - 1/(2d)$ times the Dirac measure $\delta_0 \in \mathscr{P}$ concentrated at $0 \in \mathbb{R}^d$. Then,

[1] Detailed computations are omitted from the present note due to space restrictions. They are available online at http://www.karlin.mff.cuni.cz/~nagy/ or upon request from the author.

$$
\begin{aligned}
hD(x;Q) &= \frac{1}{d} hD\left(x;P^1\right) + \left(\frac{1}{2} - \frac{1}{2d}\right) hD(x;\delta_0) \\
&= \frac{hD\left(x;P^1\right)}{d} + \left(\frac{1}{2} - \frac{1}{2d}\right) \mathbb{I}[x=0] = hD(x;P) \quad \text{for } x \in \mathbb{R}^d,
\end{aligned}
$$

where the first equality is in order because for any $x \neq 0$ the halfspace H^+ with $x \in H$ whose Q-measure is minimal is that from (5), and $0 \notin H^+$. Therefore, the depths of P and Q agree for all $\mathbb{R}^d$. It is remarkable that such a property holds true even though P and Q are quite different in nature: for all integers $d > 1$

- P and Q are mutually singular, i.e. $P(S) = P\left(\mathbb{R}^d\right)$ and $Q(S) = 0$ and
- $Q\left(\mathbb{R}^d\right) = Q(0) + P^1\left(\mathbb{R}^d\right)/d = 1/2 + 1/(2d) < 1 = P\left(\mathbb{R}^d\right)$.

In particular, we have demonstrated that from the complete knowledge of the depth $hD(\cdot;P)$ only neither the full P-measure of the sample space, nor the support of P is, in general, possible to be determined.

3 Comments on Some Positive Results

In this section, we review some important partial positive results to the characterization problem that can be found in the literature. First, in Sect. 3.1, we focus on discrete (atomic) probability measures. We identify a difficulty with the proof of the most general characterization result for discrete distributions from [2]. Measures that are absolutely continuous (with respect to the d-dimensional Lebesgue measure) are studied in Sect. 3.2. A problematic point in the main proof from [10] and a corollary from [5] are noted. Finally, it is asserted that the main characterization results from [5, 7] can be seen as special cases of a theorem stated in [13] in terms of the so-called floating bodies studied in convex geometry.

3.1 Sufficient Conditions for Discrete Distributions

For discrete distributions with a finite number of atoms, it was shown in [4, 9] that the depth characterizes probability distributions. An extension of this result to measures with infinitely many atoms from [2] appears to be problematic.

3.1.1 Characterization of Cuesta-Albertos and Nieto-Reyes

The main result in [2] concerns a continuity argument. In the key Lemma 2.2 from [2], it is asserted that

Claim Let $x \in \mathbb{R}^d$, $X \sim P \in \mathscr{P}$, and let $\{x_n\}_{n=1}^{\infty} \subset \mathbb{R}^d$ be a sequence such that $x_n \neq x$ for all n, and $\lim_{n\to\infty} x_n = x$. Then, for

$$U = \left\{u \in \mathbb{R}^d \colon \|u\|_2 = 1, \mathsf{P}\left(\langle X, u\rangle = \langle x, u\rangle\right) = P(\{x\})\right\}$$

we have that

$$\inf_{u\in U} \mathsf{P}\left(\langle X, u\rangle \leq \langle x, u\rangle\right) - P\left(\{x\}\right) \leq \liminf_{n\to\infty} \inf_{u\in U} \mathsf{P}\left(\langle X, u\rangle \leq \langle x_n, u\rangle\right). \quad (7)$$

Roughly speaking, formula (7) asserts a relaxed version of lower semi-continuity of the 'almost' halfspace depth $hD_U(x; P) = \inf_{u\in U} \mathsf{P}\left(\langle X, u\rangle \leq \langle x, u\rangle\right)$ in x (note that if P is either absolutely continuous, or finitely supported, then $hD_U(x; P) = hD(x; P)$ for all $x \in \mathbb{R}^d$). Together with the well-known fact that $hD_U(\cdot; P)$ is an upper semi-continuous mapping for any $P \in \mathscr{P}$, (7) yields for $\lim_{n\to\infty} x_n = x$, $x_n \neq x$,

$$\begin{aligned} hD_U(x; P) - P\left(\{x\}\right) &\leq \liminf_{n\to\infty} hD_U(x_n; P) \\ &\leq \limsup_{n\to\infty} hD_U(x_n; P) \leq hD_U(x; P). \end{aligned}$$

In particular, for x that is not an atom of P, it implies that $hD_U(\cdot; P)$ is continuous at x. But, this cannot be true in general. Take, for instance, P the uniform distribution on the four vertices of a square in $\mathbb{R}^2$, and x from the boundary of this square that is not its vertex. Then $hD(x; P) = hD_U(x; P) = 1/4$, yet for any sequence of points x_n that converge to x from the outside of the square, $hD(x_n; P) = hD_U(x_n; P) = 0$ for all n, and (7) is violated.[2]

Consequently, it appears that the proof of [2, Theorem 2.6] is incomplete, and the question whether two different atomic probability distributions can have the same depth is still open.

3.2 Sufficient Conditions for Absolutely Continuous Distributions

3.2.1 Characterization of Koshevoy

In [10, Theorem 5.1], it is asserted that if $P, Q \in \mathscr{M}$ are different, properly integrable absolutely continuous probability measures in $\mathbb{R}^d$, then there must exist $x \in \mathbb{R}^d$ with $hD(x; P) \neq hD(x; Q)$. The proof of this claim is quite unusual, since it involves rather advanced convex-geometric tools such as the lift zonoid or the Radon transform. Recall that a lift zonoid of a properly integrable measure $P \in \mathscr{M}$ is a deter-

[2]Inspection of the proof of [2, Lemma 2.2] shows that in the formula on page 2308, line 6 of [2], $x_{n_k^*}$ cannot be replaced by x.

ministic convex body $\hat{Z}(P) \subset \mathbb{R}^{d+1}$ constructed as the set of certain expectations of $X \sim P$. Its detailed construction and properties are described in [8]. In what follows it will be important that the lift zonoid completely characterizes P [8, Theorem 3.5]. A key step in the proof of the depth characterization result of Koshevoy is [10, Proposition 5.1], where the following is derived.

Claim Let $P, Q \in \mathcal{M}$ be absolutely continuous and compactly supported, and suppose that for some $\delta > 0$ we have $K = P_\delta = Q_\delta$. Then, for $H_\delta = \left\{x \in \mathbb{R}^{d+1} : x_1 = \delta\right\}$, it holds true that $\hat{Z}(P) \cap H_\delta = \hat{Z}(Q) \cap H_\delta$.

The proof of this statement in [10] is based on the following facts:

(a) For (any) convex body K, the set K_S of points $x \in \partial K$ with a unique unit outer normal $u(x)$ is a dense subset of the boundary of K.
(b) For any $x \in K_S$, for the halfspace $H_x^+ \subset \mathbb{R}^d$ with inner normal $u(x)$ such that $x \in H_x$ we have $P(H_x^+) = Q(H_x^+) = \delta$.
(c) Halfspaces $H^+ \subset \mathbb{R}^d$ such that $P(H^+) = \delta$ are in one-to-one correspondence with the extreme points of $\hat{Z}(P) \cap H_\delta$, and analogously for measure Q.

This argumentation does not appear to be complete. Indeed, from (a) and (b) it does not follow that the collection of the inner normals of halfspaces $\left\{H_x^+ \subset \mathbb{R}^d : x \in K_S\right\}$ is dense in the unit sphere in $\mathbb{R}^d$. We saw this in Sect. 2.1, where it was shown that for α-symmetric measures[3] with $\alpha \leq 1$ for all $\delta > 0$ there is only a finite number of distinct normals $u(x)$ at $x \in K_S$. Thus, by (a)–(c) above we are able to identify only a finite number of extreme points of $\hat{Z}(P) \cap H_\delta$ and $\hat{Z}(Q) \cap H_\delta$, which is surely not enough to guarantee that the two cuts of the lift zonoids are identical.

3.2.2 Characterization of Hassairi and Regaieg

Another intriguing characterization result for the halfspace depth from [5] asserts that if $P \in \mathcal{P}$ is absolutely continuous with density f that is positive and smooth enough in a connected open set, then the halfspace depth uniquely determines P.

This result does not appear to hold true in its full generality. Consider the example from Sect. 2.1. There, measure P is absolutely continuous, its support is $\mathbb{R}^d$ and its density is infinitely differentiable everywhere on $\mathbb{R}^d$. Thus, it satisfies all the conditions from [5]. Yet, there are different measures with the same depth.

The issue with the proof of [5, Theorem 3.1] appears to stem from a minor neglect of conditions necessary to apply the fundamental theorem of calculus in one step of the proof (formula (3.1) on page 2310). This problem could be easily fixed by imposing somewhat stronger conditions on P; for details, see [13, Sect. 8.2]. More importantly, in the latter reference it is also demonstrated that the smoothness of the density f is not enough to guarantee the technical condition (H) on page 2312 of [5]. Thus, smoothness of f alone is not sufficient for P to have a unique depth

[3] Measures from Sect. 2.1 are not compactly supported. But, also for compactly supported measures, the depth level sets may fail to be strictly convex, see [13, Example 7].

(see, again, Sect. 2.1). In fact, it turns out that condition (H) from [5] is implied by a somewhat weaker condition, stated in terms of the so-called floating bodies of measure P, recently discussed in [13].

3.2.3 Characterization Using Floating Bodies

In analogy with the research in convex geometry [3], we say that for $P \in \mathcal{M}$ and $\delta > 0$ the non-empty convex set $P_{[\delta]}$ is the *floating body* of P if for each outer supporting halfspace H^+ of $P_{[\delta]}$ (that is, $P_{[\delta]} \cap H^+ \neq \emptyset$ and $P_{[\delta]} \subset H^-$) we have $P(H^+) = \delta$. In [13, Theorem 34], the following theorem is proved.

Theorem 1 *Let $P \in \mathcal{M}$ have a connected support, and let $x_P \in \mathbb{R}^d$ be given by $hD\,(x_P;\, P) = \sup_{y \in \mathbb{R}^d} hD\,(y;\, P)$. Then the following are equivalent:*

(FB$_1$) For each $\delta \in (0, P\left(\mathbb{R}^d\right)/2)$ the floating body $P_{[\delta]}$ of P exists.
(FB$_2$) $P(H) = 0$ for any hyperplane $H \subset \mathbb{R}^d$, $P\left(\mathbb{R}^d\right) = 2 \sup_{x \in \mathbb{R}^d} hD\,(x;\, P)$, and

$$P(H^+) = \begin{cases} \sup_{x \in H} hD(x;\, P) & \text{for a hyperplane} H \text{ with } x_P \notin H^+, \\ P\left(\mathbb{R}^d\right) - \sup_{x \in H} hD(x;\, P) & \text{for a hyperplane } H \text{ with } x_P \in H^+. \end{cases} \tag{8}$$

Consequently, if (FB$_1$) is true, there is no other finite measure that satisfies (FB$_1$) with the same depth at all points in $\mathbb{R}^d$.

Existence of all floating bodies of $P \in \mathcal{M}$ is a rather strict condition. It is equivalent with the fact that all halfspaces whose measure is $\delta \leq P\left(\mathbb{R}^d\right)/2$ support P_δ. In particular, (FB$_1$) implies that P must possess a certain symmetry property. Nonetheless, (FB$_1$) is known to be satisfied if, for instance, P is elliptically symmetric, or if P is a uniform measure on a symmetric, smooth and strictly convex body in $\mathbb{R}^d$.

It can be shown that condition (H) from [5], discussed in Sect. 3.2.2, is stronger than (FB$_1$): a measure that satisfies (H) must obey also (FB$_1$). A further interesting relation of (FB$_1$) to the literature is that if all the contours of the depth $hD\,(\cdot;\, P)$ are smooth, (FB$_1$) is guaranteed to be valid for P. Therefore, the characterization theorem of Kong and Zuo [7], which states that a measure with smooth depth contours has a unique depth, is another special case of Theorem 1. For references to these results and further comments, we refer to [13, Sect. 8].

4 Characterization Problem: Summary

From our review we may conclude that the universal halfspace depth characterization conjecture holds true neither for finite measures, nor for probability distributions. This suggests the following problem.

Characterization problem
Describe the set of all (probability, or finite) Borel measures P on $\mathbb{R}^d$ such that for any $Q \neq P$ there exists a point $x \in \mathbb{R}^d$ such that $hD(x; P) \neq hD(x; Q)$.

From what we saw in Sects. 2 and 3, it is known that distinct P and Q in $\mathcal{M}$ cannot have the same halfspace depth if

- they are both finitely supported or
- if all floating bodies of both P and Q exist.

These two classes of distributions are too small for viable applications in statistics or geometry. For instance, currently it appears to be unknown whether even the uniform distribution on a triangle in $\mathbb{R}^2$ is determined by its depth!

A further, more difficult question to be addressed is the reconstruction of the measure from its depth (or its floating bodies).

Reconstruction problems
- For a Borel measure P characterized by its depth, determine the P-measure of all Borel sets of $\mathbb{R}^d$ (or, equivalently, all halfspaces in $\mathbb{R}^d$) only from its depth.
- For P, a uniform measure on a given convex body in $\mathbb{R}^d$, is it possible from a single floating body of P to recover the support of P?

Apart from immediate statistical applications, the second reconstruction problem is of great interest also in geometry and functional analysis. It is closely connected with the so-called *homothety conjecture*, or the *floating body problem*, that concerns alternative characterization of ellipsoids as the only convex bodies that are homothetic to their own floating bodies. For details see the references in [13, Sect. 8.3].

Acknowledgements The author is grateful to Jan Rataj and Dušan Pokorný for pointing out the problems with proofs in references [2, 10] and to Jan Kynčl for his contributions to the example from Sect. 2.2. This work was supported by the grant 19-16097Y of the Czech Science Foundation and by the PRIMUS/17/SCI/3 project of Charles University.

References

1. Chen, Z., Tyler, D.E.: On the behavior of Tukey's depth and median under symmetric stable distributions. J. Stat. Plan. Infer. **122**(1–2), 111–124 (2004)
2. Cuesta-Albertos, J.A., Nieto-Reyes, A.: The Tukey and the random Tukey depths characterize discrete distributions. J. Multivar. Anal. **99**(10), 2304–2311 (2008)
3. Dupin, C.: Applications de Géométrie et de Méchanique. Bachelier, Paris (1822)

4. Hassairi, A., Regaieg, O.: On the Tukey depth of an atomic measure. Stat. Methodol. **4**(2), 244–249 (2007)
5. Hassairi, A., Regaieg, O.: On the Tukey depth of a continuous probability distribution. Stat. Probab. Lett. **78**(15), 2308–2313 (2008)
6. Kong, L., Mizera, I.: Quantile tomography: using quantiles with multivariate data. Stat. Sinica **22**(4), 1589–1610 (2012)
7. Kong, L., Zuo, Y.: Smooth depth contours characterize the underlying distribution. J. Multivar. Anal. **101**(9), 2222–2226 (2010)
8. Koshevoy, G., Mosler, K.: Lift zonoids, random convex hulls and the variability of random vectors. Bernoulli **4**(3), 377–399 (1998)
9. Koshevoy, G.A.: The Tukey depth characterizes the atomic measure. J. Multivar. Anal. **83**(2), 360–364 (2002)
10. Koshevoy, G.A.: Lift-zonoid and multivariate depths. In: Developments in Robust Statistics (Vorau, 2001), pp. 194–202. Physica, Heidelberg (2003)
11. Massé, J.-C., Theodorescu, R.: Halfplane trimming for bivariate distributions. J. Multivar. Anal. **48**(2), 188–202 (1994)
12. Nagy, S.: Halfspace depth does not characterize probability distributions. Statist. Papers, 2019. To appear
13. Nagy, S., Schütt, C., Werner, E.M.: Halfspace depth and floating body. Stat. Surv. **13**, 52–118 (2019)
14. Rousseeuw, P.J., Ruts, I.: The depth function of a population distribution. Metrika **49**(3), 213–244 (1999)
15. Struyf, A., Rousseeuw, P.J.: Halfspace depth and regression depth characterize the empirical distribution. J. Multiv. Anal. **69**(1), 135–153 (1999)
16. Tukey, J.W.: Mathematics and the picturing of data. In: Proceedings of the International Congress of Mathematicians (Vancouver, B. C., 1974), Vol. 2, pp. 523–531. Canad. Math. Congress, Montreal, Que., 1975

A Component Multiplicative Error Model for Realized Volatility Measures

Antonio Naimoli and Giuseppe Storti

Abstract We propose a component Multiplicative Error Model (MEM) for modelling and forecasting realized volatility measures. In contrast to conventional MEMs, the proposed specification resorts to the use of a multiplicative component structure in order to parsimoniously parameterize the complex dependence structure of realized volatility measures. The long-run component is defined as a linear combination of MIDAS filters moving at different frequencies, while the short-run component is constrained to follow a unit mean GARCH recursion. This particular specification of the long-run component allows to reproduce very persistent oscillations of the conditional mean of the volatility process, in the spirit of Corsi's Heterogeneous Autoregressive Model (HAR). The empirical performances of the proposed model are assessed by means of an application to the realized volatility of the S&P 500 index.

Keywords Realized volatility · Component Multiplicative Error Model · Long-range dependence · MIDAS · Volatility forecasting

1 Introduction

In financial econometrics, the last two decades have witnessed an increasing interest in the development of dynamic models incorporating information on realized volatility measures. The reason is that it is believed these models can provide more accurate forecasts of financial volatility than the standard volatility models based on daily squared returns, e.g. the GARCH(1,1).

Engle and Russell [14] originally proposed the Autoregressive Conditional Duration (ACD) model as a tool for modelling irregularly spaced transaction data observed at high frequency. This model has been later generalized in the class of Multiplica-

A. Naimoli (✉) · G. Storti
University of Salerno, DISES, Via G. Paolo II, 132, 84084 Fisciano, SA, Italy
e-mail: anaimoli@unisa.it

G. Storti
e-mail: storti@unisa.it

M. La Rocca et al. (eds.), *Nonparametric Statistics*, Springer Proceedings in Mathematics & Statistics 339, https://doi.org/10.1007/978-3-030-57306-5_35

tive Error Model (MEM) by [10] for modelling and forecasting positive-valued random variables that are decomposed into the product of their conditional mean and a positive-valued i.i.d. error term with unit mean. Discussions and extensions on the properties of this model class can be found in [4–8, 18, 19], among others.

One of the most prominent fields of application of MEMs is related to the modelling and forecasting of realized volatility measures. It is well known that these variables have very rich serial dependence structures sharing the features of clustering and high persistence. The recurrent feature of long-range dependence is conventionally modelled as an Autoregressive Fractionally Integrated Moving Average (ARFIMA) process as in [3], or using regression models mixing information at different frequencies such as the Heterogeneous AR (HAR) model of [9]. The HAR model, inspired by the heterogeneous market hypothesis of [20], is based on additive cascade of volatility components over different horizons. This particular structure, despite the simplicity of the model, has been found to satisfactorily reproduce the empirical regularities of realized volatility series, including their highly persistent autocorrelation structure.

In this field, *component models* are an appealing alternative to conventional models since they offer a tractable and parsimonious approach to modelling the persistent dependence structure of realized volatility measures. Models of this type have first been proposed in the GARCH framework and are usually characterized by the mixing of two or more components moving at different frequencies. Starting from the Spline GARCH of [13], where volatility is specified to be the product of a slow-moving component, represented by an exponential spline, and a short-run component which follows a unit mean GARCH process, several contributions have extended and refined this idea. [12] introduced a new class of models called GARCH-MIDAS, where the long-run component is modelled as a MIDAS (Mixed-Data Sampling, [16]) polynomial filter which applies to monthly, quarterly or biannual financial or macroeconomic variables. [2] decomposed the variance into a conditional and an unconditional component such that the latter evolves smoothly over time through a linear combination of logistic transition functions taking time as the transition variable.

Moving to the analysis of intra-daily data, [15] developed the multiplicative component GARCH, decomposing the volatility of high-frequency asset returns into the product of three components, namely, the conditional variance is a product of daily, diurnal and stochastic intra-daily components. Recently [1] have provided a survey on univariate and multivariate GARCH-type models featuring a multiplicative decomposition of the variance into short- and long-run components.

This paper proposes a novel multiplicative dynamic component model which is able to reproduce the main stylized facts arising from the empirical analysis of time series of realized volatility. Compared to other specifications falling into the class of component MEMs, the main innovation of the proposed model can be found in the structure of the long-run component. Namely, as in [21], this is modelled as an additive cascade of MIDAS filters moving at different frequencies. This choice is motivated by the empirical regularities arising from the analysis of realized volatility measures that are typically characterized by two prominent and related features: a slowly moving long-run level and a highly persistent autocorrelation structure. For ease of reference, we will denote the parametric specification adopted for the long-run component as a Heterogeneous MIDAS (H-MIDAS) filter. Residual short-term autocorrelation is then explained by a short-run component that follows a mean reverting unit GARCH-type model. The overall model will be referred to as a H-MIDAS Component MEM model (H-MIDAS-CMEM). It is worth noting that, specifying the long-run component as an additive cascade of volatility filters as in [9], we implicitly associate this component to long-run persistent movements of the realized volatility process.

The model that is here proposed differs from that discussed in [21] under two main respects. First, in this paper, we model realized volatilities on a daily scale rather than high-frequency intra-daily trading volumes. Second, the structure of the MIDAS filters in the long-run component is based on a *pure* rolling window rather than on a *block* rolling window scheme.

The estimation of model parameters can be easily performed by maximizing a likelihood function based on the assumption of Generalized F distributed errors. The motivation behind the use of this distribution is twofold. First, nesting different distributions, the Generalized F results very flexible in modelling the distributional properties of the observed variable. Second, it can be easily extended to control the presence of zero outcomes [17].

In order to assess the relative merits of the proposed approach we present the results of an application to the realized volatility time series of the S&P 500 index in which the predictive performance of the proposed model is compared to that of the standard MEM by means of an out-of-sample rolling window forecasting experiment. The volatility forecasting performance has been assessed using three different loss functions, the Mean Squared Error (MSE), the Mean Absolute Error (MAE) and the QLIKE. The Diebold-Mariano test is then used to evaluate the significance of differences in the predictive performances of the models under analysis. Our findings suggest that the H-MIDAS-CMEM significantly outperforms the benchmark in terms of forecasting accuracy.

The remainder of the paper is structured as follows. In Sect. 2, we present the proposed H-MIDAS-CMEM model, while the estimation procedure is described in Sect. 3. The results of the empirical application are presented and discussed in Sect. 4. Finally, Sect. 5 concludes.

2 Model Specification

Let $\{v_{t,i}\}$ be a time series of daily realized volatility (RV) measures observed on day i in period t, such as in a month or a quarter. The general H-MIDAS-CMEM model can be formulated as

$$v_{t,i} = \tau_{t,i}\, g_{t,i}\, \varepsilon_{t,i}, \qquad \varepsilon_{t,i}|\mathscr{F}_{t,i-1} \overset{iid}{\sim} \mathscr{D}^{+}(1, \sigma^2), \tag{1}$$

where $\mathscr{F}_{t,i-1}$ is the sigma-field generated by the available intra-daily information until day $(i-1)$ of period t. The conditional expectation of $v_{t,i}$, given $\mathscr{F}_{t,i-1}$, is the product of two components characterized by different dynamic specifications. In particular, $g_{t,i}$ represents a daily dynamic component that reproduces autocorrelated movements around the current long-run level, while $\tau_{t,i}$ is a smoothly varying component given by the sum of MIDAS filters moving at different frequencies. This component is designed to track the dynamics of the long-run level of realized volatility.[1] In order to make the model identifiable, as in [12], the short-run component is constrained to follow a mean reverting unit GARCH-type process. Namely, $g_{t,i}$ is specified as

$$g_{t,i} = \omega^* + \sum_{j=1}^{r} \alpha_j \frac{v_{t,i-j}}{\tau_{t,i-j}} + \sum_{k=1}^{s} \beta_k \, g_{t,i-k}, \qquad \tau_{t,i} > 0 \ \ \forall_{t,i} \ . \tag{2}$$

To fulfill the unit mean assumption on $g_{t,i}$, it is necessary to set appropriate constraints on ω^* by means of a targeting procedure. In particular, taking the expectation of both sides of $g_{t,i}$, it is easy to show that

$$\omega^* = (1 - \sum_{j=1}^{r} \alpha_j - \sum_{k=1}^{s} \beta_k).$$

Positivity of $g_{t,i}$ is then ensured by setting the following standard constraints: $\omega^* > 0$, $\alpha_j \geq 0$ for $j = 1, \ldots, r$, and $\beta_k \geq 0$ for $k = 1, \ldots, s$.[2]

On the other hand, the low-frequency component is modelled as a linear combination of MIDAS filters of past volatilities aggregated at different frequencies. A general formulation of the long-run component is given by

$$\begin{aligned} log(\tau_{t,i}) = \delta + \theta_s \sum_{k=1}^{K} \varphi_k(\omega_{1,s}, \omega_{2,s}) \, log\left(VS_{t,i}^{(k)}\right) \\ + \theta_m \sum_{h=1}^{K^*} \varphi_h(\omega_{1,m}, \omega_{2,m}) \, log\left(VM_{t,i}^{(h)}\right), \end{aligned} \tag{3}$$

where $VS_{t,i}^{(k)}$ and $VM_{t,i}^{(h)}$ denote the RV aggregated over a rolling window of length equal to n_s and n_m, respectively, with $n_s > n_m$, while K is the number of MIDAS lags and $K^* = K + n_s - n_m$. In particular,

[1]The stochastic properties of the model have been derived by [21] to which the interested reader may refer for additional details.

[2]Note that strict positivity, i.e. $\alpha_j > 0$ for at least one $j \in \{1, \ldots, r\}$, is needed for identification if $s > 0$.

$$VS_{t,i}^{(k)} = \sum_{j=1}^{n_s} v_{t,i-(k-1)-j} \qquad \text{for } k = 1, \ldots, K \tag{4}$$

and

$$VM_{t,i}^{(h)} = \sum_{j=1}^{n_m} v_{t,i-(h-1)-j} \qquad \text{for } h = 1, \ldots, K^* . \tag{5}$$

In the empirical application, we choose $n_s = 125$ implying a biannual rolling window RV and $n_m = 22$, meaning that the RV is rolled back monthly. Furthermore, the long-run component is considered in terms of logarithmic specification since it does not require parameter constraints to ensure the positivity of $\tau_{t,i}$.

Finally, the weighting function $\varphi(\underline{\omega})$ is computed according to the Beta weighting scheme which is generally defined as

$$\varphi_k(\omega_1, \omega_2) = \frac{(k/K)^{\omega_1-1}(1-k/K)^{\omega_2-1}}{\sum_{j=1}^{K}(j/K)^{\omega_1-1}(1-j/K)^{\omega_2-1}}, \tag{6}$$

where the weights in Eq. (6) sum up to 1. As discussed in [16], this Beta-specification is very flexible, being able to accommodate increasing, decreasing or hump-shaped weighting schemes, where the number of lags K need to be properly chosen by information criteria to avoid overfitting problems.

This multiple frequency specification appears to be preferable to the single-frequency MIDAS filter for at least two different reasons. First, the modeller is not constrained to choose a specific frequency for trend estimation, but can determine the optimal blend of low- and high-frequency information in a fully data-driven fashion. Second, as pointed out in [9], an additive cascade of linear filters, applied to the same variable aggregated over different time intervals, can allow to reproduce very persistent dynamics such as those typically observed for realized volatilities. We have also investigated the profitability of adding more components to the specification of $\tau_{t,i}$. However, this did not lead to any noticeable improvement in terms of fit and forecasting accuracy.

3 Estimation

The model parameters can be estimated in one step by Maximum Likelihood (ML), assuming that the innovation term follows a Generalized F (GF) distribution. Alternatively, estimation could be performed by maximizing a quasi-likelihood function based on the assumption that the errors $\varepsilon_{t,i}$ are conditionally distributed as a unit Exponential distribution that can be seen as the counterpart of the standard normal distribution for positive-valued random variables [10, 11]. To save space, here we focus on ML estimation based on the assumption of GF errors.

In particular, let X be a non-negative random variable, the density function of a GF random variable is given by

$$f(x;\underline{\zeta}) = \frac{a x^{ab-1}[c + (x/\eta)^a]^{(-c-b)}\, c^c}{\eta^{ab}\, \mathscr{B}(b, c)}, \tag{7}$$

where $\underline{\zeta} = (a, b, c, \eta)', a > 0, b > 0, c > 0$ and $\eta > 0$, with $\mathscr{B}(\cdot, \cdot)$ the Beta function such that $\mathscr{B}(b, c) = [\Gamma(b)\Gamma(c)]/\Gamma(b + c)$. The GF distribution is based on a scale parameter η and three shape parameters a, b and c, and thus it is very flexible, nesting different error distributions, such as the Weibull for $b = 1$ and $c \to \infty$, the generalized Gamma for $c \to \infty$ and the log-logistic for $b = 1$ and $c = 1$. The Exponential distribution is also asymptotically nested in the GF for $a = b = 1$ and $c \to \infty$.

Note that in the presence of zero outcomes the Zero-Augmented Generalized F (ZAF) distribution [17] can be used.

In order to ensure that the unit mean assumption for $\varepsilon_{t,i}$ is fulfilled, we need to set $\eta = \xi^{-1}$, where

$$\xi = c^{1/a}\, [\Gamma(b + 1/a)\Gamma(c - 1/a)]\, [\Gamma(b)\Gamma(c)]^{-1}\, .$$

The log-likelihood function is then given by

$$\begin{aligned} \mathscr{L}(\underline{v}; \underline{\vartheta}) = \sum_{t,i} \{ & log\, a + (ab - 1)\, log\, (\varepsilon_{t,i}) + c\, log\, c - (c + b)\, log\, [c + (\xi \varepsilon_{t,i})^a] + \\ & - log(\tau_{t,i}\, g_{t,i}) - log\, \mathscr{B}(b, c) + ab\, log(\xi) \}\, , \end{aligned} \tag{8}$$

where $\varepsilon_{t,i} = \frac{v_{t,i}}{\tau_{t,i}\, g_{t,i}}$ and $\underline{\vartheta}$ is the parameter vector to be estimated.

4 Empirical Application

To assess the performance of the proposed model, in this section, we present and discuss the results of an empirical application to the S&P 500 realized volatility series.[3] The 5-min intra-daily returns have been used to compute the daily RV series covering the period between 03 January 2000 and 27 December 2018 for a total of 4766 observations. The analysis has been performed using the software *R* [23].

Graphical inspection of the S&P 500 realized volatility, displayed in Fig. 1, reveals several periods of high volatility. These essentially refer to the dot com bubble in 2002, the financial crisis starting in mid-2007 and peaking in 2008 and the crisis in Europe progressed from the banking system to a sovereign debt crisis with the highest turmoil level in the late 2011. More recently, the stock market sell-off that

[3]The data have been downloaded from the OMI realized library available at: https://realized.oxford-man.ox.ac.uk.

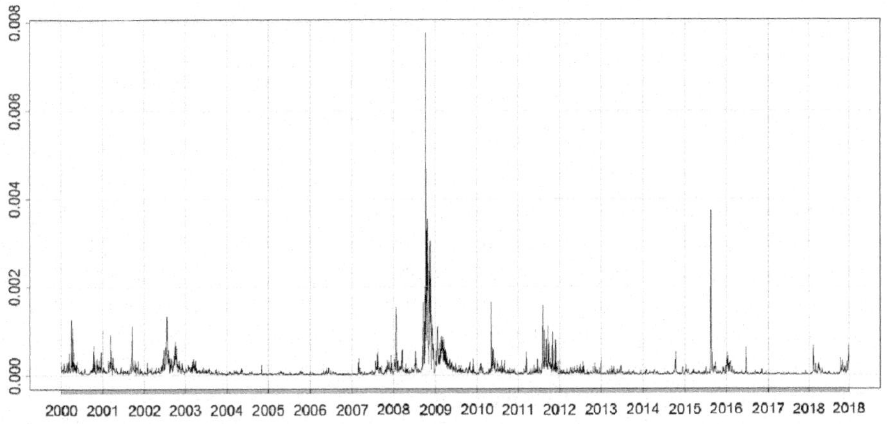

Fig. 1 S&P 500 Realized Volatility

Table 1 In sample parameter estimates for the Generalized F distribution

Model	Short run parameters			Long run parameters					GF parameters			Likelihood Information	
	ω^*	α	β	δ	θ_s	θ_m	$\omega_{2,s}$	$\omega_{2,m}$	a	b	c	$\mathscr{L}(\underline{v};\underline{\vartheta})$	BIC
MEM	0.015	0.442	0.543	-	-	-	-	-	2.015	1.877	1.767	-354.729	751.221
		0.017	0.018						0.170	0.266	0.236		
HMIDAS	0.082	0.455	0.461	2.356	-2.724	3.670	1.224	1.764	2.236	1.728	1.417	-283.363	650.253
		0.019	0.021	0.769	0.426	0.449	0.210	0.247	0.321	0.428	0.313		

Parameter estimates for MEM and H-MIDAS-CMEM. Estimation is performed on the full sample period 03 Jan 2000–27 Dec 2018 using the GF distribution. Standard errors are reported in smaller font under the parameter values. All parameters are significant at 5%

occurred between June 2015 and June 2016 is related to different events such as the Chinese stock market turbulence, but also to the uncertainty around FED interest rates, oil prices, Brexit and the U.S. presidential election. Finally, economic and political uncertainties are the most prevalent drivers of market volatility in 2018.

The model parameters have been estimated by ML, relying on the assumption of GF errors and, as a robustness check, by Exponential QML. Estimation results, based on the full sample 5-min RV, are reported in Tables 1 and 2, respectively. For ML, standard errors are based on the numerically estimated Hessian at the optimum, whereas for QML, we resort to the usual sandwich estimator. The performance of the H-MIDAS-CMEM has been compared to that of the standard MEM(1,1) specification, considered as benchmark model.

Regarding the H-MIDAS-CMEM, the short-run component follows a mean reverting unit GARCH(1,1) process, while the long-term component is specified as a combination of two MIDAS filters moving at a semiannual ($n_s = 125$) and a monthly ($n_m = 22$) frequency, with K corresponding to two MIDAS lag years. It is worth noting that, although the Beta lag structure in (6) includes two parameters, following a common practice in the literature on MIDAS models, in our empirical applications,

$\omega_{1,s}$ and $\omega_{1,m}$ have been set equal to 1 in order to have monotonically decreasing weights over the lags.

The panel of the short-term component in Table 1 shows that the intercept ω^* is slightly higher for the H-MIDAS-CMEM than the standard MEM. Furthermore, standard errors for ω^* are missing since it is estimated through the expectation targeting procedure. The parameter α takes values much larger than those typically obtained fitting GARCH models to log-returns, while the opposite holds for β. The analysis of the long-run component reveals that all the involved parameters in $log(\tau_{t,i})$ are statistically significant. In particular, the slope coefficient θ_s of the biannual filter is negative, while θ_m associated to the monthly filter is positive. Moreover, the coefficients $\omega_{2,s}$ and $\omega_{2,m}$ defining the features of the Beta weighting function take on values such that the weights slowly decline to zero over the lags. Finally, the panel referring to the error distribution parameters indicates that the GF coefficients are similar between MEM and H-MIDAS-CMEM.

From a comparison of the log-likelihoods, it clearly emerges that the value recorded for the H-MIDAS-CMEM is much larger than that of the competing model. In addition, the BIC reveals that there is a big improvement coming from the inclusion of the heterogeneous component in the MIDAS trend which allows to better capture the changes in the dynamics of the average volatility level.

In the QML case (Table 2), the estimated short-run component parameters are reasonably close to those reported for ML estimation. This is, however, not true for the parameters of the long-run component. As expected, the BIC values are always larger than the ones obtained under the GF distribution.

The out-of-sample predictive ability of the models, for the S&P 500 RV time series, has been assessed via a rolling window forecasting exercise leaving the last 500 observations as out-of-sample forecasting period, that is, 30 December 2016–27 December 2018.

The predictive performance of the examined models is evaluated by computing the Mean Squared Error (MSE), Mean Absolute Error (MAE) and QLIKE [22] loss functions, using the 5-min RV as volatility proxy, namely,

Table 2 In sample parameter estimates for the Exponential distribution

Model	Short run parameters			Long run parameters					Likelihood Information	
	ω^*	α	β	δ	θ_s	θ_m	$\omega_{2,s}$	$\omega_{2,m}$	$\mathscr{L}(\underline{v};\underline{\vartheta})$	*BIC*
MEM	0.018	0.460	0.522	-	-	-	-	-	-1561.086	3138.878
		0.038	0.041							
HMIDAS	0.069	0.504	0.427	-0.972	-0.752	1.612	4.894	5.040	-1541.576	3141.622
		0.043	0.042	0.377	0.164	0.181	0.688	0.784		

Parameter estimates for MEM and H-MIDAS-CMEM. Estimation is performed on the full sample period 03 Jan 2000–27 Dec 2018 using the Exponential distribution. Robust standard errors are reported in smaller font under the parameters value. All parameters are significant at 5%

Table 3 S&P 500 out-of-sample loss functions comparison

	Generalized F			Exponential		
	MSE	MAE	QLIKE	MSE	MAE	QLIKE
MEM	0.2525	0.2084	-0.5378	0.2517	0.2090	-0.5359
HMIDAS	0.2364	0.1933	-0.5563	0.2468	0.1992	-0.5554
DM	2.3870	6.3552	8.7803	1.0512	5.0477	9.5892
p-value	0.0174	0.0000	0.0000	0.2937	0.0000	0.0000

Top panel: loss function average values for Mean Squared Error (MSE), Mean Absolute Error (MAE) and QLIKE. Bottom panel: Diebold-Mariano test statistics (DM) with the corresponding p-values. Positive statistics are in favour of the H-MIDAS-CMEM model. Values in the table refer to models fitted using the Generalized F distribution (left panel) and the Exponential distribution (right panel). Better models correspond to lower losses

$$MSE = \sum_{t=1}^{T}\sum_{i=1}^{I}(v_{t,i} - \hat{v}_{t,i})^2;$$

$$MAE = \sum_{t=1}^{T}\sum_{i=1}^{I}|v_{t,i} - \hat{v}_{t,i}|;$$

$$QLIKE = \sum_{t=1}^{T}\sum_{i=1}^{I}log(\hat{v}_{t,i}) + \frac{v_{t,i}}{\hat{v}_{t,i}}.$$

The significance of differences in forecasting accuracy is assessed by means of the two-sided Diebold-Mariano test under the null hypothesis that MEM and H-MIDAS-CMEM exhibit the same forecasting ability.

The out-of-sample performance of the fitted models is summarized in Table 3, reporting the average values of the considered loss functions (top panel) and the Diebold-Mariano (DM) test statistics, together with the associated p-values (bottom panel). The empirical results suggest that the H-MIDAS-CMEM always returns average losses that are significantly lower than those recorded for the benchmark MEM. The only exception occurs for the MSE when models are fitted by Exponential QML. In this case, the H-MIDAS-CMEM still returns a lower average loss, but the null of equal predictive ability cannot be rejected. Finally, comparing forecasts based on models fitted by MLE and QMLE, respectively, we find that there are no striking differences between these two sets of forecasts, with the former returning slightly lower average losses.

5 Concluding Remarks

This paper investigates the usefulness of the application of the Heterogeneous MIDAS Component MEM (H-MIDAS-CMEM) for fitting and forecasting realized volatility measures. The introduction of the heterogeneous MIDAS component, specified as an additive cascade of linear filters which take on different frequencies, allows to better capture the main empirical properties of the realized volatility, such as clustering and memory persistence. The empirical analysis of the realized volatility series of the S&P 500 index points out that the H-MIDAS-CMEM outperforms the standard MEM model in fitting the S&P 500 volatility. At the same time, the out-of-sample comparison shows that, for all the loss functions considered, the H-MIDAS-CMEM significantly outperforms the benchmark in terms of predictive accuracy. These findings appear to be robust to the choice of the error distribution. Accordingly, gains in predictive ability are mainly determined by the dynamic structure of the H-MIDAS-CMEM, rather than from the estimation method (MLE versus QMLE).

Finally, although the model discussed in this paper is motivated by the empirical properties of realized volatility measures, our approach can be easily extended to the analysis of other financial variables sharing the same features, such as trading volumes, bid-ask spreads and durations.

References

1. Amado, C., Silvennoinen, A., Terasvirta, T.: Models with Multiplicative Decomposition of Conditional Variances and Correlations, vol. 2. Routledge, United Kingdom (2019)
2. Amado, C., Teräsvirta, T.: Modelling volatility by variance decomposition. J. Econ. **175**(2), 142–153 (2013)
3. Andersen, T.G., Bollerslev, T., Diebold, F.X., Labys, P.: Modeling and forecasting realized volatility. Econometrica **71**(2), 579–625 (2003)
4. Brownlees, C.T., Cipollini, F., Gallo, G.M.: Intra-daily volume modeling and prediction for algorithmic trading. J. Finan. Econ. **9**(3), 489–518 (2011)
5. Brunetti, C., Lildholdt, P.M.: Time series modeling of daily log-price ranges for chf/usd and usd/gbp. J. Deriv. **15**(2), 39–59 (2007)
6. Chou, R.Y.: Forecasting financial volatilities with extreme values: the conditional autoregressive range (carr) model. J. Money, Credit Banking 561–582 (2005)
7. Cipollini, F., Engle, R.F., Gallo, G.M.: Vector multiplicative error models: representation and inference. Technical report, National Bureau of Economic Research (2006)
8. Cipollini, F., Engle, R.F., Gallo, G.M.: Semiparametric vector mem. J. Appl. Econ. **28**(7), 1067–1086 (2013)
9. Corsi, F.: A simple approximate long-memory model of realized volatility. J. Finan. Econ. 174–196 (2009)
10. Engle, R.: New frontiers for arch models. J. Appl. Econ. **17**(5), 425–446 (2002)
11. Engle, R.F., Gallo, G.M.: A multiple indicators model for volatility using intra-daily data. J. Econ. **131**(1), 3–27 (2006)
12. Engle, R.F., Ghysels, E., Sohn, B.: Stock market volatility and macroeconomic fundamentals. Rev. Econ. Stat. **95**(3), 776–797 (2013)
13. Engle, R.F., Rangel, J.G.: The spline-garch model for low-frequency volatility and its global macroeconomic causes. Rev. Finan. Stud. **21**(3), 1187–1222 (2008)
14. Engle, R.F., Russell, J.R.: Autoregressive conditional duration: a new model for irregularly spaced transaction data. Conometrica 1127–1162
15. Engle, R.F. Sokalska, M.E.: Forecasting intraday volatility in the us equity market. multiplicative component garch. J Finan Econ *10*(1), 54–83 (2012)

16. Ghysels, E., Sinko, A., Valkanov, R.: Midas regressions: Further results and new directions. Econ. Rev. **26**(1), 53–90 (2007)
17. Hautsch, N., Malec, P., Schienle, M.: Capturing the zero: A new class of zero-augmented distributions and multiplicative error processes. J. Finan. Econ. **12**(1), 89–121 (2014)
18. Lanne, M.: A mixture multiplicative error model for realized volatility. J. Finan. Econ. **4**(4), 594–616 (2006)
19. Manganelli, S.: Duration, volume and volatility impact of trades. J. Finan. Mark. **8**(4), 377–399 (2005)
20. Müller, U.A., Dacorogna, M.M., Davé, R.D., Pictet, O.V., Olsen, R.B., Ward J.R.: Fractals and intrinsic time: a challenge to econometricians. Unpublished manuscript, Olsen & Associates, Zürich (1993)
21. Naimoli, A., Storti, G.: Heterogeneous component multiplicative error models for forecasting trading volumes. Int. J. Forecast. **35**(4), 1332–1355 (2019). https://doi.org/10.1016/j.ijforecast.2019.06.002. http://www.sciencedirect.com/science/article/pii/S0169207019301505. ISSN 0169-2070
22. Patton, A.: Volatility forecast comparison using imperfect volatility proxies. J. Econ. **160**(1), 246–256 (2011)
23. Core Team, R.: R: A Language and Environment for Statistical Computing. R Foundation for Statistical Computing, Vienna, Austria (2018)

Asymptotically Distribution-Free Goodness-of-Fit Tests for Testing Independence in Contingency Tables of Large Dimensions

Thuong T. M. Nguyen

Abstract We discuss a possibility of using asymptotically distribution-free goodness-of-fit tests for testing independence of two discrete or categorical random variables in contingency tables. The tables considered are particularly of large dimension, in which the conventional chi-square test becomes less reliable when the table is relatively sparse. The main idea of the method is to apply the new Khmaladze transformation to transform the vector of the chi-square statistic components into another vector whose limit distribution is free of the parameters. The transformation is one-to-one and hence we can build up any statistic based on the transformed vector as an asymptotically distribution-free test statistic for the problem of interest where we recommend the analogue of the Kolmogorov-Smirnov test. Simulations are used to show that the new test not only converges relatively quickly but is also more powerful than the chi-square test in certain cases.

Keywords Goodness of fit · Contingency tables · Large dimension

1 Introduction

The problem of testing independence in contingency tables can be listed as one of the most classical problems in non-parametric inference. For this problem, the conventional chi-square test had been known as the only asymptotically distribution-free goodness-of-fit test to be used for quite a long time, until a recent construction of a wider class of tests introduced in [6]. Even though the use of the chi-square test has no rule about limiting the number of cells or, in other words, about limiting the number of categories or possible values for each variable, it certainly requires some assumptions which large dimensional tables make it difficult to meet. Those assumptions for the use of the chi-square test are well known to be that the value of

T. T. M. Nguyen (✉)
School of Mathematics and Statistics, Victoria University of Wellington, Wellington 6140, New Zealand
e-mail: thuong.nguyen@vuw.ac.nz

M. La Rocca et al. (eds.), *Nonparametric Statistics*, Springer Proceedings in Mathematics & Statistics 339, https://doi.org/10.1007/978-3-030-57306-5_36

cell expectation should be 5 or more in at least 80% of the cells and no cell should have an expectation of less than 1 (see, for example, Agresti [1]).

Here we will cover the case when the tables may have several hundreds of cells while the sample size is approximately of the same order. In such situations, asymptotically distribution-free goodness of fit tests, in general, have not existed. This is the result of the fact that when many parameters need to be estimated, those estimates are not stable and the distribution-free property of the conventional chi-square test will suffer. As will be demonstrated in Sect. 4 that obstacle will not be the case for the new tests we introduce in this paper.

This work is an extension of what has been done in [6] where the author applied the new Khmaladze's transformation given in Khmaladze [2] for testing hypotheses on families of discrete random variables. The author did confirm that for testing independence in contingency tables, the new Khmaladze's transformation works for tables of reasonable dimensions like 4×6 or up to 8×8 with relatively small sample sizes. The power of the new tests like the Kolmogorov-Smirnov or omega-square test was shown to be better than that of the chi-square in several cases. In like vein, we will address the problem of testing independence in contingency tables with much larger dimensional tables and show that the method works equivalently well as already discussed in [6]. Even when the sample size is sufficiently large to be able to use the chi-square test, the new test presented here will be more powerful than the chi-square in certain scenarios.

This work, however, is not a mere extension without any practical meaning. Often we see in practice contingency tables classifying variables with no more than 10 categories, which could partly be due to limited availability of tests as already mentioned above. In various problems with diversity particularly, however, variables classified into a large number of categories should be of great desire. For example, when examining diversity relating to ethnicity, we usually neglect minor groups and classify all of them as '*Others*'. Even though that is a sensible approach as we may be able to avoid sparsity of the tables, many a time the biological and cultural diversity makes a huge contribution to the association of ethnicity to other random variables like immigrating habit, political view, etc. that we should not neglect. If we want to take diversity into account by specifying every single minor group, we will need to cope with tables of large dimension.

We need to emphasize that we do not propose only one new test for the problem of interest. Instead, we will show a construction of a class of new asymptotically distribution-free goodness-of-fit tests and investigate the behaviour of some tests as examples. Through this work we later on recommend a test from our point of view. This construction will be presented in Sect. 3 and some preliminaries for the problem will be sketched in Sect. 2. Section 4 will be devoted to demonstrate simulation results in which we discuss the distribution-free property of the new tests and how quickly the test converges to its limit, and also compare the power of the new tests with that of the chi-square for different cases.

2 Preliminaries

Our problem is to test the independence of two discrete or categorical random variables X and Y where their association is classified in a contingency table of size $(I + 1) \times (J + 1)$, which means that X and Y have $(I + 1)$ and $(J + 1)$ possible values, respectively. Within this paper we suppose that I and J are considerably large, say, both I and J are greater than 10 and can be up to 30. Meanwhile, the sample size n could be as small as of the same order as the number of cells $(I + 1) \times (J + 1)$.

Denote by $\mathbf{a} = (a_1, \cdots, a_I)^T \in \mathbb{R}^I$ and $\mathbf{b} = (b_1, \cdots, b_J)^T \in \mathbb{R}^J$ the hypothetical marginal probabilities which are freely changed provided that $a_i, b_j > 0$ for every i and j and $\sum_{i=1}^{I} a_i < 1$ and $\sum_{i=1}^{J} b_j < 1$. The marginal probabilities of the last row or column, a_{I+1} and b_{J+1}, of course, depend on $\mathbf{a}$ and $\mathbf{b}$ through the relationship $a_{I+1} = 1 - \sum_{i=1}^{I} a_i$ and $b_{J+1} = 1 - \sum_{j=1}^{J} b_j$. As we know, testing independence of two random variables belongs to the class of non-parametric testing; however, for this particular problem, we can view it as a parametric testing problem where $\mathbf{a}$ and $\mathbf{b}$ are parameters. Let ` $= (\mathbf{a}^T, \mathbf{b}^T)^T \in \mathbb{R}^d$ where $d = I + J$ then ` stands for the vector of parameters and d for its dimension. From now on, we will use `$_0$ for the true unknown parameters and n for the total number of observed values in the table, or sample size.

Suppose that the cell counts are $\{\nu_{ij}\}$ and the joint probabilities are $\{\pi_{ij}\}$. Under the hypothesis of independence between X and Y, the joint probabilities π_{ij} are defined through the marginal probabilities by $\pi_{ij} = a_i b_j$ for every i and j. These a_i and b_j can simply be estimated by

$$\hat{a}_i = \frac{\nu_{i+}}{n} = \frac{\sum_{j=1}^{(J+1)} \nu_{ij}}{n}, \quad \hat{b}_j = \frac{\nu_{+j}}{n} = \frac{\sum_{i=1}^{(I+1)} \nu_{ij}}{n}, \quad \text{for all } i, j, \tag{1}$$

which are the maximum likelihood estimators (MLE). Clearly, the estimated joint probabilities are $\hat{\pi}_{ij} = \hat{a}_i \hat{b}_j$ under the null hypothesis of independence.

Denote by $\mathbf{T}_n$ a vector of components $T_{ij} = \frac{\nu_{ij} - n\pi_{ij}(\boldsymbol{\theta}_0)}{\sqrt{n\pi_{ij}(\boldsymbol{\theta}_0)}}$. The estimates of these T_{ij} based on MLE are

$$\widehat{T}_{ij} = \frac{\nu_{ij} - n\pi_{ij}(\hat{\boldsymbol{\theta}})}{\sqrt{n\pi_{ij}(\hat{\boldsymbol{\theta}})}} = \frac{\nu_{ij} - n\hat{a}_i\hat{b}_j}{\sqrt{n\hat{a}_i\hat{b}_j}} \tag{2}$$

and let us denote $\widehat{\mathbf{T}}_n = (\widehat{T}_{ij})$. A crucial point, as stated in Khmaladze [2], is that the limit in distribution of $\widehat{\mathbf{T}}_n$, denoted by $\widehat{\mathbf{T}}$, is a Gaussian vector of the following form:

$$\widehat{\mathbf{T}} = \mathbf{V} - \sum_{\alpha=0}^{d} \langle \mathbf{V}, \mathbf{q}^{(\alpha)} \rangle \mathbf{q}^{(\alpha)}. \tag{3}$$

In this form, $\mathbf{V} = (V_{ij})$ where V_{ij} with indices $i = 1, \cdots, I+1$ and $j = 1, \cdots, J+1$ are independent and standard normal random variables and $\{\mathbf{q}^{(\alpha)}\}$ is a set of orthonormal vectors defined based on $\boldsymbol{\theta}_0$. Detail about $\mathbf{q}^{(\alpha)}$ was written in [6]. The key point is the fact that $\boldsymbol{\theta}_0$ is unknown and varied so $\mathbf{q}^{(\alpha)}$, $\alpha = 0, \ldots, d$, depends on the unknown parameter and varies from case to case as a consequence. Therefore, the limiting distribution of $\widehat{\mathbf{T}}_n$ is unspecified without the knowledge of the parameters. It is desirable to obtain a test which has a limiting distribution independent of the parameters. There is, however, so far only one test statistic which is the chi-square test statistic, as the quadratic form of $\widehat{\mathbf{T}}_n$ is asymptotically distribution-free if the sample size is significantly large enough.

The rationale behind the new Khmaladze transformation is elegant and simple and so the transformation itself is not only easy to implement but also effective. Briefly speaking, since the limiting distribution $\widehat{\mathbf{T}}$ of the vector $\widehat{\mathbf{T}}_n$ is expressed as in (3), we can transform $\widehat{\mathbf{T}}_n$ into another vector $\widehat{\mathbf{Z}}_n$ which converges in distribution to, say, $\widehat{\mathbf{Z}}$, of the form

$$\widehat{\mathbf{Z}} = \mathbf{V}' - \sum_{\alpha=0}^{d} \langle \mathbf{V}', \mathbf{r}^{(\alpha)} \rangle \mathbf{r}^{(\alpha)}, \tag{4}$$

where $\{\mathbf{r}^{(\alpha)}\}$ is a set of fixed orthonormal vectors and $\mathbf{V}'$ is of the same type as $\mathbf{V}$. This $\widehat{\mathbf{Z}}$ has the same form as $\widehat{\mathbf{T}}$ but the difference is that $\mathbf{r}^{(\alpha)}$ are fixed while $\mathbf{q}^{(\alpha)}$ are not. No matter how many vectors $\mathbf{q}^{(\alpha)}$ we have, the transformation can be done simultaneously or recursively. The transformation will remain its simplicity and transparency even for a large number of parameters d. In this text, we will show the effectiveness of the transformation for d up to 55. This transformation is used successfully for not only discrete distributions but also various cases like testing hypotheses on families of continuous distributions or with covariates as thoroughly discussed in Khmaladze [3, 4], respectively.

3 Method

The method, in fact, is no different from what is written in [6] in every detail. For readers' convenience, we will sketch some main steps on how to transform the vector $\widehat{\mathbf{T}}_n$ of components of the chi-square statistic into a vector $\widehat{\mathbf{Z}}_n$ whose limiting distribution does not depend on any parameter as given in (4). We will mainly show what is the form of the transformation and the test constructed from the transformed vector $\widehat{\mathbf{Z}}_n$.

The core unitary operator $U_{\mathbf{q},\mathbf{r}}$ used throughout the new Khmaladze's transformation which transforms a vector of unit norm $\mathbf{q}$ into another unit vector $\mathbf{r}$ is of the form

$$U_{\mathbf{q},\mathbf{r}} = \mathbf{I} - \frac{1}{1 - \langle \mathbf{q}, \mathbf{r} \rangle} (\mathbf{r} - \mathbf{q})(\mathbf{r} - \mathbf{q})^T, \tag{5}$$

where $\mathbf{I}$ is the identity operator. This operator ensures that $U_{\mathbf{q},\mathbf{r}}\mathbf{q} = \mathbf{r}$ but other properties include $U_{\mathbf{q},\mathbf{r}}\mathbf{r} = \mathbf{q}$ and $U_{\mathbf{q},\mathbf{r}}\mathbf{v} = \mathbf{v}$ for every vector $\mathbf{v}$ orthogonal to both $\mathbf{q}$ and $\mathbf{r}$. The vector $\widehat{\mathbf{T}}$ as an orthogonal projection of a standard Gaussian random vector $\mathbf{V}$ onto an orthonormal system $\{\mathbf{q}^{(\alpha)}\}$ will be transformed into a vector $\widehat{\mathbf{Z}}$ which is also an orthogonal projection of a standard Gaussian random vector $\mathbf{V}'$ onto an orthonormal system $\{\mathbf{r}^{(\alpha)}\}$ using this type of operator. This can be done by mapping the set $\mathbf{q}^{(\alpha)}$ into $\mathbf{r}^{(\alpha)}$ and in this way, $\mathbf{V}$ will be mapped into $\mathbf{V}'$ which is also a standard Gaussian random vector. More details about the transformation are presented in Khmaladze [2].

To find an explicit form of the operator that maps all $\mathbf{q}^{(\alpha)}$ into $\mathbf{r}^{(\alpha)}$ simultaneously may be complicated. We will show that finding an operator in recursive form will be much easier. The construction of the recursive form shown here is a general way for any problem with a large number of parameters. That recursive form is constructed as follows: Set the operator $U_{\mathbf{q}^{(0)},\mathbf{r}^{(0)}}$ defined as in (5) where $\mathbf{q} = \mathbf{q}^{(0)}, \mathbf{r} = \mathbf{r}^{(0)}$. Obviously, $U_{\mathbf{q}^{(0)},\mathbf{r}^{(0)}}\mathbf{q}^{(0)} = \mathbf{r}^{(0)}$, $U_{\mathbf{q}^{(0)},\mathbf{r}^{(0)}}\mathbf{r}^{(0)} = \mathbf{q}^{(0)}$. Set $\widetilde{\mathbf{q}}^{(1)} = U_{\mathbf{q}^{(0)},\mathbf{r}^{(0)}}\mathbf{q}^{(1)}$. Because $U_{\mathbf{q}^{(0)},\mathbf{r}^{(0)}}$ preserves the inner product, the images of $\mathbf{q}^{(0)}$ and $\mathbf{q}^{(1)}$ are orthogonal, which means $\mathbf{r}^{(0)} \perp \widetilde{\mathbf{q}}^{(1)}$. Therefore $U_{\widetilde{\mathbf{q}}^{(1)},\mathbf{r}^{(1)}}\mathbf{r}^{(0)} = \mathbf{r}^{(0)}$, $U_{\widetilde{\mathbf{q}}^{(1)},\mathbf{r}^{(1)}}\widetilde{\mathbf{q}}^{(1)} = \mathbf{r}^{(1)}$. In summary, by applying the composition $U_{\widetilde{\mathbf{q}}^{(1)},\mathbf{r}^{(1)}}U_{\mathbf{q}^{(0)},\mathbf{r}^{(0)}}$ on $\mathbf{q}^{(0)}, \mathbf{q}^{(1)}, \mathbf{q}^{(2)}$, we get their images $\mathbf{r}^{(0)}, \mathbf{r}^{(1)}, \widetilde{\mathbf{q}}^{(2)}$, respectively, where $\widetilde{\mathbf{q}}^{(2)} \perp \mathbf{r}^{(0)}, \mathbf{r}^{(1)}$. Continuing this process then we can define $\widetilde{\mathbf{q}}^{(\tau)}$, $\tau \geq 2$ recursively as

$$\widetilde{\mathbf{q}}^{(\tau)} = \left(\prod_{1 \leq \beta < \tau} U_{\widetilde{\mathbf{q}}^{(\beta)},\mathbf{r}^{(\beta)}} U_{\mathbf{q}^{(0)},\mathbf{r}^{(0)}} \right) \mathbf{q}^{(\tau)}.$$

The operator we wish to use is then

$$\boldsymbol{U} = \prod_{\tau=1}^{d} U_{\widetilde{\mathbf{q}}^{(\tau)},\mathbf{r}^{(\tau)}} U_{\mathbf{q}^{(0)},\mathbf{r}^{(0)}}. \tag{6}$$

It is obvious that $\boldsymbol{U}$ is a unitary operator as it is the product of unitary operators. This $\boldsymbol{U}$ satisfies $\boldsymbol{U}\mathbf{q}^{(\alpha)} = \mathbf{r}^{(\alpha)}$ for all $\alpha = 0, \cdots, d$. If the operator $\boldsymbol{U}$ transforms $\widehat{\mathbf{T}}_n$ into a $\widehat{\mathbf{Z}}_n$, i.e.

$$\widehat{\mathbf{Z}}_n = \boldsymbol{U}\widehat{\mathbf{T}}_n = \left(\prod_{\tau=1}^{d} U_{\widetilde{\mathbf{q}}^{(\tau)},\mathbf{r}^{(\tau)}} U_{\mathbf{q}^{(0)},\mathbf{r}^{(0)}} \right) \widehat{\mathbf{T}}_n, \tag{7}$$

then $\widehat{\mathbf{Z}}_n$ converges in distribution to $\widehat{\mathbf{Z}}$ of the form given in (4).

We re-emphasize that the limiting distribution of $\widehat{\mathbf{Z}}_n$ could be chosen by the users because one can freely choose any collection $\{\mathbf{r}^{(0)}, \mathbf{r}^{(1)}, \cdots, \mathbf{r}^{(d)}\}$ provided that they are orthonormal. For consistency, we choose $\mathbf{r}^{(0)} = \left(\frac{1}{\sqrt{(I+1)(J+1)}}\right)$, a vector of dimension $(I+1) \times (J+1)$ with equal components. Others $\mathbf{r}^{(\alpha)}$ with $\alpha = 1, \cdots, d$ defined for $\alpha \le I$ as

$$r^{(\alpha)}_{z_1 z_2} = \frac{1}{\sqrt{J+1}} \left[1_{\{z_1=\alpha\}} - \frac{1}{\sqrt{I+1}(1+\sqrt{I+1})} 1_{\{z_1 \neq I+1\}} - \frac{1}{\sqrt{I+1}} 1_{\{z_1=I+1\}} \right]$$

and for $\alpha \ge I+1$ as

$$r^{(\alpha)}_{z_1 z_2} = \frac{1}{\sqrt{I+1}} \left[1_{\{z_2=\alpha-I\}} - \frac{1}{\sqrt{J+1}(1+\sqrt{J+1})} 1_{\{z_2 \neq J+1\}} - \frac{1}{\sqrt{J+1}} 1_{\{z_2=J+1\}} \right],$$

where indices $z_1 = 1, \ldots, I+1$, $z_2 = 1, \ldots, J+1$. This set of $\mathbf{r}^{(\alpha)}$ is nothing else but a set of $\mathbf{q}^{(\alpha)}$ at a specific $\boldsymbol{\theta}_0$. That specific $\boldsymbol{\theta}_0$ corresponds to the discrete uniform marginal distributions of X and Y or, in other words, $\boldsymbol{\theta}_0$ with $a_i = \frac{1}{I+1}$ and $b_j = \frac{1}{J+1}$ for every i and j. Compared to the problem of testing independence of two continuous univariate random variables, this choice of $\mathbf{r}^{(\alpha)}$ will give us the vector $\widehat{\mathbf{Z}}$ such that the partial sum of coordinates of this $\widehat{\mathbf{Z}}$ will be discrete time analogue of the standard Brownian sheet (see, for example, van de Vaart and Wellner [7]).

The computation of the vector $\widehat{\mathbf{Z}}_n$ theoretically depends on the unknown vectors $\mathbf{q}^{(\alpha)}$ or, in other words, the unknown parameter $\boldsymbol{\theta}_0$. For practical use, we will substitute those unknown parameters by their estimates $\hat{a}_i$ and $\hat{b}_j$ as given in (1). The simulation result shown in the next section will use the unknown theoretical parameter $\boldsymbol{\theta}_0$ to generate realizations for the tables under the hypothesis of independence. We then use the estimates of parameters for calculating vector $\widehat{\mathbf{Z}}_n$ which is eventually used to compute the test statistics.

4 Simulation

In this section, we will use simulations to demonstrate three important points of the new tests: the distribution-free property, the rate of convergence and the power. All programmes were written and run under R version 3.5.3 (R Core Team, 2019).

First of all, let us show the form of the new statistics based on the transformed vector $\widehat{\mathbf{Z}}_n$ given in (7). For $i = 1, \ldots, I+1$ and $j = 1, \ldots, J+1$, let

$$V^Z_{n,ij} = \sum_{(z_1,z_2)\le(i,j)} \widehat{Z}_{z_1 z_2} \tag{8}$$

be the cumulative sum of the coordinates of $\widehat{\mathbf{Z}}_n$ where $(t_1, t_2) \le (i, j)$ means $t_1 \le i$ and $t_2 \le j$. The two statistics

$$KS = \max_{(1,1)\leq(i,j)\leq(I+1,J+1)} \left|V^Z_{n,ij}\right|, \tag{9}$$

$$\Omega^2 = \sum_{(1,1)\leq(i,j)\leq(I+1,J+1)} (V^Z_{n,ij})^2 \tag{10}$$

are nothing else but the discrete versions of the Kolmogorov-Smirnov (KS) statistic and the omega-square (Ω^2) statistic that will be considered as examples of the new tests.

To illustrate the distribution-free property of the new tests of the form (9) and (10), we follow the following steps: for each value of I and J, we choose two different sets of $\boldsymbol{\theta}_0$ arbitrarily. One way that we choose to generate $\boldsymbol{\theta}_0$ is to generate the marginal probabilities $(a_1, \ldots, a_{I+1})^T$ and $(b_1, \ldots, b_{J+1})^T$ as vectors of uniform realizations and then standardize those realizations to make sure that they add up to 1. By doing so, the difference between the two $\boldsymbol{\theta}_0$ is also arbitrary and can be very significant.

Then from each set of the chosen parameters, we generate around 5000 sets of realizations for the table with a fixed sample size, calculate $\widehat{\mathbf{T}}_n$ and transform it into $\widehat{\mathbf{Z}}_n$ recursively. From the transformed vector $\widehat{\mathbf{Z}}_n$, we calculate the values of the test KS or Ω^2. From these values of the test, we plot their empirical distributions corresponding to the two different $\boldsymbol{\theta}_0$ to see if they are the same or not. Figure 1 shows the result of the empirical distributions of the KS tests for 15×20 contingency tables with sample size 500 and of the omega-square tests for 30×25 tables with sample size 800. As we can see, the two curves in each plot are not easily distinguishable which indicates the distribution-free property of the new tests.

The algorithm uses finite loops so it is obviously neat and simple for users. Moreover, we observed that the algorithm runs effectively quick despite its recursive nature. For a 15×20 table with sample size 500 and 5000 iterations, the total running time for generating the empirical distribution is less than 1 hour. Hence, the time to calculate a test for each given problem should be half of a second only. When the dimension increases, i.e. the number of parameters increases to 45, we need around 7 seconds to calculate a test which is still reasonably good.

Both examples shown here are asymptotically distribution-free; however, the value of the omega-square test becomes so large as the dimension of the table increases that we are reluctant to recommend using this test. From our point of view, the KS test as the supremum of some Brownian bridge rather than its quadratic sum should be more reliable since its range is much smaller. Therefore, we will further investigate the convergence rate of the KS test to its limit and its power in the rest of this text.

Next we will demonstrate the convergence rate of the KS test. For each chosen dimension, we simulate data with different sample sizes. One sample size is very large and the other is relatively small, roughly the same as the table dimension. We then plot the empirical distributions of the test in these two cases. Figure 2 is an example with tables of dimension 25×25. In this figure, the two plots of the empirical distributions of the KS tests coincide even for two very different sample sizes, 700 and 10000. That means that the KS test converges to its limiting distribution extremely quickly. We suspect that as the number of parameters is increasing, the convergence rate is

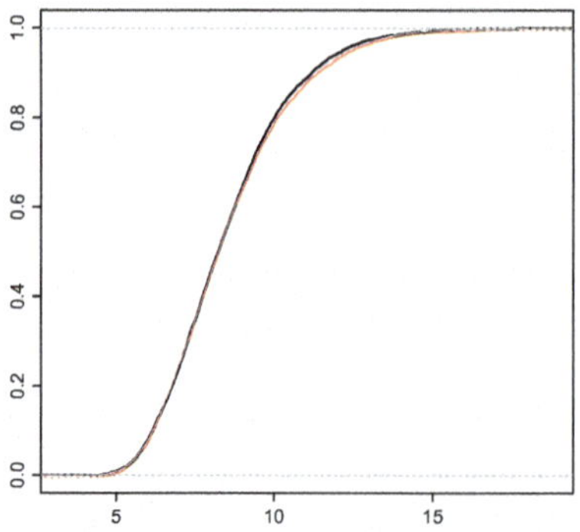

(a) *The empirical distributions of the KS statistics for* 15×20 *tables with sample size* 500 *and two different sets of parameters* θ_0

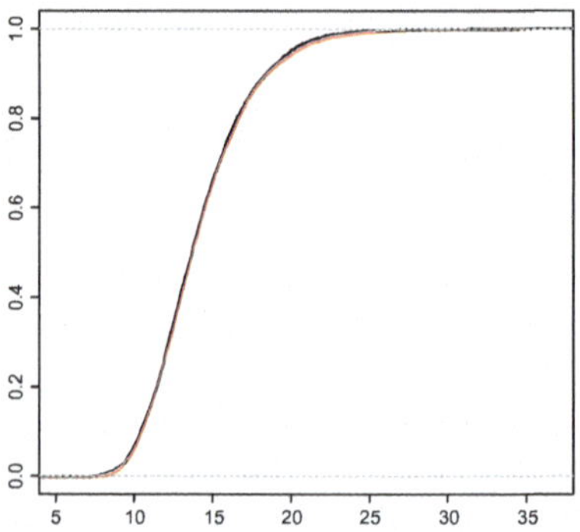

(b) *The empirical distributions of the KS statistics for* 30×25 *tables with sample size* 800 *and two different sets of parameters* θ_0

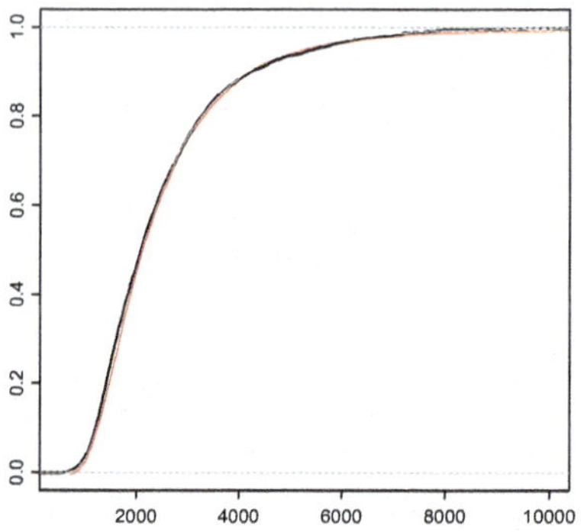

(c) *The empirical distributions of the* Ω^2 *statistics for* 15×20 *tables with sample size* 500 *and two different sets of parameters* θ_0

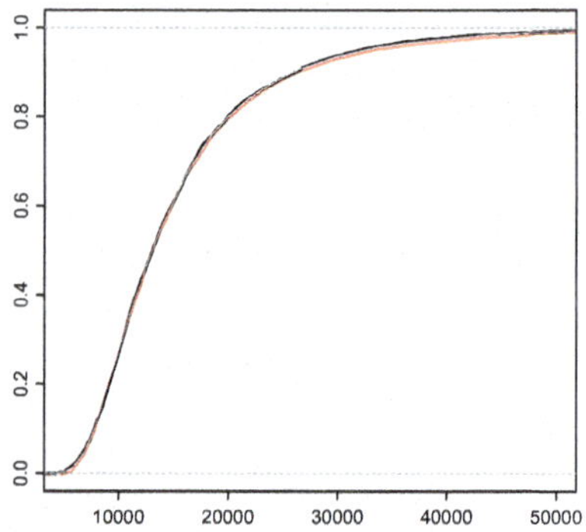

(d) *The empirical distributions of the* Ω^2 *statistics for* 30×25 *tables with sample size* 800 *and two different sets of parameters* θ_0

Fig. 1 Distribution-free property of the transformed statistics

getting better. That should be intuitively true as the asymptotic behaviour of $\mathbf{V}_n^Z$ is getting closer to the projected standard Brownian motion.

The last part of this section is to discuss the power of the new test compared to the chi-square test in case the sample size is sufficiently large enough to be able to use the chi-square. Needless to say that when the table is relatively sparse, the new test proposed here still performs well as seen above. This is certainly a case when it outperforms the chi-square test.

We will again take the KS test for comparing its statistical powers with that of the chi-square. We are not using type I and type II errors for this comparison. Instead we define the statistical power of tests under local alternatives as follows: Assume that under an alternative distribution π^a the random variables X and Y are not independent, i.e. we have $\pi^a_{ij} \neq a_i b_j$ for some i, j. Denote by F_0 and F_a the distribution functions

Fig. 2 The convergence rate of the KS test for 25×25 tables. The bold blue line is the empirical distribution with sample size 10000 and the black line is with sample size 700. These two lines, however, are not distinguishable which indicates how fast the convergence is

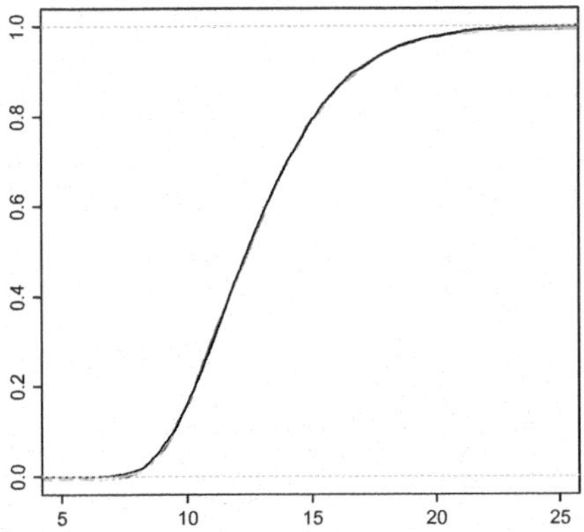

of a test statistic under the null and the alternative, respectively. The quantities

$$D = \max_{x:F_0(x)\geq 0.8} |F_0(x) - F_a(x)|$$

will be used as numerical descriptions of statistical powers of tests. Below we opt to compare $D(\chi^2_n)$ to $D(KS)$ for several scenarios of local alternatives.

Obviously there are many ways to define local alternatives, here we choose local alternatives from some families of copulas. Those families are Cuadras-Augé $C^{(1)}_\lambda$, Gumbel's bivariate exponential distribution $C^{(2)}_\lambda$ and Ali-Mikhail-Haq $C^{(3)}_\lambda$, see specific forms of these in Nelsen [5] and more explanations in [6]. Figure 3 illustrates the empirical distributions of the KS and chi-square test under the null and alternative, plotted in the same graph for each test. By looking at these figures, we can see that the KS test has better power under Gumbel's bivariate exponential distribution and Ali-Mikhail-Haq copulas and is only slightly less powerful than the chi-square under the Cuadras-Augé copula family.

In fact, for example, with tables of dimension 25×15 and the alternative distribution generated from Gumbel's bivariate exponential distribution copula with $\lambda = 0.3$, the power of the KS test is 0.914 while that of chi-square is only 0.44. We also believe that for the two scenarios when the KS test is more powerful than the chi-square, its power is getting better when the dimension of the table is getting larger.

A common characteristic of the two scenarios where the KS test is better than the chi-square test is that, under these scenarios if X and Y are discrete or ordinal categorical then there is a linear by linear relationship between X and Y. That means, in cases when X increases and Y either increases or decreases, the KS test gives us a better detection of this dependency.

In summary, we present in this text a new construction of a class of asymptotically distribution-free tests for testing independence in contingency tables, even for large tables which are relatively sparse and recommend to use the KS test. We confirm that this new test converges very quickly to its limit and also point out the cases when

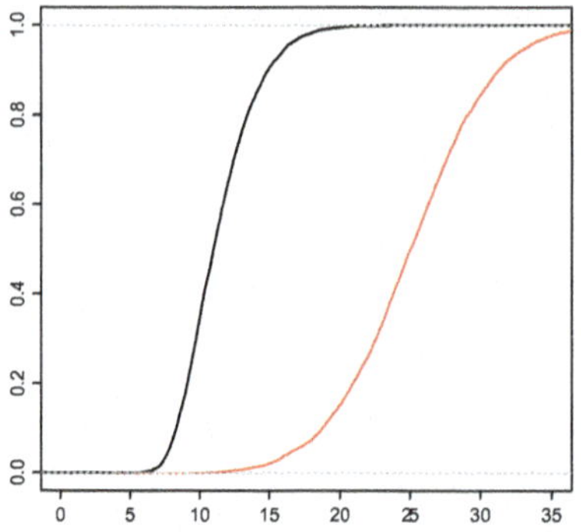

(a) *The two empirical distributions of the KS statistics under the null and the local alternative from Gumbel family with* $\lambda = 0.3$ *for* 25×20 *tables, sample size* 8000

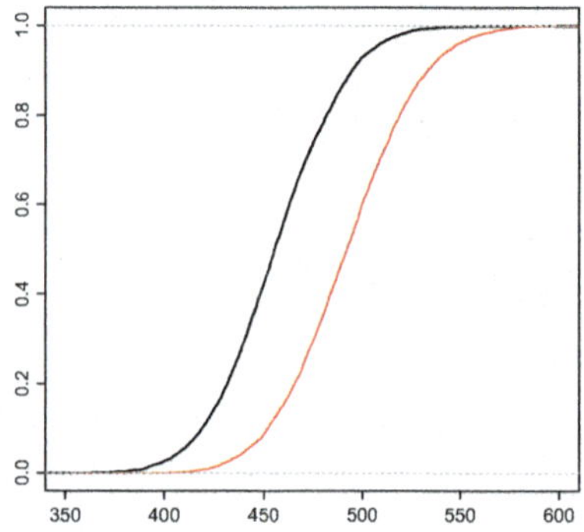

(b) *The two empirical distributions of the chi-square statistics under the null and the local alternative from Gumbel family with* $\lambda = 0.3$ *for* 25×20 *tables, sample size* 8000

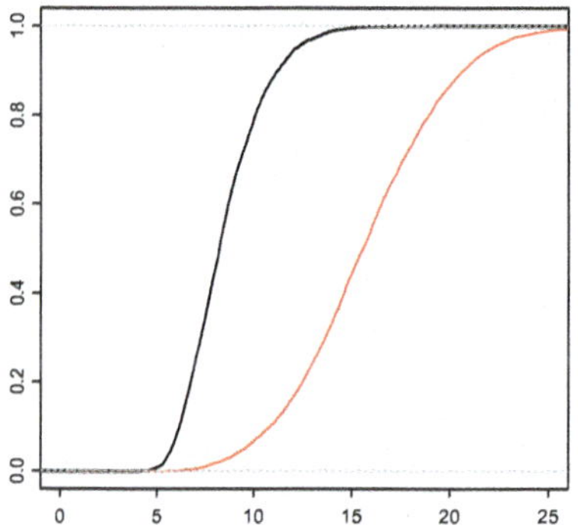

(c) *The two empirical distributions of the KS statistics under the null and the local alternative from Ali-Mikhail-Haq family with* $\lambda = 0.5$ *for* 15×20 *tables, sample size* 8000

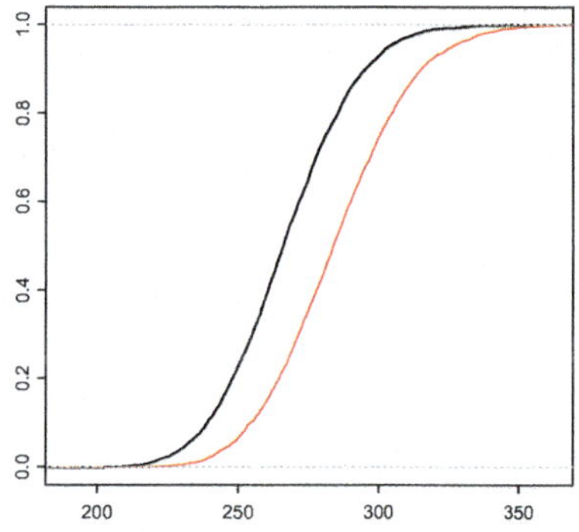

(d) *The two empirical distributions of the chi-square statistics under the null and the local alternative from Ali-Mikhail-Haq family with* $\lambda = 0.5$ *for* 15×20 *tables, sample size* 8000

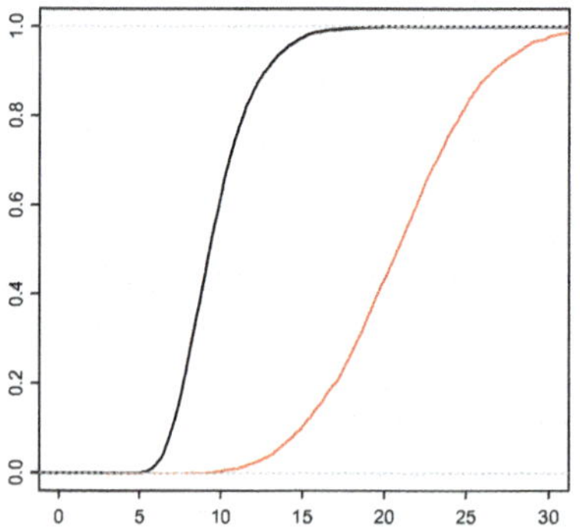

(e) *The two empirical distributions of the KS statistics under the null and the local alternative from Cuadras-Augé family with* $\lambda = 0.2$ *for* 15×25 *tables, sample size* 8000

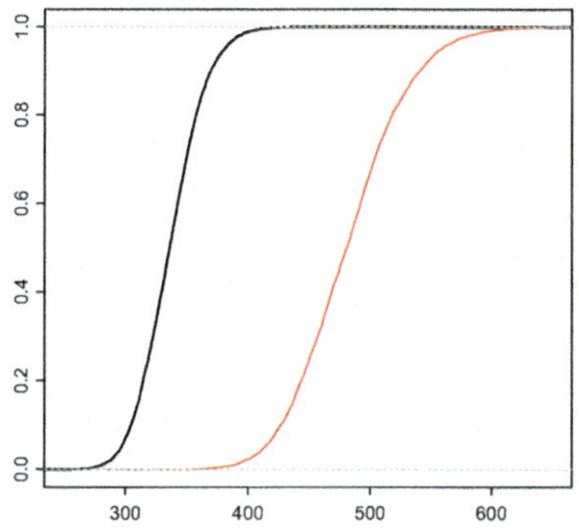

(f) *The two empirical distributions of the chi-square statistics under the null and the local alternative from Cuadras-Augé family with* $\lambda = 0.2$ *for* 15×25 *tables, sample size* 8000

Fig. 3 Statistical powers of tests under several scenarios the new test is more powerful than the chi-square. Due to restricted time for running simulations we just opt to tables of dimension 25 × 30 at most. More work could be done to see how well the new Khmaladze's transformation can do for this specific problem either by increasing the table dimension or decreasing the sample size or looking for other local alternatives where the chi-square test is less powerful than the new constructed tests.

Acknowledgements The author would like to thank Prof. Estate Khmaladze for his suggestion and encouragement on further investigation of this problem of testing independence of two random variables in sparse contingency tables. I would also like to thank the anonymous referee for their constructive suggestions and my colleague David Cox (VUW) for his careful proofreading which helped improve the exposition of the paper.

References

1. Agresti, A.: Categorical Data Analysis. John Wiley & Sons, New York (2003)
2. Khmaladze, E.: Note on distribution free testing for discrete distribution. Ann. Stat. **41**, 2979–2993 (2013)
3. Khmaladze, E.: Unitary transformations, empirical processes and distribution free testing. Bernoulli **22**, 563–588 (2016)
4. Khmaladze, E.: Distribution free testing for conditional distributions given covariates. Stat. Prob. Lett. **129**, 348–354 (2017)
5. Nelsen, R.B.: An introduction to Copulas. Springer (2006)
6. Nguyen, T.M.T.: A new approach to distribution free tests in contingency tables. Metrika **80**(2), 153–170 (2017)
7. van der Vaart, A.W., Wellner, J.A.: Weak Convergence and Empirical Processes: With Applications to Statistics. Springer (1996)

Incorporating Model Uncertainty in the Construction of Bootstrap Prediction Intervals for Functional Time Series

Efstathios Paparoditis and Han Lin Shang

Abstract A sieve bootstrap method that incorporates model uncertainty for constructing pointwise or simultaneous prediction intervals of stationary functional time series is proposed. The bootstrap method exploits a general backward vector autoregressive representation of the time series of Fourier coefficients appearing in the well-established Karhunen-Loève expansion of the functional process. The bootstrap method generates, by running backward in time, functional bootstrap samples which adequately mimic the dependence structure of the underlying process and which all have the same conditionally fixed curves at the end of every functional bootstrap sample. The bootstrap prediction error distribution is then calculated as the difference between the model-free bootstrap generated future functional pseudo-observations and the functional forecasts obtained from a model used for prediction. In this way, the estimated prediction error distribution takes into account not only the innovation and estimation error associated with prediction, but also the possible error due to model uncertainty or misspecification. Through a simulation study, we demonstrate an excellent finite-sample performance of the proposed sieve bootstrap method.

Keywords Fourier transform · Functional prediction · Prediction error · Principal components · Karhunen-Loéve expansion

1 Introduction

Functional time series consists of random functions observed at a regular or irregular time interval. Depending on whether or not the continuum is also a time variable, functional time series can be loosely grouped into two categories. On the one hand,

---

E. Paparoditis
Department of Mathematics and Statistics, University of Cyprus, Nicosia, Cyprus
e-mail: paparoditis@ucy.ac.cy

H. L. Shang (✉)
Department of Actuarial Studies and Business Analytics, Macquarie University, Sydney, NSW 2109, Australia
e-mail: hanlin.shang@mq.edu.au

M. La Rocca et al. (eds.), *Nonparametric Statistics*, Springer Proceedings in Mathematics & Statistics 339, https://doi.org/10.1007/978-3-030-57306-5_37

functional time series can arise from measurements obtained by separating an almost continuous time record into consecutive intervals, for example, days, weeks, or years (see, e.g., [6]). We refer to such data structures as sliced functional time series, examples of which include daily price curves of a financial stock, see [9] and intraday particulate matter, see [18]. On the other hand, when continuum is not a time variable, functional time series can also arise when observations in a time period can be considered as finite-dimensional realizations of an underlying continuous function; an example is the yearly age-specific mortality rates, see [1, 8].

In either case, the underlying functional process is denoted by $\{\mathcal{X}_t, t \in \mathbb{Z}\}$, where $\mathbb{Z} = \{t : t \in 0, \pm 1, \dots\}$ and where each $\mathcal{X}_t$ is a random function $\mathcal{X}_t(\tau)$ for τ within a function support range, $\tau \in \mathscr{I} \subset R$. We refer to such data structures as functional time series. Central issues in the functional time series analysis are to model the temporal dependence of the functional random variables $\{\mathcal{X}_t, t \in \mathbb{Z}\}$, to make statistical inference about a parameter ξ of interest and to predict future values of the process given an observed data $\{\mathcal{X}_1, \mathcal{X}_2, \dots, \mathcal{X}_n\}$. Not only it is vital to obtain consistent estimators, but we are also interested in estimating the variability centered around these estimators, to construct confidence or prediction intervals and to implement hypothesis tests (see, e.g., [7]). When such inference problems arise, resampling methodology, especially bootstrapping, turns out to be an important practical alternative for independent functional data or functional time series (see, e.g., [2, 5, 10, 12, 13, 17, 19]).

In the arenas of bootstrapping functional time series, from a theoretical aspect, [15] developed weak convergence results for approximate sums of weakly dependent, Hilbert space-valued random variables in a triangular array setting. They prove a central limit theorem for the stationary bootstrap. [3] also obtained weak convergence results for Hilbert space-valued random variables. The random variables are assumed to be weakly dependent in the sense of near epoch dependence, and they show the consistency of the non-overlapping block bootstrap. From a methodological aspect, [16] extended the stationary bootstrap method of [15] to functional time series. [14] extended the moving block and the tapered block bootstrap to functional time series, while [19] extended the maximum entropy bootstrap method. [12] proposed a sieve bootstrap method for functional time series. [11] proposed a residual-based bootstrap method for functional autoregressions, while [4] applied a residual-based bootstrap method to construct confidence intervals for the regression function in a nonparametric functional regression.

We focus on the problem of constructing pointwise or simultaneous prediction intervals for functional time series. To elaborate, suppose that for every $t \in \mathbb{Z}$ the random element $\mathcal{X}_t$ is generated as

$$\mathcal{X}_t = f(\mathcal{X}_{t-1}, \mathcal{X}_{t-2}, \dots) + \varepsilon_t,$$

with some appropriate function f and a zero mean independent and identically distributed (I.I.D.) innovation process $\{\varepsilon_t\}$ with finite second moments; we write for simplicity $\varepsilon_t \sim$ I.I.D.$(0, C_\varepsilon)$, where $C_\varepsilon = E(\varepsilon_t \otimes \varepsilon_t)$ and $E(\cdot)$ denotes expectation. Suppose further that a "model"

$$\mathcal{X}_t = g(\mathcal{X}_{t-1}, \mathcal{X}_{t-2}, \ldots, \mathcal{X}_{t-k}) + v_t \tag{1}$$

is used for prediction, where $k < n$ is some fixed integer, g is an unknown function, and $v_t \sim$ I.I.D.$(0, C_v)$. Given the observed functional time series $\{\mathcal{X}_1, \mathcal{X}_2, \ldots, \mathcal{X}_n\}$, one-step-ahead prediction of $\mathcal{X}_{n+1}$ based on model (1) is obtained as

$$\widehat{\mathcal{X}}_{n+1} = \widehat{g}(\mathcal{X}_n, \mathcal{X}_{n-1}, \ldots, \mathcal{X}_{n-k+1}),$$

where $\widehat{g}$ is an estimator of the unknown function g. The prediction error $\mathcal{E} = \mathcal{X}_{n+1} - \widehat{\mathcal{X}}_{n+1}$ of the one-step-ahead prediction can then be decomposed as follows:

$$\begin{aligned}
\mathcal{E}_{n+1} &= \mathcal{X}_{n+1} - \widehat{\mathcal{X}}_{n+1} \\
&= \varepsilon_{n+1} \\
&\quad + [f(\mathcal{X}_n, \mathcal{X}_{n-1}, \ldots) - g(\mathcal{X}_n, \mathcal{X}_{n-1}, \ldots, \mathcal{X}_{n-k+1})] \\
&\quad + [g(\mathcal{X}_n, \mathcal{X}_{n-1}, \ldots, \mathcal{X}_{n-k+1}) - \widehat{g}(\mathcal{X}_n, \mathcal{X}_{n-1}, \ldots, \mathcal{X}_{n-k+1})] \\
&= \mathcal{E}_{I,n+1} + \mathcal{E}_{M,n+1} + \mathcal{E}_{E,n+1},
\end{aligned}$$

with an obvious notation for $\mathcal{E}_{I,n+1}$, $\mathcal{E}_{M,n+1}$, and $\mathcal{E}_{E,n+1}$. Note that $\mathcal{E}_{I,n+1}$ is the error due to the I.I.D. innovation ε_{t+1}, $\mathcal{E}_{M,n+1}$ is the model specification error, and $\mathcal{E}_{E,n+1}$ the error due to estimation of the unknown function g used for prediction. An appropriate bootstrap method for constructing prediction intervals is one which is able to take all three sources of prediction error into account and to estimate consistently the conditional distribution function

$$\Pr(\mathcal{E}_{n+1}(\tau_i) \leq x | \mathcal{X}_n, \mathcal{X}_{n-1}, \ldots, \mathcal{X}_{n-k+1}), \qquad x \in \mathbb{R}, \tag{2}$$

for a set of values $\tau_i \in \mathcal{I}$.

However, even in the most studied real-valued case, a common approach to estimate the prediction error distribution is to ignore the model specification error, i.e., to construct the prediction intervals based on the innovation and the estimation error only. The corresponding bootstrap methods use the same model for prediction *and for generating the functional pseudo-time series.*

2 Bootstrapping Prediction Intervals

2.1 The Bootstrap Method

The basic idea of the bootstrap method proposed is to generate bootstrap samples of functional time series $(\mathcal{X}_1^*, \mathcal{X}_2^*, \ldots, \mathcal{X}_n^*, \mathcal{X}_{n+1}^*)$ which imitate the dependence structure of the original functional time series and at the same time satisfy the following condition:

$$\mathcal{X}^*_{n-k+1} = \mathcal{X}_{n-k+1}, \quad \mathcal{X}^*_{n-k+2} = \mathcal{X}_{n-k+2}, \quad \ldots, \quad \mathcal{X}^*_n = \mathcal{X}_n. \tag{3}$$

This requirement is essential since as we have seen, our interest is focused on estimating the distribution of the prediction error, that is, the conditional distribution of $\mathcal{E}_{n+1}$ given $\mathcal{X}_n, \mathcal{X}_{n-1}, \ldots, \mathcal{X}_{n-k+1}$.

To achieve this and motivated by the functional sieve bootstrap proposed by [12], we use the Karhunen-Loève expansion and first decompose the random element $\mathcal{X}_t$ as

$$\mathcal{X}_t = \sum_{j=1}^{n} \xi_{j,t} \nu_j = \underbrace{\sum_{k=1}^{m} \xi_{j,t} \nu_j}_{X_{t,m}} + \underbrace{\sum_{j=m+1}^{\infty} \xi_{j,t} \nu_j}_{U_{t,m}}. \tag{4}$$

Here, $\xi_{j,t} = \langle \mathcal{X}_t, \nu_j \rangle$ where $(\nu_j, j = 1, 2, \ldots)$ are the orthonormal eigenvectors corresponding to the eigenvalues $(\lambda_1 > \lambda_2 > \ldots)$, in descending order of the variance operator $C_0 = E(\mathcal{X}_t \otimes \mathcal{X}_t)$. Based on decomposition (4) the main idea is to consider $X_{t,m}$ as the main driving component of the functional random element $\mathcal{X}_t$ and to treat the "remainder" $U_{t,m}$ as a white noise component.

Since, in practice, we do not observe the eigenvectors v_j and the scores $\xi_{j,t}$, we use their sample estimates. As in [12], the dependence structure of the estimated scores, $\widehat{\boldsymbol{\xi}}_t = (\widehat{\xi}_{1,t}, \ldots, \widehat{\xi}_{m,t})$, is modeled by a forward vector autoregressive (VAR) process,

$$\widehat{\boldsymbol{\xi}}_t = \sum_{j=1}^{p} \widehat{A}_{j,p} \widehat{\boldsymbol{\xi}}_{t-j} + \widehat{\boldsymbol{e}}_t, \qquad t = p+1, p+2, \ldots, n, \tag{5}$$

where p denotes the order of the VAR model and $\widehat{\boldsymbol{e}}_t$ denotes the residuals of the VAR fit. To select the optimal number of components m and the order of VAR model p, we implement the method proposed by ([12], Sect. 5).

Based on the decomposition (4) and to generate the functional pseudo-time series $\mathcal{X}^*_1, \mathcal{X}^*_2, \ldots, \mathcal{X}^*_n$ we also fit a VAR(p) process to the backward series of estimated scores, i.e.,

$$\widehat{\boldsymbol{\xi}}_t = \sum_{j=1}^{p} \widehat{B}_{j,p} \widehat{\boldsymbol{\xi}}_{t+j} + \widehat{\boldsymbol{v}}_t, \qquad t = 1, 2, \ldots, n-p, \tag{6}$$

where the $\widehat{B}_j$'s denote the estimated coefficient matrices. Using the backward vector autoregressive representation allows for the generation of a time series of pseudo-scores $\xi^*_1, \xi^*_2, \ldots, \xi^*_n$ which satisfies the condition $\xi^*_t = \xi_t$ for $t = n-k+1, n-k+2, \ldots, n$. This is important in order to achieve that our bootstrap time series $\{\mathcal{X}^*_1, \mathcal{X}^*_2, \ldots, \mathcal{X}^*_n\}$ fulfills condition (3). Using (6), we generate

$$\boldsymbol{\xi}^*_t = \sum_{k=1}^{p} \widehat{B}_{j,p} \boldsymbol{\xi}^*_{t+k} + \boldsymbol{v}^*_t, \qquad \text{for} \quad t = n-k, n-k-1, \ldots, 1. \tag{7}$$

Here, $(\boldsymbol{v}_1^*, \boldsymbol{v}_2^*, \ldots, \boldsymbol{v}_{n-k}^*)$ are obtained by

$$\boldsymbol{v}_t^* = \widehat{B}_p(L^{-1})\widehat{A}_p^{-1}(L)\boldsymbol{e}_t^*, \tag{8}$$

where $\widehat{A}_p(z) = I - \sum_{j=1}^{p} \widehat{A}_{j,p}z^j$, $\widehat{B}_p(z) = I - \sum_{j=1}^{p} \widehat{B}_{j,p}z^j$, $z \in \mathbb{C}$, and $\boldsymbol{e}_t^*$ are I.I.D. resampled with replacement from the empirical distribution of the centered residual $\widehat{e}_t$ in (5) for $t = p+1, p+2, \ldots, n$.

If the order p of the VAR model is larger than the number k of past observations used in the time series model applied for prediction, then we generate for $l = 1, 2, \ldots, p-k$ random vectors $\boldsymbol{\xi}_{n+l}^+ = \sum_{j=1}^{p} \widehat{A}_{j,p}\boldsymbol{\xi}_{n+l-j}^+ + \boldsymbol{e}_{n+l}^+$, where $\boldsymbol{\xi}_t^+ = \widehat{\boldsymbol{\xi}}_t$ for $t \leq n$ and $\boldsymbol{e}_{n+l}^+$ are I.I.D. resampled with replacement from the empirical distribution of the centered residuals $\widehat{\boldsymbol{e}}_t$, $t = p+1, p+2, \ldots, n$.

To generate a functional pseudo-time series $\{\mathcal{X}_1^*, \mathcal{X}_2^*, \ldots, \mathcal{X}_n^*, \mathcal{X}_{n+1}^*\}$, we first set

$$\mathcal{X}_t^* = \mathcal{X}_t, \qquad t = n, n-1, \ldots, n-k+1.$$

Using the backward obtained pseudo-series $(\xi_1^*, \xi_2^*, \ldots, \xi_{n-k}^*)$ calculate

$$\mathcal{X}_t^* = \sum_{j=1}^{m} \xi_{j,t}^* \widehat{v}_j + U_{t,m}^*, \qquad t = n-k, n-k-1, \ldots, 1,$$

where $U_{t,m}^*$ are I.I.D. resampled with replacement from $\{\widehat{U}_{t,m} - \overline{U}_n, t = 1, 2, \ldots, n\}$, $\overline{U}_n = n^{-1}\sum_{t=1}^{n} \widehat{U}_{t,m}$ and $\widehat{U}_{t,m} = \mathcal{X}_t - \sum_{j=1}^{m} \widehat{\xi}_{j,t}\widehat{v}_j$. Finally, let

$$\begin{aligned}
\mathcal{X}_{n+1}^* &= \sum_{j=1}^{m} \xi_{j,n+1}^* \widehat{v}_j + U_{n+1,m}^* \\
&= \sum_{j=1}^{m} \Big(\sum_{l=1}^{p} \widehat{A}_{l,p} \xi_{j,n+1-l}^* + \boldsymbol{e}_{n+1}^* \Big) \widehat{v}_j + U_{n+1,m}^*,
\end{aligned}$$

where $\boldsymbol{e}_{n+1}^*$ and $U_{n+1,m}^*$ are I.I.D. as $\boldsymbol{e}_t^*$ and $U_{t,m}^*$, respectively.

Using the bootstrapped functional time series $(\mathcal{X}_1^*, \mathcal{X}_2^*, \ldots, \mathcal{X}_n^*)$, we calculate the forecast $\widehat{\mathcal{X}}_{n+1}^*$ as

$$\widehat{\mathcal{X}}_{n+1}^* = \widehat{g}^* \left(\mathcal{X}_n^*, \mathcal{X}_{n-1}^*, \ldots, \mathcal{X}_{n-k+1}^* \right),$$

where $\widehat{g}^*$ is the same estimator as $\widehat{g}$ but obtained using the functional pseudo-time series $\{\mathcal{X}_1^*, \mathcal{X}_2^*, \ldots, \mathcal{X}_n^*\}$.

We approximate the distribution of $\mathcal{E}_{n+1}$ by the distribution of $\mathcal{E}_{n+1}^* = \mathcal{X}_{n+1}^* - \widehat{\mathcal{X}}_{n+1}^*$. Notice that by generating B replicates of $\mathcal{E}_{n+1}^*$, say

$$\left(\mathcal{E}_{n+1,1}^*, \mathcal{E}_{n+1,2}^*, \ldots, \mathcal{E}_{n+1,B}^* \right), \tag{9}$$

we can use the empirical distribution of the pseudo-prediction errors $\mathcal{E}^*_{n+1,b}$ to estimate the unknown distribution of $\mathcal{E}^*_{n+1}$. From (9), we can estimate for any $\tau_i \in \mathcal{I}$, the quantiles of $\mathcal{E}^*_{n+1}(\tau_i)$, say $c^*_{\alpha}(\tau_i)$ and $c^*_{1-\alpha}(\tau_i)$, i.e.,

$$\Pr\left[c^*_{\alpha}(\tau_i) \leq \mathcal{E}^*_{n+1}(\tau_i) \leq c^*_{1-\alpha}(\tau_i)\right] = 1 - 2\alpha, \quad i = 1, \ldots, \kappa,$$

where κ is the number of discretized data points and 2α denotes the level of significance. Using these quantiles, we can, for instance, construct a pointwise prediction interval for $\mathcal{X}_{n+1}$ which is given by

$$\left[\widehat{\mathcal{X}}_{n+1}(\tau_i) + c^*_{\alpha}(\tau_i),\ \widehat{\mathcal{X}}_{n+1}(\tau_i) + c^*_{1-\alpha}(\tau_i)\right].$$

3 Numerical Studies

We utilize Monte Carlo methods to evaluate the performance of the proposed sieve bootstrap method. The ultimate goal of our simulation study is to provide an assessment and comparison based on interval forecast accuracy.

To define the data generating process that we considered, let $\{\mathcal{X}_t(\tau), \tau \in [0, 1]\}$ be simulated series from Brownian motions with zero mean and variance $1/(N-1)$, where N denotes the number of discrete data points. Let $B_t(\tau)$ be simulated series from Brownian motions with zero mean and variance $0.05 \times 1/(N-1)$. We generate functional data according to

$$\mathcal{X}_t(\tau) = \int_0^1 \psi(\tau, \gamma)\mathcal{X}_{t-1}(\gamma)d\gamma + b \times \mathcal{X}_{t-2}(\tau) + B_t(\tau), \qquad t = 1, 2, \ldots, 100, \tag{10}$$

where $\psi(\tau, \gamma) = 0.07 \exp^{\frac{1}{2}(t^2+s^2)}$. The choice of the constant in the definition of $\psi(\cdot, \cdot)$ is performed so that $\|\psi\|_2 \approx 0.1$. When $b \neq 0$ in (10), the model is FAR(1) model; when $b = 0.8$, the model is FAR(2) model. As an illustration, Figure 1 displays simulated functional time series.

We consider two sample sizes $n = 100$ and $n = 250$. Using the first 80% of the data as the initial training sample, we compute the one-step-ahead prediction interval. Then, we increase the training sample by one and again compute the one-step-ahead prediction interval. This procedure continues until the training sample reaches the sample size. With the 20% of the data as the testing sample, we compute interval forecast accuracy for the one-step-ahead prediction. To measure the interval forecast accuracy, we consider the coverage probability deviance (CPD) as an absolute difference measure between the nominal coverage probability and empirical coverage probability. The empirical coverage probability is defined as

$$\text{Empirical coverage} = \frac{1}{n_{\text{test}} \times \kappa} \sum_{\eta=1}^{n_{\text{test}}} \sum_{i=1}^{\kappa} [\mathbb{1}\{\mathcal{X}_\eta(\tau_i) < \widehat{\mathcal{X}}^{\text{ub}}_\eta(\tau_i)\} + \mathbb{1}\{\mathcal{X}_\eta(\tau_i) > \widehat{\mathcal{X}}_{\eta^{\text{lb}}}(\tau_i)\}],$$

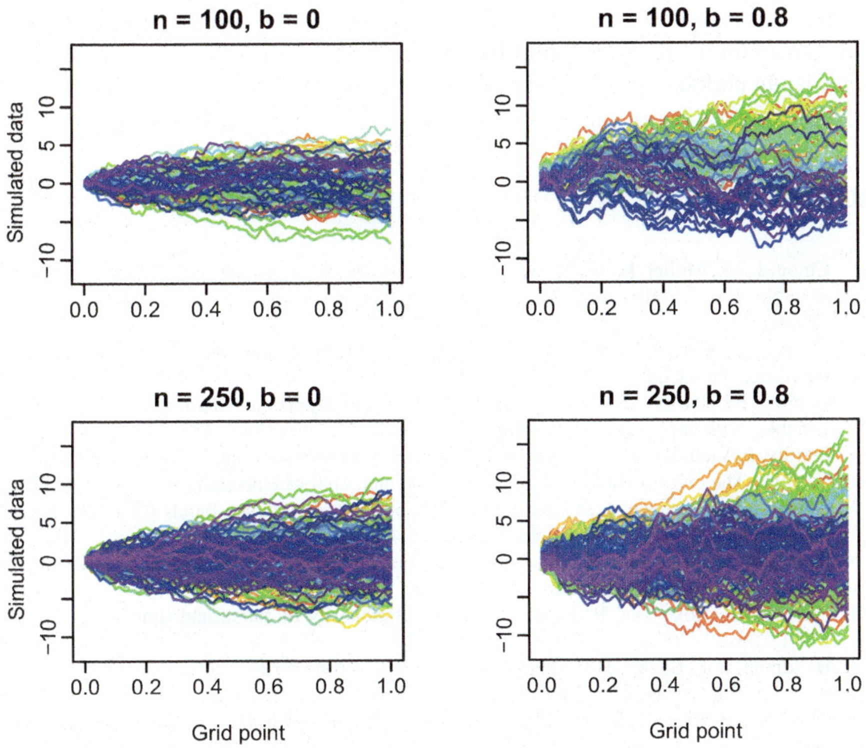

Fig. 1 Simulated functional time series

where n_{test} denotes the number of curves in the forecasting period. The CPD is defined as

$$\text{CPD} = |\text{Empirical coverage probability} - \text{Nominal coverage probability}| .$$

As pointed out by [19], we implement the FAR(1) bootstrap and compare its performance with our sieve bootstrap that uses the FAR(1) method to produce one-step-ahead forecasts. In Table 1, we present the CPD for the two bootstrap methods,

Table 1 CPD for the two bootstrap methods, where the sieve bootstrap method uses $k = 1$

		$n = 100$		$n = 250$	
Method	Statistic	$b = 0$	$b = 0.8$	$b = 0$	$b = 0.8$
FAR(1) bootstrap	Median	0.0359	0.0782	0.0238	0.0520
Sieve bootstrap	Median	**0.0349**	**0.0337**	**0.0228**	**0.0250**

where we choose $k = 1$ for the sieve bootstrap method. The sieve bootstrap performs better than the FAR(1) bootstrap with a smaller median of the CPD values in the forecasting period.

References

1. Chiou, J.-M., Müller, H.-G.: Modeling hazard rates as functional data for the analysis of cohort lifetables and mortality forecasting. J. Am. Stat. Assoc.: Appl. Case Stud. **104**(486), 572–585 (2009)
2. Cuevas, A., Febrero, M., Fraiman, R.: On the use of the bootstrap for estimating functions with functional data. Comput. Stat. Data Anal. **51**(2), 1063–1074 (2006)
3. Dehling, H., Sharipov, S.O., Wendler, M.: Bootstrap for dependent Hilbert space-valued random variables with application to von Mises statistics. J. Multivar. Anal. **133**, 200–215 (2015)
4. Ferraty, F. Vieu, P.: Kernel regression estimation for functional data. In: Ferraty, F., Romain, Y. (Eds.) The Oxford Handbook of Functional Data. Oxford University Press, Oxford (2011)
5. Goldsmith, J., Greven, S., Crainiceanu, C.: Corrected confidence bands for functional data using principal components. Biometrics **69**(1), 41–51 (2013)
6. Hörmann, S., Kokoszka, P.: Functional time series. Handbook of Statistics, vol. 30, pp. 157–186. North Holland, Amsterdam (2012)
7. Horvàth, L., Kokoszka, P., Rice, G.: Testing stationarity of functional time series. J. Econ. **179**(1), 66–82 (2014)
8. Hyndman, R.J., Shang, H.L.: Forecasting functional time series (with discussion). J. Korean Stat. Soc. **38**(3), 199–221 (2009)
9. Kokoszka, P., Rice, G., Shang, H.L.: Inference for the autocovariance of a functional time series under conditional heteroscedasticity. J. Multivar. Anal. **162**, 32–50 (2017)
10. McMurry, T., Politis, D.N.: Resampling methods for functional data. In: Ferraty, F., Romain, Y. (Eds.) The Oxford Handbook of Functional Data Analysis. Oxford University Press, New York, pp. 189–209 (2011)
11. Nyarige, E.G.: The bootstrap for the functional autoregressive model FAR(1), Ph.d. thesis, Technische Universität Kaiserslautern. (2016) https://kluedo.ub.uni-kl.de/frontdoor/index/index/year/2016/docId/4410
12. Paparoditis, E.: Sieve bootstrap for functional time series. Annals of Statistics. **46**(6B), 3510–3538 (2018)
13. Paparoditis, E., Sapatinas, T.: Bootstrap-based testing of equality of mean functions or equality of covariance operators for functional data. Biometrika **103**(3), 727–733 (2016)
14. Pilavakis, D., Paparoditis, E., Sapatinas, T.: Moving block and tapered block bootstrap for functional time series with an application to the K-sample mean problem. Bernoulli (2018). in press
15. Politis, D.N., Romano, J.P.: The stationary bootstrap. J. Am. Stat. Assoc.: Theory Methods **89**(428), 1303–1313 (1994)
16. Raña, P., Aneiros-Perez, G., Vilar, J.M.: Detection of outliers in functional time series. Environmetrics **26**(3), 178–191 (2015)
17. Shang, H.L.: Resampling techniques for estimating the distribution of descriptive statistics of functional data. Commun. Stat. - Simul. Comput. **44**(3), 614–635 (2015)
18. Shang, H.L.: Functional time series forecasting with dynamic updating: An application to intraday particulate matter concentration. Econ. Stat. **1**, 184–200 (2017)
19. Shang, H.L.: Bootstrap methods for stationary functional time series. Stat. Comput. **28**(1), 1–10 (2018)

Measuring and Estimating Overlap of Distributions: A Comparison of Approaches from Various Disciplines

Judith H. Parkinson and Arne C. Bathke

Abstract In this work, we will compare three approaches on measuring the overlap of datasets. Different research areas lead to differing definitions and interpretations of overlap. We will discuss the differences, advantages and disadvantages of three methods which were all introduced in different research fields. Coming from a medical, a cryptographical and a statistical background, all three methods show interesting aspects of overlap. Even though quite differently defined, all three show reasonably interpretable results in simulations and data example.

Keywords overlap measures · overlap coefficient · cryptosets · probabilistic interpretation · comparison

1 Introduction

The volume of overlap of sets is measured in many different research areas for various reasons. An exchange of those approaches between the research fields seldom occurs as the terminology tends to diverge, and it may not seem clear that a method from a different discipline may be fitting for ones own concept of set overlap. Even though the terminology is different and maybe even daunting, a look at the methods of colleagues in other research areas can be beneficial for ones own research.

In this paper, we will compare three approaches of determining the volume of overlap with different interpretations coming from various disciplines. Each of them is designed for different types of datasets, yet we will apply them to one dataset and modifications of it and show how they perform under different conditions. This paper is not about showing that one method is better than the other, as all of them have their

J. H. Parkinson (✉) · A. C. Bathke
Department of Mathematics, University of Salzburg, Hellbrunner Straße 34, 5020 Salzburg, Austria
e-mail: judith.parkinson-schwarz@tum.de

A. C. Bathke
e-mail: arne.bathke@sbg.ac.at

M. La Rocca et al. (eds.), *Nonparametric Statistics*, Springer Proceedings in Mathematics & Statistics 339, https://doi.org/10.1007/978-3-030-57306-5_38

own strengths and weaknesses. This paper is about showing that an interdisciplinary approach can have positive effects on the research.

The first approach based on a publication by [2] uses the overlap coefficient (OVL) as first mentioned in [9]. The OVL measures the similarity of two probability distributions or two populations represented by such distributions. In medicine and biomathematics it is commonly used, and various estimates exist for this measure. [2] focus on the overlap of two normally distributed populations using maximum-likelihood estimators and even analyse their asymptotic behaviour. As [2] claim, one can construct asymptotic valid confidence intervals using the results they presented.

The second paper finds application in cryptography combined with a medical motivation. [8] proposed an approach to measure the overlap between private datasets with cryptosets. Due to sensitive information, certain datasets can only be accessed after passing ethical and legal barriers. Whether the time and money was worth investing cannot easily be assessed in advance. Being able to analyse whether a new private dataset contains new informations thus proves profitable. Using certain encryptions the original multivariate data is mapped to a single value such that no critical informations are accessible. Based on the cryptographically secured data one then can estimate the overlap, the volume of the duplicated data in the private dataset.

The third and last approach considered here is a stochastic interpretation of overlap motivated by a question from ecology. Using a receiver operating characteristic (ROC) curve and rank estimator, [5] calculated the overlap of two (ecological) niches. They further provided asymptotic results and shortcut variance estimators leading to confidence intervals with a high expected coverage probability (ECP). Their mathematical definition of overlap has a probabilistic interpretation. Further they propose a method to measure the overlap in an arbitrary dimension d.

In the following section, we will introduce those three methods in a more detailed manner and show their most important properties. Section 3 then provides results of a data example and of some simulations based on the dataset followed by a short discussion of the methods.

2 Methodology

In this section, we will introduce all three methods in a more detailed manner. We will point out their properties and how the resulting estimates should be interpreted.

2.1 Estimator of the OVL

As mentioned in the previous section, the OVL measures the similarity of two probability distributions. Let f_1 and f_2 be two density functions on $\mathbb{R}$. Then the OVL is defined to be

$$OVL = \int_{\mathbb{R}} \min\{f_1(t), f_2(t)\}\, dt\ . \tag{1}$$

A version for discrete probability distributions can be defined analogously. We will, however, focus on the continuous case in this paper, as the method by [2] was designed for normal distributions which are continuous.

The OVL has some basic properties that are useful when interpreting the result. To start with, the OVL can only take values in [0, 1]. The boundaries can only be obtained in special cases, such that $OVL = 0$ if and only if the supports of $f_1(t)$ and $f_2(t)$ are disjoint, and $OVL = 1$ if and only if $f_1 \equiv f_2$. Additionally, the measure is invariant under transformations of the form $t \to g(t)$ with g a continuous differentiable function defined on $\mathbb{R}$ that is one-to-one and preserves order. This may be helpful in situations where a normalizing transformation can be applied. Thus, the estimator of [2] has obtained good performance results even in instances where the original data was not normally distributed.

However, in the derivation of the OVL estimator, only normal distributions, $f_1(t, \mu_1, \sigma_1^2)$ and $f_2(t, \mu_2, \sigma_2^2)$, were considered. The case $\sigma_1^2 = \sigma_2^2 = \sigma^2$ was discussed intensively in the original paper while the more general case $\sigma_1^2 \neq \sigma_2^2$ was only briefly mentioned.. The focus was on the simpler version of equal variances as the authors said it had attracted more attention in the past.

Denoting the standard normal distribution function by $\Phi(\cdot)$, we can express the OVL in the simple case as $2\Phi(-|\mu_1 - \mu_2|/(2\sigma)) = 2\Phi(-|\delta|/2)$ with $\delta := (\mu_1 - \mu_2)/\sigma$. To estimate the OVL, [2] used the canonical estimators of μ_1, μ_2, namely,

$$\hat{\mu}_1 = \frac{1}{n_1}\sum_{i=1}^{n_1} X_{1i},\ \hat{\mu}_2 = \frac{1}{n_2}\sum_{i=1}^{n_2} X_{2i},$$

with n_1 and n_2 the sample sizes, X_{1i} drawn from the distribution with density $f_1(t, \mu_1, \sigma^2)$ and X_{2i} from $f_2(t, \mu_2, \sigma^2)$, and the estimator of the variance

$$S^2 = \frac{1}{n_1 + n_2}\left(\sum_{i=1}^{n_1}(X_{1i} - \hat{\mu}_1)^2 + \sum_{i=1}^{n_2}(X_{2i} - \hat{\mu}_2)^2\right),$$

the pooled maximum-likelihood estimator for σ^2. A straightforward estimator of OVL is then given in the simple case of $\sigma_1^2 = \sigma_2^2 = \sigma^2$ by

$$O\hat{V}L = 2\Phi\left(-\frac{|\hat{\mu}_1 - \hat{\mu}_2|}{2S}\right).$$

Even though the above expression of the OVL resembles $P(X \leq Y)$ for X and Y two independent random normal variables with mean μ_1 and μ_2 and common variance σ^2, there is no direct relation between those two values.

Further the sampling distribution of this estimator can be linked to a non-central F-distribution. More precisely

$$F_{O\hat{V}L}(p) = P(O\hat{V}L \leq p) = P\left(\frac{4n_1n_2(n_1+n_2-2)}{(n_1+n_2)^2}\left[\Phi^{-1}\left(\frac{p}{2}\right)\right]^2 \leq F\right),$$

where F has a non-central F-distribution with 1 and $n_1 + n_2 - 2$ degrees of freedom and non-centrality parameter $\lambda = \delta^2 n_1 n_2/(n_1 + n_2)$.

Simulations reported in [2] showed that the relative error was small unless sample sizes were small and additionally $|\delta|$ was large. For small sample sizes, a bias reduced estimator of OVL was proposed as

$$O\tilde{V}L = 2\Phi\left(-\frac{\left(\frac{n_1+n_2-2}{n_1+n_2}\right)^{1/2} |\hat{\mu}_1 - \hat{\mu}_2|}{2S}\right).$$

In the unequal variance case, due to the properties of the normal distribution, the two densities $f_1(t, \mu_1, \sigma_1^2)$ and $f_2(t, \mu_2, \sigma_2^2)$ cross at exactly two points. Denoting the smaller of the two points with t_1 and the other with t_2, then the OVL is given by

$$OVL = \Phi\left(\frac{t_1 - \mu_1}{\sigma_1}\right) + \Phi\left(\frac{t_2 - \mu_2}{\sigma_2}\right) - \Phi\left(\frac{t_1 - \mu_2}{\sigma_2}\right) - \Phi\left(\frac{t_2 - \mu_1}{\sigma_1}\right) + 1.$$

To estimate the OVL in this scenario, maximum-likelihood estimators for μ_1, μ_2, σ_1 and σ_2 were substituted in the above formula. For an approximation to the sampling distribution, [2] refer to the dissertation [1] of one of the authors.

2.2 *Overlap of Private Datasets Using Cryptosets*

The method proposed by [8] not only deals with estimating overlap, but also with the problem of encrypting data without losing too much information. In a first step, for two private datasets, a common transformation to so-called hashs needs to be defined. Every entry in the private dataset gets assigned a private ID that is constructed analogously for all entries. Those private IDs do not have to be unique, but duplicates shall be held at a minimum. A cryptographic hash function $H(\cdot)$ maps those IDs to an integer space ranging from 0 to $L - 1$, which are the public IDs. The length of the cryptoset L is typically chosen between 500 and 10,000, with smaller L leading to a higher security of the sensible data, while larger L lead to a higher information rate. The number of public IDs with value $'i'$ is denoted by A_i and B_i for the two private datasets. The total number of entries in the two datasets are then given by $A = \sum_{i=0}^{L-1} A_i$ and $B = \sum_{i=0}^{L-1} B_i$. Further let $\mathcal{A} := (A_i)_{i=0}^{L-1}$ and $\mathcal{B} := (B_i)_{i=0}^{L-1}$ be the two resulting cryptosets, based on which the overlap is estimated, and A and B denote the number of entries in the two datasets. For the method to perform well, one shall use a hash function that maps in a manner indistinguishable from an uniform, random process. The elements of $\mathcal{A}$ and $\mathcal{B}$ can be modelled each as the sum of two independent, random variables that are Poisson distributed, as claimed by [8]. More precisely we can denote the counts derived from items only in the first dataset with $A^{'}$, from the second dataset with $B^{'}$ and counts derived from items in both datasets with $A \cap B$, such that $A = A^{'} + A \cap B$ and $B = B^{'} + A \cap B$. Then the elements of the cryptosets are Poisson distributed with rates $(A^{'} + A \cap B)/L$ and, respectively, $(B^{'} + A \cap B)/L$. [8] claim that, using the attributes of the Poisson distribution, it holds that

$$A \cap B = Cov(\mathcal{A}, \mathcal{B})L + \epsilon,$$

with ϵ a random error term with zero mean. For $R_{\mathcal{A},\mathcal{B}}$ the Pearson correlation, $A \cap B$ can be written as $R_{\mathcal{A},\mathcal{B}}\sqrt{AB} + \epsilon$. [8] then defined the overlap proportion to be given by

$$\frac{A \cap B}{\min\{A, B\}} = R_{\mathcal{A},\mathcal{B}}\sqrt{\eta} + \epsilon^*, \tag{2}$$

where $\eta = \max\{A, B\}/\min\{A, B\}$ and ϵ^* a zero-mean error term. Similar to the method by [2] one should be able to construct confidence intervals based on the proposed results by [8].

In addition to the results on overlap estimation, [8] pointed out the security of their cryptosets and stated that together with a high accuracy of the proposed method made it favourable in comparison to bloom filters, which is a commonly used method for encrypting. For further discussion on this topic, we refer to the paper itself as it is explained in detail there and validated using some simulations.

2.3 Overlap Volume in a Probabilistic Interpretation

The method presented by [5] deals with the question of how a true overlap volume between distributions (or sets) should be defined, as well as the actual estimation of the overlap. The work was motivated by a question from ecology (see also [3, 4]), but the authors derived an approach with a probabilistic interpretation.

Consider two random variables X and Y with some continuous cumulative distribution functions F and G. For each $\alpha \in [0, 1]$ we are interested in the probability that measure G assigns to the central $(1-\alpha)$-portion of F. More precisely, we define an overlap function

$$h: [0, 1] \mapsto [0, 1], \ \alpha \mapsto G(F^{-1}(1-\alpha/2)) - G(F^{-1}(\alpha/2)),$$

where F^{-1} is the quantile function of F. The asymmetric overlap volume is then defined as

$$I_2 := \int_0^1 h(\alpha)\, d\alpha.$$

We call it *asymmetric* because switching the roles of F and G yields a different value, as opposed to, for example, the OVL as defined in Sect. 2.1. For given datasets containing observations $X_1, \ldots, X_n \sim F$ and $Y_1, \ldots, Y_m \sim G$, the overlap function and thus the asymmetric overlap volume can be estimated consistently using the empirical cdfs. Similar to the OVL, I_2 fulfils some nice properties. For example, if F and G are equal then $I_2 = 1/2$. If $I_2 = 0$ holds, this is equivalent to the fact that the probability mass of G puts no weight within the range of F. Alternatively, $I_2 = 1$ is equivalent to the fact that all probability mass of G lies between the quantiles $F^{-1}(1/2)$ and $F^{-1}(1/2+)$, where $F^{-1}(1/2+)$ is the right side limit. Further, one can apply strictly monotone, continuous transformations to X and Y simultaneously without changing the value of I_2. For example, one may take logarithms or square roots, if applicable, and the value of I_2 remains unaffected. Now consider the second asymmetric overlap volume I_1, which is defined analogously with the roles of F and G switched. As long as the cdfs F and G are continuous, the sum of these two asymmetric overlap volumes will lie between 0 and 1, with $I_1 + I_2 = 1$ if and only if $F \equiv G$.

As for the probabilistic interpretation of I_2, we will split up X into two random variables $X_1 \sim F_1$ and $X_2 \sim F_2$ with

$$F_1(t) = \begin{cases} 22\, F(t), & t < F^{-1}(1/2), \\ 1, & t \geq F^{-1}(1/2), \end{cases} \quad F_2(t) = \begin{cases} 0, & t < F^{-1}(1/2), \\ 22\, F(t) - 1, & t \geq F^{-1}(1/2). \end{cases} \tag{3}$$

Then, I_2 can be expressed as $P(X_1 < Y < X_2)$, which can be loosely interpreted as the chance that an observation of Y will lie between two random observations of X already knowing that one of them is bigger and the other one is smaller than the median of X.

Instead of using the empirical cdfs of F and G to estimate I_2, which looks rather complex at first sight, [5] propose an approach which uses ranks and is easy to calculate. We can rank all $m + n$ observations, using midranks in case of ties. Denote the ranks of the X-observations below or equal to their median by $R_1^{X<}, \ldots, R_K^{X<}$ and the ones above by $R_{K+1}^{X>}, \ldots, R_n^{X>}$. Then I_2 can be estimated by

$$\frac{2}{mn}\left(\sum_{i=1}^{K} R_i^{X<} - \sum_{j=K+1}^{n} R_j^{X>}\right) + \frac{1}{2}c,$$

with $c = -n/m$ for n even and $c \approx -n/m$ for n odd and n and m large. It was shown in [5] that this estimator of I_2 follows asymptotically a normal distribution with the expectation given by the true value of I_2. Using a shortcut formula for the variance estimation in special cases and a bootstrap approach for the variance estimation in the general case, [5] give explicit confidence intervals.

They further define a symmetric overlap volume NO as $4I_1I_2$ and also analyse its properties. Among the most important properties is that the symmetric overlap volume lies between 0 and 1, with 1 indicating identical distribution functions, that is, $F \equiv G$. For more detailed informations on the derived results and substantial simulation results, we refer to the original paper.

3 Data Examples and Comparison

While [2, 5] check how much the distributions of two datasets overlap, [8] focus on how many of the observations are probably contained in both datasets. Even though this seems similar, bigger differences could be noticed in the simulations. The method by [8] produced average overlap estimates far lower than the other two methods when using the same distributions for both datasets. Their method is designed to identify true duplicates. As the choice of simulation setting will influence the performance of the individual methods immensely we will focus on illustrating the differences and similarities between those approaches using a data example. The dataset we will thus use is 'Finch2' as given in the R-package 'dynRB' ([7]). It contains 81 observations of the *Geospiza fuliginosa pavula* and 22 observations of the *Geospiza fortis fortis*. In the first setting, we will consider two datasets where each represents one of the species. By the definition of the method of [8] their method should obtain a value close to zero as no observation is contained in both datasets. The other two methods should obtain some kind of overlap when considering special traits of the two species of finches. Note that the latter two methods were both analysed in an univariate setting. That is, each trait was considered separately, while the method by [8] was devised as multivariate. We consider the four traits 'tarsus length', 'body length', 'wing length' and 'tail length'.

In Figure 1, we can see the distribution of the individual four traits of the two species. Figure 2 shows a scatterplot matrix of those four traits. In the following, we will refer to the methods introduced in [2], in [8] and in [5] as OVL, CS and Rank method, respectively.

As one can see in Tables 1 and 2, OVL and Rank methods both return reasonably interpretable values for the overlaps. Both measures agree that the overlap of the tail length is the highest whereas wing length has the lowest overlap, and that the distributions of those two finches somewhat overlap. As for the confidence intervals, those of the OVL method are much smaller, yet recall that an assumption of their method is that the data are normally distributed, which is likely violated for certain traits in this dataset considering Fig. 1. For calculating the overall overlap over all four traits of the OVL method, we have used the same approach as for the rank methods, as [2] did not explicitly state how one could expand their method to a higher dimension. When calculating the Pearson correlation for the CS method, we obtain a slightly negative value, implying that there is no overlap at all. This is indeed correct as the two datasets don't contain duplicates. A small overlap of distributions is thus not sufficient to trigger a noticeable overlap with this method, so that in the

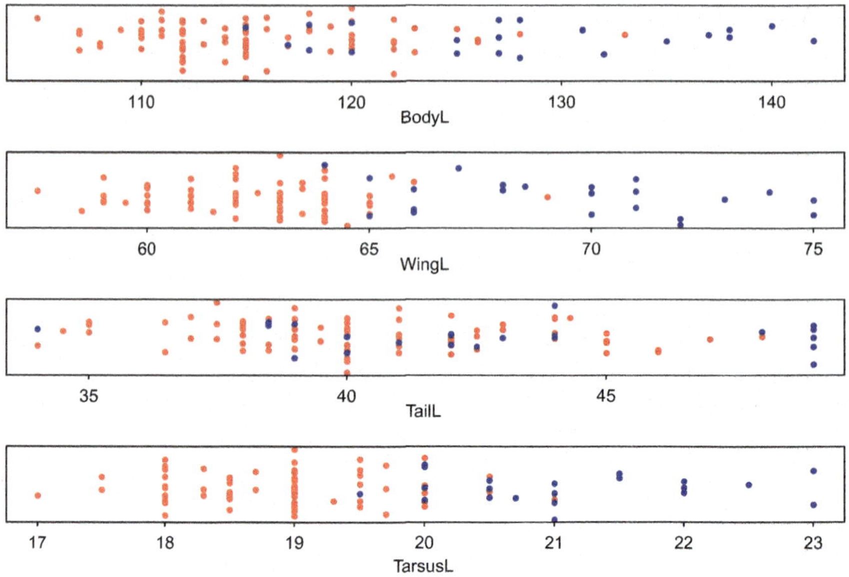

Fig. 1 The four considered traits of the Galapagos finches. The red observations represent *Geospiza fuliginosa parvula* and the blue represents *Geospiza fortis fortis*.

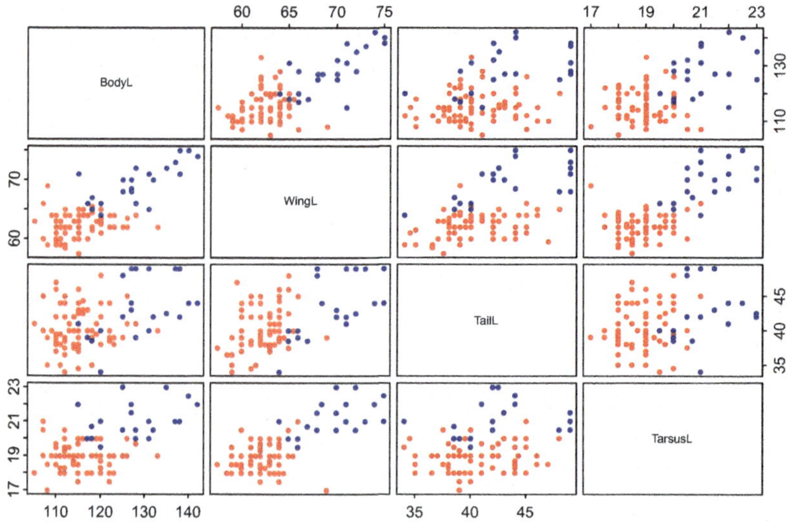

Fig. 2 Scatterplot of four of the nine traits for two species of finches, *Geospiza fuliginosa parvula* (red) and *Geospiza fortis fortis* (blue)

Table 1 Estimated values for the OVL along with the estimated variance and a confidence interval

	OVL	Var	CI
BodyL	0.276	0.004	0.250, 0.303
WingL	0.123	0.002	0.107, 0.140
TailL	0.696	0.008	0.657, 0.734
TarsusL	0.169	0.002	0.149, 0.189
Overall	0.251		

Table 2 Estimated asymmetric overlap I_1 and I_2 together with their confidence interval using a bootstrap approach, as well as the resulting symmetric overlap estimation

	I_1	CI_1	I_2	CI_2	NO
BodyL	0.177	0.071, 0.31	0.167	0.063, 0.282	0.118
WingL	0.039	0.007, 0.093	0.039	0.006, 0.089	0.006
TailL	0.383	0.241, 0.527	0.470	0.301, 0.608	0.720
TarsusL	0.061	0.016, 0.126	0.058	0.014, 0.117	0.014
Overall	0.113		0.115		0.052

Table 3 Expected overlap (variance) return with respect to the number of duplicates 5, 10, 15, 20 and 22

	5	10	15	20	22
OVL	0.352 (5.9e-4)	0.424 (4.9e-4)	0.478 (2.9e-4)	0.521 (7.6e-5)	0.536 (–)
I_1	0.177 (1.6e-4)	0.224 (1.6e-4)	0.265 (1.2e-4)	0.299 (3.8e-5)	0.312 (–)
I_2	0.147 (2.9e-5)	0.172 (3.0e-5)	0.193 (2.1e-5)	0.212 (6.7e-6)	0.218 (–)
NO	0.104 (7.4e-5)	0.155 (1.1e-4)	0.205 (1.1e-4)	0.253 (4.2e-5)	0.272 (–)
CS	0.227 (9.6e-3)	0.518 (1.3e-2)	0.796 (1.1e-2)	1.065 (3.5e-3)	1.170 (–)
$\# CS$	5.05 (4.72)	11.38 (6.53)	17.50 (5.13)	23.46 (1.81)	26 (–)

scenario it was introduced, patients may have observed values that are identically distributed, yet they will be registered as different individuals.

In a second run, we added some observations of *Geospiza fortis* to the data of *Geospiza fuliginosa*, thereby duplicating 5, 10, 15, 20 and 22 observations, and compared the overall overlap measures of all three methods with each other. The results can be found in Table 3. For matter of lucidity, we only show the overall overlap and not the ones of the individual traits. For each number of duplicates, we drew 10,000 different samples from the first species and added them to the second species. The results shown are the averages of the simulations and the variances.

Results from an analogous simulation, where from the second species we added observations to the first species, can be found in Table 4. Looking at both tables in combination, we do notice a few differences between the three methods.

Table 4 Expected overlap (variance) return with respect to the number of duplicates 5, 20, 40, 75, 81

	5	20	40	75	81
OVL	0.407 (3.8e-4)	0.634 (3.8e-4)	0.754 (2.0e-4)	0.840 (2.1e-5)	0.848 (–)
I_1	0.195 (1.8e-4)	0.300 (1.8e-4)	0.356 (1.1e-4)	0.396 (1.2e-5)	0.401 (–)
I_2	0.214 (1.0e-3)	0.484 (8.0e-4)	0.538 (3.2e-4)	0.525 (2.3e-5)	0.523 (–)
NO	0.167 (7.1e-4)	0.579 (1.4e34)	0.765 (5.8e-4)	0.831 (1.8e-5)	0.838 (–)
CS	0.087 (7.5e-3)	0.359 (9.9e-3)	0.540 (5.1e-3)	0.847 (4.1e-4)	0.926 (–)
$\# CS$	2.17 (5.83)	15.09 (17.74)	33.50 (19.69)	68.59 (2.78)	75 (–)

All three methods pick up the fact that there is an increase of overlap as we add duplicates. The OVL method obtains values close to 1 for a high number of duplicates, yet does not obtain this value as long as there are data points in one dataset that are not obtained in the other. For smaller numbers of duplicates, it clearly is increased in comparison of the case with no duplicates, as the duplicates have a great influence on the underlying density function. Especially for small sample sizes this effect is quite strong, see Table 4. The rank method, on the other hand, does not seem to pick up the effect of duplicates quite as fast as the OVL. This could be based on the fact that if the duplicates contain outliers or observations that are extreme values they get dropped during the calculations quite early and have a very limited influence. Yet for larger numbers of duplicates it performs adequately. Obtaining an almost perfect overlap in the case of 81 duplicates. Recall that a perfect overlap, that is, identical underlying distribution functions, leads to $I_1 = I_2 = 0.5$. Last but not least, let us discuss the CS method. As the overlap as defined by [8] is the number of duplicates we not only give the overlap percentage but also the corresponding estimated number of duplicates in the last row of the two tables. In Table 3, we notice that the method overestimates the number of duplicates slightly, while in Table 4 it underestimates it quite a lot. It seems reasonable to assume that this is caused by the sample size correction that is being applied to the Pearson correlation. Otherwise, it clearly detects duplicates and is not triggered by an identical distribution itself.

All three methods proved to be reliable and to perform well for what they were designed for. While estimation of the OVL is the most common of the three proposed approaches, it lacks some flexibility, as the method by [2] only considers normal distributions. While there exist approaches for kernel-density-based estimation of the OVL, see [6], those often are burdened with computational complexity and require higher sample sizes to perform well, as the densities need to be estimated explicitly. When using simplified methods that do not estimate the densities themselves one has to carefully analyse whether the needed assumptions are met. The methods by [5, 8] were published more recently. The method by [5] clearly benefits from its easy and fast non-parametric estimation and its probabilistic interpretation, which might even make it easier to understand for people without a strong mathematical background. The method by [8] is the method that stands out the most in comparison with the other two. Due to its design, it is more likely limited to certain application areas, yet it covers those areas better than either of the others could. Overall the three methods discussed here have their advantages and disadvantages.

References

1. Inman, H.F.: Behavior and Properties of the Overlapping Coefficient as a Measure of Agreement between Distributions. PhD thesis, University of Alabama in Birmingham, 1984
2. Inman, H.F., Bradley Jr., E.L.: The overlapping coefficient as a measure of agreement between probability distributions and point estimation of the overlap of two normal densities. Commun. Stat. - Theory Methods **18**(10), 1989
3. Junker, R.R., Kuppler, J., Bathke, A.C., Schreyer, M.L., Trutschnig, W.: Dynamic range boxes–a robust nonparametric approach to quantify size and overlap of n-dimensional hypervolumes. Methods Ecol. Evol. **7**(12), 1503–1513 (2016)
4. Kuppler, J., Höfers, M.K., Trutschnig, W., Bathke, A.C., Eiben, J.A., Daehler, C.C., Junker, R.R.: Exotic flower visitors exploit large floral trait spaces resulting in asymmetric resource partitioning with native visitors. Function. Ecol. **31**(12), 2244–2254 (2017)
5. Parkinson, J.H., Kutil, R., Kuppler, J., Junker, R., Trutschnig, W., Bathke A.: A fast and robust way to estimate overlap of niches, and draw inference. Int. J. Biostat. **06** (2018)
6. Schmid, Friedrich, Schmidt, Axel: Nonparametric estimation of the coefficient of overlapping - theory and empirical application. Comput. Stat. Data Anal. **50**, 1583–1596 (2006)
7. Schreyer, M., Trutschnig, W., Junker, R.R., Kuppler, J., Bathke, A.: Maintainer Manuela Schreyer. Package “dynrb”. 2018
8. Swamidass, S.J., Matlock, M., Rozenblit, L.: Securely measuring the overlap between private datasets with cryptosets. PLOS ONE (2015)
9. Weitzmann, M.S.: Measures of overlap of income distributions of white and negro families in the united states. Technical report, Department of Commerce, Bureau of Census, Washington (U.S.), 1970

Bootstrap Confidence Intervals for Sequences of Missing Values in Multivariate Time Series

Maria Lucia Parrella, Giuseppina Albano, Michele La Rocca, and Cira Perna

Abstract This paper is aimed at deriving some specific-oriented bootstrap confidence intervals for missing sequences of observations in multivariate time series. The procedure is based on a spatial-dynamic model and imputes the missing values using a linear combination of the neighbor contemporary observations and their lagged values. The resampling procedure implements a residual bootstrap approach which is then used to approximate the sampling distribution of the estimators of the missing values. The normal based and the percentile bootstrap confidence intervals have been computed. A Monte Carlo simulation study shows the good empirical coverage performance of the proposal, even in the case of long sequences of missing values.

Keywords Spatio-temporal models · Bootstrap · Missing values

1 Introduction

Imputing missing data from a data set is still a challenging issue both in theoretical and applied statistics. Recent approaches to the problem include Multiple Imputation (MI) and Maximum Likelihood (ML) techniques (see, for a review, [5] or [7]). They have been proved to be superior with respect to traditional techniques based on simple deletion, averaging, and regression estimation.

In MI, missing data are imputed several times by using an appropriate model to produce several different complete-data estimates of the parameters. The parameter estimates from each imputation are then combined across the missing value samples to obtain an overall estimate of the complete-data parameters as well as reasonable estimates of the standard errors. MI assumes that data are Missing Completely

M. L. Parrella (✉) · G. Albano · M. La Rocca · C. Perna
University of Salerno, Via Giovanni Paolo II n.132, Fisciano, Italy
e-mail: mparrella@unisa.it

G. Albano
e-mail: pialbano@unisa.it

C. Perna
e-mail: perna@unisa.it

M. La Rocca et al. (eds.), *Nonparametric Statistics*, Springer Proceedings
in Mathematics & Statistics 339, https://doi.org/10.1007/978-3-030-57306-5_39

At Random (MCAR) or Missing At Random (MAR) and requires that the missing data should be imputed under a particular model that in most applications is the multivariate normal model ([1]). ML-based algorithms have the advantage of being theoretically unbiased under both MCAR and MAR conditions, since they implicitly account for the dependencies of missingness on other variables in the data set ([5]). Like MI, this method delivers unbiased parameter estimators along with their standard errors. Moreover, the ML approach does not require a careful selection of the variables used to impute values as necessary in MI schemes. However, even if limited to linear models, it often presents a high computational cost since estimates could have no closed form.

In the context of multivariate time series, the problem of missing data becomes even more challenging due to the dependence structure which is present in the data. For example, when dealing with environmental data, PM_{10} data are simultaneously collected by monitoring stations for different sites and different time points. As a consequence, missing data, which often occurs due to equipment failure or measurement errors, can be imputed using the time series dependency structure and measurements from nearby stations. Several approaches, proposed in the recent literature, do not jointly account for cross-correlation among variables and serial correlation (see, for example, [6, 8, 10]). However, in the context of environmental data, [2] have proposed a multivariate hidden dynamic geostatistical model which is able to reveal dependencies and spatio-temporal dynamics across multiple variables. In the same framework, [9] have proposed a procedure that aims to reconstruct the missing sequences by exploiting the spatial correlation and the serial correlation of the multivariate time series, simultaneously. The approach is based on a spatial-dynamic model and imputation of the missing sequences in the observed time series is based on a linear combination of the neighbor contemporary observations and their lagged values. This latter approach has several advantages. Firstly, it takes into account both the serial correlation and the spatial correlation simultaneously, in a single stage. Secondly, it does not depend on any tuning parameter or ad hoc choices made by the user. In addition, it has a low computational burden so it nicely scales up to high dimensional multivariate time series. Finally, it can also be applied in those cases where the number of time observations is equal or smaller than the number of variables/locations.

This paper is aimed at deriving some specific-oriented bootstrap confidence intervals for missing sequences in the framework of the model proposed in [9]. They are essentially based on a residual bootstrap scheme to approximate the sampling distribution of the missing value estimators. The performance of the normal-based bootstrap and the percentile bootstrap confidence intervals are compared through a Monte Carlo experiment. The paper is organized as follows. In Sect. 2, after a brief review of the underlying model and the estimation of the parameters, the iterative imputation procedure is proposed and discussed. In Sect. 3, the sampling distribution of the missing value estimator is approximated by the residual bootstrap and the confidence intervals for the imputed value are derived. In Sect. 4, Monte Carlo simulation study is implemented to evaluate and compare the empirical performance of the proposed confidence intervals. Some remarks close the paper.

2 The Model and the Iterative Imputation Procedure

Let $\mathbf{y}_t$ be a multivariate stationary process of dimension p, assumed for simplicity with zero mean value, collecting the observations at time t from p different variables. Following [4], we assume that the process can be modeled by the following *Spatial-Dynamic Panel Data* (*SDPD*) model

$$\mathbf{y}_t = D(\boldsymbol{\lambda}_0)\mathbf{W}\mathbf{y}_t + D(\boldsymbol{\lambda}_1)\mathbf{y}_{t-1} + D(\boldsymbol{\lambda}_2)\mathbf{W}\mathbf{y}_{t-1} + \boldsymbol{\varepsilon}_t, \tag{1}$$

where $D(\cdot)$ denotes a diagonal matrix with diagonal coefficients from the vectors $\boldsymbol{\lambda}_0, \boldsymbol{\lambda}_1$ and $\boldsymbol{\lambda}_2$, respectively, and the error process $\boldsymbol{\varepsilon}_t$ is serially uncorrelated. Model (1) belongs to the family of *spatial econometric models*, so it is particularly oriented to model spatio-temporal data. The matrix $\mathbf{W}$ is called *spatial matrix* and collects the weigths used in the *spatial regression* of each time series observation with simultaneous or delayed observations of neighboring data. In particular, note that the term $D(\boldsymbol{\lambda}_0)\mathbf{W}\mathbf{y}_t$ captures the pure spatial effects, since it only considers contemporary observations, the component $D(\boldsymbol{\lambda}_1)\mathbf{y}_{t-1}$ captures the pure dynamic effects, since it involves lagged observations, while $D(\boldsymbol{\lambda}_2)\mathbf{W}\mathbf{y}_{t-1}$ captures the spatial-dynamic effects. However, if one uses a correlation based matrix $\mathbf{W}$ to measure variable distances, instead of using physical distances, one can use model (1) to analyze any kind of multivariate time series, not necessarily of a strictly spatial nature.

In the following, we assume that $\mathbf{y}_1, \ldots, \mathbf{y}_T$ are realizations from the stationary process defined by (1). Then, we denote with $\boldsymbol{\Sigma}_j = Cov(\mathbf{y}_t, \mathbf{y}_{t-j}) = E(\mathbf{y}_t\mathbf{y}'_{t-j})$ the autocovariance matrix of the process at lag j, where the prime subscript denotes the transpose operator.

The parameters of model (1) can be estimated following [4]. In particular, given stationarity, from (1) we derive the Yule-Walker equation system

$$(\mathbf{I} - D(\boldsymbol{\lambda}_0)\mathbf{W})\boldsymbol{\Sigma}_1 = (D(\boldsymbol{\lambda}_1) + D(\boldsymbol{\lambda}_2)\mathbf{W})\boldsymbol{\Sigma}_0,$$

where $\mathbf{I}$ is the identity matrix of order p. The i-th row of the equation system is

$$(\mathbf{e}'_i - \lambda_{0i}\mathbf{w}'_i)\boldsymbol{\Sigma}_1 = (\lambda_{1i}\mathbf{e}'_i + \lambda_{2i}\mathbf{w}'_i)\boldsymbol{\Sigma}_0, \quad i = 1, \ldots, p, \tag{2}$$

with $\mathbf{w}_i$ the i-th row vector of $\mathbf{W}$ and $\mathbf{e}_i$ the i-th unit vector. Replacing $\boldsymbol{\Sigma}_1$ and $\boldsymbol{\Sigma}_0$ by the sample (auto)covariance matrices

$$\widehat{\boldsymbol{\Sigma}}_1 = \frac{1}{T}\sum_{t=1}^{T-1}\mathbf{y}_{t+1}\mathbf{y}'_t \quad \text{and} \quad \widehat{\boldsymbol{\Sigma}}_0 = \frac{1}{T}\sum_{t=1}^{T}\mathbf{y}_t\mathbf{y}'_t,$$

the vector $(\lambda_{0i}, \lambda_{1i}, \lambda_{2i})'$ is estimated by the generalized Yule-Walker estimator, available in closed form,

$$(\widehat{\lambda}_{0i}, \widehat{\lambda}_{1i}, \widehat{\lambda}_{2i})' = (\widehat{\mathbf{X}}'_i\widehat{\mathbf{X}}_i)^{-1}\widehat{\mathbf{X}}'_i\widehat{\mathbf{Y}}_i, \quad i = 1, 2, \ldots, p, \tag{3}$$

where $\widehat{\mathbf{X}}_i = \left(\widehat{\boldsymbol{\Sigma}}_1'\mathbf{w}_i, \widehat{\boldsymbol{\Sigma}}_0\mathbf{e}_i, \widehat{\boldsymbol{\Sigma}}_0\mathbf{w}_i\right)$ and $\widehat{\mathbf{Y}}_i = \widehat{\boldsymbol{\Sigma}}_1'\mathbf{e}_i$.

The estimated model (1) can be used to reconstruct sequences of missing values as follows.

Let us assume that $\widetilde{\mathbf{y}}_1, \ldots, \widetilde{\mathbf{y}}_T$ are realizations from a stationary process as in (1), not necessarily with zero mean value. In case of processes with no zero mean, model (1) can be still used for parameter estimation after a pre-processing step which centers the observed time series. Let $\boldsymbol{\delta}_t = (\delta_{t1}, \ldots, \delta_{tp})$ be a vector of zeroes/ones that identifies all the missing values in the observed vector $\widetilde{\mathbf{y}}_t$, so that $\delta_{ti} = 0$ if the observation $\widetilde{y}_{ti}$ is missing, otherwise it is $\delta_{ti} = 1$.

The imputation procedure starts, at iteration 0, by initializing the mean centered vector $\mathbf{y}_t^{(0)}$, for $t = 1, \ldots, T$, as

$$\mathbf{y}_t^{(0)} = \boldsymbol{\delta}_t \circ \left(\widetilde{\mathbf{y}}_t - \bar{\mathbf{y}}^{(0)}\right), \qquad \text{with } \bar{\mathbf{y}}^{(0)} = \sum_{t=1}^{T} \boldsymbol{\delta}_t \circ \widetilde{\mathbf{y}}_t / \sum_{t=1}^{T} \boldsymbol{\delta}_t, \tag{4}$$

where the operator $\circ$ denotes the Hadamard product (which substantially implies replacing the missing values with zero) and the ratio between the two vectors in the formula of $\bar{\mathbf{y}}^{(0)}$ is made component-wise.

Then, the generic iteration s of the procedure, with $s \geq 1$, requires that:

(a) we estimate $(\widehat{\boldsymbol{\lambda}}_0^{(s-1)}, \widehat{\boldsymbol{\lambda}}_1^{(s-1)}, \widehat{\boldsymbol{\lambda}}_2^{(s-1)})$ as in Eq. (3), using the centered data $\{\mathbf{y}_1^{(s-1)}, \ldots, \mathbf{y}_T^{(s-1)}\}$;

(b) we compute, for $t = 1, \ldots, T$,

$$\widehat{\mathbf{y}}_t^{(s)} = D(\widehat{\boldsymbol{\lambda}}_0^{(s-1)})\mathbf{W}\,\mathbf{y}_t^{(s-1)} + D(\widehat{\boldsymbol{\lambda}}_1^{(s-1)})\mathbf{y}_{t-1}^{(s-1)} + D(\widehat{\boldsymbol{\lambda}}_2^{(s-1)})\mathbf{W}\,\mathbf{y}_{t-1}^{(s-1)} \tag{5}$$

$$\bar{\mathbf{y}}^{(s)} = \frac{1}{T}\sum_{t=1}^{T}\left(\boldsymbol{\delta}_t \circ \widetilde{\mathbf{y}}_t + (\mathbf{1} - \boldsymbol{\delta}_t) \circ (\widehat{\mathbf{y}}_t^{(s)} + \bar{\mathbf{y}}^{(s-1)})\right) \tag{6}$$

$$\mathbf{y}_t^{(s)} = \boldsymbol{\delta}_t \circ (\widetilde{\mathbf{y}}_t - \bar{\mathbf{y}}^{(s)}) + (\mathbf{1} - \boldsymbol{\delta}_t) \circ \widehat{\mathbf{y}}_t^{(s)}, \tag{7}$$

where $\mathbf{1}$ is a vector of ones.

(c) We iterate steps (a) and (b) with increasing $s = 1, 2, \ldots$, until

$$\|\mathbf{y}_t^{(s)} - \mathbf{y}_t^{(s-1)}\|_2^2 \leq \gamma, \tag{8}$$

with γ sufficiently small.

At the end of the procedure, the reconstructed multivariate time series is given by $\widetilde{\mathbf{y}}_t^{(s)} = \mathbf{y}_t^{(s)} + \bar{\mathbf{y}}^{(s)}$, $t = 1, 2, \ldots, T$, with the original missing data replaced by the estimated values.

3 Bootstrap Confidence Intervals for Missing Values

We use a resampling procedure based on the residual bootstrap approach to approximate the sampling distribution of the estimators of the missing value. The theoretical properties of the following residual bootstrap scheme for time series can be derived following [3]. The bootstrap algorithm can be implemented as follows.

- Denote with $\mathcal{Y} = (\widetilde{\mathbf{y}}_1, \ldots, \widetilde{\mathbf{y}}_T)$ the observed time series. The bootstrap resampled time series $\mathcal{Y}^* = (\widetilde{\mathbf{y}}_1^*, \ldots, \widetilde{\mathbf{y}}_T^*)$ is built as follows.

1. Compute the residuals $\widehat{\boldsymbol{\varepsilon}}_t^{(s)} = \mathbf{y}_t^{(s)} - \widehat{\mathbf{y}}_t^{(s)}$, where $\mathbf{y}_t^{(s)}$ is computed by the (7) and $\widehat{\mathbf{y}}_t^{(s)}$ is computed by the (5). The value for the index s is taken from the last iteration of the imputation procedure described in the previous section.
2. Obtain the bootstrap error series $\{\boldsymbol{\varepsilon}_t^*\}$ by drawing T samples independently and uniformly, with replacement, from the centered residuals $\tilde{\boldsymbol{\varepsilon}}_t^{(s)} = \widehat{\boldsymbol{\varepsilon}}_t^{(s)} - \bar{\boldsymbol{\varepsilon}}_T^{(s)}$.
3. Generate the bootstrap series $\widehat{\mathbf{y}}_t^*$, for $t = 1, \ldots, T$, as

$$\widehat{\mathbf{y}}_t^* = (\mathbf{I}_p - D(\widehat{\boldsymbol{\lambda}}_0^{(s)})\mathbf{W})^{-1} \left[\left(D(\widehat{\boldsymbol{\lambda}}_1^{(s)}) + D(\widehat{\boldsymbol{\lambda}}_2^{(s)})\mathbf{W} \right) \mathbf{y}_{t-1}^{(s)} + \boldsymbol{\varepsilon}_t^* \right].$$

- Repeat the previous steps 1–3 for B times and derive the empirical distribution of the bootstrap replications $\widetilde{\mathbf{y}}_t^{*(b)} = \widehat{\mathbf{y}}_t^{*(b)} + \bar{\mathbf{y}}^{(s)}$, for $b = 1, \ldots, B$.

Confidence intervals are derived as individual intervals, for each missing value of the time series, with nominal confidence level $(1 - \alpha)$ using the normal approximation with bootstrap standard errors and the percentile method. See [3] for theoretical references on such bootstrap intervals for multivariate autoregressive time series.

The normal-based bootstrap confidence intervals are constructed assuming normality for the error process. For all t and i such that $\delta_{ti} = 0$, this confidence intervals are

$$\mathcal{I}_{ti,s,1-\alpha}^{B,norm} = \left[m_B(\widehat{y}_{ti}^*) - z_{1-\alpha/2} sd_B(\widehat{y}_{ti}^*);\ m_B(\widehat{y}_{ti}^*) + z_{1-\alpha/2} sd_B(\widehat{y}_{ti}^*) \right], \tag{9}$$

where z_α is the α-th percentile of the standard normal distribution, whereas $m_B(\widehat{y}_{ti}^*)$ and $sd_B(\widehat{y}_{ti}^*)$ are the estimated mean and standard deviation for $\widehat{y}_{ti}^*$, derived as the mean and standard deviation of the bootstrap replications $\widehat{y}_{ti}^{*(1)}, \ldots, \widehat{y}_{ti}^{*(B)}$, respectively.

The percentile bootstrap confidence intervals are more general since they do not depend on the normality assumption for the error process. They are given by

$$\mathcal{I}_{ti,s,1-\alpha}^{B,perc} = \left[\widetilde{y}_{ti,\alpha/2}^{*(B)};\ \widetilde{y}_{ti,1-\alpha/2}^{*(B)} \right], \qquad \forall t, i : \delta_{ti} = 0, \tag{10}$$

where $\widetilde{y}_{ti,\alpha}^{*(B)}$ is the estimated α-th percentile for the bootstrap distribution of $\widetilde{y}^*$, derived from the bootstrap replications $\widetilde{y}_{ti}^{*(b)}$, for $b = 1, \ldots, B$.

Further bootstrap confidence intervals of better performance (i.e., second-order correct), such as the calibrated bootstrap confidence intervals or the t-percentile bootstrap intervals, can be obtained by implementing a double bootstrap procedure to perceive an additive correction of the original nominal coverage level for the percentile bootstrap intervals. However, these procedures are not considered in this paper since they are computationally heavy and therefore unfeasible in the high-dimensional setup.

4 A Monte Carlo Simulation Study

To validate the empirical performance of the confidence intervals proposed in Eqs. (9) and (10), we have implemented a Monte Carlo simulation study to compute empirically the coverage of the intervals and compare it with the nominal coverage.

We have considered multivariate time series of dimension $p = 30$ and length $T = (50, 100, 500)$. The true weight matrix $\mathbf{W}_1$ has been randomly generated as a full rank symmetric matrix and has been row-normalized. The parameters of model (1) have been randomly generated in the interval $[-0.9, 0.9]$. In order to guarantee the consistency of the normal-based bootstrap confidence intervals in (9), the error component $\boldsymbol{\varepsilon}_t$ has been generated from a multivariate normal distribution, with mean vector zero and diagonal variance-covariance matrix, with heteroscedastic variances $(\sigma_1^2, \ldots, \sigma_p^2)$. In particular, the standard deviations $(\sigma_1, \ldots, \sigma_p)$ have been generated randomly from a Uniform distribution $U(0.5; 1.5)$. All bootstrap estimates have been computed by using $B = 999$ replicates.

Figure 1 shows an example of a simulated time series with dimension $p = 10$ and length $T = 100$. For brevity, we have only considered time series with zero mean value. However, results similar to those reported here can also be shown in the general case where the mean of the time series is not zero (see [9] for some examples).

In the implementation of the estimation procedure, we have considered two cases for the spatial matrix $\mathbf{W}$. In the first part of the simulation study, we have assumed the matrix as known and we have derived the coverage performance of the intervals using the true weight matrix in the estimation formulas. In the second part of the simulation study, instead, we have assumed that the spatial matrix is unknown. Therefore, we have plugged in the estimation formulas an estimated weight matrix derived as the (row-normalized) sample correlation matrix. In this way, the performance of our inferential procedure has been evaluated in those cases where no information about the spatial weights is available.

We have simulated $N = 500$ replications of the model and, for each one, we have removed 50 values and considered them as missing values. Of course, we kept a record of the true values. In particular, we have simulated a missing sequence of length 30 for location 2 (i.e., the first 30 values of this location have been removed and considered as missing). The other 20 missing values have been generated randomly at other locations. The plot on the bottom of Fig. 1 shows an example of time series with the simulated missing values.

For each missing value, indexed by $i = 1, \ldots, 50$, the empirical coverage of the relative confidence interval has been evaluated as the proportion of times that the

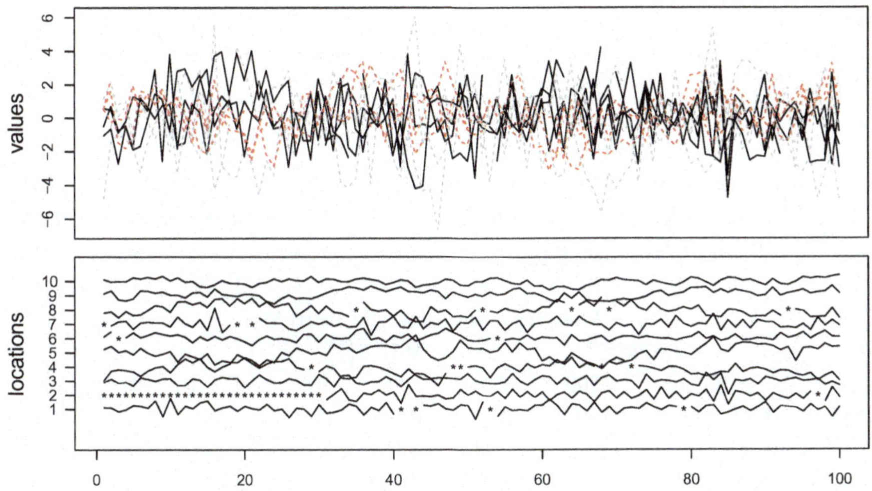

Fig. 1 Plot of a simulated multivariate time series of dimension $p = 10$ and length $T = 100$. On the top, the 10 time series are plotted in different colors and lines to show the variability of the observed values over time and over space. On the bottom, the time plots of the series are shifted in order to highlight (through stars) the simulated missing sequence, at location 2, and the missing values, at other locations

interval includes the true value of the missing, over the 500 simulated series. Such estimated proportion is denoted by $\widehat{\gamma}_{i,1-\alpha}$. Then, for all the different intervals, we have computed the mean squared coverage error as

$$MSCE_{1-\alpha} = \sqrt{\frac{1}{50}\sum_{i=1}^{50}\left[\widehat{\gamma}_i^{(1-\alpha)} - (1-\alpha)\right]^2}.$$

The results for the normal-based bootstrap intervals in (9) and for the percentile bootstrap intervals in (10) are reported in Table 1. To facilitate comparisons between the two types of intervals, the last column of the table reports the ratios of the two mean squared coverage errors. Clearly, the ratios are approximately equal to one and the MSCEs constantly decrease over T. So, the percentile bootstrap confidence intervals produces almost equivalent results, with respect to the normal-based bootstrap confidence intervals, even if the normality assumption is made for the error process. Such comments are valid both when using the true weight matrix **W** and when it is consistently estimated.

To see the results in detail, Fig. 2 shows the estimated proportions $\widehat{\gamma}_{i,1-\alpha}$ for the 50 missing values, for increasing values of $T = (50, 100, 500)$. The labels on the x-axis denote the locations where the missing values have been simulated (note that the first 30 values, on the left of the vertical dashed line, represent a missing sequence for location 2, see also Fig. 1). The black circles show the results for the percentile

Table 1 Mean Squared Coverage Errors $MSCE_{1-\alpha}$ for $T = 50, 100, 500$ and their ratio using the true (on the top) and the estimated (on the bottom) weight matrix **W** for several values of the confidence level $1-\alpha$. R denotes the ratio $MSCE_{1-\alpha}^{norm}/MSCE_{1-\alpha}^{perc}$

Using the true weight matrix **W**									
	$MSCE_{1-\alpha}^{norm}$			$MSCE_{1-\alpha}^{perc}$			R		
$1-\alpha$	$T = 50$	100	500	$T = 50$	100	500	$T = 50$	100	500
0.99	0.228	0.204	0.142	0.238	0.205	0.143	0.957	0.995	0.995
0.95	0.256	0.163	0.145	0.249	0.16	0.144	1.027	1.02	1.005
0.90	0.246	0.171	0.145	0.244	0.171	0.146	1.009	1.002	0.995
Using the estimated weight matrix $\widehat{\mathbf{W}}$									
	$MSCE_{1-\alpha}^{norm}$			$MSCE_{1-\alpha}^{perc}$			R		
$1-\alpha$	$T = 50$	100	500	$T = 50$	100	500	$T = 50$	100	500
0.99	0.250	0.153	0.14	0.237	0.148	0.14	1.055	1.033	0.999
0.95	0.242	0.168	0.135	0.213	0.156	0.135	1.134	1.075	1
0.90	0.274	0.179	0.128	0.254	0.169	0.128	1.080	1.064	1

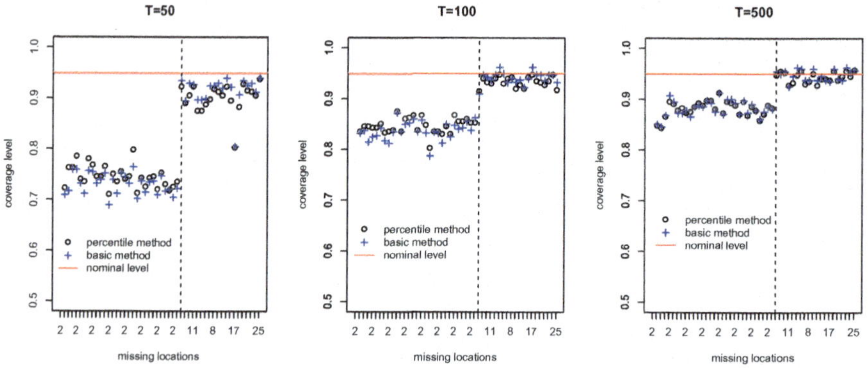

Fig. 2 Estimated proportions $\widehat{\gamma}_{i,1-\alpha}$ for the $i = 1, \ldots, 50$ missing values, for increasing values of $T = (50, 100, 500)$, using the true weight matrix **W**. The labels on the x-axis denote the locations where the missing values have been simulated (the first 30 values, on the left of the vertical dashed line, represent a missing sequence for location 2). The empirical coverages for the percentile bootstrap confidence intervals (black "o") are slightly closer to the nominal level (red line) compared to the normal-based bootstrap intervals (blue "+")

bootstrap confidence intervals whereas the red "+" represents the empirical coverage for the normal-based bootstrap intervals. Note how the empirical coverage for the percentile bootstrap intervals is generally closer to the true nominal coverage (red line). As expected, the coverage error is higher for the missing sequence values, since the confidence intervals in this case are built on the basis of the estimated sequence of lagged values (for isolated missing values, instead, the confidence intervals are based on observed lagged values). Also, note that the missing sequence is at the beginning of the observed time series and this makes estimating the missing sequence even more difficult. Finally, note that when $T = 50$ we have more than 50% of the time series which is missing for location 2. However, the coverage error for the missing sequence rapidly converges to zero when the length of the time series increases.

5 Concluding Remarks

In this paper, bootstrap confidence intervals for single or sequences of missing values have been derived in the context of multivariate time series. Confidence intervals are derived as individual intervals, one for each missing value of the (sequence of) time series, with nominal confidence level $(1 - \alpha)$. The construction of joint confidence regions for the missing sequence is left to future work. In particular, starting from the generalized spatial-dynamic autoregressive model proposed in [4] and applied to environmental data in [9], we approximated the sampling distribution of the missing value estimators by residual bootstrap. The normal based and the percentile bootstrap method have been considered for the constructions of approximated confidence intervals. Their performance has been evaluated and compared in terms of empirical coverage in a Monte Carlo simulation study, for different time series lengths and different nominal coverages.

The results show that the residual bootstrap delivers satisfactory coverage results even for short time series with sequences of missing data. In all experiments, the percentile method is substantially equivalent to the normal-based one, even if the normality of the error process is assumed. For all the values of T and all the values $1 - \alpha$, the mean squared coverage errors obtained by the percentile method appears slightly better than the corresponding values of the normal method. Moreover, the empirical coverage for the percentile is slightly closer to the true nominal coverage. As expected, both the empirical coverage errors go to zero as the time series length increases.

References

1. Allison, P.D.: Missing data (Sage University Papers Series on Quantitative Applications in the Social Sciences, 07–136). Sage, Thousand Oaks, CA (2002)
2. Calculli, C., Fassò, A., Finazzi, F., Pollice, A., Turnone, A.: Maximum likelihood estimation of the multivariate hidden dynamic geostatistical model with application to air quality in Apulia. Italy. Environmetrics **26**, 406–417 (2015)
3. Choi, E., Hall, P.: Bootstrap confidence regions computed from autoregressions of arbitrary order. J. R. Statist. Soc., Ser. B **623**, 461-477 (2000)
4. Dou, B., Parrella, M.L., Yao, Q.: Generalized Yule-Walker estimation for Spatio-Temporal models with unknown diagonal coefficients. J. Econ. **194**, 369–382 (2016)
5. Enders, C.K.: A primer on maximum likelihood algorithms for use with missing data. Struct. Equ. Model. **8**(1), 128–141 (2001)
6. Liu, S., Molenaar, P.C.: iVAR: a program for imputing missing data in multivariate time series using vector autoregressive models. Behav. Res. Method. **46**(4), 1138–1148 (2014)

7. Newman, D.A.: Longitudinal modeling with randomly and systematically missing data: A simulation of ad hoc, maximum likelihood, and multiple imputation techniques. Organ. Res. Methods **6**, 328–362 (2003)
8. Oehmcke, S., Zielinski, O., Kramer O.: kNN Ensembles with Penalized DTW for Multivariate Time Series Imputation.In: International Joint Conference on Neural Networks (IJCNN), IEEE, (2016)
9. Parrella, M.L., Albano, G., La Rocca, M., Perna, C.: Reconstructing missing data sequences in multivariate time series: an application to environmental data. Stat. Methods Appl. 1–25, ISSN:1613-981X (2018)
10. Pollice, A., Lasinio, G.J.: Two approaches to imputation and adjustment of air quality data from a composite monitoring network. J. Data Scie. **7**, 43–59 (2009)

On Parametric Estimation of Distribution Tails

Igor Rodionov

Abstract The aim of this work is to propose a method for estimating the parameter of the continuous distribution tail based on the largest order statistics of a sample. We prove the consistency and asymptotic normality of the proposed estimator. Note especially that we do not assume the fulfillment of the conditions of the extreme value theorem.

Keywords Distribution tail · Parametric estimation · Extreme value theory · Weibull-tail index · Super-heavy tails.

1 Introduction

In certain situations it is of interest to draw inference not about the whole distribution, but only about its tail. Of interest to researchers are both the parametric case, when the distribution tail belongs to a certain parametric class, and the nonparametric case. Such situations appear, in particular, in fields related to computer science and telecommunications, Internet traffic, finance and economics, and are the subject of statistics of extremes.

The Fisher-Tippet-Gnedenko theorem [11] is a central result in the extreme value theory. It states that if there exist sequences of constants $a_n > 0$ and b_n, such that the cumulative distribution function (cdf) of the normalized maximum $M_n = \max(X_1, \ldots, X_n)$ tends to some non-degenerate cdf G, i.e.,

$$\lim_{n\to\infty} P(M_n \leq a_n x + b_n) = G(x), \tag{1}$$

then there exist constants $a > 0$ and b such that $G(ax + b) = G_\gamma(x)$, where

I. Rodionov (✉)
V.A. Trapeznikov Institute of Control Sciences of Russian Academy of Sciences, Moscow, Russia

Moscow Institute of Physics and Technology, Moscow, Russia
e-mail: vecsell@gmail.com

M. La Rocca et al. (eds.), *Nonparametric Statistics*, Springer Proceedings
in Mathematics & Statistics 339, https://doi.org/10.1007/978-3-030-57306-5_40

$$G_\gamma(x) = \exp\left(-(1+\gamma x)^{-1/\gamma}\right), \quad 1+\gamma x > 0, \tag{2}$$

$\gamma \in \mathbb{R}$, and for $\gamma = 0$ the right-hand side should be understood as $\exp(-e^{-x})$. The parameter γ is called the extreme value index [8] (EVI). The cdf of the sample $(X_1, \ldots, X_n)$ is said to belong to the Fréchet (Weibull, Gumbel, respectively) maximum domain of attraction (MDA) if (1) holds for $\gamma > 0$ ($\gamma < 0$, $\gamma = 0$, respectively). Distributions with tails heavier than those of distributions in the Fréchet MDA are referred to as distributions with super-heavy tails; these distributions do not belong to any maximum domain of attraction (for details, see [4], Sects. 8.8 and 8.15, [17], Sect. 5).

The problem of tail estimation is central to statistics of extremes. Now the most popular method of tail estimation is based on Pickands' theorem [16]. The theorem states that if the cdf F of the random variable X belongs to the MDA of G_γ (2) (hereafter $F \in D(G_\gamma)$) for some $\gamma \in \mathbb{R}$, then for x, such that $0 < x < (\min(0, -\gamma))^{-1}$, it holds

$$\lim_{t \uparrow x^*} P\left(\frac{X-t}{f(t)} > x \,|X > t\right) = (1+\gamma x)^{-1/\gamma},$$

where $x_F^* = \sup\{x : F(x) < 1\}$, $f(t)$ is some function (for details, see [8]), and for $\gamma = 0$ $(1+\gamma x)^{-1/\gamma}$ should be understood as e^{-x}. So, it is enough to estimate EVI in order to estimate the distribution tail. The estimators of EVI were discussed in [6, 7, 16], among others, whereas the problem of the tail index estimation, i.e., estimation of $\gamma > 0$, was considered in [5, 14, 15], see also [8], Chap. 3.

This approach works well for distributions belonging to Fréchet or Weibull MDA, since these domains are fairly accurately described by EVI. However, one cannot distinguish between the tails within the Gumbel MDA using this approach, since $\gamma = 0$ for the whole domain. Additionally, the rate of convergence in the Gnedenko's limit theorem for distributions belonging to the Gumbel MDA is extremely slow (namely, it is logarithmic, see [13]). Next, the conditions of Gnedenko's theorem are not satisfied for super-heavy-tailed distributions, since their tail index $\alpha = 1/\gamma = 0$. Therefore, the tail estimation method using estimators of γ is not applicable for such distributions. It is worthwhile either to solve the problem of tail estimation in each of the above cases separately, or develop a general method of tail estimation, that could be applied to all these cases.

A number of authors follow the first way and consider a particular problem of parametric estimation of distribution tails belonging to some wide class of distributions from the Gumbel MDA. The Weibull and log-Weibull (with $\theta > 1$, see below) classes of distributions are the examples of such classes. We say that a cdf F is of Weibull-type, if there exists $\theta > 0$ such that for all $\lambda > 0$ we have

$$\lim_{x \to \infty} \frac{\ln(1 - F(\lambda x))}{\ln(1 - F(x))} = \lambda^\theta. \tag{3}$$

Clearly, the parameter θ, called the Weibull-tail index, governs the tail behavior, with larger values indicating faster tail decay. In the analysis of exponentially decreasing tails, the estimation of θ and the subsequent estimation of extreme quantiles assume a central position. If for some cdf F the distribution function $F(e^x)$ belongs to the Weibull class with $\theta > 0$, then one says that F is of log-Weibull-type. Note, that distributions of log-Weibull-type with $\theta > 1$ belong to Gumbel MDA. If $0 < \theta < 1$, then such distributions have super-heavy tails. The estimators of the Weibull-tail index were proposed in [2, 3, 9, 10], among others, whereas the problem of log-Weibull-tail index estimation defined similarly to (3), for our best knowledge, has not yet be considered.

The purpose of our work is to propose a general method to estimate the parameter of the tail in case of its continuity. The proposed method is appropriate to estimate, on the one hand, the Weibull and log-Weibull-tail indices and, on the other hand, the parameters of super-heavy-tailed distributions. We emphasize that the proposed method is independent of whether the distribution tail belongs to some MDA or not.

2 Main Results

Let $X_1, \ldots, X_n$ be independent identically distributed (iid) random variables with a cumulative distribution function (cdf) F with $x_F^* = +\infty$. We call $\overline{G}(x) = 1 - G(x)$, the tail of a cdf G. We say, that cdfs G and H with $x_G^* = x_H^* = +\infty$ have the same tail, if $\overline{G}(x)/\overline{H}(x) \to 1$ holds as $x \to +\infty$. Assume, that the tail of F belongs to the parametric class of distribution tails $\{F_\theta, \theta \in \Theta\}$, $\Theta \in \mathbb{R}$. Let $X_{(1)} \leq X_{(2)} \leq \ldots \leq X_{(n)}$ be the order statistics corresponding to the sample $(X_1, \ldots, X_n)$. Before considering the problem of estimation of the parameter θ, discuss how to find the parametric class of distribution tails to which the tail of the cdf F belongs.

Definition 1 [18] We say, that cdfs H and G satisfy the condition B(H,G) (written B-condition), if for some $\varepsilon \in (0, 1)$ and x_0

$$\frac{(1 - H(x))^{1-\varepsilon}}{1 - G(x)} \text{is non increasing as } x > x_0.$$

It is easy to see, that under this condition the tail of H is lighter than the tail of G, i.e.,

$$\frac{1 - H(x)}{1 - G(x)} \to 0$$

as $x \to \infty$.

Consider two distribution classes A_0 and A_1 such that the tails of distributions lying in A_0 are lighter than the distribution tails lying in A_1. We say, that the classes A_0 and A_1 are *separable*, if there exist the "separating" cdf $\widetilde{F}$ such that the tail of $\widetilde{F}$ is not lighter than the tail of G for all $G \in A_0$ and the condition $B(\widetilde{F}, H)$ is satisfied with some ε and x_0 for all $H \in A_1$.

So, suppose that two classes A_0 and A_1 of continuous distribution tails are separable via the cdf $\widetilde{F}$. Consider the null hypothesis $H_0 : F \in A_0$ and the alternative

$H_1 : F \in A_1$. Propose the test to check the hypothesis H_0. For this purpose consider the following statistic

$$R_{k,n} = \ln(1 - \widetilde{F}(X_{(n-k)})) - \frac{1}{k} \sum_{i=n-k+1}^{n} \ln(1 - \widetilde{F}(X_{(i)})). \tag{4}$$

Theorem 1 *[18] Let $X_1, \ldots, X_n$ be iid random variables with a common continuous cdf F with $x_F^* = +\infty$. Assume that the sequence $k = k(n)$ is such that*

$$k \to \infty, \ \frac{k}{n} \to 0 \ \text{as} \ n \to \infty. \tag{5}$$

Then the test

$$\text{if } R_{k,n} > 1 + \frac{u_{1-\alpha}}{\sqrt{k}}, \text{ then reject } H_0, \tag{6}$$

has the asymptotical significance level α, here $u_{1-\alpha}$ is the $(1-\alpha)$-quantile of $N(0, 1)$. Moreover, the test is consistent.

The proposed test allows us to distinguish such classes of distribution tails as super-heavy-tailed distributions, heavy-tailed distributions, Weibull- and log-Weibull-type distributions, see for details [18]. In addition, the tests to distinguish between the distributions belonging to the Gumbel MDA are proposed in [12, 19, 20].

Let us return to the problem of estimation the parameter of the distribution tail based on the largest order statistics of a sample. Consider the generalization of the statistic (4)

$$R_{k,n}(\theta) = \ln(1 - F_\theta(X_{(n-k)})) - \frac{1}{k} \sum_{i=n-k+1}^{n} \ln(1 - F_\theta(X_{(i)})), \tag{7}$$

here, as before, $\{F_\theta, \theta \in \Theta\}$, $\Theta \in \mathbb{R}$, is the parametric class of distribution tails to which the tail of the cdf F of the sample $X_1, \ldots, X_n$ belongs. We say that the parametric class $\{F_\theta, \theta \in \Theta\}$ is *ordered*, if the condition $B(F_{\theta_1}, F_{\theta_2})$ is satisfied $\forall \theta_1, \theta_2 \in \Theta$, $\theta_1 < \theta_2$, or the condition $B(F_{\theta_2}, F_{\theta_1})$ is satisfied $\forall \theta_1, \theta_2 \in \Theta$, $\theta_1 < \theta_2$. We propose the following estimator of the parameter θ

$$\widehat{\theta}_{k,n} = \arg\{\theta : R_{k,n}(\theta) = 1\}. \tag{8}$$

The next theorem states that the estimator (8) is consistent. The consistent estimators of the scale and location parameters of the distribution tails are proposed in [1].

Theorem 2 *Let $X_1, \ldots, X_n$ be iid random variables with a cdf $F = F_{\theta_0}$, the class of distribution tails $\{F_\theta, \theta \in \Theta\}$ is ordered and $F_\theta(x)$ is continuous in both x and θ. Assume that the sequence $k = k(n)$ satisfies the condition (5). Then*

$$\widehat{\theta}_{k,n} \to \theta_0 \;\; P_{\theta_0} - a.s. \tag{9}$$

Provide some examples of families of distribution tails for which the conditions of Theorem 2 are satisfied. All the distributions below are assumed to be continuous.

(1) We can describe the class of Weibull-type distributions as $\{F_{\theta,\ell}(x), \theta > 0\}$, where $F_{\theta,\ell}(x) = 1 - \exp\{-x^{\theta}\ell(x)\}$ as $x \to \infty$ and $\ell(x)$ is slowly varying at infinity. One can see, that this class is ordered with respect to the parameter θ. For practitioners, it is sufficient to use $F_{\theta}(x) = 1 - \exp\{-x^{\theta}\}$, $x > 0$ in (7) to obtain the estimate of the Weibull-tail index, see Sect. 3 for details. For the same reasons, the class of log-Weibull-type distributions satisfies the conditions of Theorem 2.

(2) Another example is the class of regularly varying distributions with $\gamma > 0$. Recall that the right endpoint for all distributions in this class equals $+\infty$. Indeed, we have $1 - F(x) = x^{-1/\gamma}\ell(x)$ as $x \to +\infty$ for all cdfs belonging to this class, where γ is EVI and $\ell(x)$ is slowly varying at infinity. We see, that this class is ordered with respect to γ.

(3) The class of logarithm-tailed distributions with $1 - F(x) = (\ln x)^{-\theta}(1 + o(1))$ as $x \to +\infty$, $\theta > 0$ is ordered with respect to θ and therefore satisfies the conditions of Theorem 2. Recall, that such distributions, as the distributions of log-Weibull-type with $\theta \in (0, 1)$, have super-heavy tails and do not satisfy the conditions of the Fisher-Tippet-Gnedenko theorem. This example emphasizes that the proposed method can be applied even if the cdf of a sample does not belong to any of the maximum domains of attraction.

Remark 1 It follows from Theorems 1 and 2 [18], that under the assumptions of Theorem 2 the equation $R_{k,n}(\theta) = 1$ has only one solution a.s. as $n \to \infty$, therefore, the estimator $\widehat{\theta}_{k,n}$ is defined correctly.

Proof of Theorem 2. Assume that under the conditions of Theorem 2 $B(F_{\theta_1}, F_{\theta_2})$ holds for all $\theta_1, \theta_2 \in \Theta$, $\theta_1 < \theta_2$. We write $X \ll Y$, if the random variable X is stochastically smaller than Y. From Theorem 1, [18],

$$R_{k,n}(\theta_0) \to 1 \;\; P_{\theta_0} - \text{a.s.}$$

under the conditions (5). Next, assume that $\theta_1 > \theta_0$ and the condition $B(F_{\theta_0}, F_{\theta_1})$ holds with some ε. Let $E_1, \ldots, E_k$ be independent random variables, exponentially distributed with a parameter $1 - \varepsilon$. From Theorem 2, [18], under the conditions (5), it holds

$$R_{k,n}(\theta_1) \gg \frac{1}{k}\sum_{i=1}^{k} E_i \xrightarrow{\text{a.s.}} \frac{1}{1-\varepsilon}.$$

Similarly, if $\theta_1 < \theta_0$ and the condition $B(F_{\theta_1}, F_{\theta_0})$ holds with some δ, then under conditions (5) it holds

$$R_{k,n}(\theta_1) \ll \frac{1}{k}\sum_{i=1}^{k} E_i' \xrightarrow{\text{a.s.}} 1 - \delta,$$

here $\{E'_i\}_{i=1}^k$ are independent random variables, exponentially distributed with the parameter $\frac{1}{1-\delta}$. Thereby, (9) holds, if the condition $B(F_{\theta_1}, F_{\theta_2})$ is satisfied for all $\theta_1, \theta_2 \in \Theta,\ \theta_1 < \theta_2$, the proof of the second case is similar. The statement of Theorem 2 follows.

The next theorem states the asymptotic normality of the estimator (8) under some additional conditions imposed on the class $\{F_\theta, \theta \in \Theta\}$. Denote $S(x, \theta) = -\ln(1 - F_\theta(x))$. It is easy to see, that the function $S(x, \theta)$ is differentiable with respect to θ if and only if cdf $F_\theta(x)$ is differentiable with respect to θ. Let us denote

$$I(q, \theta) = \frac{1}{1 - F_\theta(q)} \int_q^\infty \frac{\partial S(x, \theta)}{\partial \theta} dF_\theta(x) - \frac{\partial S(q, \theta)}{\partial \theta}.$$

Consider the following regularity conditions.

A1 The function $\frac{\partial S(x,\theta)}{\partial \theta}$ is continuous in (x, θ) as $x > x_0$;
A2 There exists $x_1 = x_1(\theta_0)$ for all $\theta_0 \in \Theta$ such that the function $\frac{\partial S(x,\theta)}{\partial \theta}$ is monotone in θ in some neighborhood of θ_0 as $x > x_1(\theta_0)$.
A3 There exists $x_2(\theta_0)$ for all $\theta_0 \in \Theta$ such that $|I(x, \theta_0)| < \infty$ as $x > x_2(\theta_0)$.

Theorem 3 *Let the conditions* **A1-A3** *be fulfilled. Then under the conditions of Theorem 2,*

$$\sqrt{k} I(X_{(n-k)}, \widehat{\theta}_{k,n})(\widehat{\theta}_{k,n} - \theta_0) \xrightarrow{d_{\theta_0}} \xi \sim N(0, 1)$$

for all $\theta_0 \in \Theta$.

Proof of Theorem 3 From (8), we have $\sqrt{k}(R_{k,n}(\widehat{\theta}_{k,n}) - 1) = 0$. Denote

$$\tau = \sqrt{k} I(X_{(n-k)}, \theta_0)(\widehat{\theta}_{k,n} - \theta_0)$$

and expand $\sqrt{k}(R_{k,n}(\widehat{\theta}_{k,n}) - 1)$ in a Taylor's series in θ_0 with a Lagrange form of the reminder,

$$0 = \sqrt{k}(R_{k,n}(\widehat{\theta}_{k,n}) - 1) = \sqrt{k}(R_{k,n}(\theta_0) - 1) + \frac{\tau}{I(X_{(n-k)}, \theta_0)} \frac{\partial}{\partial \theta} R_{k,n}(\widetilde{\theta}) \quad (10)$$

where $\widetilde{\theta}$ is between θ_0 and $\widehat{\theta}_{k,n}$. It follows from Theorem 1 [18], that

$$\sqrt{k}(R_{k,n}(\theta_0) - 1) \xrightarrow{d_{\theta_0}} N(0, 1) \quad (11)$$

under the conditions of Theorem, and, additionally, that the distribution of the statistic $R_{k,n}(\theta_0)$ does not depends on $X_{(n-k)}$.

Consider $\tau(I(X_{(n-k)}, \theta_0))^{-1} \frac{\partial}{\partial \theta} R_{k,n}(\widetilde{\theta})$ given $X_{(n-k)} = q$. Note, that if $\{\xi_n\}$ is some sequence of random variables such that $\xi_n \to C$ a.s. as $n \to \infty$, $C \in \mathbb{R}$, then it follows from criterion for almost sure convergence, that for all $\varepsilon > 0$ there exists the sequence $c_n > 0$, $c_n \to 0$ as $n \to \infty$, such that for some $n = n(\varepsilon)$ it holds

$$P(\forall l > n : \; |\xi_l - C| > c_l) < \varepsilon. \tag{12}$$

Let $\varepsilon > 0$ be fixed. Since $\widetilde{\theta}$ is between θ_0 and $\widehat{\theta}_{k,n}$, then $\widetilde{\theta} \to \theta_0$ P_{θ_0}-a.s. under the conditions of Theorem. Using (12) and monotonicity of $\frac{\partial S(x,\theta)}{\partial \theta}$ (without loss of generality suppose that $\frac{\partial S(x,\theta)}{\partial \theta}$ increases monotonically with θ), we have with a probability of $1-\varepsilon$ as $n > n(\varepsilon)$,

$$\frac{1}{k}\sum_{i=n-k+1}^{n} \frac{\partial}{\partial\theta} S(X_{(i)}, \theta_0 - c_n) - \frac{\partial}{\partial\theta} S(X_{(n-k)}, \theta_0 + c_n) \le \frac{\partial}{\partial\theta} R_{k,n}(\widetilde{\theta}) \le$$

$$\frac{1}{k}\sum_{i=n-k+1}^{n} \frac{\partial}{\partial\theta} S(X_{(i)}, \theta_0 + c_n) - \frac{\partial}{\partial\theta} S(X_{(n-k)}, \theta_0 - c_n) \tag{13}$$

From lemma 3.4.1 [8], the joint distribution of $\{X_{(i)}\}_{i=n-k+1}^{n}$ given $X_{(n-k)} = q$ agrees with the joint distribution of the set of order statistics $\{X^*_{(j)}\}_{j=1}^{k}$ of the sample $\{X^*_i\}_{j=1}^{k}$ with the cdf

$$F_q(x) = \frac{F_{\theta_0}(x) - F_{\theta_0}(q)}{1 - F_{\theta_0}(q)}, \quad q < x.$$

We have given $X_{(n-k)} = q$

$$\frac{1}{k}\sum_{i=n-k+1}^{n} \frac{\partial}{\partial\theta} S(X_{(i)}, \theta_0 - c_n) - \frac{\partial}{\partial\theta} S(q, \theta_0 + c_n) \overset{d_{\theta_0}}{=} \frac{1}{k}\sum_{j=1}^{k} \frac{\partial}{\partial\theta} S(X^*_{(j)}, \theta_0 - c_n) -$$

$$\frac{\partial}{\partial\theta} S(q, \theta_0 + c_n) = \frac{1}{k}\sum_{j=1}^{k} \frac{\partial}{\partial\theta} S(X^*_j, \theta_0 - c_n) - \frac{\partial}{\partial\theta} S(q, \theta_0 + c_n).$$

From the Law of large numbers,

$$\frac{1}{k}\sum_{j=1}^{k} \frac{\partial}{\partial\theta} S(X^*_j, \theta_0 - c_n) - \frac{1}{1 - F_{\theta_0}(q)} \int_q^{\infty} \frac{\partial}{\partial\theta} S(x, \theta_0 - c_n) dF_{\theta_0}(x) \xrightarrow{P_{\theta_0}} 0.$$

Therefore, we obtain

$$\frac{1}{I(q,\theta_0)} \left(\frac{1}{k}\sum_{j=1}^{k} \frac{\partial}{\partial\theta} S(X^*_j, \theta_0 - c_n) - \frac{\partial}{\partial\theta} S(q, \theta_0 + c_n) \right) \xrightarrow{P_{\theta_0}} \frac{I(q,\theta_0)}{I(q,\theta_0)} = 1.$$

Similarly we derive, that the expression in the right-hand side of the relation (13) given $X_{(n-k)} = q$ converges to 1 is probability. Since the value of ε is arbitrary, then it holds $\frac{\partial}{\partial\theta} R_{k,n}(\widetilde{\theta}) \xrightarrow{P_{\theta_0}} 1$ given $X_{(n-k)} = q$. Therefore, from (10) and (11) we have

$$\tau = \sqrt{k} I(q, \theta_0)(\widehat{\theta}_{k,n} - \theta_0) \xrightarrow{d_{\theta_0}} N(0, 1).$$

Using the regularity conditions **A1**, **A3** and Slutsky's theorem, we obtain given $X_{(n-k)} = q$

$$\sqrt{k} I(q, \widehat{\theta}_{k,n})(\widehat{\theta}_{k,n} - \theta_0) \xrightarrow{d_{\theta_0}} N(0, 1).$$

Note, that the last relation holds for all q, the statement of Theorem 2 follows.

3 Simulation Study

In our simulation study, we focus on the performance of the proposed method applied to estimating the Weibull-tail index in comparison with performance of two other Weibull-tail index estimators. We also show the performance of our approach adapted to estimating the log-Weibull-tail index.

In [3], the following estimator of the Weibull-tail index is proposed,

$$\widehat{\theta}^{(1)} = \frac{X_{(n-k+1)}}{\ln(n/k)} \Bigg/ \sum_{i=1}^{k-1} (X_{(n-i+1)} - X_{(n-k+1)}) .$$

We also consider another estimator of the Weibull-tail index introduced in [10],

$$\widehat{\theta}^{(2)} = \sum_{i=1}^{k-1} (\ln_2(n/i) - \ln_2(n/k)) \Bigg/ \sum_{i=1}^{k-1} (\ln X_{(n-i+1)} - \ln X_{(n-k+1)}) ,$$

where $\ln_2(x) = \ln(\ln x)$, $x > 1$. The latter estimator is a normalization of the Hill tail index estimator, [14]. We compare the finite sample performance of the estimators $\widehat{\theta}^{(1)}$, $\widehat{\theta}^{(2)}$ and $\widehat{\theta}^{(0)} = \widehat{\theta}_{k,n}$ (for the latter we select $F_\theta(x) = 1 - \exp(-x^\theta)$, $x > 0$) on 4 different distributions: $Exp(1)$, $\Gamma(3, 1)$, $N(0, 1)$ and Weibull distribution $W(3)$ with cdf $F(x) = 1 - \exp(-x^3)$, $x > 0$. In each case, $m = 300$ samples $(\mathbf{X}_i)_{i=1}^m$ of size $n = 1000$ are simulated. On each sample $\mathbf{X}_i$, the estimates $\widehat{\theta}_i^{br}$, $\widehat{\theta}_i^{g}$ and $\widehat{\theta}_i$ are computed for $k = 5, 10, \ldots, 150$. Finally, we build the Hill-type plots by drawing the points $\left(k, \frac{1}{m}\sum_{i=1}^m \widehat{\theta}_i^{(j)}\right)$, $j = 0, 1, 2$ for samples generated from $\Gamma(3, 1)$, $N(0, 1)$ and $W(3)$. We also present the MSE plots obtained by plotting the points $\left(k, \frac{1}{m}\sum_{i=1}^m (\widehat{\theta}_i^{(j)} - \theta)^2\right)$, j=0,1,2 for observations sampled from standard exponential distribution.

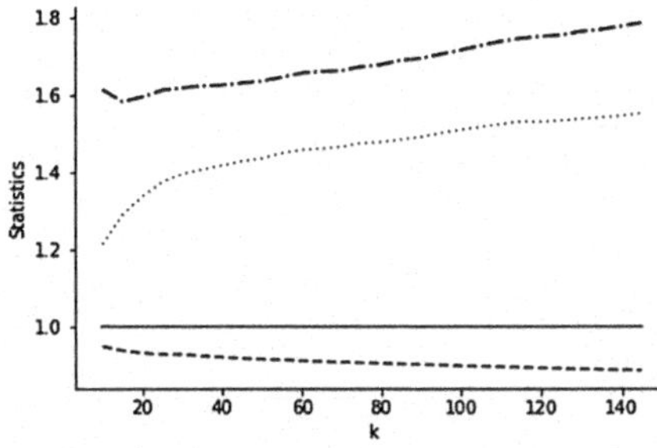

(a) $\Gamma(3, 1)$

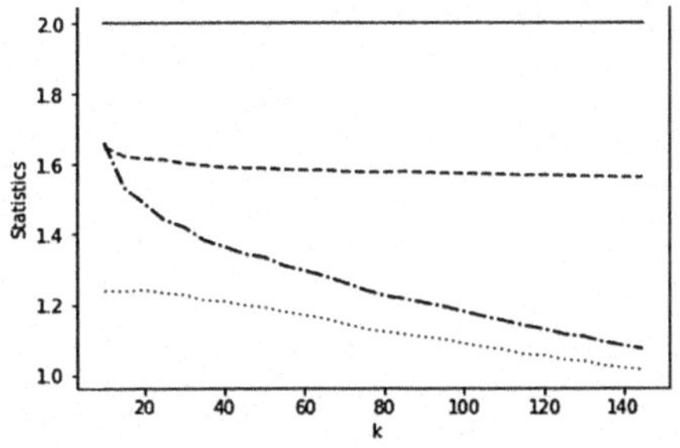

(b) $N(0, 1)$

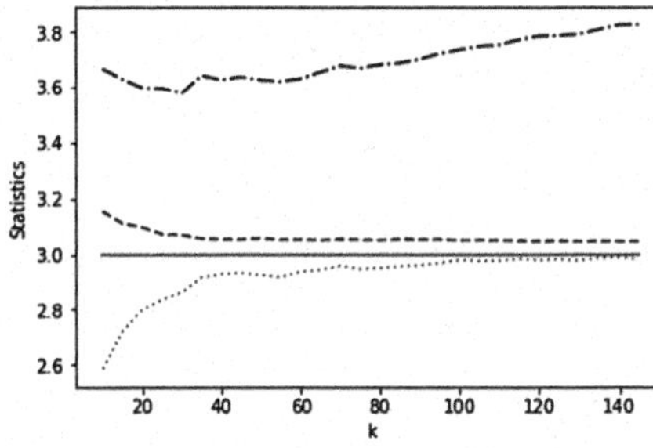

(c) $W(3)$

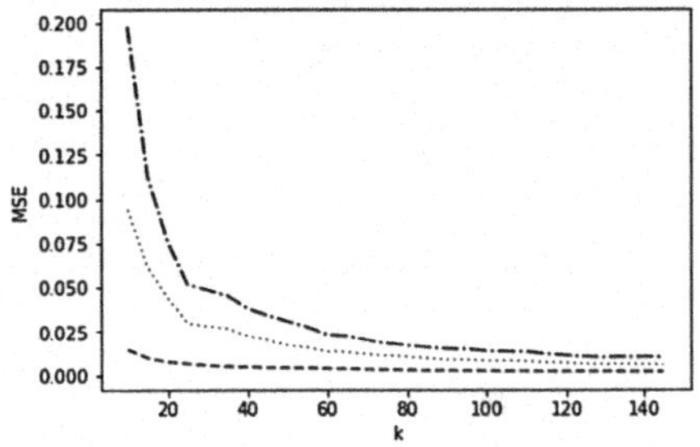

(d) $Exp(1)$

Fig. 1 Comparison of the estimators $\hat{\theta}^{(j)}$, $j = 0, 1, 2$ for various distributions. Solid line: true value, dashed line: $\hat{\theta}^{(0)}$, dot-dash line: $\hat{\theta}^{(1)}$, dotted line: $\hat{\theta}^{(2)}$.

Results are presented on Fig. 1. It appears that, in all the simulated cases, the estimator $\widehat{\theta}^{(0)}$ is better than others. One can also note, that all considered estimators do not provide the best performance for the standard normal distribution. Our simulation results correspond to the numerical results provided in [10].

The numerical performance of the estimator $\widehat{\theta_{k,n}}$ of log-Weibull-tail index is investigated for two classes, $\mathbb{F}^0 = \{F_\theta^0, \theta > 0\}$ with $F_\theta^0(x) = 1 - \exp(-(\ln x)^\theta)$, $x > 1$, and $\mathbb{F}^1 = \{F_\theta^1, \theta > 1\}$ with $F_\theta^1(x) = 1 - x^{-1}\exp(-(\ln x)^\theta)$, $x > 1$. The observations are sampled in each case from the standard log-normal distribution. One can see that the log-normal distribution does not belong to any of the considered classes. As before, we present Hill-type plots built on $m = 300$ samples of size $n = 1000$ for $k = 5, 10, \ldots, 150$.

Results for estimating the log-Weibull-tail index are presented on Fig. 2. One can see, that the choice of the class of distribution tails within the proposed approach is significant enough for the problem of log-Weibull-tail index estimation.

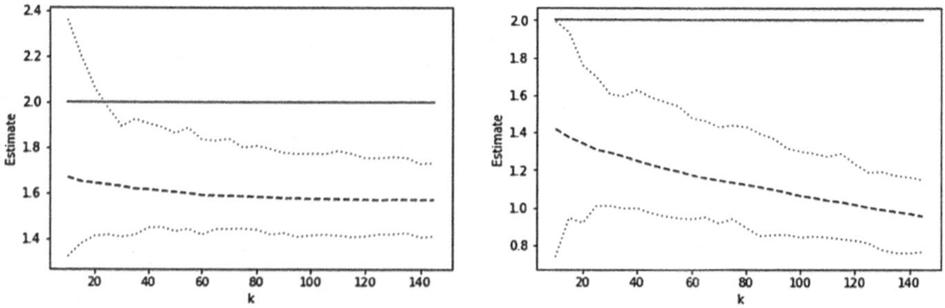

(a) $\mathbb{F}^0$ (b) $\mathbb{F}^1$

Fig. 2 Comparison of the estimates $\hat{\theta}_{k,n}$ built on classes $\mathbb{F}^0$ and $\mathbb{F}^1$. Solid line: true value, dashed line: $\hat{\theta}_{n,k}$, dotted lines: empirical 95% confidence interval

Acknowledgments The author appreciates the partial financial support by the Russian Foundation for Basic Research, grant 19-01-00090.

References

1. Akhtyamov, P.I., Rodionov, I.V.: On estimation of scale and location parameters of distribution tails. Fundam. Appl. Math. **23(1)**, (2019), in press
2. Balakrishnan, N., Kateri, M.: On the maximum likelihood estimation of parameters of Weibull distribution based on complete and censored data. Stat. Probabil. Lett. **78**, 2971–2975 (2008)
3. Beirlant, J., Broniatowski, M., Teugels, J.L., Vynckier, P.: The mean residual life function at great age: Applications to tail estimation. J. Statist. Plann. Inference **45**, 21–48 (1995)
4. Bingham, N.H., Goldie, C.M., Teugels, J.L.: Regular Variation. Encyclopedia of Mathematics and its Application, vol. 27. Cambridge University Press, Cambridge (1987)
5. Danielsson, J., Jansen, D.W., de Vries, C.G.: The method of moments ratio estimator for the tail shape parameter. Commun. Stat. Theory **25**, 711–720 (1986)
6. Dekkers, A.L.M., Einmahl, J.H.J., Haan, L. de.: A moment estimator for the index of an extreme-value distribution. Ann. Statist. **17**, 1833–1855 (1989)
7. Drees, H., Ferreira, A., de Haan, L.: On maximum likelihood estimation of the extreme value index. Ann. Appl. Probab. **14**, 1179–1201 (2003)
8. Ferreira, A., Haan, L. de: Extreme value theory. An introduction. Springer, Springer Series in Operations Research and Financial Engineering, New York (2006)
9. Gardes, L., Girard, S., Guillou, A.: Weibull tail-distributions revisited: a new look at some tail estimators. J. Stat. Plann. Inf. **141**(4), 429–444 (2009)
10. Girard, S.: The Hill-type of the Weibull tail-coefficient. Comm. Stat. Theor. Meth. **33**(2), 205–234 (2004)
11. Gnedenko, B.V.: Sur la distribution limite du terme maximum d'une serie aleatoire. Ann. Math. **44**, 423–453 (1943)
12. Goegebeur, J., Guillou, A.: Goodness-of-fit testing for Weibull-type behavior. J. Stat. Plann. Inf. **140**(6), 1417–1436 (2010)
13. de Haan, L., Resnick, S.: Second-order regular variation and rates of convergence in extreme value theory. Ann. Probab. **24**(1), 97–124 (1996)
14. Hill, B.M.: A simple general approach to inference about the tail of a distribution. Ann. Statist. **3**, 1163–1174 (1975)

15. Paulauskas, V., Vaiciulis, M.: On the improvement of Hill and some others estimators. Lith. Math. J. **53**, 336–355 (2013)
16. Pickands III, J.: Statistical inference using extreme order statistics. Ann. Stat. **3**, 119–131 (1975)
17. Resnick, S.: Point processes, regular variation and weak convergences. Adv. Appl. Probab. **18**(1), 66–138 (1986)
18. Rodionov, I.V.: On discrimination between classes of distribution tails. Probl. Inform. Transm. **54**(2), 124–138 (2018)
19. Rodionov, I.V.: A discrimination test for tails of weibull-type distributions. Theory Probab. Appl. **63**(2), 327–335 (2018)
20. Rodionov, I.V.: Discrimination of close hypotheses about the distribution tails using higher order statistics. Theory Probab. Appl. **63**(3), 364–380 (2019)

An Empirical Comparison of Global and Local Functional Depths

Carlo Sguera and Rosa E. Lillo

Abstract A functional data depth provides a center-outward ordering criterion that allows the definition of measures such as median, trimmed means, central regions, or ranks in a functional framework. A functional data depth can be global or local. With global depths, the degree of centrality of a curve x depends equally on the rest of the sample observations, while with local depths the contribution of each observation in defining the degree of centrality of x decreases as the distance from x increases. We empirically compare the global and the local approaches to the functional depth problem focusing on three global and two local functional depths. First, we consider two real data sets and show that global and local depths may provide different data insights. Second, we use simulated data to show when we should expect differences between a global and a local approach to the functional depth problem.

1 Introduction

The theory of statistics for functional data is a well-established field with a great amount of applications and ongoing research. See, for instance, [5, 11, 12, 16] for overviews of Functional Data Analysis (FDA).

In this paper we deal with the notion of data depth in the functional framework. A functional depth provides a center-outward data ordering criterion that, for example, allows the definition of the functional median or ranks. Behind any implementation of the idea of data depth there is an explicit or implicit approach to the depth problem. For example, in the multivariate framework, where the notion of depth originated, we find a well-established classification of depths in global and local measures. A multivariate global depth provides a data ordering based on the behavior of each observation relative to the complete sample. Several implementations of this notion have been proposed in the literature, e.g., [20] proposed the halfspace depth, [13]

C. Sguera (✉)
UC3M-Santander Big Data Institute, Getafe, Spain
e-mail: carlo.sguera@uc3m.es; sguera.carlo@gmail.com

R. E. Lillo
UC3M Department of Statistics, UC3M-Santander Big Data Institute, Getafe, Spain

M. La Rocca et al. (eds.), *Nonparametric Statistics*, Springer Proceedings in Mathematics & Statistics 339, https://doi.org/10.1007/978-3-030-57306-5_41

introduced the simplicial depth, while [17] defined the spatial depth. On the contrary, a multivariate local depth provides a data ordering based on the behavior of each observation relative to a certain neighborhood. Existing multivariate local depths are the kernelized spatial depth proposed by [4], local versions of the halfspace and simplicial depths introduced by [1] and the β-local depth defined by [15]. Multivariate local depths try to deal with data that have some complex or local features, and they are usually able to capture the underlying structure of data in nonstandard scenarios.

Global and local depths have been proposed also in FDA, but there are no studies that provide researchers and practitioners with guidance on differences between them. Therefore, the main aim of this paper is to point out the structural differences between global and local functional data depths and help users to decide which type of functional data depth to use.

As global-oriented depths, in this paper we consider the Fraiman and Muniz depth (FMD, [10, 10]), which measures how long a curve remains in the middle of a sample of functional observations, the modified band depth (MBD, [14, 14]), which is based on a measure of how much a curve is inside the bands defined by all the possible pairs of curves of a sample, and the functional spatial depth (FSD, [3, 3]), which is a functional version of the multivariate spatial depth. As local-oriented depths, we consider the h-modal depth (HMD, [6, 6]), which measures how densely a curve is surrounded by other curves in a sample, and the kernelized functional spatial depth (KFSD, [18, 18]), which represents an explicit local version of the functional spatial depth.

To compare global and local functional depths, we first consider two real data examples that involve the presence of functional local features such as bimodality, presence of isolated observations and potential outliers, or asymmetry (Sect. 2). Then, we focus on the relationship between FSD and KFSD (Sect. 3). Finally, we use a simulation study to analyze the behavior of global and local depths under the presence of complex features (Sect. 4) and draw some conclusions (Sect. 5).

2 Comparing Global and Local Depths: Real Data Examples

This section represents an introductory part of our empirical study. Here will illustrate the differences between global depths (FSD, FMD, and MBD) and local depths (KFSD and HMD) using two real functional data sets: phonemes data (Sect. 2.1) and nitrogen oxides data (Sect. 2.2).

2.1 *Phonemes Data*

The phonemes data set, available in the *fda.usc* R package [8, 8], consists in log-periodograms corresponding to recordings of speakers pronouncing specific

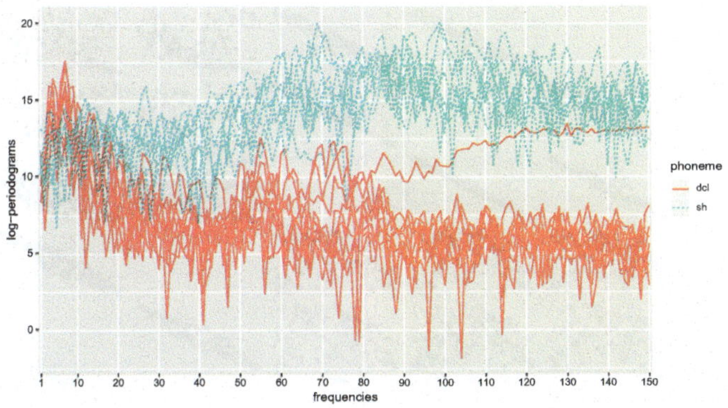

Fig. 1 Log-periodograms of 10 speakers pronouncing *sh* and 10 speakers pronouncing *dcl*

phonemes. A detailed description of the data set which contains information about five speech frames corresponding to five phonemes can be found in [9]. In this subsection we consider 50 observations for the phoneme *sh* as in she and 50 observations for the phoneme *dcl* as in dark. For illustrative purposes, Fig. 1 shows 10 randomly chosen log-periodograms for each phoneme. As in [9], we consider the first 150 frequencies from each recording.

Treating this data set as a unique sample, we obtain data that show bimodality, in particular starting from frequencies around 40, and a central region where fall few isolated observations (see Fig. 1). Our first goal is to show that global and local depths may behave differently in presence of such complex data features, and since the center-outward ordering of curves is possibly the main by-product of any depth analysis, we evaluate depth measures considering the ranks associated to its values.[1] We first consider all the possible pairs of depths, and then we focus on the pair FSD-KFSD due to their direct relationship (see Sect. 3 for more details).

Figure 2 shows the scatter plots of the ten possible pairs of depth-based ranks, and we observe strong relationships between either global or local depths and relatively weaker relationships between global and local depths.

In Table 1 we report the Spearman's rank correlation coefficients corresponding to Fig. 2, and they confirm the visual inspection of the figure: the coefficients are never less than 0.96 between either global or local depths, while they are never greater than 0.26 between a global and a local depth.

In Fig. 3 we focus on the scatter plot of the FSD-based and KFSD-based ranks to compare more in detail the behaviors of a global and a local depth: it is clear that there are important differences in terms of ranks, except for low FSD-based ranks (lower than 20). Therefore, the functional phonemes data set represents a clear example of global and local depths showing very different behaviors.

[1]Note that the higher the depth values, the higher the associated ranks.

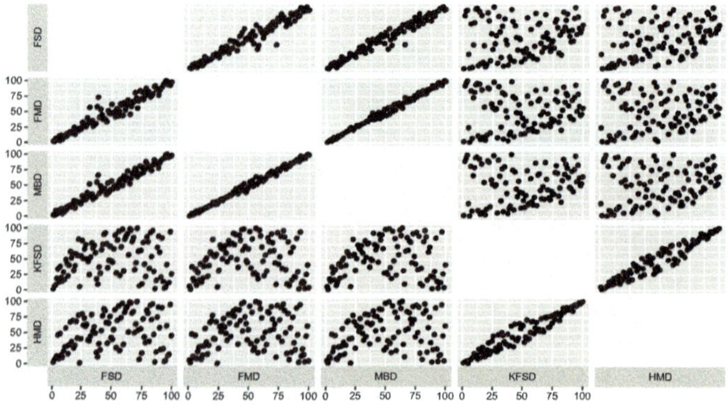

Fig. 2 Scatter plots of the ten possible pairs of depth-based ranks for the phonemes data set

Table 1 Spearman's rank correlation matrix of FSD, FMD, MBD, KFSD, and HMD values for the phonemes data set

	FSD	FMD	MBD	KFSD	HMD
FSD	1.00	0.97	0.98	0.17	0.26
FMD	0.97	1.00	0.99	0.02	0.13
MBD	0.98	0.99	1.00	0.08	0.19
KFSD	0.17	0.02	0.08	1.00	0.96
HMD	0.26	0.13	0.19	0.96	1.00

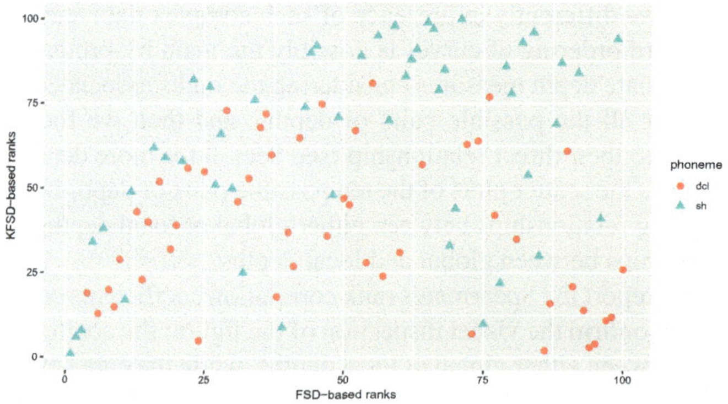

Fig. 3 Scatter plot of the FSD-based and KFSD-based ranks for the phonemes data set

2.2 Nitrogen Oxides (NO_x) Data

The nitrogen oxides (NO_x) data set, also available in the *fda.usc* R package, consists in nitrogen oxides (NO_x) emission daily levels measured in a Barcelona area between 2005-02-23 and 2005-06-29. More details about this data set can be found in [7], where it is used to implement functional outlier detection techniques after

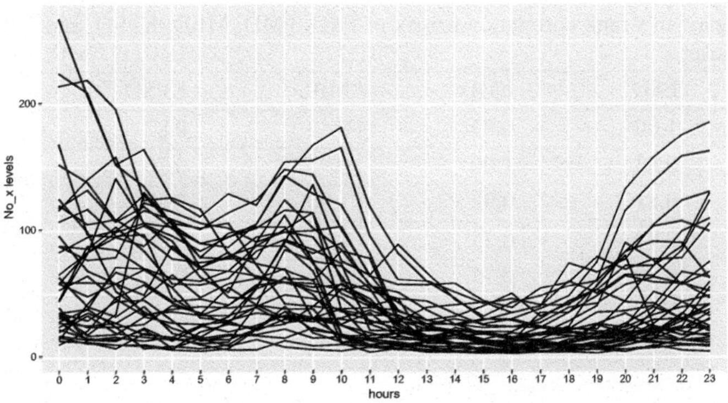

Fig. 4 NO_x levels (nanograms per cubic meter) measured every hour of 39 nonworking days between 23/02/2005 and 26/06/2005 close to an industrial area in Poblenou, Barcelona, Spain

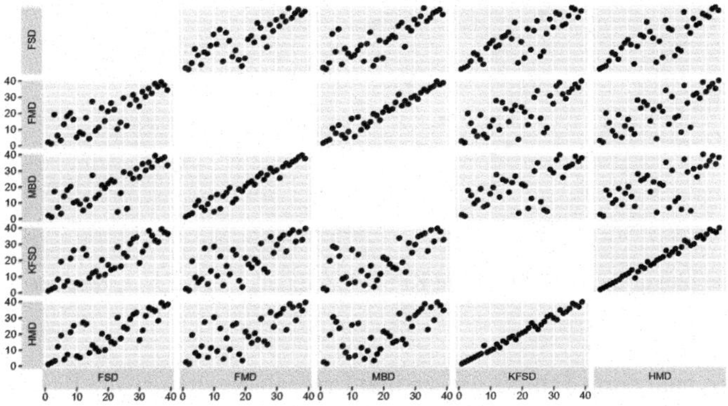

Fig. 5 Scatter plots of the ten possible pairs of depth-based ranks for the NO_x data set

splitting the whole data set in two samples referring to working and nonworking days, respectively. In this subsection we consider the nonworking days sample (see Fig. 4).

Observing Fig. 4 we notice at least two features that can be described as complex and/or local: first, the data set contains NO_x levels having a potential atypical behavior; second, the data set shows partial asymmetry, i.e., between roughly 10 and 24 24 h there are many relatively low NO_x levels and few relatively high NO_x levels. Therefore, it seems interesting to compare the behavior of global and local functional depths using this sample affected by potential outliers and asymmetry.

Figures 5 and 6 and Table 2 mimic Figs. 2 and 3 and Table 1 for this new data set: when comparing all the depths between each other in Fig. 5, we see that the juxtaposition between global and local depths exists but it appears less strong than in the phonemes data set.

Table 2 Spearman's rank correlation matrix of FSD, FMD, MBD, KFSD, and HMD values for the NO_x data set

	FSD	FMD	MBD	KFSD	HMD
FSD	1.00	0.83	0.82	0.82	0.80
FMD	0.83	1.00	0.97	0.75	0.73
MBD	0.82	0.97	1.00	0.67	0.64
KFSD	0.82	0.75	0.67	1.00	0.99
HMD	0.80	0.73	0.64	0.99	1.00

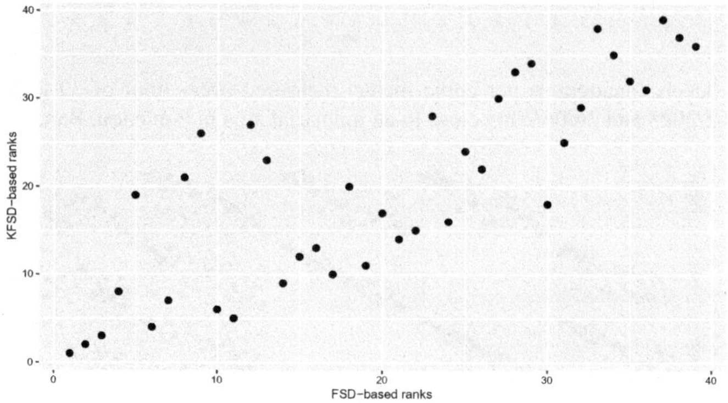

Fig. 6 Scatter plot of the FSD-based and KFSD-based ranks for the NO_x data set

However, analyzing Table 2 we see that for each depth measure the highest Spearman's rank correlation coefficient is still observed with a depth measure of the same nature, and therefore global and local depths show different behaviors also when they are used to analyze a data set affected by the complex features identified in the NO_x data set.

Focusing on FSD and KFSD, in Fig. 6 we observe that these depths have a stronger relationship than in the phonemes data set. However, there are several observations for which the FSD-based ranks differ significantly from the KFSD-based ranks. For example, it is easily seen a group of five observations having FSD-based ranks roughly between 5 and 15 and KFSD-based ranks roughly between 20 and 30.

The real data examples of this section show that we may expect different behaviors from global and local depths. In the next section we give an idea about the why, focusing on FSD and KFSD.

3 Comparing Global and Local Depths: FSD Versus KFSD

For reasons of space, we present the global and local approach to the functional depth problem focusing on the relationship between FSD and KFSD since the second is a direct local version of the first, and we refer to the original articles for the definitions of FMD, MBD, and HMD.

[3] introduced an extension of the multivariate spatial depth, the Functional Spatial Depth (FSD), with the aim of considering the geometry of the data to assign depth values. Let $\mathbb{H}$ be an infinite-dimensional Hilbert space. The FSD of $x \in \mathbb{H}$ with respect to the functional sample $Y_n = \{y_1, \ldots, y_n\}$ is defined as

$$FSD(x, Y_n) = 1 - \frac{1}{n} \left\| \sum_{i=1; y_i \neq x}^{n} \frac{x - y_i}{\|x - y_i\|} \right\|, \tag{1}$$

where $\| \cdot \|$ is the norm induced by the inner product $\langle \cdot, \cdot \rangle$ defined in $\mathbb{H}$.

[18] introduced the kernelized functional spatial depth modifying FSD in the following way:

$$KFSD(x, Y_n) = 1 - \frac{1}{n} \left\| \sum_{i=1}^{n} \frac{\phi(x) - \phi(y_i)}{\|\phi(x) - \phi(y_i)\|} \right\|, \tag{2}$$

where $\phi : \mathbb{H} \rightarrow \mathbb{F}$ is an embedding map and $\mathbb{F}$ is a feature space. Since ϕ can be defined implicitly by a positive definite and stationary kernel through $\kappa(x, y) = \langle \phi(x), \phi(y) \rangle$, $x, y \in \mathbb{H}$, and after some standard calculations, the kernel-based definition of KFSD is given by

$$KFSD(x, Y_n) = 1 - \frac{1}{n} \sqrt{\left(\sum_{\substack{i,j=1; \\ y_i \neq x; y_j \neq x}}^{n} \frac{\kappa(x,x) + \kappa(y_i, y_j) - \kappa(x, y_i) - \kappa(x, y_j)}{\sqrt{\kappa(x,x) + \kappa(y_i, y_i) - 2\kappa(x, y_i)}\sqrt{\kappa(x,x) + \kappa(y_j, y_j) - 2\kappa(x, y_j)}} \right)}. \tag{3}$$

Note that KFSD has been used for classification [18, 18] and outlier detection [2, 2, 19, 19].

As stated before, the pair FSD-KFSD represents the unique case where one functional depth (KFSD) is a direct local version of another (FSD), and therefore in what follows we briefly explain the why.

On the one hand, $FSD(x, Y_n)$ depends equally on all the possible deviations of x from y_i, for $i = 1, \ldots, n$. Therefore, behind FSD there is an approach based on the following fundamental assumption: each y_i should count equally in defining the degree of centrality of x. This is the feature that turns FSD into a global-oriented functional depth. A similar approach is behind FMD and MBD.

On the other hand, as a modification of FSD, KFSD aims to substitute the equally dependence of the depth value of x on the y_i's with a kernel-based dependence producing that y_i's closer to x influence more the depth value of x than y_i's that are more distant. Therefore, the alternative approach behind KFSD suggests that the contribution of each y_i in defining the degree of centrality of x should decrease for y_i's distant from x. This is the feature that turns KFSD in a local-oriented functional depth. A similar approach is behind HMD.

Moreover, the choice of the kernel makes KFSD and HMD flexible tools as it allows the practitioner to implement her/his preferences about the form of the neighborhoods of x. Additionally, the kernel bandwidth allows to tune the size of the neighborhood of x. In this paper we implement KFSD and HMD using a Gaussian kernel and setting the kernel bandwidth equal to the 25% percentile of the empirical distribution of $\{\|y_i - y_j\|, i = 1, \ldots, n; i < j \leq n\}$. Such a bandwidth defines fairly local versions of KFSD and HMD.

In the next section we complete our empirical study using simulated data.

4 Simulation Study

In Sect. 2 we have anticipated that global and local depths may behave differently when used to analyze real functional data sets. In this section, using the results of a simulation study, our goal is to establish when we should expect that a local functional depth may provide alternative data insights with respect to the ones that would arise using a global functional depth.

We are interested in models capable to replicate specific data features such as:

- absence of complex/local features;
- presence of atypical observations;
- asymmetry;
- bimodality and presence of isolated observations.

To do this, we consider models based on truncated Karhunen-Loève expansions to which we add an error term. For example, the curves generating process defining the first model (Model 1) is given by

$$x(t) = \mu(t) + \xi_1\phi_1(t) + \xi_2\phi_2(t) + \epsilon(t), \tag{4}$$

where $t \in \left\{\frac{s-0.5}{50}, s = 1, \ldots, 50\right\}$, $\mu(t) = 2t$, $\xi_1 \sim N(0, \lambda_1)$ and $\lambda_1 = 1.98$, $\xi_2 \sim N(0, \lambda_2)$ and $\lambda_2 = 0.02$, $\phi_1(t) = 1$, $\phi_2(t) = \sqrt{7}\left(20t^3 - 30t^2 + 12t\right)$ and $\epsilon(t) \sim N(0, \sigma^2 = 0.01)$. Figure 7 shows a simulated data set under Model 1 with sample size $n = 100$. We use this sample size along the whole simulation study.

Model 1 generates functional data that strongly depend on the realizations of ξ_1. Since ξ_1 is normal, Model 1 represents a scenario where complex data features are absent. Models 2, 3, and 4 will be defined modifying the distribution of ξ_1 and they

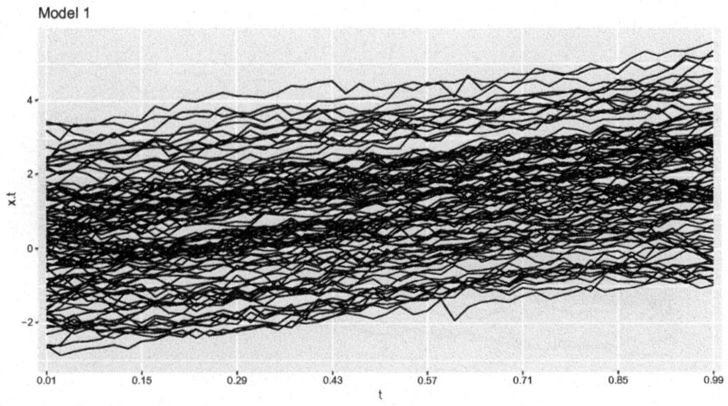

Fig. 7 Simulated data set from Model 1

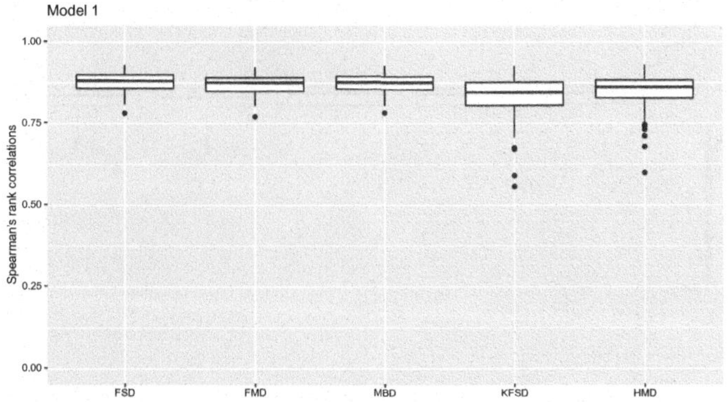

Fig. 8 Model 1: boxplots of the Spearman's rank correlation coefficients between the FSD, FMD, MBD, KFSD and HMD values and the $f(\xi_1)$ values

will reproduce other data features of our interest. The design of our simulation study allows the attainment of two goals: first, the considered models will both replicate and isolate specific data features, and, second, the theoretical densities of the realizations of ξ_1, say $f(\xi_1)$, will represent a natural benchmark to evaluate the performances of the functional depths under study.

We generate 100 samples from Model 1, and we evaluate the performance of a functional depth with each generated data set from Model 1 looking at the Spearman's rank correlation coefficient between depth and $f(\xi_1)$ values. Figure 8 shows the five boxplots obtained under Model 1. The boxplots illustrate that in absence of complex features there are very mild differences in favor of global depths, which behave similarly among them. Local depths show similar but slightly more variable performances.

Model 2 is obtained modifying the distribution of ξ_1.

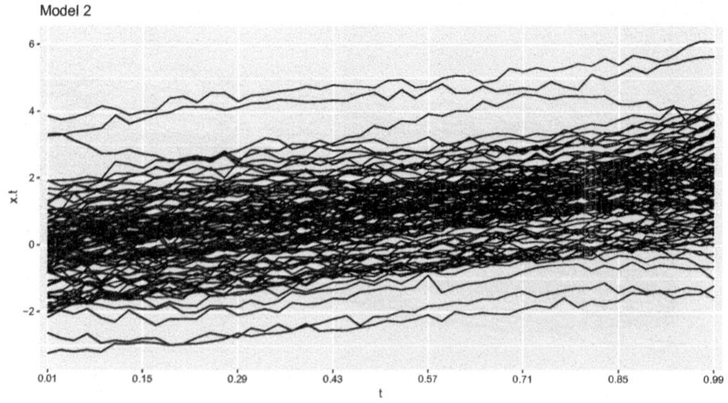

Fig. 9 Simulated data set from Model 2

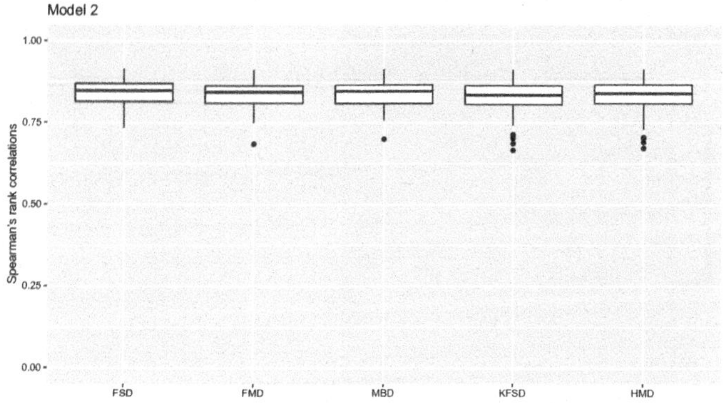

Fig. 10 Model 2: boxplots of the Spearman's rank correlation coefficients between the FSD, FMD, MBD, KFSD, and HMD values and the $f(\xi_1)$ values

Under Model 2, $\xi_1 \sim \sqrt{\lambda_1 \frac{3}{5}} X$ and $X \sim t_5$. Note that this change allows us to obtain functional data sets potentially contaminated by atypical observations (see Fig. 9 for an example).

Figure 10 replicates Fig. 8 for Model 2. According to this new figure, in presence of a complex feature such as the existence of potential outliers both classes of depths behave very similarly. We claim that this result is due to the fact that both global and local depths analyze reasonably well those functional samples symmetrically contaminated by curves that are outlying because of their relative levels.

To obtain Model 3 we consider an alternative modification of the distribution of ξ_1.

Under Model 3, $\xi_1 \sim \sqrt{\lambda_1 \frac{1}{10}} X$ and $X \sim \chi^2_5$. In this case the change allows to obtain asymmetric functional data sets, i.e., for all t, Model 3 generates many relatively low $x(t)$ and fewer relatively high $x(t)$ (see Fig. 11 for an example).

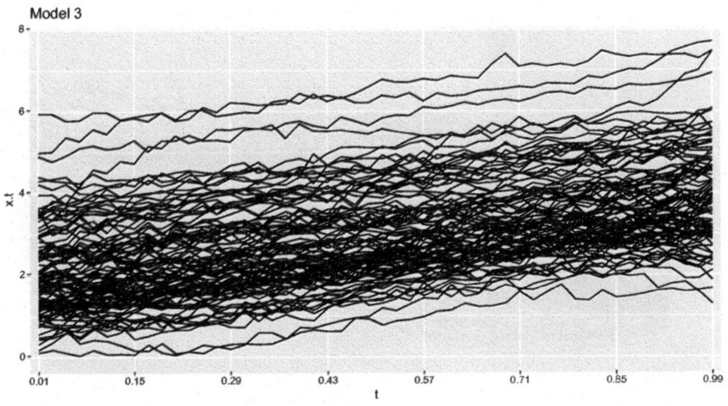

Fig. 11 Simulated data set from Model 3

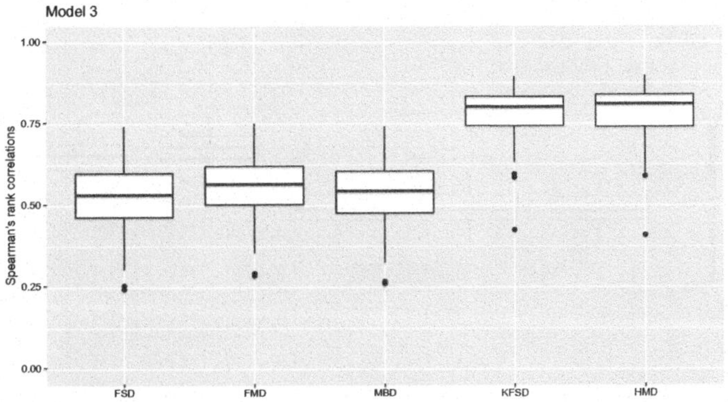

Fig. 12 Model 3: boxplots of the Spearman's rank correlation coefficients between the FSD, FMD, MBD, KFSD, and HMD values and the $f(\xi_1)$ values

Figure 12 shows the boxplots obtained under Model 3, and it is clear that asymmetry represents a complex feature that affects the performances of all the depths under study, but in general local-oriented KFSD and HMD show a stronger association with the benchmark $f(\xi_1)$.

Finally, we consider a mixture of two normals to obtain Model 4: with equal probability, $\xi_1 \sim N\left(-\sqrt{\lambda_1 - \frac{1}{10}}, \frac{1}{10}\right)$ or $\xi_1 \sim N\left(\sqrt{\lambda_1 - \frac{1}{10}}, \frac{1}{10}\right)$. We employ Model 4 to obtain data showing bimodality and potential presence of isolated observations lying between the two main groups of curves (see Fig. 13 for an example).

Due to the behaviors of FSD, FMD, and MBD under Model 4, when reporting the boxplots in Fig. 14, we use $[-1, 1]$ as range for the vertical axis. The information provided by Fig. 14 suggests that the ranking of whole bimodal data sets represents a problem that is hard to be handled in an unsupervised way by the functional depths under study. However, the local-oriented KFSD and HMD show a generally positive association with the benchmark $f(\xi_1)$, whereas for the global-oriented FSD, FMD,

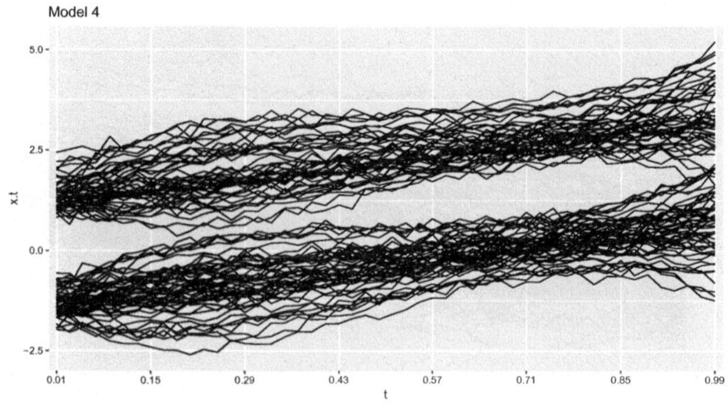

Fig. 13 Simulated data set from Model 4

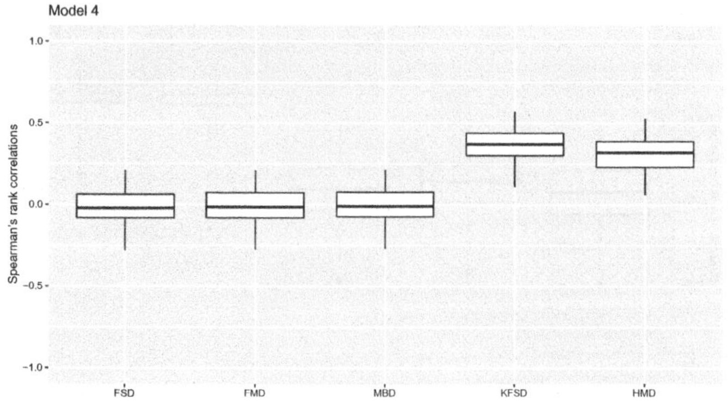

Fig. 14 Model 4: boxplots of the Spearman's rank correlation coefficients between the FSD, FMD, MBD, KFSD, and HMD values and the $f(\xi_1)$ values

and MBD we observe Spearman's rank correlation coefficients that vary symmetrically around 0.

The results of the simulation study presented in this section have shown that the behaviors of global and local functional depths can be fairly similar under some circumstances (e.g., absence of complex data features and presence of a particular type of outliers), but quite different under others (e.g., asymmetry and bimodality).

5 Conclusions

With the aim of extending to the functional context the knowledge about the differences between a global and a local approach to the depth problem, in this paper we have presented an empirical study that studied and compared the behavior of three global and two local functional depths. We have illustrated that local functional depths may behave differently with respect to their global alternatives. Indeed, using real and simulated data sets, we have observed that analyses based on local depths may be an alternative under specific scenarios. In this work we have identified at least two: first, in presence of asymmetry (see Model 3 and NO_x analyses); second, in presence of bimodality and isolated observations (see Model 4 and phonemes analyses).

References

1. Agostinelli, C., Romanazzi, M.: Local depth. J. Stat. Plan. Inference **141**, 817–830 (2011)
2. Azcorra, A., Chiroque, L.F., Cuevas, R., Anta, A.F., Laniado, H., Lillo, R.E., Romo, J., Sguera, C.: Unsupervised scalable statistical method for identifying influential users in online social networks. Sci. Rep. **8**(1), 6955 (2018)
3. Chakraborty, A., Chaudhuri, P.: On data depth in infinite dimensional spaces. Ann. Inst. Stat. Math. **66**, 303–324 (2014)
4. Chen, Y., Dang, X., Peng, H., Bart, H.L.: Outlier detection with the kernelized spatial depth function. IEEE Trans. Pattern Anal. Mach. Intell. **31**, 288–305 (2009)
5. Cuevas, A.: A partial overview of the theory of statistics with functional data. J. Stat. Plan. Inference **147**, 1–23 (2014)
6. Cuevas, A., Febrero, M., Fraiman, R.: On the use of the bootstrap for estimating functions with functional data. Comput. Stat. Data Anal. **51**, 1063–1074 (2006)
7. Febrero, M., Galeano, P., González-Manteiga, W.: Outlier detection in functional data by depth measures, with application to identify abnormal nox levels. Environmetrics **19**, 331–345 (2008)
8. Febrero-Bande, M., Oviedo de la Fuente, M.: Statistical computing in functional data analysis: the r package fda.usc. J. Stat. Softw. **51**, 1–28 (2012)
9. Ferraty, F., Vieu, P.: Nonparametric Functional Data Analysis: Theory and Practice. Springer, New York (2006)
10. Fraiman, R., Muniz, G.: Trimmed means for functional data. TEST **10**, 419–440 (2001)
11. Horváth, L., Kokoszka, P.: Inference for Functional Data With Applications. Springer, New York (2012)
12. Kokoszka, P., Reimherr, M.: Introduction to Functional Data Analysis. CRC Press (2017)
13. Liu, R.Y.: On a notion of data depth based on random simplices. Ann. Stat. **18**, 405–414 (1990)
14. López-Pintado, S., Romo, J.: On the concept of depth for functional data. J. Am. Stat. Assoc. **104**, 718–734 (2009)
15. Paindaveine, D., Van Bever, G.: From depth to local depth: a focus on centrality. J. Am. Stat. Assoc. **108**, 1105–1119 (2013)
16. Ramsay, J.O., Silverman, B.W.: Functional Data Analysis. Springer, New York (2005)
17. Serfling, R.: A depth function and a scale curve based on spatial quantiles. In: Dodge, Y. (ed.) Statistical Data Analysis Based on the L1-Norm and Related Methods, pp. 25–38. Birkhaüser, Basel (2002)
18. Sguera, C., Galeano, P., Lillo, R.: Spatial depth-based classification for functional data. TEST **23**, 725–750 (2014)
19. Sguera, C., Galeano, P., Lillo, R.: Functional outlier detection by a local depth with application to nox levels. Stoch. Environ. Res. Risk Assess. **30**, 1115–1130 (2016)
20. Tukey, J.W.: Mathematics and the picturing of data. Proc. Int. Congr. Math. **2**, 523–531 (1975)

AutoSpec: Detecting Exiguous Frequency Changes in Time Series

David S. Stoffer

Abstract Most established techniques that search for structural breaks in time series may not be able to identify slight changes in the process, especially when looking for frequency changes. The problem is that many of the techniques assume very smooth local spectra and tend to produce overly smooth estimates. The problem of over-smoothing tends to produce spectral estimates that miss slight frequency changes because frequencies that are close together will be lumped into one frequency. The goal of this work is to develop techniques that concentrate on detecting slight frequency changes by requiring a high degree of resolution in the frequency domain.

1 Introduction

Many time series are realizations of nonstationary random processes, hence estimating their time varying spectra may provide insight into the physical processes that give rise to these time series. For example, EEG time series are typically nonstationary, and estimating the time varying spectra based on the EEG of epilepsy patients may lead to methods capable of predicting seizure onset; e.g., see [2]. Similarly, analyzing the time varying spectrum of the Southern Oscillation Index (SOI) may further our knowledge of the frequency of the El Niño Southern Oscillation (ENSO) phenomenon and its impact on global climate; e.g., see [3].

Most established techniques that search for structural breaks in time series, however, may not be able to identify slight frequency changes at the resolution of interest. Of course, the resolution depends on the particular application. The problem is that many of the techniques assume very smooth local spectra and tend to produce overly smooth estimates. The problem of assuming very smooth spectra produces spectral

Supported in part by NSF DMS-1506882.

D. S. Stoffer (✉)
Department of Statistics, University of Pittsburgh, Pittsburgh, PA, USA
e-mail: stoffer@pitt.edu

M. La Rocca et al. (eds.), *Nonparametric Statistics*, Springer Proceedings in Mathematics & Statistics 339, https://doi.org/10.1007/978-3-030-57306-5_42

estimates that may miss slight frequency changes because frequencies that are close together will be lumped into one frequency; see Sect. 5 for further details. The goal of this work is to develop techniques that concentrate on detecting slight frequency changes by requiring a high degree of resolution in the frequency domain.

The basic assumptions here are that, conditional on the location and number of segments, the time series is piecewise stationary with each piece having a spectral density. A detailed description of the model is given in Sect. 2. In addition to representing time series that have regime changes, the model can be used to approximate slowly varying processes; e.g., see [1], which uses dyadic segmentation to find the approximate location of breakpoints. The approach taken in [8] was to fit piecewise AR models using minimum description length and a genetic algorithm for solving the difficult optimization problem. [13] proposed nonparametric estimators based on dyadic segmentation and smooth local exponential functions. [15] estimated the log of the local spectrum using a Bayesian mixture of splines. The basic idea of this approach is to first partition the data into small sections. It is then assumed that the log spectral density of the evolutionary process in any given partition is a mixture of individual log spectra. A mixture of smoothing splines model with time varying mixing weights is used to estimate the evolutionary log spectrum. Later, [16] improved on the technique of [15] by adaptively selecting breakpoints.

For background, note that spectral analysis has to do with partitioning the variance of a stationary time series, $\{X_t\}$, into components of oscillation indexed by frequency ω, and measured in cycles per unit of time, for $-1/2 < \omega \leq 1/2$. Given a time series sample, $\{X_t;\ t = 1, \ldots, n\}$, that has been centered by its sample mean, the sample spectral density (or *periodogram*) is defined in terms of frequency ω:

$$I_n(\omega) = \left| n^{-1/2} \sum_{t=1}^{n} X_t \, e^{-2\pi i \omega t} \right|^2 . \tag{1}$$

The periodogram is essentially the squared-correlation of the data with sines and cosines that oscillate at frequency ω.

The spectral density, $f(\omega)$, of a stationary time series can be defined as the limit $(n \to \infty)$ of $E[I_n(\omega)]$, provided that the limit exists; details can be found in [17, Chap. 4]. It is worthwhile to note that $f(\omega) \geq 0$, $f(\omega) = f(-\omega)$, and

$$\int_{-1/2}^{1/2} f(\omega)\, d\omega = 2 \int_{0}^{1/2} f(\omega)\, d\omega = \sigma^2, \tag{2}$$

where $\sigma^2 = \text{var}(X_t) < \infty$. Thus, the spectral density can be thought of as the variance density of a time series relative to frequency of oscillation. That is, for positive frequencies between 0 and 1/2, the proportion of the variance that can be attributed to oscillations in the data at frequency ω is roughly $2f(\omega)d\omega$. If the time series X_t is *white noise*, that is, $E(X_t)$ is independent of time t, and $\text{cov}(X_s, X_t) = 0$ for all $s \neq t$, then $f(\omega) \equiv \sigma^2$. The designation white originates from the analogy with white light and indicates that all possible periodic oscillations are present with equal strength.

2 Model and Existing Methods

Let a time series $\{X_t;\ t = 1, \ldots, n\}$ consist of an unknown number of segments, m, and let ξ_j be the unknown location of the end of the jth segment, $j = 0, 1, \ldots, m$, with $\xi_0 = 0$ and $\xi_m = n$. Then conditional on m and $\boldsymbol{\xi} = (\xi_0, \ldots, \xi_m)'$, we assume that the process $\{X_t\}$ is piecewise stationary. That is,

$$X_t = \sum_{j=1}^{m} X_{t,j}\, \delta_{t,j}\,, \tag{3}$$

where for $j = 1, \ldots, m$, the processes $X_{t,j}$ have spectral density $f_j^\theta(\omega)$ that may depend on parameters θ, and $\delta_{t,j} = 1$ if $t \in [\xi_{j-1} + 1, \xi_j]$ and 0 otherwise.

Consider a realization $\boldsymbol{x} = (x_1, \ldots, x_n)'$ from process (3), where the number and locations of the stationary segments is unknown. Let n_j be the number of observations in the jth segment. We assume that each n_j is large enough for the local Whittle likelihood (see [20]) to provide a good approximation to the likelihood. Given a partition of the time series $\boldsymbol{x}$, the jth segment consists of the observations $\boldsymbol{x}_j = \{x_t : \xi_{j-1} + 1 \le t \le \xi_j\}$, $j = 1, \ldots, m$, with underlying spectral densities f_j^θ and periodograms I_j, evaluated at frequencies $\omega_{k_j} = k_j/n_j, 0 \le k_j \le n_j - 1$. For a given partition $\boldsymbol{\xi}$, the approximate likelihood of the time series is given by

$$L(f_1^\theta, \ldots, f_m^\theta \mid \boldsymbol{x}, \boldsymbol{\xi}) \approx \prod_{j=1}^{m} (2\pi)^{-n_j/2} \prod_{k_j=0}^{n_j-1} \exp\Big\{-\frac{1}{2}\Big[\log f_j^\theta(\omega_{k_j}) + I_j(\omega_{k_j})/f_j^\theta(\omega_{k_j})\Big]\Big\}. \tag{4}$$

Note that in setting up the model, most items depend on the number of regimes, m. For ease, that dependence is understood and dropped from the notation.

2.1 AdaptSpec

The frequency domain approach used in [16] is a Bayesian method that incorporates (4) with a linear smoothing spline prior on the $\log f_j^\theta(\omega)$ for $j = 1, \ldots, m$. In addition, a uniform prior is placed on the breakpoints, $\Pr(\xi_j = t \mid m) = 1/p_j$, for $j = 1, \ldots, m-1$, where p_j is the number of available locations for split point ξ_j, as is the prior on the number of segments, $\Pr(m = k) = 1/M$ for $k = 1, \ldots, M$ and M is some large but fixed number. The approach uses reversible jump Markov chain Monte Carlo (RJ-MCMC) methods to evaluate the posteriors. The technique is available in an R package called `BayesSpec`.

2.2 AutoParm

Although this method, which is described in [8], is a time domain approach, the authors argue that AR models are dense in the space of bounded spectral densities (e.g., see [17, Sect. 4.5]) and can thus be used as a frequency domain approach. In this case, each time series $\{X_{t,j}\}$ is piecewise stationary with AR(p_j) behavior in each segment, $j = 1, \ldots, m$. Then, Minimum Description Length (MDL) as described in [14] is used to find the best combination of the number of segments, m, the breakpoints ξ_j (or segment sizes n_j), and the orders/estimates of the piecewise AR processes. The idea is to minimize the Code Length (CL) necessary to store the data (i.e., the amount of memory required to encode the data), which leads to a BIC-type criterion to find the model that minimizes

$$\sum_j \left(\ell_j + \log p_j + \frac{p_j + 2}{2} \log n_j \right) + \log m + (m+1) \log n, \tag{5}$$

where $\ell_j = -\log \hat{L}_j(\mu_j, \phi_1, \ldots, \phi_{p_j}, \sigma_j^2 \mid x, \xi_m)$ and $\hat{L}_j$ maximized value of the usual Gaussian AR(p_j) likelihood for segment $j = 1, \ldots, m$,

$$L_j(\cdot \mid \cdot) = (2\pi)^{-n_j/2} |\Gamma_j|^{-1/2} \exp\left\{ -\tfrac{1}{2} (\boldsymbol{x}_j - \mu_j \mathbf{1})' \Gamma_j^{-1} (\boldsymbol{x}_j - \mu_j \mathbf{1}) \right\}, \tag{6}$$

where μ_j is the segment's constant mean value, $\mathbf{1}$ is the corresponding vector of ones, and $\Gamma_j = \sigma_j^2 V_j$ is the usual AR(p_j) variance-covariance matrix corresponding to $\boldsymbol{x}_j$. Because of the Markov structure of AR models, the likelihood has a simple form; see [6, Prob. 8.7] for details. Fitting the model has to be done via numerical optimization, which is accomplished via a genetic algorithm (a derivative free smart search for minimization based on evolutionary biology concepts). Basic information on genetic algorithms may be obtained from the Mathworks site [11].

2.3 The Problem Is Resolution

The problem with the aforementioned techniques is that they tend to over-smooth the spectral density estimate so that small frequency shifts cannot be detected. Resolution problems were thoroughly discussed in the literature in the latter half of the twentieth century. Some details of the history of the problem as well as a simulation example are given in Sect. 5. For our analysis, we focus on El Niño–Southern Oscillation (ENSO). The Southern Oscillation Index (SOI) measures changes in air pressure related to sea surface temperatures in the central Pacific Ocean. The central Pacific warms every three to seven years due to the El Niño effect, which has been blamed for various global extreme weather events. It has become taken as fact that sometime after 1980, the frequency of the El Niño–La Niña (or ENSO) cycle has increased with the global warming; e.g., see [19].

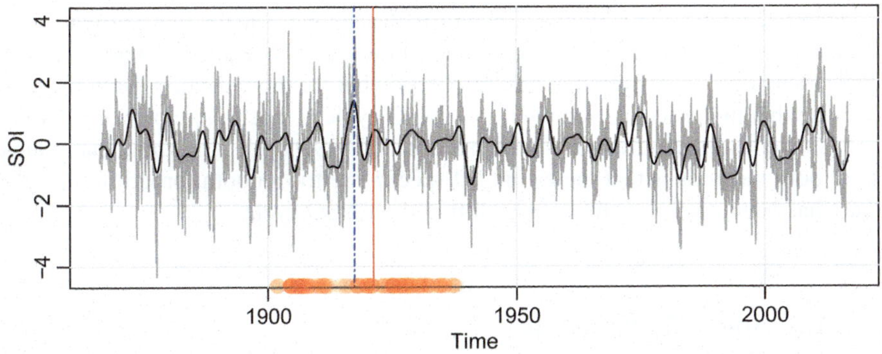

Fig. 1 Monthly values of the SOI for years 1866–2017 with breakpoints (vertical lines) determined by AutoParm (- - -) and by AdaptSpec (|). The solid smooth line is the filtered series that exhibits the ENSO cycle. For AdaptSpect, Pr(break | data) = 0.3 indicates there is probably not a breakpoint

Monthly values of the SOI are displayed in Fig. 1 for years 1866–2017 [7]; additionally, the data have been filtered to exhibit the ENSO cycle. Also shown in Fig. 1 are the AutoParm results (vertical dashed line) and AdaptSpec results (vertical solid line) when applied to the SOI series. AutoParm prefers a breakpoint around 1920, whereas AdaptSpec is indicating there are no breakpoints because Pr(break | data) = 0.3. However, assuming that there is one structural break, the posterior distribution of the breakpoints (with a vertical line at the mean) is displayed in the figure.

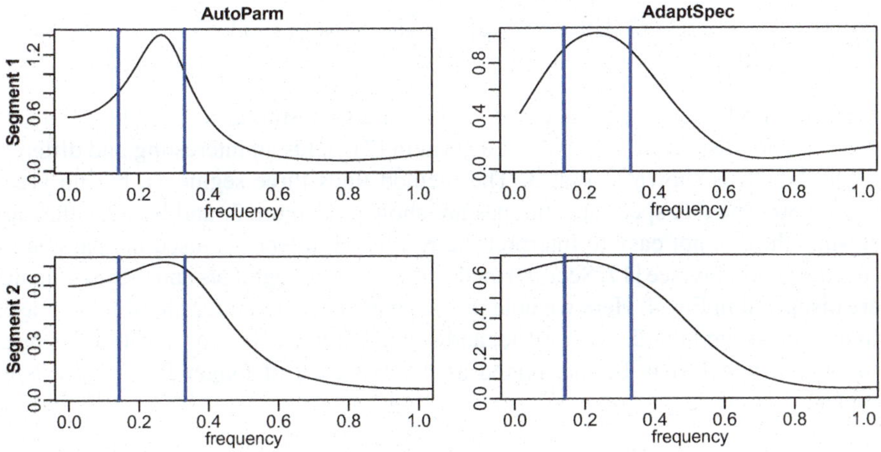

Fig. 2 The segmented spectral estimates using AutoParm and AdaptSpec. The vertical lines show the 3–7 7 year cycle known ENSO cycle range. The frequency scale is in years and is truncated at the annual cycle

Figure 2 shows the estimated spectra for each segment for the AutoParm and AdaptSpec techniques. The vertical lines show the 3–7 year cycle known ENSO cycle (see https://www.weather.gov/mhx/ensowhat for more details). Both methods indicate that in the second segment, the ENSO cycle is much more broad, including both slower and faster frequencies than the usual ENSO cycle. One thing that is clear from both methods is that the estimated spectra are too smooth (broad) to reveal if there has been a decisive frequency shift in the ENSO cycle.

3 AutoSpec—Parametric

Since the interest is in spectra, an obvious extension of AutoParm is to replace the Gaussian AR likelihood in MDL with Whittle likelihood. That is, in (6), replace L_j with the Whittle likelihood (4) but with

$$f_j^{\theta}(\omega) = \sigma_j^2 \, |\phi_j(e^{2\pi i \omega})|^{-2}, \tag{7}$$

where $\phi_j(z)$ is the AR polynomial of order p_j given by.

$$\phi_j(z) = 1 - \phi_1 z - \cdots - \phi_{p_j} z^{p_j},$$

for $j = 1, \ldots, m$. The basic idea is use peridograms as the data in a move from the time domain to the frequency domain. Another similar method would be to use the Bloomfield EXP model [5],

$$f_j^{\theta}(\omega) = \sigma_j^2 \exp\Big(2 \sum_{\ell=1}^{p_j} \theta_{\ell,j} \cos(2\pi \ell \omega)\Big).$$

However, EXP yields spectral estimates that are very similar to the AR spectra, so these are not displayed. Using the AR form in (7) yields an interesting and different result, which is shown in Fig. 3. The method finds three segments, but no break near 1980. The AR spectral estimates are shown in Figure 3, and because they are so smooth, it is not easy to interpret the results. However, we used nonparametric spectral estimates (see [17, Sect. 4]) on the three chosen segments and those estimates are displayed in Fig. 4. Here we note that segments two and three show an increased frequency around the 2.5 year cycle, and segment three shows an additional slower frequency, which may be interpreted as the existence of longer El Niño/La Niña cycles.

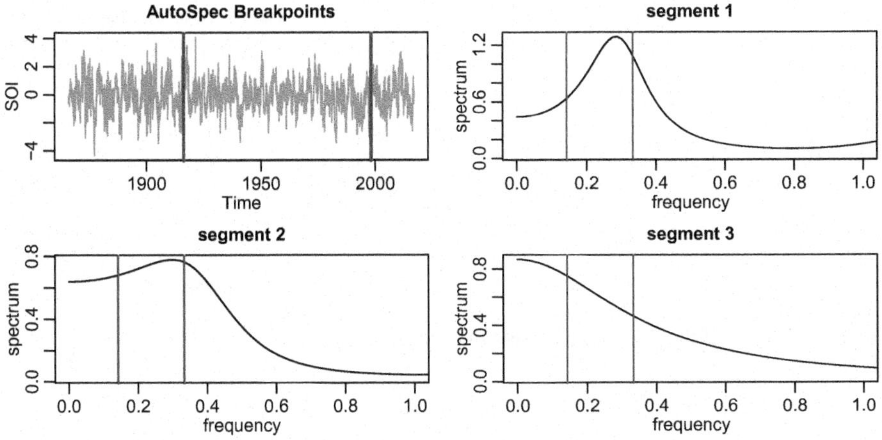

Fig. 3 The SOI series with the two breakpoints found using AutoSpec. The individual AR spectral estimates for each of the three segments. The vertical lines show the 3–7 year cycle known ENSO cycle range

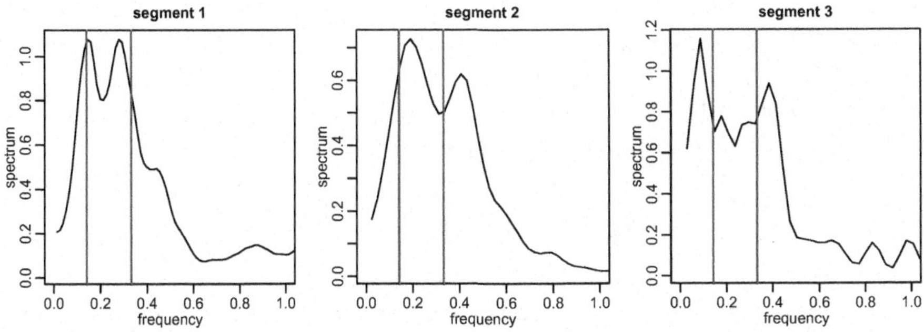

Fig. 4 Nonparametric estimates of the spectra in the three segments identified by AutoSpec; compare to Figure 3. The vertical lines show the 3–7 year cycle known ENSO cycle range

4 AutoSpecNP—Nonparametric

Because of the success of the nonparametric approach in the previous section, the natural next step would be to develop a fully nonparametric technique. To this end, consider a triangular kernel (Bartlett window), $\{W(\ell);\ \ell = 0, \pm 1, \dots, \pm b\}$ with $W(\ell) \propto 1 - |\ell|/(b+1)$ such that $\sum W(\ell) = 1$, with the bandwidth b chosen by MDL to smooth the periodogram of the fully tapered data and then used the Whittle likelihood for the given spectral estimate. That is, to nonparametrically evaluate the likelihood (4) in each segment $j = 1, \dots, m$, use

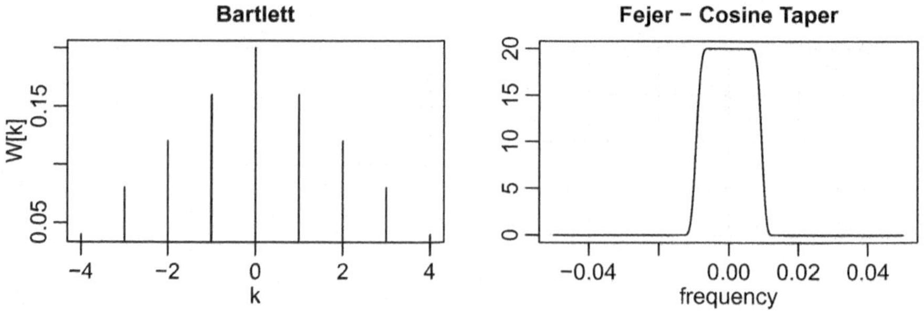

Fig. 5 Example of the Bartlett kernel and the corresponding Fejér spectral window when a taper is applied

$$\hat{f}_j(\omega_{k_j}) = \sum_{\ell=-b_j}^{b_j} W(\ell)\, I_j^{\text{taper}}(\omega_{k_j} + \ell), \tag{8}$$

where the $b_j = 1, 2, \ldots$ are chosen by MDL (similar to AR orders in AutoParm). Here, $I_j^{\text{taper}}(\cdot)$ represents the periodogram of the fully cosine tapered data in segment j for $j = 1, \ldots, m$.

For example, if $\{x_t\}$ represents the data in a segment, then they are preprocessed as $y_t = h_t x_t$ where h_t is the cosine bell taper favored by [4],

$$h_t = .5\left[1 + \cos\left(\frac{2\pi(t - \bar{t})}{n}\right)\right],$$

where $\bar{t} = (n+1)/2$ and n is the number of observations in that segment. In this case, the periodogram is of the preprocessed data, y_t. Figure 5 shows an example of the Bartlett window with $b = 4$; the corresponding spectral window (see [17, Sect. 4]) of the Bartlett kernel is not very good unless the data are tapered. The spectral window corresponding to the Bartlett kernel with tapering is also displayed in Figure 5.

Figure 6 shows the results of the fully nonparametric method. The figure displays the SOI series along with the estimated breakpoints. The fully nonparametric method finds a breakpoint near 1980 as has been suggested by climatologists, initially in [18]. Although the differences in each segment are subtle, this method has enough resolution to distinguish between minor frequency shifts. We may interpret the findings as follows. From 1866 to 1905, the ENSO cycle was a 3–7 year cycle. After that, there appears to be a shift to a higher ENSO cycle of about 2.5 years in addition to the usual 3–7 year cycle. Finally, after 1980, there appears to be a slower cycle introduced into the system. That is, after 1980, the ENSO cycle included a much slower cycle that indicates that El Niño events tend to be *longer*, but not faster than 2.5 years.

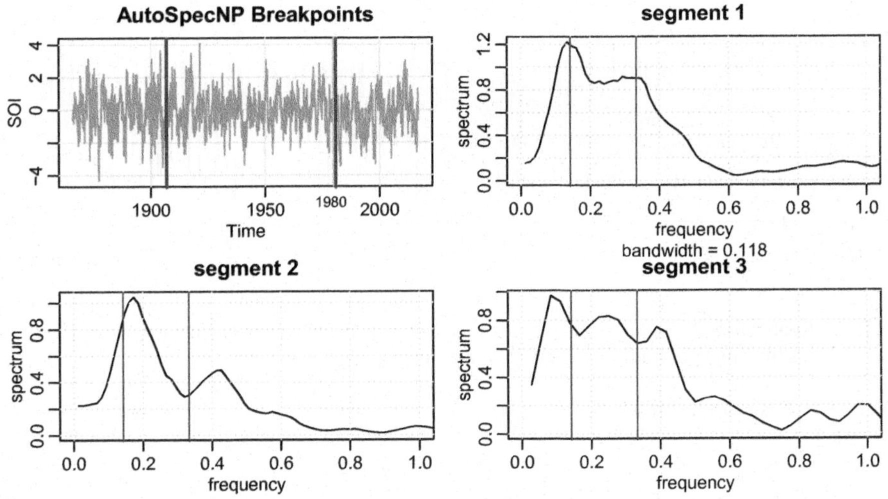

Fig. 6 Nonparametric estimates of the spectra in the three segments identified by AutoSpecNP; compare to Figs. 3 and 4. The vertical lines show the 3–7 year cycle known ENSO cycle range

5 Spectral Resolution and a Simulation

As previously stated, the problem of resolution was discussed in the literature in the latter half of the twentieth century; e.g., [9, 10]. The basic rule of thumb is that the achievable frequency resolution, $\Delta\omega$ should be approximately the reciprocal of the observational time interval, Δt of the data, $\Delta\omega \approx 1/\Delta t$. Two signals can be as close as $1/\Delta t$ apart before there is significant overlap in the transform and the separate peaks are no longer distinguishable.

For example, we generated a time series of length 2000 where

$$X_t = \begin{cases} X_{1t} = 2\cos(2\pi\omega\, t)\cos(2\pi\delta\, t) + Z_{1t}\,, & 1 \le t \le 1000\,, \\ X_{2t} = \cos(2\pi\omega\, t) + Z_{2t}\,, & 1001 \le t \le 2000\,, \end{cases} \tag{9}$$

$\omega = 1/25$, $\delta = 1/150$, and Z_{it} for $i = 1, 2$ are independent i.i.d. standard normals. The difference between the two halves of the data is that X_{1t} is a modulated version of X_{2t}. Modulation is a common occurrence in many signal processing applications, e.g., EEG (see [12]). In addition, note that

$$X_{1t} = \cos(2\pi[\omega + \delta]\, t) + \cos(2\pi[\omega - \delta]\, t) + Z_{1t}\,,$$

so that X_{1t} is distinguishable by twin peaks in the frequency domain.

Figure 7 shows a realization of X_t with the changepoint marked. The figure also displays the breakpoint $t = 1087$ identified by AutoSpecNP. We note, however, that AutoSpec and AutoParm do not identify any breakpoints.

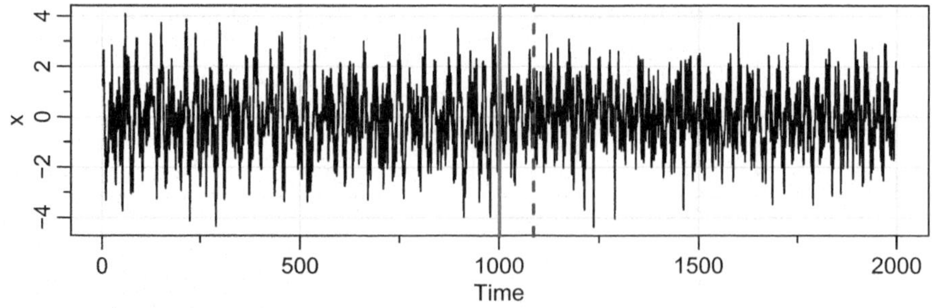

Fig. 7 Realization of (9) showing the true breakpoint as a solid vertical line; the dashed vertical line shows the breakpoint identified by AutoSpecNP

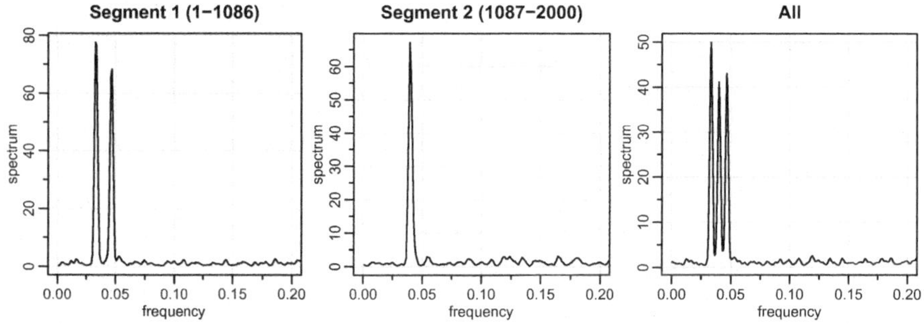

Fig. 8 The results of running AutoSpecNP on the data X_t, which finds one breakpoint at $t = 1087$. The figures are the estimated AutoSpecNP spectra for each identified segment and the estimate (8) with $b = 4$ on all the data

Figure 8 shows the results of running AutoSpecNP described in Sect. 4 on the data X_t. As seen from the figure, the procedure is able to distinguish between the two processes (with a breakpoint at $t = 1087$). The method works because the procedure allows very limited to no smoothing of the periodogram. In addition to showing the spectral estimates of each segment, the figure also displays the estimate (8) with $b = 4$ on the entire realization of X_t. This figure helps in realizing why the method works.

References

1. Adak, Sudeshna: Time-dependent spectral analysis of nonstationary time series. J. Am. Stat. Assoc. **93**, 1488–1501 (1998)
2. Aksenova, T., Volkovych, V., Villa, A.: Detection of spectral instability in EEG recordings during the preictal period. J. Neural Eng. **4**, 173–178 (2007)
3. An, S.-I., Wang, B.: Interdecadal change of the structure of the ENSO mode and its impact on the ENSO frequency. J. Clim. **13**, 2044–2055 (2000)

4. Blackman, R.B., Tukey, J.W.: The measurement of power spectra from the point of view of communications engineering. Bell Syst. Tech. J. **37**(1), 185–282 (1958)
5. Bloomfield, P.: An exponential model for the spectrum of a scalar time series. Biometrika **60**(2), 217–226 (1973)
6. Brockwell P.J., Davis, R.A.: *Time Series: Theory and Methods*, 2nd edn. Springer Science & Business Media (2013)
7. University of East Anglia Climatic Research Unit. Southern Oscillation Index (SOI) (2018). https://crudata.uea.ac.uk/cru/data/soi/
8. Davis, R.A., Lee, T.C.M., Rodriguez-Yam, G.A.: Structural breaks estimation for nonstationary time series models. J. Am. Stat. Assoc. **101**, 223–239 (2006)
9. Kay, S.M., Marple, S.L.: Spectrum analysis—a modern perspective. Proc. IEEE **69**(11), 1380–1419 (1981)
10. Marple Jr., S.L.: Frequency resolution of Fourier and maximum entropy spectral estimates. Geophysics **47**(9), 1303–1307 (1982)
11. Mathworks. Genetic Algorithm—MATLAB (2018). https://www.mathworks.com/discovery/genetic-algorithm.html
12. Novak, P., Lepicovska, V., Dostalek, C.: Periodic amplitude modulation of EEG. Neurosci. Lett. **136**(2), 213–215 (1992)
13. Ombao, H.C., Raz, J.A., Von Sachs, R., Malow, B.A.: Automatic statistical analysis of bivariate nonstationary time series. J. Am. Stat. Assoc. **96**, 543–560 (2001)
14. Rissanen, J.: A universal prior for integers and estimation by minimum description length. Ann. Stat. 416–431 (1983)
15. Rosen, O., Wood, S., Stoffer, D.S.: Local spectral analysis via a Bayesian mixture of smoothing splines. J. Am. Stat. Assoc. **104**, 249–262 (2009)
16. Rosen, O., Wood, S., Stoffer, D.S.: Adaptspec: adaptive spectral estimation for nonstationary time series. J. Am. Stat. Assoc. **107**(500), 1575–1589 (2012)
17. Shumway, R.H., Stoffer, D.S.: Time Series Analysis and Its Applications: With R Examples, 4th edn. Springer, New York (2017)
18. Timmermann, A., Oberhuber, J., Bacher, A., Esch, M., Latif, M., Roeckner, E.: Increased El Niño frequency in a climate model forced by future greenhouse warming. Nature **398**(6729), 694 (1999)
19. Wang, G., Cai, W., Gan, B., Wu, L., Santoso, A., Lin, X., Chen, Z., McPhaden, M.J.: Continued increase of extreme El Niño frequency long after 1.5 °C warming stabilization. Nat. Clim. Change **7**(8), 568 (2017)
20. Whittle, P.: Curve and periodogram smoothing. J. R. Stat. Soc. B **19**, 38–47 (1957)

Bayesian Quantile Regression in Differential Equation Models

Qianwen Tan and Subhashis Ghosal

Abstract In many situations, nonlinear regression models are specified implicitly by a set of ordinary differential equations. Often, mean regression may not adequately represent the relationship between the predictors and the response variable. Quantile regression can give a more complete picture of the relationship, can avoid distributional assumptions and can naturally handle heteroscedasticity. However, quantile regression driven by differential equations has not been addressed in the literature. In this article, we consider the problem and adopt a Bayesian approach. To construct a likelihood without distributional assumptions, we consider all quantile levels simultaneously. Because of the lack of an explicit form of the regression function and the indeterminate nature of the conditional distribution, evaluating the likelihood and sampling from the posterior distribution are very challenging. We avoid the computational bottleneck by adopting a "projection posterior" method. In this approach, the implicit parametric family of regression function of interest is embedded in the space of smooth functions, where it is modeled nonparametrically using a B-spline basis expansion. The posterior is computed in the larger space based on a prior without constraint, and a "projection" on the parametric family using a suitable distance induces a posterior distribution on the parameter. We illustrate the method using both simulated and real datasets.

Keywords Differential equation model · Projection posterior · Quantile regression · B-splines · Finite random series prior

1 Introduction

We consider a nonlinear quantile regression model, where for a specific τ, the τth quantile regression function is given implicitly through a set of Ordinary Differential

---

Q. Tan · S. Ghosal (✉)
Department of Statistics, North Carlina State University, 2311 Stinson Dr., Raleigh, NC 27695, USA
e-mail: sghosal@ncsu.edu

Q. Tan
e-mail: qtan@ncsu.edu

M. La Rocca et al. (eds.), *Nonparametric Statistics*, Springer Proceedings in Mathematics & Statistics 339, https://doi.org/10.1007/978-3-030-57306-5_43

Equations (ODE). Quantile regression Koenker and Bassett [11] is popularly used as a robust and more flexible alternative to the standard mean regression. The likelihood function may be constructed without any distributional assumption from the specification of quantiles of all levels u, $0 < u < 1$; see Tokdar and Kadane [14], Das and Ghosal [7]. However, since the τth quantile regression function is given only implicitly through an ODE and the τth quantile alone does not determine the entire conditional distribution, evaluation of the likelihood function of the parameters of the ODE is very challenging. We adopt a Bayesian approach, but avoid the computational barrier by using a "projection posterior". This posterior distribution is obtained by modeling all quantile regression functions nonparametrically as a function of both quantile level u and predictor variableś value t, without any functional constraint on the τth quantile regression function and then inducing a distribution on the parameter through a map obtained by minimizing the distance to the parametric family specified by the ODE. For mean regression driven by an ODE, this type of Bayesian approach was proposed by Bhaumik and Ghosal [1–3] building on the two-step method of Varah [15] and Brunel [5], who used the classical approach instead based on the least-squares method of estimation. In this paper, we modify the Bayesian two-step procedure for quantile regression, by first considering a nonparametric approach to simultaneous quantile regression for all quantiles $0 < u < 1$, and then inducing the projection posterior on the parameter θ of the ODE by minimizing a distance based on the ODE between the τth quantile in the nonparametric model with the parametric family described by the ODE. It may be noted that, even though we are primarily interested in the inference on θ, which is linked with the specific τth quantile, the evaluation of the likelihood function without a parametric model for the residual is possible only by considering the simultaneous quantile regression of all levels $0 < u < 1$.

The paper is organized as follows, Sect. 2 contains the modeling assumptions and prior specifications. A block-Metropolis–Hastings MCMC algorithm for the computation of the posterior and the details about the projection step are also described there. In Sect. 3, we carry out a simulation study for a quantile regression model governed by a set of ordinary differential equations describing the dynamics of the relations between prey and predator populations. We analyze a real-life data on stock prices in Sect. 4 using a simplified hyperbolic diffusion equation model Bibby and Sørensen [4].

2 Methodology and Computation

Consider observations Y_i of the ith measurement taken at a point $t_i \in [0, 1]$, $i = 1, \ldots, n$, where the predictor variable typically stands for time, and $Y = (Y_i : i = 1, \ldots, n)$. We assume, without loss of generality, that each observation of response variables has been monotonically transformed into the unit interval. Let $Q(u|t)$ stand for the uth quantile of Y at the value t of the predictor, $0 \leq u \leq 1$. We assume that, for a specific fixed τ, $0 < \tau < 1$, the τth quantile $Q(\tau|t)$ is given by a nonlinear function $f_\theta(t)$ depending on an unknown parameter $\theta \in \Theta$, where Θ is a compact subset of $\mathbb{R}^d$. However, the function $f_\theta(t)$ is only implicitly given, as the solution of the ODE

$$\frac{df_\theta(t)}{dt} = F(t, f_\theta(t), \theta), \quad t \in [0, 1], \quad \theta \in \Theta, \tag{1}$$

where F is a given smooth function of its co-ordinates.

The first step of the two-step method is to embed the implicit parametric nonlinear quantile regression model in a nonparametric model for simultaneous quantile regressions and put a prior distribution in the bigger nonparametric model which does not have any restriction on its τth quantile function. To this end, we follow the approach of Das and Ghosal [8] of using a finite random series based on tensor products of B-splines; see Shen and Ghosal [12] and Ghosal and van der Vaart [10] for a systematic development of the finite random series prior for function estimation. Let $\{B_s(\cdot) : s = 1, \dots, J\}$ denote B-spline basis functions of order m on $k-1$ equidistant interior knots, $J = m + k - 1$; see de Boor [9] for an introduction to B-splines. Expanding the quantile function in a series

$$Q(u|t) = \sum_{s_1=1}^{J_1} \sum_{s_2=1}^{J_2} \beta_{s_1 s_2} B_{s_1}(u) B_{s_2}(t), \quad u, t \in [0, 1], \tag{2}$$

we put prior on the coefficients $(\beta_{s_1 s_2} : s_1 = 1, \dots, J_1,\ s_2 = 1, \dots, J_2)$, given $J_1,\ J_2$. Note that $Q(0|t) = 0$, $Q(1|t) = 1$, and $Q(u|t)$ is increasing in u for every t. These constraints are addressed by the condition $0 = \beta_{1s_2} < \beta_{2s_2} < \cdots < \beta_{J_1 s_2} = 1$ for every $s_2 = 1, \dots, J_2$. We put a prior of these coefficients by considering the vector of the spacings $(\gamma_{s_1 s_2} = \beta_{(s_1+1)s_2} - \beta_{s_1 s_2} : s_1 = 1, \dots, J_1 - 1)$, which lives on the $(J_1 - 1)$-unit simplex for all $s_2 = 1, \dots, J_2$. We put the uniform prior on each of these J_2 many $(J_1 - 1)$-simplex blocks. Ideally, an infinitely supported prior with geometric-like tail should be put of J_1, J_2. However, to reduce computational cost, we do not put any prior on J_1 and J_2, but in the end we choose them based on the data using the Akaike Information Criterion (AIC).

The likelihood $\prod_{i=1}^n p(Y_i|t_i)$ in the nonparametric model is derived from the quantile function through the relation $p(Y_i|t_i) = \left(\frac{\partial}{\partial u} Q(u|t_i)\big|_{u=u_{t_i}(Y_i)}\right)^{-1}$, $i = 1, \dots, n$, where $u_{t_i}(Y_i)$ solves the equation

$$Y_i = Q(u|t_i) = \sum_{s_1=1}^{J_1} \sum_{s_2=1}^{J_2} \beta_{s_1 s_2} B_{s_1}(u) B_{s_2}(t_i). \tag{3}$$

Since $Q(u|t_i)$ is monotonically increasing in u and is a piece-wise polynomial, the solution of (3) is unique and can be easily computed. In particular, if we use piece-wise quadratic B-spline (i.e., $m = 3$), (3) reduces to a quadratic equation and hence can be solved analytically. Now the B-spline representation, $\frac{\partial}{\partial u} Q(u|t_i) = \sum_{s_1=1}^{J_1} \sum_{s_2=1}^{J_2} \beta_{s_1 s_2} \dot{B}_{s_1}(u) B_{s_2}(t_i)$, where $\dot{B}_s(\cdot)$ is the derivative of the B-spline basis functions and is itself a spline function. Hence the log-likelihood is given by

$$\sum_{i=1}^{n} \log p(Y_i|t_i) = -\sum_{i=1}^{n} \log \left\{ \sum_{s_1=1}^{J_1} \sum_{s_2=1}^{J_2} \beta_{s_1 s_2} \dot{B}_{s_1}(u_{t_i}(Y_i)) B_{s_2}(t_i) \right\}. \quad (4)$$

Note that like the two-step method for mean regression, the likelihood is based on the nonparametric model only, and can be used to make inference on all quantile functions $\{Q(u|t) : 0 \leq u, t \leq 1\}$ independent of the ODE. With fixed J_1, J_2, the likelihood may also be regarded as a function of the array $((\beta_{s_1 s_2}))$.

The posterior distribution in the nonparametric model may be computed by the following block-Metropolis–Hastings Markov chain Monte Carlo (MCMC) algorithm. Following Das and Ghosal [7], we propose Metropolis–Hastings moves as follows. For a given $s_2 = 1, \ldots, J_2$, we generate independent sequence $(U_{s_1} : s_1 = 1, \ldots, J_1)$ from $\mathrm{Unif}(1/r, r)$ for some $r > 1$. Let $V_{s_1 s_2} = \gamma_{s_1 s_2} U_{s_1}$ and propose a move $\gamma_{s_1 s_2} \mapsto \gamma^*_{s_1 s_2}$, where $\gamma^*_{s_1 s_2} = V_{s_1 s_2} / \sum_{s_1'=1}^{J_1} V_{s_1' s_2}$, $s_1 = 1, \ldots, J_1$. The conditional density of $\gamma^*_{\cdot s_2} = (\gamma^*_{s_1 s_2} : s_1 = 1, \ldots, J_1)$ given $\gamma_{\cdot s_2} = (\gamma_{s_1 s_2} : s_1 = 1, \ldots, J_1)$ with respect to the Lebesgue measure on the simplex is then

$$h(\gamma^*_{\cdot s_2}|\gamma_{\cdot s_2}) = \left(\frac{r}{r^2 - 1}\right)^{J_1} \left(\prod_{s_1=1}^{J_1} \gamma_{s_1 s_2}\right)^{-1} \left(\frac{D_1 - D_2}{J_1}\right),$$

where $D_1 = (\max\{r\gamma_{s_1 s_2}/\gamma^*_{s_1 s_2} : 1 \leq s_1 \leq J_1\})^{J_1}$ and $D_2 = (\min\{r\gamma_{s_1 s_2}/\gamma^*_{s_1 s_2} : 1 \leq s_1 \leq J_1\})^{J_1}$.

Denote the likelihood at the parameter values $\gamma_{\cdot s_2}$ and $\gamma^*_{\cdot s_2}$ by $L(\gamma_{\cdot s_2})$ and $L(\gamma^*_{\cdot s_2})$ respectively. Then the acceptance probability of a single block update for $s_2 = 1, \ldots, J_2$ is $R_{s_2} = \min(1, L(\gamma^*_{\cdot s_2}) h(\gamma_{\cdot s_2}|\gamma^*_{\cdot s_2}) / L(\gamma_{\cdot s_2}) h(\gamma^*_{\cdot s_2}|\gamma_{\cdot s_2}))$.

The parameter r works as a tuning parameter of the MCMC procedure. A smaller value of r yields sticky movement with a higher acceptance rate while a larger value of r results in bigger jumps with a lower acceptance rate. We maintain the acceptance rate within $[0.15, 0.45]$ by tuning r, following the standard recipe for the Metropolis–Hastings algorithm.

Since the parameter space is very large consisting of several simplexes, a good starting point in the center of the posterior distribution for the MCMC is essential. We use the Maximum Likelihood Estimator (MLE) of $\{\beta_{s_1 s_2} : 1 \leq s_1 \leq J_1, 1 \leq s_2 \leq J_2\}$ given by maximizing (4) for the purpose. The optimization of the likelihood function is itself challenging since there is no convexity or explicit expression for the derivative function. We use the Greedy Coordinate Descent of Varying Step sizes on Multiple Simplexes (GCDVSMS) algorithm of Das [6] specifically designed for simplex parameter spaces, which seemed to work well for nonparametric simultaneous quantile regression Das and Ghosal [7].

Now we describe the projection step on the parameter space Θ. Let $w(\cdot)$ be a continuous weight function with $w(0) = w(1) = 0$ and be positive on $(0, 1)$, such that $w(t) = t(1 - t)$. For a given function f, its derivative $\dot{f}$ and $\eta \in \Theta$, define

$$R_f(\eta) = \sqrt{\int_0^1 |\dot{f}(t) - F(t, f(t), \eta)|^2 w(t)dt} \tag{5}$$

and a functional $\psi(f) = \operatorname{argmin}\{R_f(\eta) : \eta \in \Theta\}$ from the space of smooth functions on $[0, 1]$ to Θ. Now for each sampled array of the coefficients $\beta = (\beta_{s_1 s_2} : s_1 = 1, \ldots, J_1, s_2 = 1, \ldots, J_2)$, we obtain its projection on Θ as $\psi(f(\cdot; \beta))$, where $f(t; \beta) = \sum_{s_1=1}^{J_1} \sum_{s_2=1}^{J_2} \beta_{s_1 s_2} B_{s_1}(\tau) B_{s_2}(t)$.

3 Simulation Study

Let $f_\theta(t) = (f_{\theta,1}(t), f_{\theta,2}(t))$ be the solution of the system of ODE given by the Lotka–Volterra equations

$$\begin{aligned} \frac{df_{\theta,1}(t)}{dt} &= \theta_1 f_{\theta,1}(t) - \theta_2 f_{\theta,1}(t) f_{\theta,2}(t), \\ \frac{df_{\theta,2}(t)}{dt} &= -\theta_3 f_{\theta,2}(t) + \theta_4 f_{\theta,1}(t) f_{\theta,2}(t), \end{aligned} \tag{6}$$

where $t \in [0, 1]$ and the parameters are $\theta = (\theta_1, \theta_2, \theta_3, \theta_4)$ and with initial conditions $f_{\theta,1}(0) = 1$, $f_{\theta,2}(0) = 0.5$. These equations are often used to model the average size of prey and predator populations respectively in natural habitat. For a specific $0 < \tau < 1$, we assume that the τth quantiles of response variables Y and Z are given, respectively, by $f_\theta(t) = (f_{\theta,1}(t), f_{\theta,2}(t))$, where $\theta = (\theta_1, \theta_2, \theta_3, \theta_4)$. The true value θ_0 of the vector of these parameters is taken to be $(10, 10, 10, 10)$ in our simulation study, and $f_{\theta_0,1}$ and $f_{\theta_0,2}$ are computed by numerically solving (6). We generate the data from independent gamma distributions $Y_i \sim \text{Gamma}(\alpha_1(t_i), \lambda_1(t_i))$ and $Z_i \sim \text{Gamma}(\alpha_2(t_i), \lambda_2(t_i))$, where we choose the parameters such that the vector of medians of (Y, Z) at a given time t is $f_{\theta_0}(t)$, i.e., $\tau = 0.5$ and $(Q_\tau(Y|t), Q_\tau(Z|t)) = (f_{\theta_0,1}(t), f_{\theta_0,2}(t))$. More specifically, we choose the shape functions $\alpha_1(t) = t(1 - t)/4 + 5$ and $\alpha_2(t) = 5t(1 - t) + 5$, and determine the scale parameters $\lambda_j(t) = f_{\theta_0,j}(t)/Q_\tau(\text{Gamma}(\alpha_j(t), 1))$, $j = 1, 2$, by back-calculation using the scaling property of the gamma distribution $\text{Gamma}(\alpha, \lambda) \stackrel{d}{=} \lambda \times \text{Gamma}(\alpha, 1)$. For a sample of size n, the observation points $t_1, \ldots, t_n$ are chosen as $t_i = (2i - 1)/2n$ for $i = 1, \ldots, n$. Samples of sizes $n = 200, 500$, and 1000 are considered. We simulate 1000 replications for each case.

Under each replication, in the first step, we first transform the prey and predator population sizes into the unit interval. We use the gamma distribution function to transform the population sizes into the unit interval. For the prey population we choose the parameters of the transforming gamma distribution to be $\alpha_1 = 2.9620$, $\beta_2 = 0.3871$; and for predator population we choose the values as $\alpha_2 = 2.6097$, $\beta_2 = 0.4135$. We choose these parameter values to match the mean and variance of data we generated for prey and predator populations.

In the simulation setting, although the observations are two-dimensional, the methodology described in Sect. 2 still applies on each component separately, and the joint inference on θ is obtained in the projection step.

We use piece-wise quadratic B-splines to expand $Q(u|t)$ in both u and t (i.e., $m_1 = m_2 = 3$), so that the solution of (3) can be obtained analytically. Using k equidistant knots for the B-spline functions used to expand $Q(u|t)$ in both u and t, we let $k = 3, \ldots, 10$, and choose the best model using the AIC. We obtain 10000 MCMC samples and discard the first 1000 iterations as burn-in. After the quantile curves are estimated, the corresponding inverse transformation is performed on the response variables to their original scales.

To obtain the projection posterior using the two-step method with the distance given by (5), where the weight function is chosen as $w(t) = t(1-t), t \in [0, 1]$, and consider the functional $\psi(f)$. For each posterior draw of B-spline coefficients β, we obtain the induced posterior of $\theta = \psi(f(\cdot; \beta))$ and then calculate the Monte Carlo bias and Mean Squared Error (MSE) of the Bayes estimator of θ. To compare, we also consider a non-Bayesian estimator $\hat{\theta} = \psi(\hat{f})$ of θ, where $\hat{f}(t)$ is a nonparametric estimate of the vector of median regression function of (Y, Z) at t, obtained by applying a smoothing method. In this application, we also apply the B-spline smoothing method to estimate f by minimizing $\sum_{i=1}^{n} |Y_i - \sum_{s_2=1}^{J_2} \gamma_{1,s_2} B_{s_2}(t_i)|$ and $\sum_{i=1}^{n} |Z_i - \sum_{s_2=1}^{J_2} \gamma_{2,s_2} B_{s_2}(t_i)|$ with respect to $(\gamma_{1,s_2} : s_2 = 1, \ldots, J_2)$ and $(\gamma_{2,s_2} : s_2 = 1, \ldots, J_2)$, and plugging in the series expansions $f_1(t) = \sum_{s_2=1}^{J_2} \gamma_{1,s_2} B_{s_2}(t)$ and $f_2(t) = \sum_{s_2=1}^{J_2} \gamma_{2,s_2} B_{s_2}(t)$, respectively. Convergence properties of this estimator and numerical performance were studied in Chap. 3 of Tan [13]. We use 1000 replications to calculate the Monte Carlo bias and MSE for this estimator. The Monte Carlo bias

Table 1 Monte Carlo bias and MSE of two-step estimators based on Bayesian nonparametric simultaneous quantile regression (TSBNPSQR) and quantile regression (TSQR). The data are generated from a gamma population with median regression curves satisfying the Lotka–Volterra equations

		TSBNPSQR				TSQR			
n		θ_1	θ_2	θ_3	θ_4	θ_1	θ_2	θ_3	θ_4
200	Bias	−0.424	0.822	0.481	−0.848	−0.874	−0.950	−0.924	−1.081
		(0.046)	(0.061)	(0.046)	(0.062)	(1.010)	(1.011)	(1.011)	(1.010)
	MSE	2.467	3.025	2.627	3.184	4.850	4.950	5.184	4.896
		(0.089)	(0.090)	(0.089)	(0.091)	(1.010)	(1.011)	(1.011)	(1.010)
500	Bias	−0.186	0.363	0.174	−0.353	−0.375	−0.731	−0.389	−0.810
		(0.024)	(0.031)	(0.024)	(0.031)	(0.042)	(0.055)	(0.043)	(0.058)
	MSE	1.063	1.490	1.129	1.597	1.877	1.955	2.034	1.960
		(0.078)	(0.085)	(0.081)	(0.085)	(0.087)	(0.087)	(0.088)	(0.088)
1000	Bias	−0.111	0.193	0.101	−0.248	−0.199	−0.674	−0.270	0.752
		(0.022)	(0.022)	(0.022)	(0.021)	(0.024)	(0.051)	(0.021)	(0.057)
	MSE	0.406	0.417	0.503	0.581	0.509	0.919	1.010	1.577
		(0.045)	(0.046)	(0.048)	(0.053)	(0.048)	(0.078)	(0.078)	(0.085)

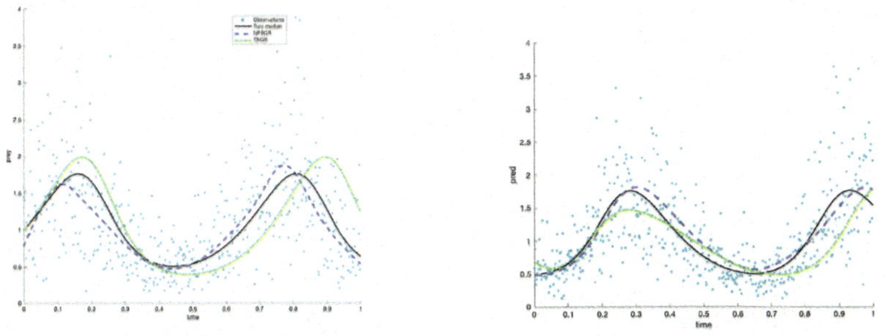

(a) Prey population (b) Predator population

Fig. 1 Estimated median regression curves using nonparametric simultaneous quantile regression (TSBNPSQR) and two-step quantile regression (TSQR) based on $n = 500$ observations from a gamma population with median regression curves satisfying the Lotka–Volterra equations

and MSE of the two-step Bayesian nonparametric estimator based on simultaneous quantile regression and the two-step (non-Bayesian) quantile regression estimator based on median regression are given in Table 1, where the two methods are referred to as TSBNPSQR and TSQR, respectively. The corresponding Monte Carlo errors of the Monte Carlo bias and MSE are given inside parentheses. It is evident from the table that the Bayesian method TSBNPSQR based on simultaneous quantile regression is considerably more accurate than the two-step non-Bayesian method TSQR based on nonparametric median regression. The higher accuracy is also reflected in the estimated median regression curves for both populations by these two methods, as shown in Figure 1.

The proposed method also gives estimated quantile regression curves at any quantile level as a by-product. In Fig. 2, we present the true and the estimated quantiles of prey and predator populations at the quantile levels $u = 0.05, 0.1, 0.2, 0.3, 0.4, 0.5, 0.6, 0.7, 0.8, 0.9, 0.95$ for $n = 500$ using the two-step approach for nonparametric simultaneous quantile regression with the data points generated from a gamma distribution. Note that for low quantile levels, the nonparametric simultaneous quantile regression method performs quite well. For high quantile levels, i.e., $u = 0.9$ and $u = 0.95$, as there are very few data points above 2.5 for both prey and predator populations, it is very difficult to estimate these high-level quantile functions.

4 Realdata Analysis

Bibby and Sørensen [4] introduced a hyperbolic diffusion model for stock prices, where the log-price is a diffusion process with coefficient depending on the instantaneous stock price in a particular way, plus a linear trend. We consider a simplified version of the model for the trend, ignoring the temporal correlation. The model can be described by the ordinary differential equation

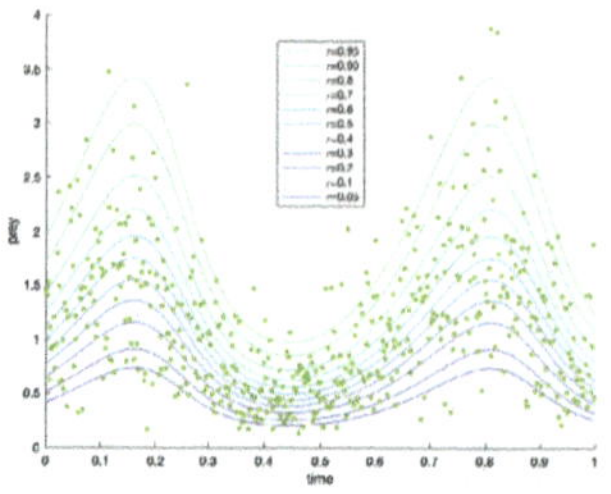

(a) True prey quantiles

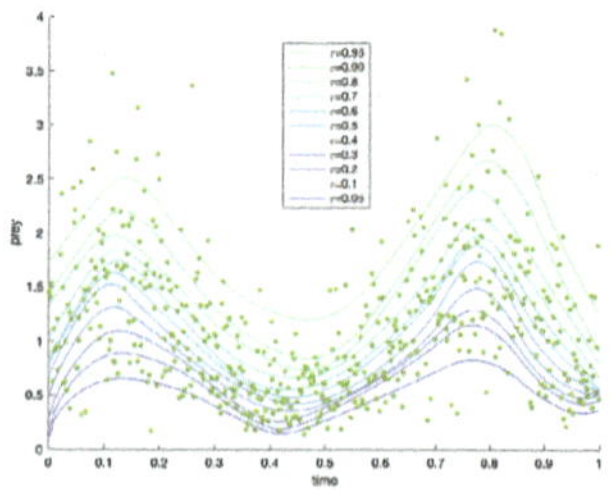

(b) Estimated prey quantiles

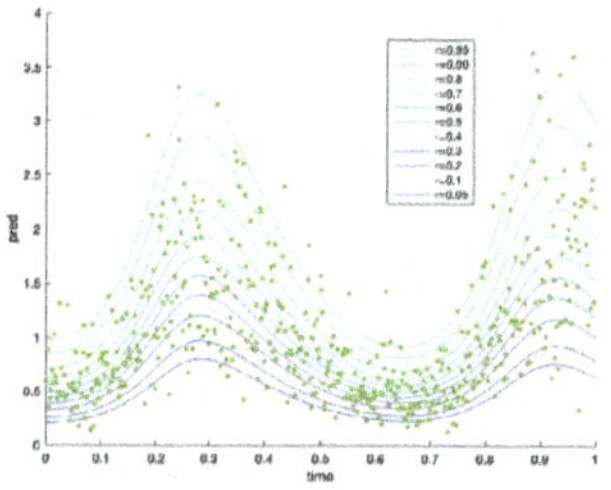

(c) True predator quantiles

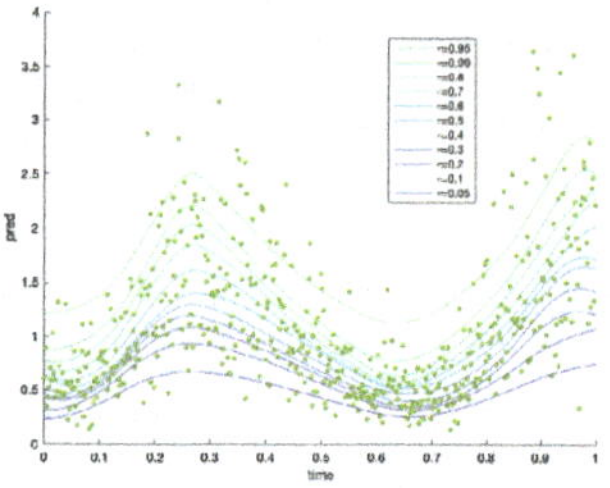

(d) Estimated predator quantiles

Fig. 2 True and estimated quantiles using NPSQR with sample size $n = 500$ at levels $u = 0.05, 0.1, 0.2, 0.3, 0.4, 0.5, 0.6, 0.7, 0.8, 0.9, 0.95$ of prey and predator populations

$$\frac{d}{dt} f_\theta(t) = \kappa + \frac{\sigma^2}{2} \exp\left\{\alpha\sqrt{\delta^2 + (f_\theta(t) - \kappa t - \mu)^2} - \beta(f_\theta(t) - \kappa t - \mu)\right\}$$

for the median regression function $f_\theta(t)$ involving unknown parameters $\theta = (\kappa, \sigma, \alpha, \delta, \mu, \beta)$. We apply the proposed methodology on this model to analyze the bi-weekly Walmart stock-price data during the period 1999–2011 obtained from Yahoo! Finance. We use the closing price, adjusted for both dividends and splits. First, the explanatory variable time is linearly transformed to the unit interval such that the years 1999 and 2011 are mapped to 0 and 1, respectively. To transform the stock prices into the unit interval, we use the log-normal distribution function $F(y) = \Phi\big((\log y - \mu)/\sigma\big)$, where Φ is the standard normal cumulative distribution function. We choose $\mu = 3.6812$, $\sigma = 0.1070$ to match the mean and variance of the stock prices. After transforming both explanatory and response variables into the unit interval, we use nonparametric simultaneous quantile regression method to estimate the simultaneous quantiles of the stock price S_t at time t. We start the MCMC procedure with a warm starting point using the GCDVSMS algorithm. We obtain 10000 posterior samples discarding the first 1000 samples as burn-in. The number of equidistant knots to be used for B-spline basis functions is selected using the AIC. After the quantile curves are estimated, the corresponding inverse transformations

(a) Walmart stock-price quantiles 1999–2011.

(b) Trace and density plots of the posterior of σ, α, δ and β.

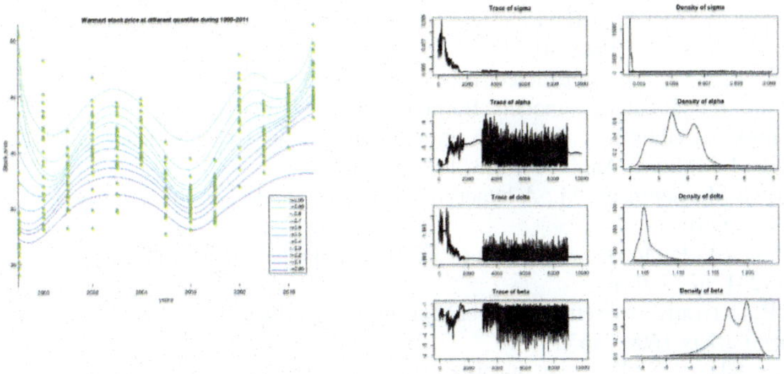

Fig. 3 Estimated quantiles of Walmart stock prices, and trace and density plots of posterior distributions of model parameters

are performed on the response and the explanatory variables before plotting them. In Fig. 3, we note that the upper quantiles of the stock prices have changed more dramatically over time compared to the lower quantiles. We note a periodic pattern in the quantiles at all levels $u = 0.8, 0.9, 0.95$.

To fit this median regression model, we apply the projection method based on the distance (5) with weight function $w(t) = t(1-t)$. Since the ordinary differential equation is a very complicated nonlinear function in parameters, like Bibby and Sørensen [4], we first estimate the slope κ and intercept μ parameters from the linear regression model $\log S_t = \mu + \kappa t$. For the Walmart stock price we found the value $\kappa = 0.0135$ and $\mu = -23.3375$. Then the optimization step is simplified to finding the location of the minimum of

$$\int \left| f'(t) - \kappa - \frac{\sigma^2}{2} \exp\left(\alpha\sqrt{\delta^2 + (f(t) - \kappa t - \mu)^2} - \beta(f(t) - \kappa t - \mu)\right)\right|^2 w(t)dt$$

with respect to $(\sigma, \alpha, \delta, \beta)$ only.

Based on the data, we find that the 95% credible interval for σ is $(0.0047, 0.0073)$, for α is $(4.4875, 6.8396)$, for δ is $(1.1840, 1.1949)$, and for β is $(-3.8412, -1.1496)$. Note that since the value of σ is close to zero, the squared diffusion coefficient $v^2(x) = \sigma^2 \exp\left\{\alpha\sqrt{\delta^2 + (x - \kappa t - \mu)^2} - \beta(x - \kappa t - \mu)\right\}$ is a very small number for all log-prices over the time period. Figure 3 also shows the trace plots and posterior densities of the parameters σ, α, δ and β assuring convergence of MCMC after the burn-in period.

References

1. Bhaumik, P., Ghosal, S.: Bayesian two-step estimation in differential equation models. Electron. J. Stat. **9**(2), 3124–3154 (2015)
2. Bhaumik, P., Ghosal, S.: Efficient Bayesian estimation and uncertainty quantification in ordinary differential equation models. Bernoulli **23**, 3537–3570 (2017)
3. Bhaumik, P., Ghosal, S.: Bayesian inference for higher-order ordinary differential equation models. J. Multivar. Anal. **157**, 103–114 (2017)
4. Bibby, B.M., Sørensen, M.: A hyperbolic diffusion model for stock prices. Financ. Stoch. **1**, 25–41 (1996)
5. Brunel, N.J.: Parameter estimation of ODE's via nonparametric estimators. Electron. J. Stat. **2**, 1242–1267 (2008)
6. Das, P.: Derivative-free efficient global optimization on high-dimensional simplex (2016). *arXiv preprint* arXiv:1604.08636
7. Das, P., Ghosal, S.: Bayesian quantile regression using random B-spline series prior. Comput. Stat. Data Anal. **109**, 121–143 (2017)
8. Das, P., Ghosal, S.: Bayesian non-parametric simultaneous quantile regression for complete and grid data. Comput. Stat. Data Anal. **127**, 172–186 (2018)
9. de Boor, C.: A Practical Guide to Splines. Springer-Verlag, New York (1978)
10. Ghosal, S., van der Vaart, A.: Fundamentals of Nonparametric Bayesian Inference. Cambridge University Press, Cambridge, UK (2017)
11. Koenker, R., Bassett Jr, G.: Regression quantiles. Econometrica J. Econ. Soc. 33–50 (1978)
12. Shen, W., Ghosal, S.: Adaptive Bayesian procedures using random series prior. Scandinavian J. Stat. **42**, 1194–1213 (2015)
13. Tan, Q. (2017). *Two-step Methods for Differential Equation Models.* Ph.D. Dissertation, North Carolina State University. Available for download at https://repository.lib.ncsu.edu/handle/1840.20/34722
14. Tokdar, S.T., Kadane, J.B.: Simultaneous linear quantile regression: a semiparametric Bayesian approach. Bayesian Anal. **7**(1), 51–72 (2012)
15. Varah, J.: A spline least squares method for numerical parameter estimation in differential equations. SIAM J. Sci. Stat. Comput. **3**, 28–46 (1982)

Predicting Plant Threat Based on Herbarium Data: Application to French Data

Jessica Tressou, Thomas Haevermans, and Liliane Bel

Abstract Evaluating formal threat criteria for every organism on earth is a tremendously resource-consuming task which will need many more years to accomplish at the actual rate. We propose here a method allowing for a faster and reproducible threat prediction for the 360,000+ known species of plants. Threat probabilities are estimated for each known plant species through the analysis of the data from the complete digitization of the largest herbarium in the world using machine learning algorithms, allowing for a major breakthrough in biodiversity conservation assessments worldwide. First, the full scientific names from the Paris herbarium database were matched against all the names from the international plant list using a text mining open-source search engine called Terrier. A series of statistics related to the accepted names of each plant were computed and served as predictors in a statistical learning algorithm with binary output. The training data was built based on the International Union for Conservation of Nature (IUCN) global Redlisting plants assessments. For each accepted name, the probability to be of least concern (LC, not threatened) was estimated with a confidence interval and a global misclassification rate of 20%. Results are presented on the world map and according to different plant traits.

Keywords Machine learning · Threatened species · Digitization · The plant list · Random uniform forest algorithm

---

J. Tressou (✉) · L. Bel
UMR MIA-Paris, Université Paris-Saclay, AgroParisTech, INRAE, 16 rue Claude Bernard 75231, Paris 05, France
e-mail: jessica.tressou@inrae.fr

L. Bel
e-mail: liliane.bel@agroparistech.fr

T. Haevermans
Institut de Systématique Evolution Biodiversité, MNHN, CNRS, EPHE, UA, SU, Muséum national d'histoire naturelle, CP39, 75231, Paris 05, France
e-mail: thomas.haevermans@mnhn.fr

M. La Rocca et al. (eds.), *Nonparametric Statistics*, Springer Proceedings in Mathematics & Statistics 339, https://doi.org/10.1007/978-3-030-57306-5_44

Introduction

In the last years, the French National Museum of Natural History (MNHN) started a huge work of digitalization of its herbarium, one of the largest collections in the world. Based on this and the increasing development of machine learning in all fields, botanists from the MNHN met the statisticians from INRAE to develop an algorithm allowing to predict plant threat and constitute a list of threatened species similar to what is currently performed in the IUCN Red List. These assessments require a lot of time and resources and many plants still have not been assessed (20,000 out of nearly 400,000 have been assessed).

We propose in this work an original approach combining the French Paris (P) herbarium data to international public data to classify plants according to their threat level, first in a binary model (threatened vs not threatened) and then in a multinomial model (merging some of the IUCN threat categories). A huge part of the work concerned the matching of the different databases and the construction of predictors based on the available data and the knowledge of what is actually used for performing red list assessments for the IUCN. Then based on the 20, 000 plants already classified by IUCN, a uniform random forest algorithm is trained to be able to predict the threat category of all 400,000 plants. The end goal of the present work is to provide a tool that can rapidly and at a less cost predict roughly the threat level for a large amount of plants so that it may help botanists prioritize which plant should be assessed in detail next. A side result is also the analysis of the features determining whether a plant is endangered or not.

The paper is organized as follows. First, we give a brief description of the available databases that were combined in the analysis. Then we describe the proposed methodology with a focus first on the matching between the different databases and then on the modeling approach based on random uniform forests. In the last section, we expose some of the results and discuss the perspectives of this work.

1 Data Description

The data used come from three main different sources: the data collected by the Herbarium of the National Museum of Natural History in Paris (MNHN, available at https://science.mnhn.fr/institution/mnhn/collection/p/item/search/form); international public data from The Plant List (TPL, http://www.theplantlist.org/), and previously assessed plants from the IUCN redlisting data (http://www.iucnredlist.org).

1.1 Herbarium

The specimens stored at Paris MNHN Herbarium were fully digitized constituting one of the largest collections in the world, see [5] for the construction of this huge database. For this study, we extracted the following information: taxonomic information (family, genus, species, names of authors), geographic sector information, and collected information when available (ISO code of the country where the plant was collected, year when collected). External data was used to add the area of each country with respect to its ISO code. The raw data is constituted of 6,104,130 records, associated with 5,318,001 physical distinct herbarium sheets. Each of these sheets is identified by a barcode. Each barcode is associated with at least one plant name (and up to 8 due to synonymy) and a geographic sector (ASI, AME, EUR, FRA, etc.). A total of 613,313 collections were described covering 1,463,754 of the records (24.0%).

1.2 TPL and International Databases

The Plant List is a working list of all known plant species. It aims to be comprehensive for species of Vascular plants (flowering plants, conifers, ferns, and their allies) and of Bryophytes (mosses and liverworts). It was created jointly by the Royal Botanic Gardens, Kew, and Missouri Botanical Garden. It provides the Accepted Latin name for most species, with links to all synonyms by which that species has been known. Around 20% of names are unresolved indicating that the data sources included provided no evidence or view as to whether the name should be treated as accepted or not, or there were conflicting opinions that could not be readily resolved. See http://www.theplantlist.org/ for summary statistics by the family of plants. Our extract from the TPL database contained 1,298,042 records: each record has an identifier, a scientific name (family, genus, species, authors), the associated accepted name (ANID in the following), and the year of publication. 393,585 names are recognized as accepted names, 356,106 if we narrow the database to vascular plants only. This database has been supplemented with geographical, climate, and plant life information from the Royal Botanical Gardens Kew, World Checklist of Selected Plant Families http://apps.kew.org/wcsp/, 684,477 records). Geographical information (9 continent codes, 53 region codes, 388 sub-region codes or "area" called TDWG code and used in Figure 1) of the collection site of the plant is available for 168,725 distinct plants (among which 130,726 have accepted names). Lifeform data was available for 126,730 plants (among which 113,264 have accepted names) and climate information for 136,783 plants (among which 122,346 have accepted names). The 258 described lifeforms were summarized into 25 lifeform binary criteria (such as phanerophyte, epiphyte, annual, climbing, hydrophyte, ...). Similarly, the 23 described climates were summarized into 5 climate binary criteria (tropical, aquatic, temperate, dry, altitude). To group plants at a more aggregated level than the family level (about 500 distinct categories), the order (around 70 categories) was considered as well as the "super order" (8 distinct categories: Gymnosperms, Magnoliids, Monocots, other Angiosperms, other Eudicots, Pteridophyta, Superasterids, Superrosids).

1.3 IUCN Redlist

The IUCN Red List consisted of 19,200 assessments that can be extracted by country or by taxon ranking plants as LC for *Least Concern* (28.3%), NT for *Near Threatened* (9.1%), VU for *Vulnerable* (27.0%), EN for *Endangered* (16.4%), CR for *Critically Endangered* (10.8%), EW for *Extinct in the Wild* (0.2%), EX for *Extinct* (0.5%), or DD for *Data Deficient* (7.7%). These 19,200 rankings correspond to 18,826 accepted names (all vascular plants): when several evaluations relate to the same plant in the sense of the accepted name, the "highest" ranking was retained, considering the following order LC > NT > VU > EN > CR > EW > EX > DD. In the following, the plants classified as DD are excluded from the training data, yielding a training sample of size 15,824.

2 Methods

The main idea of the proposed methodology was to predict the red list status of each plant based on training data (IUCN data) and the available information from the French herbarium and general information (TPL mainly). A first step to this approach is to match the different databases which all have taxonomic information but no common identifier. Then the information available in the Herbarium and TPL had to be summarized at the accepted name level, that is, each and every synonym of a plant will have the same red list status prediction. We first considered the binary problem of predicting whether a species is of least concern (LC) or not. A natural extension is to predict each of the nine statuses or at least to work with some groupings of these, isolating the three categories of endangered species that are CR, EN, and VU in a group.

2.1 Text Mining

The matching of the three main databases described in the previous section was done manually concerning IUCN and TPL and from an open-source search engine called Terrier [6] for the Herbarium and TPL. For each row of the databases, a "document" is created by concatenating the text (in lower case) of the family, genus, species, and different fields of authors. Then, a similarity score is calculated between each "Herbarium" document and the list of "TPL" documents that constitute our reference/dictionary. This allows us to identify for each record of the Herbarium the closest record in TPL. To ensure a certain efficiency of the procedure, the records of the herbarium with too many missing or indeterminate values were deleted, leaving 5,589,233 records to be matched to the TPL reference. The quality of this matching was evaluated by calculating the concordance rate of different fields. Some random checks were also performed by the botanists to ensure the number of errors resulting from this approach which remains low.

2.2 Dealing with Synonyms

When several names (synonymy) coexist for the same plant, one of these synonyms is retained as the accepted name of the plant that will serve as an identifier (ANID for "accepted name identifier"). Each TPL line is either an accepted name or a synonym pointing to an accepted name or an "unresolved" (i.e., the name has not been critically evaluated yet and is thus neither an accepted name nor a synonym). The 1,298,052 lines correspond to 393,585 separate ANIDs, (356,106 ANID excluding non-vascular plants). In the Herbarium, after matching the names to the TPL reference, the 5,589,233 records finally correspond to 167,891 ANID, (167,355 excluding non-vascular plants). For IUCN, the 19,200 lines correspond to 17,098 distinct ANIDs (15,824 excluding those classified as DD). Non-vascular plants are excluded from the analysis because they are absent from our learning base (IUCN).

2.3 Construction of the Predictors

Predictors were constructed by summarizing the available information at the accepted name id level. For example, for each ANID, the variable `N_LINE` counts the number of herbarium records related to the ANID, `N_CB` counts the number of barcodes linked to the ANID, `NB_SYN_SONNERAT` counts the number of synonyms linked to the ANID, `NB_SECTOR` counts the number of distinct geographical sectors, the number of occurrences in each sector being stored in the variables `ASI` for Asia, `AME` for America, `EUR` for Europe, `AFT` for Tropical and South Africa, `AFM` for Africa and Madagascar, `OCE` for Oceania, etc. `N_ISO` counts the number of distinct ISO codes. In TPL, in addition to the year of publication of ANID (`YEAR_TPL`), the minimum and maximum year of publication associated with the ANID via the dates of publication of the synonyms were calculated, as well as the difference between the two (`DELTA_YEAR_TPL`). The number of synonyms in TPL was also calculated for each ANID (`NB_SYN_TPL`) and used to compute the ratio of the number of synonyms in the Herbarium to the number of synonyms in TPL (`RATIO_SYN`). The number of distinct continents, regions, and areas (`N_CONTINENT`,`N_REGION`, `N_AREA`) from the checklist data were computed for each ANID including the synonyms or not (suffix `_ANID` added when synonyms are not included).

A total of 38 quantitative variables and 31 qualitative variables (`SUPER_ORDER`, 5 on climate, and 25 on lifeforms) were constructed following this principle. Other variables were not included in the model but constructed for the presentation of results such as the number of ANIDs associated with a TDWG code or ISO code, or the lifeform and climate most frequently associated with a given code. Due to the predictor construction process, a large number of data is missing, some are missing from the original database and some are inherently missing due to the fact, for example, that some ANID do not appear in the French herbarium.

2.4 Random Uniform Forests

Several approaches have been tested. The most classic approach is logistic regression (binary case) or multinomial regression (to classify into three or more categories). They are well known and very popular methods among botanists but they lack robustness when dealing with a high number of covariates or/and factors with many levels. Our choice then turned to a method based on regression trees [2] of the CART type that allows a non-parametric modeling of the link between predictors and response, and the interpretation of the decision rules in a graphical form. However, the simplest approaches in this family are generally too close to the training data and present a high risk of overlearning. Methods where individuals and/or variables are randomly resampled are more robust, hence the use of boosting [4] or random forests [1]. Missing values can be dealt with using imputation. We have retained uniform random forests because of their low sensitivity to tuning parameters, the possibility of including/comparing different methods for the imputation of missing values (FastImpute, AccurateImpute), and its native handling of categorical variables using a randomization mechanism at the node level (see [3] for details). Furthermore, the associated R package includes the calculation of the generalization error (OOB prediction for "out of bag"), and the graph showing the influence of the different predictors. It is referred to hereinafter as the RUF algorithm. The principle of this algorithm is to combine the responses of several regression trees, presenting a very low correlation that is obtained by randomly choosing the variables to be included in each tree and by choosing from the uniform distribution the cut-points which determine the branches of the tree. Each tree is grown on a random subsample of the training observations; the rest of them is used to evaluate the generalization error (OOB) similar to what cross-validation allows to do. The missing value imputation can either be performed within the R package by FastImpute (missing values are replaced with the median value of the observed) or AccurateImpute (after initialization with FastImpute, a RUF learning algorithm is run on the observed values of each variable using the remaining ones as predictors).

3 Main Results

Text mining

The text analysis of plant names allowed to determine that the Herbarium of Paris covers about 42.7% of the plant species in terms of accepted names, and even 47% if we exclude non-vascular plants.

Table 1 OOB prediction errors for different tuning parameters of the RUF model (`ntree` is the number of trees to grow, `mtry` is the number of variables randomly sampled with replacement as candidates at each split)

Missing imputation method	`ntree`	`mtry`	OOB error (%)	Time
Fast	100	69	19.8	2.5 min
Accurate	100	69	5.9	6.6 min
Fast	200	69	19.7	5.1 min
Fast	500	69	19.6	12.8 min
Fast	1000	69	19.7	37.6 min
Fast	50	69	20.5	1.3 min
Fast	100	50	20.3	1.9 min
Fast	100	100	19.9	3.2 min

Generalization error

We run the RUF algorithm using the default parameters with the 69 predictors (31 categorical variables). We obtained an OOB prediction error of 19.8% on the training dataset of size 15,824. This OOB prediction error was compared to the misclassification error obtained by cross-validation: from the 15,824 training observations, we built 40 test sets (sampled with replacement) of size 1,000 (or 5,000) and used the remaining observations to train the model. For each test set, the misclassification error is calculated: it varies from 17.9 to 22.3% for the 1,000 size, and from 19.1% to 21.3% for the 5,000 size. The OOB prediction error is therefore a good proxy of the generalization error.

Tuning the RUF algorithm

Table 1 illustrates how the OOB prediction error varies when modifying the tuning parameters that are the missing value imputation method, `ntree`, the number of trees to grow, `mtry`, the number of variables randomly sampled with replacement as candidates at each split, and the nested missing values treatment. Modifying `ntree` and `mtry` does not reduce the OOB error but can substantially increase the running time. Using accurateImpute rather than fastImpute reduces the OOB error (from 20 to 6%). However, further tests should be performed as the proportion of missing values is high and the risks of overlearning by accurately imputing missing values here are also high as a consequence.

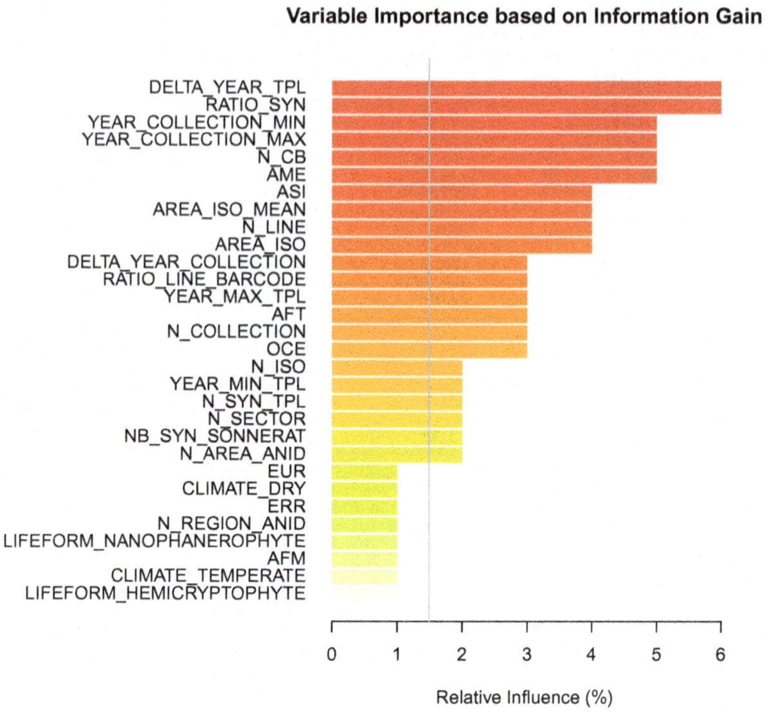

Fig. 1 Variable importance in predicting the binary response LC vs not LC. (see the construction of the predictors' section for the meaning of variables names)

Important variables

We chose to keep the simplest model with the default parameters of the RUF algorithm (Fast, `ntree`= 100, `mtry`= 69) and the full set of variables (69 in total). Figure. 1 lists the most influential variables for the prediction.

We observe that the variables that are the most influential are `DELTA_YEAR_TPL` and `RATIO_SYN` as well as the collection year (min or max) or the number of herbarium sheets stored at the herbarium (`N_CB`). Graphics showing the links between these most influential variables and the response probability to be LC were drawn (not shown here). For example, we observed that the larger `DELTA_YEAR_TPL`, the larger the probability to be LC, meaning that plants with synonyms having very different dates of publication in TPL tend to not be threatened. The important conclusion is that, as expected, the more specimens of a plant were collected, the greater the probability to be of least concern, but this relationship is highly nonlinear as some point rare plants tend to be specifically searched for while more frequent plants may be ignored. These results will be further detailed and commented on in a publication aiming at the botanists' audience.

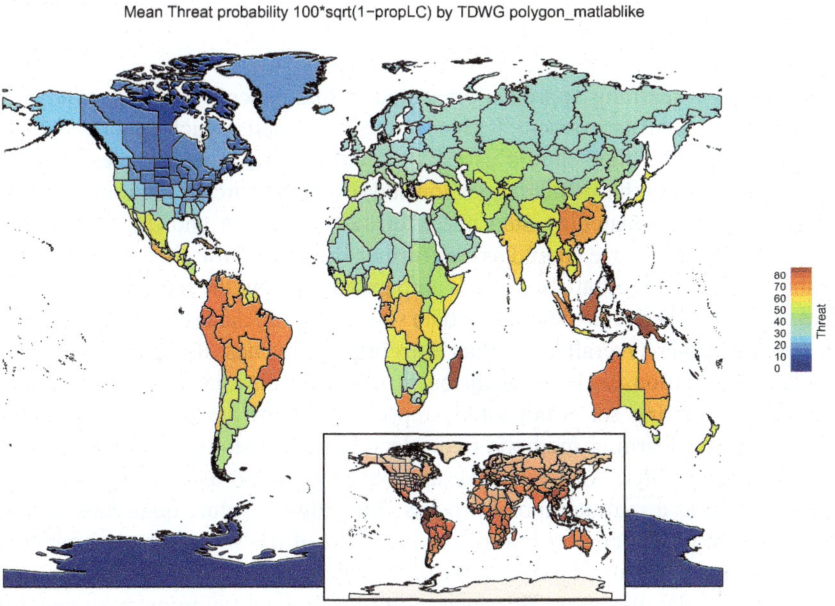

Fig. 2 Threat map (main map) with enclosed richness map. The threat is calculated as the square root of the mean probability to be "not LC" among plants within each polygon; the richness is the number of plants per polygon in our database

Visualization of the results

For each of the 356,106 vascular plants of TPL, we can use this model to predict whether the plant is LC or non-LC as well as the probability and associated confidence interval, based on the distribution of the votes of the different trees of the forest. We also trained the model with a three class response (LC-NT, CR-EN-VU, EX-EW), yielding an OOB prediction error of 26.5% with the default tuning parameters (and a running time of 2.9 min).

Globally, in the binary model, we find a mean probability to be of least concern (LC) of 29.1%, ranging from 19.4% for Magnoliids to 39.8% for Monocots. In the three class model, we find a mean probability to be LC/NT of 38.2%, ranging from 28.6% for Magnoliids to 50.1% for Gymnosperms. More detailed results will be published soon at the family level or at the ANID level.

By aggregating the results at the level of the TDWG codes (based only on the 130,408 ANID for which the information was available), we obtain in Fig. 2 a threat map (main map), the red zones containing the most endangered species, to be linked to the map of the number of species per polygon (small map called richness map).

4 Perspectives

The statistical learning approach presented in this paper is quite innovative in the field of plant threat assessment. It gives interesting results which could help botanists choose what plant they should assess in detail next. It is nevertheless only a first attempt at tackling this difficult question and several research directions merit further study. The initial matching step needs further validation and due to the way the predictors are built, we should assess further the role of missing value imputation. Although descriptive statistics were compared to rule out the representativeness bias that could exist between the training data set and the full data set, this could definitely be studied further. Overall, more than the representativeness itself it has to be checked whether the relationship between the outcome and the covariates is still well estimated even if the training set is not totally representative of the whole set. In addition, other machine learning methods (e.g., deep learning) should be tested to confirm the obtained results. An alternative approach would be a direct modeling of the phenomenon as a spatiotemporal process, allowing to capture quantities such as the area covered by the convex hull of the locations of the specimens of a plant or the evolution of the density of points along time, which are some of the main determinants of the IUCN classification. This type of approach would eliminate the aggregating step in the data preparation.

Acknowledgements All the datasets used are freely available online and an upload of TPL and the checklist was made possible through a Data Transfer Agreement between RBG Kew and Paris MNHN. The authors would like to thank Dr. Alan Paton, head of Science (Collections) at Royal Botanic Gardens, Kew for providing the TPL data and for his insights about the use of the data in this work.

References

1. Breiman, L.: Random forests. Mach. Learn. **45**(1), 5–32 (2001)
2. Breiman, L., Friedman, J., Olshen, R., Stone, C.: Classification and regression trees (chapman y hall, eds.). Monterey, CA, EE. UU.: Wadsworth International Group (1984)
3. Ciss, S.: randomuniformforest-package: random uniform forests for classification, regression and unsupervised learning (2015)
4. Friedman, J., Hastie, T., Tibshirani, R., et al.: Additive logistic regression: a statistical view of boosting (with discussion and a rejoinder by the authors). Ann. Stat. **28**(2), 337–407 (2000)
5. Le Bras, G., Pignal, M., Jeanson, M.L., Muller, S., Aupic, C., Carré, B., Flament, G., Gaudeul, M., Gonçalves, C., Invernón, V.R., et al.: The french muséum national d'histoire naturelle vascular plant herbarium collection dataset. Sci. Data **4**, 170,016 (2017)
6. Macdonald, C., McCreadie, R., Santos, R., Ounis, I.: From puppy to maturity: Experiences in developing terrier. Open Source Inf. Retr. **60** (2012)

Monte Carlo Permutation Tests for Assessing Spatial Dependence at Different Scales

Craig Wang and Reinhard Furrer

Abstract Spatially dependent residuals arise as a result of missing or misspecified spatial variables in a model. Such dependence is observed in different areas, including environmental, epidemiological, social and economic studies. It is crucial to take the dependence into modelling consideration to avoid spurious associations between variables of interest or to avoid wrong inferential conclusions due to underestimated uncertainties. An insight about the scales at which spatial dependence exist can help to comprehend the underlying physical process and to select suitable spatial interpolation methods. In this paper, we propose two Monte Carlo permutation tests to (1) assess the existence of overall spatial dependence and (2) assess spatial dependence at small scales, respectively. A p-value combination method is used to improve statistical power of the tests. We conduct a simulation study to reveal the advantages of our proposed methods in terms of type I error rate and statistical power. The tests are implemented in an open-source R package `variosig`.

Keywords Spatial data · Combining p-values · Empirical Brown's method · Variogram · Nonparametric

1 Introduction

Independent and identically distributed residuals are a key assumption in many statistical analysis models. When analyzing spatial, i.e. geolocated data, this assumption is violated if one fails to account for the existence of spatial dependence in the modelling components. Such violation can lead to biased parameter estimates and spurious associations between the dependent variable and its covariates. Therefore, it

C. Wang (✉) · R. Furrer
Department of Mathematics, University of Zurich, Zurich, Switzerland
e-mail: craig.wang@math.uzh.ch

R. Furrer
e-mail: reinhard.furrer@math.uzh.ch

R. Furrer
Department of Computational Science, University of Zurich, Zurich, Switzerland

M. La Rocca et al. (eds.), *Nonparametric Statistics*, Springer Proceedings in Mathematics & Statistics 339, https://doi.org/10.1007/978-3-030-57306-5_45

is important to take spatial dependence into modelling consideration when it exists. Spatially dependent data arise in many research domains, for example in spatial epidemiology where researchers are interested in the relationship between disease prevalence and risk factors [9, 23], in economics where it is of interest to identify regions of housing externalities [15] or in ecological studies where species distribution needs to be mapped [10, 17]. The scales of spatial dependence also play an important role in understanding the underlying physical and biological processes [10, 17].

The assessment of spatial dependence and its scales is often done by plotting and modelling the empirical semi-variogram estimates, based on extracted residuals after first-stage statistical modelling. If the empirical semi-variogram indicates spatial dependence, then the model needs to be adjusted to account for the remaining dependence. However, semi-variogram estimates can be sensitive to outliers, choice of distance binning and sampling design. Several robust variogram estimators [4, 7] and methods to quantify uncertainty of variogram estimates [3, 5, 12] are available. One can use a maximum likelihood estimator and its uncertainty estimate to assess the spatial dependence, or use parametric bootstrap [16] by firstly fitting a variogram model and simulate new values based on the estimated model to obtain additional variogram estimates hence an uncertainty estimate. However, both approaches require a pre-defined variogram model and a sufficient sample size. An alternative approach to assess the existence of spatial dependence is to use a Monte Carlo permutation test. The permutation test is a nonparametric approach that does not make any assumptions on the distribution of residuals. Walker et al. [21] introduced a permutation test to permute the residual values across spatial locations, in order to simulate under the null hypothesis of complete spatial randomness. Diblasi and Bowman [5] compared the performance of a permutation test on their proposed test statistics based on the assumption of normally distributed residuals. Viladomat et al. [20] used a permutation method in a two-step procedure to test the correlation of two variables when they both exhibit spatial dependence.

In this paper, we propose two Monte Carlo permutation tests to assess the existence of overall spatial dependence and spatial dependence specifically at small scales, respectively. We demonstrate that our proposed methods have more accurate type I error rate compared to the standard permutation test in [21] and achieve good statistical power at the same time.

2 Assessing Spatial Dependence at Different Scales

We assume that trend components have been taken out of data, so we work with residuals. Let $Y(s) : s \in D \subseteq R^2$ be a zero-mean second-order stationary spatial process that is observed at coordinates s, then the semi-variogram is defined as

$$\gamma(h) = \frac{1}{2}\text{Var}\left(Y(s+h) - Y(s)\right), \tag{1}$$

where h is the spatial lag, with an estimator as the empirical semi-variogram

$$\hat{\gamma}(h) = \frac{1}{2|N(h)|} \sum_{N(h)} (Y(s_i) - Y(s_j))^2, \tag{2}$$

where $N(h)$ denotes the set of all pairs whose spatial locations are separated by distance between $[h - \delta, h + \delta]$. In practice, different distance binnings $h_1, \ldots, h_k$ are used to obtain empirical semi-variogram estimates $\hat{\gamma}(h_d)$ for $d = 1, \ldots, k$.

2.1 *Permutation Test for Overall Spatial Dependence*

The permutation test for overall spatial dependence has been described in [21]. Under the null hypothesis of complete spatial randomness, residuals are permuted randomly over all locations. With such permutation, the spatial dependence at any scale is destroyed. There are $n!$ number of possible permutations for n locations, hence the Monte Carlo method with a fixed number of permutations is often used to save computation time. Pointwise p-values based on semi-variogram estimates can be computed as

$$p_d = \frac{1}{n_{mc}} \sum_{i=1}^{n_{mc}} I_{\{\hat{\gamma}_i(h_d) \leq \hat{\gamma}(h_d)\}} \tag{3}$$

for the dth distance binning, where n_{mc} is the number of Monte Carlo iterations and $\hat{\gamma}_i(h_d)$ is the dth semi-variogram estimates from the ith permuted samples. Walker et al. [21] compared the p-values in Eq. (3) with a type I error rate α, and deemed that the null hypothesis is rejected if any p-value is below α. This approach implicitly used the p-value combination method proposed in [19], which takes the overall p-value $\Psi_T = \min(p_d)$ and compares it with α. When the evidence of spatial dependence is relatively strong, the p-values tend to be small. In such cases, the null hypothesis will be rejected as long as one of the p-values is smaller than α. However, when none of the p-values are smaller than α under settings of weak spatial dependence, the rejection region of using overall p-value Ψ_T is smaller than using other p-value combination methods (e.g. [6, 11]). This will lead to smaller statistical power. In addition, using the minimum p-value can inflate the type I error which leads to spurious spatial dependence. To mitigate these problems, we propose a modified permutation test for overall spatial dependence.

2.2 Modified Permutation Test for Overall Spatial Dependence

We describe a modified version of the permutation test, which still uses p-values obtained via Monte Carlo permutations but combines them more appropriately. Under the null hypothesis of no spatial dependence, p-values are uniformly distributed. Fisher's method [6] was proposed to combine p-values based on independent test statistics into a single χ^2-distributed test statistic. If we assume the test statistics $\gamma(h_d)$ are mutually independent, then the Fisher's method states

$$\Psi_F = -2\sum_{d=1}^{k} \log(P_d) \sim \chi^2_{2k}. \tag{4}$$

However, the semi-variogram estimates are not mutually independent since each estimate is based on an overlapping set of residual values from the same data. Failing to account for positive correlation between test statistics tends to give under-estimated combined p-value. Conversely, failing to account for negatively correlated test statistics gives an over-estimated combined p-value.

We propose a modified version of the permutation test for overall spatial dependence using the empirical Brown's method [13], an extension of the Fisher's method, for combining dependent p-values from multivariate normal test statistics. Under the null hypothesis of no spatial dependence, we assume that the residual value at each location is normally distributed as $Y(s) \sim N(0, \sigma^2)$. For locations $s_1, \ldots, s_n$, let $\mathbf{Y} = (Y(s_1), \ldots, Y(s_n))^\top$. The dth semi-variogram then follows a scaled χ^2-distribution with r_d degrees of freedom, i.e.

$$\gamma(h_d) = \frac{1}{N(h)}\mathbf{Y}^\top \mathbf{A}_d \mathbf{Y} \sim \chi^2_{r_d}/N(h), \tag{5}$$

if the matrix $\mathbf{A}_d$ is idempotent and has rank r_d [1]. The matrix $\mathbf{A}_d$ is the spatial design matrix of the data at lag d [7]. For non-gridded locations, the matrix is close to idempotent if the number of semi-variogram estimate is not too small, which yields Eq. (5) as a good approximation. For moderate to large ranks, $\chi^2_{r_d}/N(h) \xrightarrow{d} N(r_d/N(h), 2r_d/N(h)^2)$. Therefore, we can approximate the vector of test statistics $[\gamma(h_1), \gamma(h_2), \ldots, \gamma(h_k)]^\top$ with a multivariate normal distribution. The Brown's method [2] allows us to derive a scaled χ^2-distribution to replace χ^2_{2k} from the Fisher's method. The overall test statistic stays as Ψ_F; however, the distribution under the null hypothesis becomes a scaled chi-squared distribution $c\chi^2_{2f}$, with

$$f = \frac{2k^2}{2k+s}, \qquad c = 1 + \frac{s}{2k}, \quad \text{where}$$
$$s = \sum_{i<j} \mathrm{cov}(-2\log P_i, -2\log P_j). \tag{6}$$

The evaluation of the covariance terms in Eq. (6) requires numerical integration, and can be approximated using either a polynomial regression based on the correlation between test statistics [8] or the empirical Brown's method to obtain approximated samples of P_i and P_j [13]. The latter has been shown to be more robust compared to polynomial approximation when there is deviation from normality in the test statistics. In our modified permutation test, we use the empirical Brown's method to combine the p-values generated by Monte Carlo permutations into an overall test statistic and compare it with $c\chi^2_{2f}$.

2.3 Permutation Test for Spatial Dependence at Small Scales

Sometimes the existence of scale-specific spatial dependence is a more meaningful hypothesis to test against. We propose a permutation test to permute the residuals in a way such that only small-scale dependence is destroyed.

Instead of randomly permuting the residuals over all spatial locations, we first apply a clustering algorithm on the locations to divide them into small clusters. The clustering algorithm should not result in clusters that have a high variance in size. Popular algorithms such as k-means or hierarchical clustering can be used, or simply hex-binning when the locations are evenly distributed over the spatial domain. After clusters are defined, the residuals are randomly permuted only within each cluster. The null hypothesis then concerns only the first few semi-variogram estimates at small scale. The clustering algorithm should be tuned depending on the scale of interest. Since there are still correlations among the pointwise p-values, we use the empirical Brown's method to combine them to get an overall test statistic.

Figure 1 shows semi-variogram estimates based on a simulated set of residuals in 200 locations in a unit square and an exponential covariance function $\gamma(h) = 0.5\exp(-0.05h) + 0.5$. In this case, spatial dependence only exists at small scales. The 95% confidence band shown in light blue is based on random permutation where all of the spatial dependence is destroyed. The 95% confidence band in red is based on cluster permutation where only the small-scale spatial dependence is destroyed. The latter allows more powerful hypothesis testing focusing only on a small scale, as we will show in the simulation results.

3 Simulation Study

We conduct a simulation study to compare the permutation tests in terms of type I error rate and statistical power of detecting spatial dependence. The null hypothesis of random permutation is that there is no spatial dependence, while the null hypothesis of cluster permutation is that there is no spatial dependence at small scales.

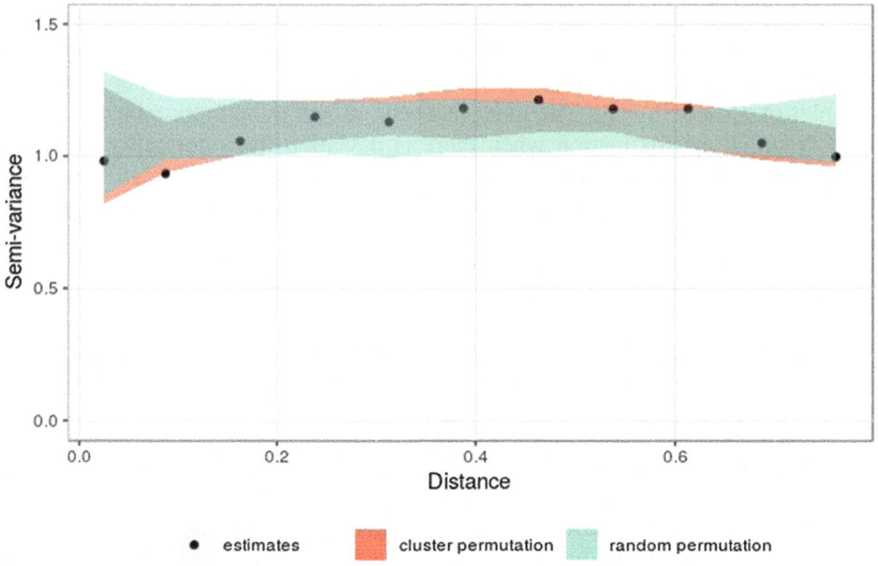

Fig. 1 Semi-variogram estimates based on simulated residuals. Ribbons indicate 95% confidence band of permutation samples. The original and modified permutation test for overall spatial dependence provided p-values of 0.001 and 0.078, respectively, while the cluster permutation method provided an overall p-value of 0.027

3.1 Simulation Setup

We simulate a Gaussian process on n uniformly distributed locations within a $[0, 1] \times [0, 1]$ domain. Without loss of generality, we assume the Gaussian process has an exponential covariance function from the Matérn family, i.e. $\mathrm{Cov}(Y(s_i), Y(s_j)) = \sigma^2 \exp(||s_i - s_j||/\phi) + \tau^2$, where $||s_i - s_j||$ denotes the Euclidean distance between locations s_i and s_j. The magnitude of spatial covariance is represented by σ^2, the magnitude of noise is τ^2. The spatial range ϕ controls the covariance decay over distance, which corresponds to an effective range of 3ϕ for the exponential covariance function. Hence, 95% of the spatial correlation disappears at distance 3ϕ.

We simulate a total of 10 different spatial dependence structures each with 5 different sample sizes, where $n = \{50, 100, 200, 300, 500\}$, $\phi = \{0.05, 0.2\}$ and $\sigma^2/\tau^2 = \{0/0.5, 0.25/1, 0.5/1, 0.5/0.5, 1/0.5\}$. Different σ^2/τ^2 represents differing strengths of spatial dependence. We choose $n_{mc} = 1000$ and repeat each scenario 1000 times. For the cluster permutation, we use k-means clustering with 5 clusters when $n = 50$, and with 10 clusters for all other sample sizes. The null hypothesis is that the first two semi-variogram estimates are 0, which corresponds to no spatial dependence at a distance smaller than approximately 0.1. All the simulation results are obtained in R version 3.5 [14].

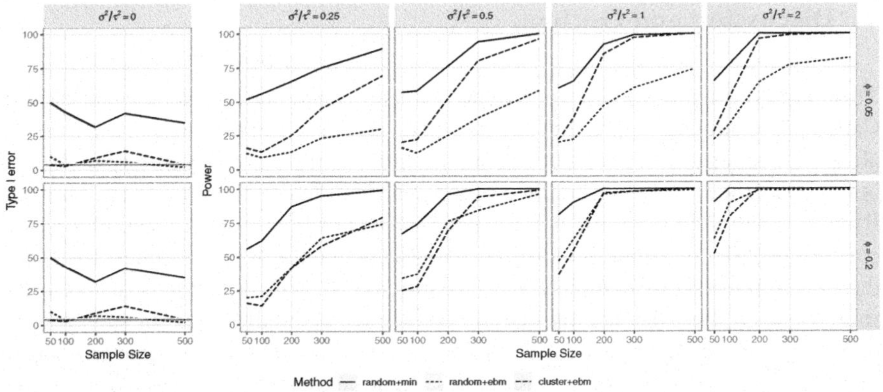

Fig. 2 Simulation results showing type I error rate (first column) and statistical power (other columns) of three different permutation tests for different scenarios. The horizontal red line in the first column shows the nominal type I error rate at 5%

3.2 Simulation Results

Figure 2 shows the simulation results. The permutation test for overall spatial dependence (denoted as random+min) using minimum p-value is shown in solid lines; the modified permutation test for overall spatial dependence with empirical Brown's method (denoted as random+ebm) is shown in dotted lines; the permutation test for spatial dependence at small scales with empirical Brown's method (denoted as cluster+ebm) is shown in dashed lines.

The first permutation test has the highest power, since it uses the minimum p-value across semi-variogram estimates without any adjustment. This makes the null hypothesis easy to reject. As a result of this, the type I error rate is inflated to around 37% at a nominal level of 5%. After combining the individual p-values using empirical Brown's method, our proposed modified permutation test obtained close-to-nominal type I error rate. This comes at a cost of losing some statistical power. When the main interest is to test the existence of spatial dependence at small scale, the clustering-based permutation test boosts the statistical power compared to the modified permutation test while maintaining the correct type I error rate. It can be observed from the first row, where spatial dependence exists at small scales with $\phi = 0.05$, the power of the clustering-based permutation test is higher than the modified permutation test.

4 Discussion and Outlook

This paper presents two new approaches of testing spatial dependence using Monte Carlo permutation tests. The first is a modified version of the permutation test for overall spatial dependence. Instead of using the minimum p-values at different distance binnings as the overall p-value, we propose a modified version that uses empirical Brown's method to combine the p-values into a new test statistic. The second is a clustering-based permutation test for spatial dependence at small scales. Instead of the null hypothesis of complete spatial randomness, sometimes it is of interest to focus only on the existence of spatial dependence at small scales. In such a situation, our proposed approach can improve the statistical power compared to using an overall permutation test. Both approaches are implemented in the open-source software package `variosig` [20] available on the Comprehensive R Archive Network (CRAN, https://cloud.r-project.org/).

Our simulation study shows that the type I error rate is maintained by the modified permutation test for overall spatial dependence and the clustering-based permutation test for spatial dependence at small scales. The clustering-based permutation test has increased statistical power compared to the modified permutation test when the sample size is not too small. When the interest is spatial dependence at small scales, the clustering-based permutation test should be used.

In addition to permutation tests, our proposed clustering-based permutation method can also be used in conjunction with a functional boxplot [18] to obtain a visual inspection of the spatial dependence at small scales. The permutation tests can also be applied to large spatial datasets, since it is computationally efficient and can be tuned using the number of permutations. Finally, the result of our permutation tests can help to inform about subsequent analysis such as spatial interpolation, scale decomposition and regression modelling.

Acknowledgements This work was supported by the Swiss National Science Foundation (grant no. 175529).

References

1. Anderson, T.W.: An Introduction to Multivariate Statistical Analysis, vol. 2. Wiley, New York (1958)
2. Brown, M.B.: A method for combining non-independent, one-sided tests of significance. Biometrics **31**(4), 987–992 (1975)
3. Clark, R.G., Allingham, S.: Robust resampling confidence intervals for empirical variograms. Math. Geosci. **43**(2), 243–259 (2011)
4. Cressie, N., Hawkins, D.M.: Robust estimation of the variogram: I. J. Int. Assoc. Math. Geol. **12**(2), 115–125 (1980)
5. Diblasi, A., Bowman, A.W.: On the use of the variogram in checking for independence in spatial data. Biometrics **57**(1), 211–218 (2001)
6. Fisher, R.A.: Statistical Methods for Research Workers, 4th edn. Oliver & Boyd (1932)
7. Genton, M.G.: Highly robust variogram estimation. Math. Geol. **30**(2), 213–221 (1998)

8. Kost, J.T., McDermott, M.P.: Combining dependent P-values. Stat. Probab. Lett. **60**(2), 183–190 (2002)
9. Lee Elizabeth, C., Asher Jason, M., Sandra, Goldlust., Kraemer John, D., Lawson Andrew, B., Shweta, Bansal: Mind the scales: harnessing spatial big data for infectious disease surveillance and inference. J. Infect. Dis. **214**, S409–S413 (2016)
10. Leiterer, R., Furrer, R., Schaepman, M.E., Morsdorf, F.: Forest canopy-structure characterization: a data-driven approach. Forest Ecol. Manag. **358**, 48–61 (2015)
11. Liptak, T.: On the combination of independent tests. Magyar Tudomanyos Akademia Matematikai Kutato Intezenek Kozlomenyei **3**, 127–141 (1958)
12. Marchant, B.P., Lark, R.M.: Estimating variogram uncertainty. Math. Geol. **36**(8), 867–898 (2004)
13. Poole, W., Gibbs, D.L., Shmulevich, I., Bernard, B., Knijnenburg, T.A.: Combining dependent P-values with an empirical adaptation of Brown's method. Bioinformatics **32**(17), 430–436 (2016)
14. Core Team, R.: R: A Language and Environment for Statistical Computing. R Foundation for Statistical Computing, Vienna, Austria (2018)
15. Redding, S.J., Rossi-Hansberg, E.: Quantitative spatial economics. Ann. Rev. Econ. **9**(1), 21–58 (2017)
16. Ribeiro Jr., P.J., Diggle, P.J.: geoR: a package for geostatistical analysis. R News **1**(2), 11–15 (2001)
17. Schneider, F.D., Felix, M., Bernhard, S., Petchey, O.L., Andreas, H., Schimel, D.S., Schaepman, M.E.: Mapping functional diversity from remotely sensed morphological and physiological forest traits. Nat. Commun. **8**(1), 1441 (2017)
18. Sun, Y., Genton, M.G.: Functional boxplots. J. Comput. Graph. Stat. **20**(2), 316–334 (2011)
19. Tippett, L.H.C.: Methods of Statistics. Williams Norgate, London (1931)
20. Júlia, V., Rahul, M., Alex, M., McCauley Douglas, J., Trevor, H.: Assessing the significance of global and local correlations under spatial autocorrelation: a nonparametric approach. Biometrics **70**(2), 409–418 (2014)
21. Walker, D.D., Loftis, J.C., Mielke, J.P.W.: Permutation methods for determining the significance of spatial dependence. Math. Geol. **29**(8), 1011–1024 (1997)
22. Wang, C., Furrer, R.: Variosig: Spatial dependence based on empirical variograms. R package version 0.3 (2018). https://CRAN.R-project.org/package=variosig
23. Wang, C., Puhan, M.A., Furrer, R.: Generalized spatial fusion model framework for joint analysis of point and areal data. Spatial Stat. **23**, 72–90 (2018)

Introduction to Independent Counterfactuals

Marcin Wolski

Abstract The aim of this contribution is to introduce the idea of independent counterfactuals. The technique allows to construct a counterfactual random variable which is independent from a set of given covariates, but it follows the same distribution as the original outcome. The framework is fully nonparametric, and under error exogeneity condition the counterfactuals have causal interpretation. On an example of a stylized linear process, I demonstrate the main mechanisms behind the method. The finite-sample properties are further tested in a simulation experiment.

Keywords Statistical independence · Probability theory · Random variable · Counterfactual

1 Introduction

Estimation of counterfactual designs has become a focal point for policymakers and practitioners in the fields of policy evaluation and impact assessment. Counterfactual distributions are an important landmark in the methodology, as they allow to measure not only average effects but, under some regularity conditions, they also capture the relationship for any point across the distribution of interest [1].

In the context of a counterfactual analysis, one is interested in approximating the dynamics of an outcome variable Y under a new, possibly unobserved, scenario. Typically, the construction of such a scenario assumes a shift of a set of covariates from X to, say, X'. For instance, a policymaker may want to investigate the effects of a tariff change on local food prices where the relevant covariates (taxes, fees or other policy instruments) increase or decrease by some amount.

The vast majority of counterfactual scenarios are user-designed, suffering from an over-simplification and potential model misspecification biases. Nevertheless, the recent advances in counterfactual distributions aim at providing possibly assumption-free inference techniques. [1] offers a complete toolbox to study counterfactual dis-

M. Wolski (✉)
European Investment Bank, 98-100 Boulevard Konrad Adenauer, 2950 Luxembourg, Luxembourg
e-mail: M.Wolski@eib.org

M. La Rocca et al. (eds.), *Nonparametric Statistics*, Springer Proceedings
in Mathematics & Statistics 339, https://doi.org/10.1007/978-3-030-57306-5_46

tributions through a prism of regression methods. [5] extends the approach to a fully nonparametric setup and demonstrates that nonparametric estimation has superior Mean Squared Error (MSE) performance in the case of (functional) model misspecification. [6] further extends the nonparametric approach to cover partial distributional effects.

Capitalizing on [8], I propose an alternative identification strategy which defines the counterfactual scenario as independent from a given set of covariates. Using an example from above, a policymaker may be interested in approximating the behaviour of food prices under no policy intervention, exemplifying the overall distortions created by relevant taxes or fees. In this simple case, one would consider independent counterfactuals as dropping the entire policy instrument rather than estimating a counterfactual distribution of food prices at a zero tax rate. Setting a covariate to zero does not have to uniquely identify the independence criterion. If the taxation becomes effective only above some minimum threshold, there may be multiple choices for the counterfactual designs. Similarly, the true relation between the outcome and the covariates may be actually undefined, or not directly interpretable, for zero-valued arguments. In such cases, independent counterfactuals offer an attractive alternative to a standard toolkit.

The framework requires to take a somehow broader perspective on the interpretation of counterfactuals. More specifically, it asks what would be the realization of an outcome variable for which there would be no evidence against the independence condition given the realizations of the covariates. As such, the distribution of the counterfactual coincides with the distribution of the observed variable, spanning over the same information set, but the dependence link versus the covariates is removed.

The framework has desired asymptotic properties, allowing to apply standard statistical inference techniques. It also advertises the use of nonparametric methods, utilizing a smooth version of kernel density/distribution estimates. This, in fact, turns out to generate substantial efficiency gains over the step-wise estimators [8].

The purpose of this contribution is to offer the basic concepts behind independent counterfactual random variables. The extended description of the framework, covering also an idea of conditionally independent counterfactuals, together with an extensive numerical exercise and empirical study, is offered by [8]. Section 2 introduces the methodology, which is further illustrated numerically and compared against the standard linear framework in Sect. 3. A brief numerical study is described in Sect. 4. Finally, Sect. 5 concludes.

2 Framework

Assume two random variables $Y \in \mathbb{R}$ and $X \in \mathbb{R}^{d_X}$, where $d_X \geq 1$, with a joint Cumulative Distribution Function (CDF) denoted by $F_{Y,X}(y, x)$, which is r-times differentiable and strictly monotonic.

Filtering out the effects between X and Y means constructing a counterfactual random variable $Y' =^D Y$ that is independent of X. (Clearly, in case, Y and X are independent, Y' would be simply equal to Y.)

In terms of CDFs, one can write the independence condition as

$$F_{Y'|X}(y|x) = F_Y(y) \tag{1}$$

for all y and x.

The random variable Y' can be obtained directly from Eq. (1) by assuming that, for any point along the X marginal, there is an increasing functional ϕ, such that $Y' = \phi(Y, X)$, which is invertible in Y for all x in the support of X, for which Eq. (1) holds. The realizations of the counterfactual random variable Y' are given by $y' = \phi(y, x)$. [8] shows that Eq. (1) is satisfied by

$$Y' = F_Y^{-1}(F_{Y|X}(Y|x)), \tag{2}$$

where $F_Y^{-1}(q) = \inf\{y : F_Y(y) \geq q\}$ is the quantile function of Y, under the assumption that F_Y is invertible around the argument. The invertability assumption is satisfied by the monotonicity of $F_Y(y)$, which also guarantees that the relation is uniquely identified for any y and x.[1]

The relation between Eqs. (2) and (1) follows from

$$F_{Y'|X}(y|x) = P(\phi(Y, X) \leq y|X = x) = P(Y \leq \phi^{-1}(y, X)|X = x) = F_{Y|X}(\phi^{-1}(y, x)|x),$$

which makes $\phi^{-1}(y, x) = F_{Y|X}^{-1}(F_Y(y)|x)$, or equivalently $\phi(y, x) = F_Y^{-1}(F_{Y|X}(y|x))$, under the assumptions outlined above.

For the moment, the setup is designed for real-valued Y. In principle, the framework may be extended to multivariate outcome variables, under additional regularity conditions on the corresponding CDF and conditional CDF. This topic is, however, beyond the scope of this manuscript.

2.1 Estimation

A major challenge in estimating the function in Eq. (2) results from its nested structure. [8] provides a set of necessary conditions under which the kernel-based estimator of Eq. (2) is asymptotically tight. In fact, the crucial condition is the Donsker property of the quantile and conditional CDF estimators, respectively.

---

[1]One can define the independence condition in Eq. (1) in terms of PDFs. However, even though Eq. (2) would still satisfy such a condition, it would not be a unique solution to the PDF condition for some processes.

In the setup below I take that Y is univariate and X is potentially multivariate with $d_X \geq 1$. The kernel CDF and conditional CDF estimators are given by[2]

$$\hat{F}_Y(y) = n^{-1} \sum_{i=1}^{n} \bar{K}_{\mathbf{H}_0^Y}(y - Y_i), \quad (3)$$

and

$$\hat{F}_{Y|X}(y|x) = \frac{\sum_{i=1}^{n} \bar{K}_{\mathbf{H}_0^{Y|X}}(y - Y_i) K_{\mathbf{H}^{Y|X}}(x - X_i)}{\sum_{i=1}^{n} K_{\mathbf{H}^{Y|X}}(x - X_i)}, \quad (4)$$

where $\bar{K}_{\mathbf{H}_0}(w) = \int_{-\infty}^{w} K(\mathbf{H}_0^{-1/2} u) du$ is an integrated kernel function. Matrices $\mathbf{H}$ contain smoothing parameters, dubbed as bandwidths, with subscript 0 marking the CDF marginal and superscripts determining the corresponding distribution of interest. To simplify the presentation, I take $\mathbf{H}_0^Y = h_{0Y}^2$, $\mathbf{H}_0^{Y|X} = h_{0YX}^2$ and $\mathbf{H}^{Y|X} = \text{diag}(h_{1YX}^2, ..., h_{d_X YX}^2)$. Expression

$$K_{\mathbf{H}}(\boldsymbol{w}) = (\det \mathbf{H})^{-1/2} K(\mathbf{H}^{-1/2} \boldsymbol{w}) \quad (5)$$

is the scaled kernel with 'det' denoting the determinant and K being a generic multiplicative d_W-variate kernel function

$$K(w_1, ..., w_{d_W}) = \prod_{j=1}^{d_W} k(w_j), \quad (6)$$

satisfying for each marginal j

$$\begin{aligned} \int k(w_j) dw_j &= 1, \\ \int w_j^c k(w_j) dw_j &= 0 \quad \text{for} \quad c = 1, ..., r-1, \\ \int w_j^c k(w_j) dw_j &= \kappa_r < \infty \quad \text{for} \quad c = r, \end{aligned} \quad (7)$$

and $k(w)$ being symmetric and r-times differentiable [4].

The convergence properties of estimators in Eqs. (3) and (4) can be tuned by the rates of convergence of the smoothing parameters, i.e. h_{0Y} and h_{jYX} for $j = 0, ..., d_X$. Following [3], to guarantee that Eqs. (3) and (4) are uniformly tight, the sequences of bandwidths $h \equiv h(n)$ need to satisfy

---

[2]The quantiles of Y distribution can be directly extracted from the CDF estimates by solving for the argument. Although asymptotic properties of the quantiles and CDF correspond, the extraction of the quantiles through the CDF performs better in applied settings.

$$\lim_{n\to\infty} n^{1/2} h_{0Y}^{r} = 0, \quad \lim_{n\to\infty} n^{\alpha_1} h_{0Y} = \infty,$$
$$\lim_{n\to\infty} n^{1/2} h_{0YX}^{r} = 0, \quad \lim_{n\to\infty} n^{\alpha_2} h_{0YX} = \infty, \tag{8}$$

for some $\alpha_1, \alpha_2 > 0$ and

$$\lim_{n\to\infty} n^{1/2} \max_{j\in 1,\dots,d_X} (h_{jXY})^r = 0, \quad \lim_{n\to\infty} \frac{\log(n)}{n^{1/2}\Pi_{j=1}^{d_X} h_{jXY}} = 0. \tag{9}$$

If the support of Y is a compact set on $\mathbb{R}$, the functionals in Eqs. (3) and (4) are Donsker, and under an additional assumption that F_Y^{-1} is Hadamard differentiable, the fitted values of $y' \equiv \hat{y}'$ are asymptotically tight [7].

If one represents the sequence of bandwidth as $h = Cn^{-\beta}$, for some constant $C > 0$, Eq. (8) implies that $\beta > 1/(2r)$ for h_{0Y} and h_{0YX}, and from Eq. (9) it follows that $\beta \in (1/(2r), 1/(2d_X))$ for h_{jYX} where $j = 1, \dots, d_X$. These conditions are satisfied for the basic setup with the second-order kernels and $d_X = 1$. In fact, if one extends dimensionality of X to $d_X > 1$, condition Eq. (9) requires a higher order kernel.

A plug-in estimator of Eq. (2) becomes

$$\hat{y}' = \hat{F}_Y^{-1}(\hat{F}_{Y|X}(y|x)), \tag{10}$$

for fixed realizations $(Y, X) = (y, x)$. By rearranging the terms and substituting the kernel estimators from Eqs. (3) and (4), one may obtain $\hat{y}'$ by solving

$$n^{-1}\sum_{i=1}^{n} \bar{K}_{\mathbf{H}_0^Y}\left(\hat{y}' - Y_i\right) = \frac{\sum_{i=1}^{n} \bar{K}_{\mathbf{H}_0^{Y|X}}(y - Y_i) K_{\mathbf{H}^{Y|X}}(x - X_i)}{\sum_{i=1}^{n} K_{\mathbf{H}^{Y|X}}(x - X_i)}. \tag{11}$$

[8] shows that under the data assumptions outlined above and if $\hat{F}_Y$ and $\hat{F}_{Y|X}$ are Donsker then

$$\sqrt{n}\left(\hat{y}' - y'\right) \xrightarrow{d} N(0, \sigma^2), \tag{12}$$

where σ^2 is given by

$$\sigma^2 = \frac{F_{Y|X}(y|x)(1 - F_{Y|X}(y|x))}{f_Y\left(F_Y^{-1}(F_{Y|X}(y|x))\right)} + \frac{\int K(u)^2 du / f_X(x)}{\Pi_{j=1}^{d_X} h_{jXY}} \frac{F_{Y|X}(y|x)(1 - F_{Y|X}(y|x))}{f_Y\left(F_Y^{-1}(F_{Y|X}(y|x))\right)}. \tag{13}$$

The first term in σ^2 is the variance of the standard quantile estimator evaluated at the known quantity $F_{Y|X}(y|x)$. The second term results from the fact that the quantity $F_{Y|X}(y|x)$ is, in fact, estimated.

3 Interpretation

Removing the dependence between X and Y cannot be directly interpreted as a causal relation from X to Y. Reverse causality effects are also present in the joint distribution of (Y, X), and so are in the conditional distribution of $Y|X = x$. Nevertheless, the effects of X onto Y have causal interpretation under the so-called exogeneity assumption, or selection on observables. The assumption requires that there is no dependence between the covariates and the unobserved error component, $X \perp\!\!\!\perp \varepsilon$.

To introduce the concept formally, imagine that ε describes a (possibly discrete) policy option assigned between different groups of individuals. With the aim to study the causal effects of a policy e on the outcome Y, denote the set of potential outcomes by $(Y_e^* : \varepsilon \sim F_\varepsilon(e))$. The identification problems arise as Y is observed only conditional on $\varepsilon = e$. If the error term e is not randomly assigned (for instance, a policymaker discriminates between groups what policy e they receive), the observed Y conditional on $\varepsilon = e$ may not be equal to the true variable Y_e^*. On the other hand, if e is assigned randomly, variables Y_e^* and $Y|\varepsilon = e$ coincide. The exogeneity assumption may be extended by a set of conditioning covariates X. Under conditional exogeneity, the independent counterfactuals have also causal interpretation such that if conditional on X, the error component e is randomly assigned to Y, variables $Y_e^*|X$ and $Y|X, \varepsilon = e$ agree. Since the observed conditional random variable has causal interpretation, so has the independent counterfactual for which the X conditional effects have been integrated out (for more discussion see [1]).

Exogeneity assumption allows also to relate independent counterfactuals to the distribution of the error term. Consider a general nonseparable model

$$Y = m(X, \varepsilon), \tag{14}$$

where m is the general functional model and ε is an unobserved continuous error term. For identification purposes, let us assume that $m(x, .)$ is strictly increasing in e and continuous for all $x \in$ supp(X), so that its inverse exists and is strictly increasing and continuous.

Under exogeneity, one finds that after removing the effects of X onto Y, the counterfactual random variable Y' is identified at the $F_\varepsilon(\varepsilon)$ quantiles of Y. Note that

$$\begin{aligned} Y' &= F_Y^{-1}(P(m(X, \varepsilon) \le Y|X = x)) \\ &= F_Y^{-1}(P(\varepsilon \le m^{-1}(X, Y)|X = x)) \\ &= F_Y^{-1}(F_{\varepsilon|X}(\varepsilon|x)) \\ &= F_Y^{-1}(F_\varepsilon(\varepsilon)). \end{aligned} \tag{15}$$

By the inverse transformation method, one can also readily observe that the distribution of Y' coincides with the distribution of Y, i.e. $F_{Y'}(y) = F_Y(y)$ for all y. This is not surprising as a sample from a null hypothesis of independence can be often constructed by permutation methods [2].[3] Permutations are, however, not uniquely defined, as for a sample $\{Y_i, X_i\}_{i=1}^n$, for any fixed point $X = X_i$ any outcome Y_i may be assigned in the permutation process. Therefore, although permutations are a powerful tool in hypothesis testing, they cannot be applied as an identification strategy. Independent counterfactuals offer an alternative in this respect, for which the counterfactual realization is identified at the quantiles determined by the realization of the error term. It follows that

$$F_{Y'}(y') = F_Y(y') = F_{Y|X}(y|x) = F_Y(y)\delta(y, x), \tag{16}$$

where I substituted $\delta(y, x) \equiv F_{Y,X}(y, x)/(F_Y(y)F_X(x))$.

With endogenous error terms, the counterfactual Y' is still identified by the data but the dependence filtering is contaminated by the relation between X and ε. In such a case, the independent counterfactual removes the causal relation from X onto Y, but also from Y onto X, such that the random variables Y' and $F_Y^{-1}(F_\varepsilon(\varepsilon))$ do not necessarily agree. To illustrate it analytically, let us consider a simple linear framework.

3.1 Exogenous Linear Model

Consider a stylized process with the first-moment dependence between X and Y

$$\begin{aligned} x &= e_X, \\ y &= ax + \sqrt{1 - a^2}e_Y, \end{aligned} \tag{17}$$

where $a \in (0, 1)$ is a tuning parameter. Error terms ε_X and ε_Y follow standard normal distributions and are mutually independent. (Note that the setup ensures that the marginal of Y follows also a standard normal distribution.) The closed form expression for transformation in Eq. (2) can be derived as

$$\begin{aligned} F_Y^{-1}(q) &= \Phi^{-1}(q) \quad q \in (0, 1), \\ F_{Y|X}(y|x) &= \Phi\left(\frac{y - ax}{\sqrt{1 - a^2}}\right), \end{aligned} \tag{18}$$

where Φ is the standard normal CDF. Putting the expressions together, for the linear mean-dependent process in Eq. (17) I arrive at

[3]For an i.i.d sample from a dependent process, one may permute the data along each marginal to construct a sample from an independent process. In this context, permutation preserves the marginal distributions but breaks the dependence structure between covariates.

$$
\begin{aligned}
y' \equiv \phi(y, x) &= F_Y^{-1}(F_{Y|X}(y|x)) \\
&= \Phi^{-1}\left(\Phi\left(\frac{y - ax}{\sqrt{1 - a^2}}\right)\right) = \frac{y - ax}{\sqrt{1 - a^2}} = e_Y.
\end{aligned}
\tag{19}
$$

Equation (19) confirms Eq. (15). In the proposed stylized setup, the distribution of Y' corresponds to the distribution of errors so that the independent counterfactuals are asymptotically equal to the residuals from the standard Ordinary Least Squares (OLS) regression applied to the process from Eq. (17). In more general nonseparable models, the distribution of the error component would be scaled, by the inverse transformation method, to match the scale of the dependent variable.

3.2 Endogenous Linear Model

Consider now a similar process as in Eq. (17) but with inverse causality structure

$$
\begin{aligned}
y &= e_Y, \\
x &= ay + \sqrt{1 - a^2} e_X,
\end{aligned}
\tag{20}
$$

with similar stationarity conditions as before. Clearly, the exogeneity condition is violated as $X|\varepsilon_Y = e_Y \sim N(ae_Y, 1 - a^2)$. Having pointed this out, the identification in independent counterfactuals removes the entire dependence structure between the variables, which is exactly the same as in Eq. (17), such that

$$
y' = \frac{y - ax}{\sqrt{1 - a^2}} = \sqrt{1 - a^2} e_Y - a e_X. \tag{21}
$$

In this extreme example, because of reverse causality, the counterfactual variable Y' does not correspond to the potential outcome variable, which in this case is given by ε_Y. Nevertheless, the independence condition between Y' and X is satisfied as both variables are transformations of independent random variables and, since the distributions of Y' and Y coincide, $F_{Y'|X}(y|x) = F_{Y'}(y) = F_Y(y)$.

4 Illustration

To present the setup graphically, I choose the linear model given in Eq. (17), with additive and exogenous errors. For transparency, I fix the X marginal at $x = 1$, and I set the dependence parameter at $a = 0.75$, such that $Y|X = 1 \sim N(0.75, 1 - 0.75^2)$. The unconditional distribution of Y and the distribution of ε follow standard normal distributions.

Table 1 Average MSE and number of fails of fitted independent counterfactuals from Eq. (17). The numbers are aggregated over 1000 runs

	n = 50	n = 100	n = 200	n = 500	n = 1000	n = 2000
MSE($\hat{Y}'$)	0.116	0.08	0.056	0.035	0.024	0.017
Fails	0.001	0.001	0.001	0.001	0.002	0.002

The strategy is as follows. I randomly draw samples from the joint distribution (Y, X) and from the conditional distribution $Y|X = 1$ for different sample lengths n. Each realization from the conditional distribution sample is then transformed by Eq. (10), estimated over the joint distribution. The bandwidth parameters are set by the rule of thumb at $h_{0Y} = 1.59\hat{\sigma}_Y n^{-1/3}$, $h_{0XY} = 1.59\hat{\sigma}_Y n^{-1/3}$ and $h_{1XY} = 1.06\hat{\sigma}_X n^{-1/3}$, where $\hat{\sigma}_Y$ and $\hat{\sigma}_X$ correspond to standard deviation of samples $\{Y_i\}$ and $\{X_i\}$, respectively. Quantiles of Y are evaluated over the support $[-3.7, 3.7]$ to meet the compactness condition. If the value falls beyond that interval, I record it as a fail, and set $\hat{Y}_i' = Y_i$.

The results are presented in two ways. Firstly, for different sample sizes, I plot the histograms of random realizations of independent counterfactuals against the true densities of Y and $Y|X = 1$. The outcomes are depicted in Fig. 1.

Secondly, I calculate the MSE of the fitted independent counterfactuals as

$$\mathrm{MSE}(\phi(Y, 1)) = n^{-1} \sum_{i=1}^{n} \left(\hat{F}_Y^{-1}(\hat{F}_{Y|X}^{-i}(Y_i|X = 1)) - F_Y^{-1}(F_{Y|X}(Y_i|X = 1)) \right)^2, \tag{22}$$

where the superscript $-i$ stands for the leave-one-out kernel aggregate. The numbers are aggregated over 1000 runs of process in Eq. (17). The MSE results, together with the average estimation fails, are given in Table 1.

The simulation results suggest that as the sample size increases the independent counterfactuals converge to the true unconditional realizations of ε. The number of estimation fails appears to be contained at negligible levels, and clearly would be even lower for wider quantile support.

5 Conclusions

The purpose of this study is to familiarize the Reader with a novel dependence filtering framework. Under mild regularity conditions, and without assuming any specific parametric structure, the method allows to construct a counterfactual random variable which is independent from the effects of given covariates. Under error exogeneity assumption such a counterfactual has causal interpretation, and moreover, one can directly relate the counterfactuals with the distribution of the error component through the probability integral transform.

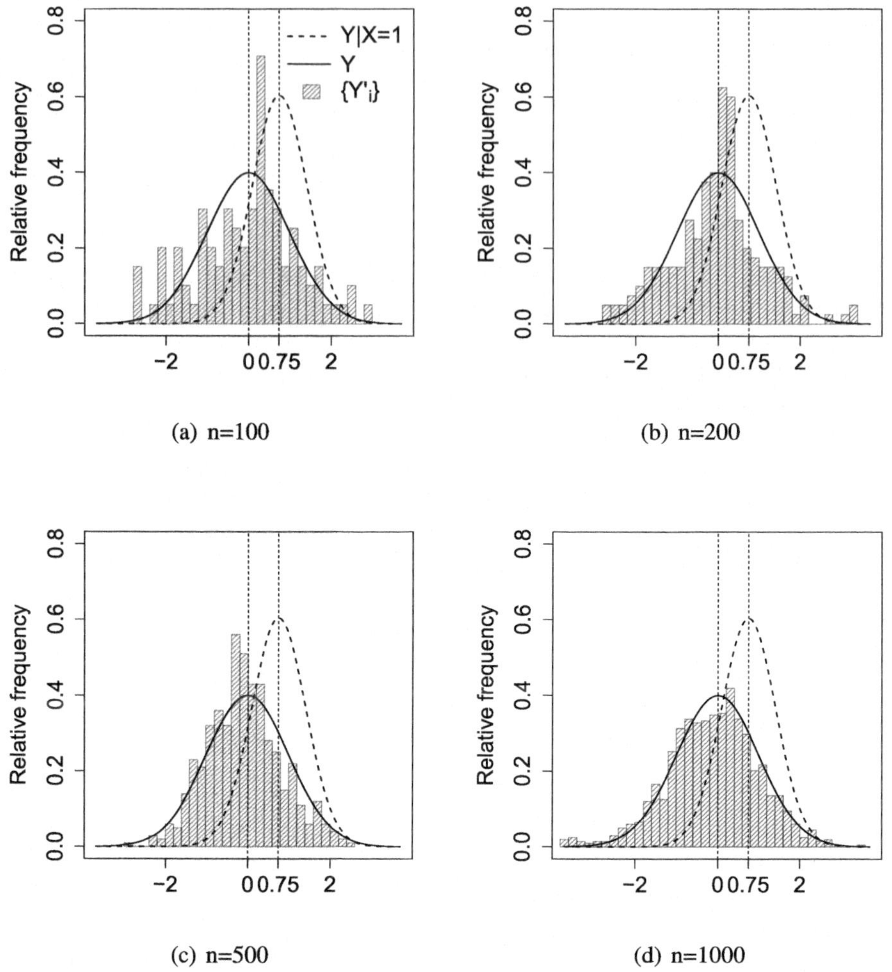

(a) n=100

(b) n=200

(c) n=500

(d) n=1000

Fig. 1 Independent counterfactuals. The plots show the true densities of random variables Y and $Y|X = 1$ under process from Eq. (17), together with a histogram of a counterfactual sample $\{Y_i'\}$ of an independent counterfactual random variable Y'. Vertical lines correspond to the expectations of Y and $Y|X = 1$

In settings where a no-dependence scenario can be expressed by specific values of the covariates, for instance, $X = 0$, independent counterfactuals can be related to the literature on counterfactual distributions [1, 5, 6]. Whenever $X = 0$ is not directly interpretable as independence, the proposed framework offers an attractive alternative to a standard toolkit.

I demonstrate how independent counterfactuals perform in a simple linear model with exogenous and endogenous error terms. In a simulation study, I also show the finite-sample consistency of the method.

The framework offers an easy extension to conditionally independent counterfactuals, along the lines proposed by [8]. It can be also applied to support identification in nonseparable models, statistical tests of independence between the variables or tests of error exogeneity.

Acknowledgements The author would like to thank Cees Diks, Laurent Maurin, Michiel van de Leur, Debora Revoltella, Christoph Rothe and participants of the 23rd International Conference Computing in Economics and Finance in New York, the 26th Annual Symposium of the Society for Nonlinear Dynamics and Econometrics in Tokyo and 4th Conference of the International Society for Nonparametric Statistics in Salerno for useful comments. The opinions expressed herein are those of the author and do not necessarily reflect those of the European Investment Bank.

References

1. Chernozhukov, V., Fernandez-Val, I., Melly, B.: Inference on counterfactual distributions. Econometrica **81**(6), 2205–2268 (2013). https://doi.org/10.3982/ECTA10582
2. Diks, C.: Nonparametric tests for independence. In: Meyers, R. (ed.) Encyclopedia of Complexity and Systems Science. Springer Verlag, New York (2009)
3. Ferraty, F., Laksaci, A., Tadj, A., Vieu, P.: Rate of uniform consistency for nonparametric estimates with functional variables. J. Stat. Plan. Infer. **140**(2), 335–352 (2010). https://doi.org/10.1016/j.jspi.2009.07.019
4. Li, Q., Racine, J.S.: Nonparametric Econometrics: Theory and Practice. Princeton University Press, Princeton and Oxford (2007)
5. Rothe, C.: Nonparametric estimation of distributional policy effects. J. Econom. **155**(1), 56–70 (2010). https://doi.org/10.1016/j.jeconom.2009.09.001
6. Rothe, C.: Partial distributional policy effects. Econometrica **80**(5), 2269–2301 (2012). https://doi.org/10.3982/ECTA9671
7. van der Vaart, A.W.: Asymptotic Statistics. Cambridge University Press, Cambridge (2000)
8. Wolski, M.: Sovereign risk and corporate cost of borrowing: evidence from a counterfactual study. Working Paper 2018/05, European Investment Bank (2018)

The Potential for Nonparametric Joint Latent Class Modeling of Longitudinal and Time-to-Event Data

Ningshan Zhang and Jeffrey S. Simonoff

Abstract Joint latent class modeling (JLCM) of longitudinal and time-to-event data is a parametric approach of particular interest in clinical studies. JLCM has the flexibility to uncover complex data-dependent latent classes, but it suffers high computational cost, and it does not use time-varying covariates in modeling time-to-event and latent class membership. In this work, we explore in more detail both the strengths and weaknesses of JLCM. We then discuss the sort of nonparametric joint modeling approach that could address some of JLCM's weaknesses. In particular, a tree-based approach is fast to fit, and can use any type of covariates in modeling both the time-to-event and the latent class membership, thus serving as an alternative method for JLCM with great potential.

Keywords Biomarker · Recursive partitioning · Survival data

1 Introduction

Clinical studies often collect three types of data on each patient: the time to the event of interest (possibly censored), the longitudinal measurements on a continuous response (for example, some sort of biomarker viewed as clinically important), and an additional set of covariates (possibly time-varying) about the patient. The clinical studies then focus on analyzing the relationship between the time-to-event and the longitudinal responses, using the additional covariates. A common approach is the shared random effects model, which jointly models the time-to-event by a survival model while modeling the longitudinal responses using a linear mixed-effects model, with the two models sharing random effects, with both the survival and the linear mixed-effects models potentially making use of the additional covariates [16].

N. Zhang (✉) · J. S. Simonoff
Leonard N. Stern School of Business, New York University, 44 West 4th Street, New York, NY 10012, USA
e-mail: nzhang@stern.nyu.edu

J. S. Simonoff
e-mail: jsimonof@stern.nyu.edu

M. La Rocca et al. (eds.), *Nonparametric Statistics*, Springer Proceedings in Mathematics & Statistics 339, https://doi.org/10.1007/978-3-030-57306-5_47

A different interesting line of work in this area assumes that the heterogeneous population consists of several homogeneous "latent classes," within which subjects follow the same time-to-event and longitudinal relationships, and uses parametric approaches to model the latent class membership together with time-to-event and longitudinal outcomes. This method is called the joint latent class model (JLCM) [12, 14, 15]. The idea of latent class membership in JLCM is of particular interest in clinical studies, since the latent class can be used to describe disease progression [7, 12, 14]. It is well known that many diseases have different stages; examples include dementia, AIDS, cancer, and chronic obstructive pulmonary disease (COPD) [6]. From a clinical point of view, it is important to identify those stages, since treatment could change with those different stages [8]. Currently the clinical definitions of stages of a disease consists of using diagnostic findings (such as biomarkers) to produce clusters of patients. However, it is possible that by jointly studying biomarker trajectories and survival experiences, one can find data-dependent latent classes that uncover new, meaningful stages.

Like most frequentist parametric approaches, JLCM uses maximum likelihood estimation to estimate the parameters, which comes with high computational cost. In addition, JLCM enforces certain restrictions in its model, which could limit its performance. In this work, we first give a brief introduction to JLCM in Sect. 2, and review its strengths and weaknesses in Sects. 3 to 5. In particular, we use simulations to examine JLCM's modeling flexibility and running time. In Sect. 6 we discuss why a nonparametric approach could address some of those weaknesses, and conclude that such a nonparametric approach is an alternative method for joint latent class modeling with great potential.

2 The JLCM Setup

In this section, we give a brief description of JLCM. Assume there are N subjects in the sample. For each subject i, we observe n_i repeated measurements of a longitudinal outcome at times $\mathbf{t}_i = (t_{i1}, \ldots, t_{in_i})'$. We denote the vector of longitudinal outcomes by $\mathbf{Y}_i = \left(y_{it_{i1}}, \ldots, y_{it_{in_i}}\right)'$. In addition, for each subject i we observe a vector of p covariates at each measurement time $t \in \mathbf{t}_i$: $\mathbf{X}_{it} = (x_{it1}, \ldots, x_{itp})$. These covariates can be either time-invariant or time-varying. We will introduce four subsets of $\mathbf{X}_{it}$ for modeling the three components of JLCM: $\mathbf{X}^{\mathrm{f}}_{it}$($\mathbf{X}^{\mathrm{r}}_{it}$) for the fixed (random) effects in modeling longitudinal outcomes, $\mathbf{X}^{\mathrm{s}}_{it}$ for modeling the (survival) time-to-event, and $\mathbf{X}^{\mathrm{g}}_{it}$ for modeling the latent class membership. The four subsets can either be identical, or share common covariates, or share no covariates at all. In addition, JLCM enforces the restriction that $\mathbf{X}^{\mathrm{s}}$ and $\mathbf{X}^{\mathrm{g}}$ do not contain any time-varying covariates. This is a major drawback of JLCM, which we will discuss in more detail in Sect. 5. Finally, each subject i is associated with a time-to-event tuple (T_i, δ_i), where T_i is the time of the event, and δ_i is the censoring indicator with $\delta_i = 0$ if subject i is censored at T_i, and $\delta_i = 1$ otherwise.

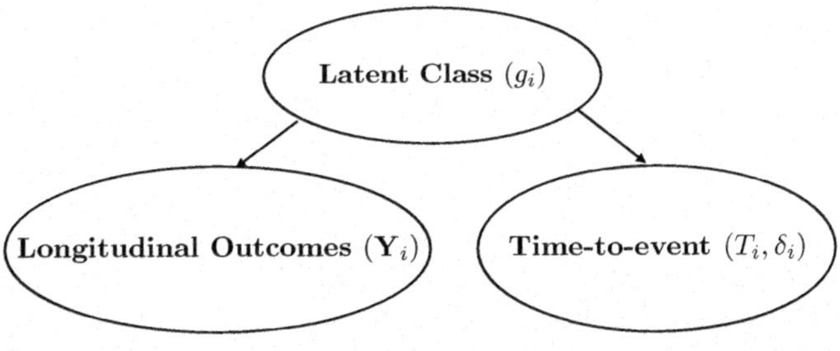

Fig. 1 Causal graph describing how JCLM assumes that time-to-event and longitudinal outcomes are independent conditioning on the latent class membership

For each subject i, we assume the latent class membership $g_i \in \{1, \ldots, G\}$ for subject i is determined by the set of covariates $\mathbf{X}_i^{\mathrm{g}}$ (since $\mathbf{X}_{it}^{\mathrm{g}}$ must be time-invariant, we drop the time indicator t), through the following probabilistic model:

$$\pi_{ig} = \Pr(g_i = g|\mathbf{X}_i^{\mathrm{g}}) = \frac{\exp\{\xi_{0g} + \mathbf{X}_i^{\mathrm{g}}\xi_{1g}\}}{\sum_{l=1}^{G} \exp\{\xi_{0l} + \mathbf{X}_i^{\mathrm{g}}\xi_{1l}\}},$$

where ξ_{0g}, ξ_{1g} are class-specific intercept and slope parameters for class $g = 1, \cdots, G$.

Figure 1 uses a causal graph to describe the key assumption made by JLCM: a subject's time-to-event (T_i, δ_i) and longitudinal outcomes $(\mathbf{Y}_i)$ are independent conditioning on his or her latent class membership (g_i). Without controlling the latent class membership g_i, time-to-event and longitudinal outcomes may appear to be correlated because each is related to the latent class, but given g_i the two are independent of each other, and therefore the longitudinal outcomes have no prognostic value for time-to-event given the latent class. The modeling of (T_i, δ_i) and $\mathbf{Y}_i$ are therefore separated conditioning on g_i.

The longitudinal outcomes are described by the following linear mixed-effects model [11]:

$$\begin{aligned} y_{it}|_{g_i=g} &= \mathbf{X}_{it}^{\mathrm{f}}\mathbf{u}_g + \mathbf{X}_{it}^{\mathrm{r}}\mathbf{v}_{ig} + \varepsilon_{it}, \\ \mathbf{v}_{ig} &= \mathbf{v}_i|_{g_i=g} \sim N(\boldsymbol{\mu}_g, \mathbf{B}_g),\ \varepsilon_{it} \sim N(0, \sigma^2). \end{aligned}$$

Here we assume the longitudinal outcomes depend on two subsets of $\mathbf{X}_{it}$, where $\mathbf{X}_{it}^{\mathrm{f}}$ is the set of covariates associated with a class-specific fixed effect vector $\mathbf{u}_g$, and $\mathbf{X}_{it}^{\mathrm{r}}$ is the set of covariates associated with a class and subject-specific random effect vector $\mathbf{v}_{ig}$. The random effect vector $\mathbf{v}_{ig}$ is independent across latent classes and subjects, and normally distributed with mean $\boldsymbol{\mu}_g$ and variance-covariance matrix $\mathbf{B}_g$. Finally, the errors $varepsilon_{it}$ are assumed to be independent and normally distributed with mean 0 and variance σ^2, and independent of all of the random effects as well. Let

$f(\mathbf{Y}_i|g_i = g)$ denote the likelihood of longitudinal outcomes $\mathbf{Y}_i$ given that subject i belongs to latent class g.

The time-to-event T_i is considered to follow the proportional hazards model:

$$h_i(t|g_i = g) = h_{0g}(t; \zeta_g)e^{\mathbf{X}_i^s \eta_g},$$

where ζ_g parameterizes the class-specific baseline hazard h_{0g}, and η_g is associated with the set of covariates $\mathbf{X}_i^s$ (we drop the time indicator t from $\mathbf{X}_{it}^s$ since it must be time-invariant). Let $S_i(t|g_i = g)$ denote the survival probability at time t if subject i belongs to latent class g.

Let $\theta_G = \big(\xi_{0g}, \xi_{1g}, \mathbf{u}_g, \mathbf{v}_{ig}, \mu_g, \mathbf{B}_g, \sigma, \zeta_g, \eta_g : g = 1, \cdots, G, i = 1, \cdots, N\big)$ be the entire vector of parameters of JLCM. These parameters are estimated together via maximizing the log-likelihood function

$$L(\theta_G) = \sum_{i=1}^{N} \log\left(\sum_{g=1}^{G} \pi_{ig} f(\mathbf{Y}_i|g_i = g; \theta_G) h_i(T_i|g_i = g; \theta_G)^{\delta_i} S_i(T_i|g_i = g; \theta_G)\right). \tag{1}$$

The log-likelihood function above uses the assumption that conditioning on the latent class (g_i), longitudinal outcomes ($\mathbf{Y}_i$) and time-to-event (T_i, δ_i) are independent.

3 Latent Class Membership of JLCM

JLCM uses a multinomial logistic regression to model latent classes. To study the modeling flexibility of multinomial logistic regression, we simulate various scenarios of true latent class membership, and then use multinomial logistic regression to predict the membership. Simulation results indicate that multinomial logistic regression is, in fact, quite flexible in modeling latent class membership, and thus JLCM has the potential of uncovering complex clustering of the population. We give more details below.

We consider four underlying structures of latent class membership: partition by a tree, linear separation, nonlinear separation, and partition by an asymmetric tree, which are demonstrated in Fig. 2. The class membership is determined by the values of X_1, X_2. For each latent class membership structure, we generate samples of various sizes (100, 1000, 10000), where X_1, X_2 are drawn i.i.d. from uniform [0, 1]. In addition, three superfluous covariates X_3, X_4, X_5 are drawn i.i.d. from uniform [0, 1] as well. We fit multinomial logistic regression to predict the true latent class membership using $X_1, \cdots, X_5$. For observation i, we denote by g_i and $\widehat{g_i}$ the true and predicted class membership, respectively. We use the proportion of incorrect predictions to measure the prediction error: $\frac{1}{N}\sum_{i=1}^{N} 1_{g_i \neq \widehat{g_i}}$.

The above procedure is repeated 20 times, and we summarize the out of sample prediction error in Table 1. Multinomial logistic regression gives very accurate predictions when the underlying partitions are trees or linear separations, and the error

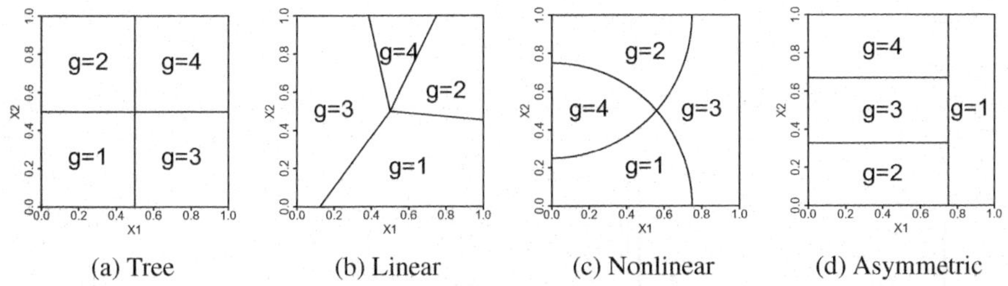

Fig. 2 Four structures of latent class membership based on X_1 and X_2: **a** Tree partition, **b** Linear partition, **c** Nonlinear partition, **d** Asymmetric tree partition

Table 1 The out of sample prediction error (standard deviation in parentheses) of multinomial logistic model predicting latent class membership

Sample size	Tree	Linear	Nonlinear	Asymmetric
100	0.0810 (0.0404)	0.0780 (0.0311)	0.0885 (0.0344)	0.1240 (0.0472)
1000	0.0060 (0.0052)	0.0082 (0.0033)	0.0403 (0.0069)	0.0188 (0.0065)
10000	0.0000 (0.0000)	0.0012 (0.0004)	0.0374 (0.0021)	0.0002 (0.0003)

decreases as the sample size increases. When the underlying partition is nonlinear, multinomial logistic regression can still give reasonably good approximations, but it is not working as effectively as it does for other partition structures.

Among the four structures, only the linear partition (Fig. 2b) is consistent with the multinomial logistic regression model, while the tree partitions (Figs. 2a and d) are consistent with the model only in limits, as certain slope parameters in the regression model approach 0 or ∞. The fact that the multinomial logistic regression can accurately predict membership under the tree partitions demonstrates its flexibility and robustness in modeling unknown latent class membership.

4 Running Time of JLCM

The log-likelihood function of JCLM (1) is a complicated and non-convex function of all the parameters θ_G, and it therefore can take a significant amount of time to find an optimal solution. In this section, we run simulations to examine JCLM's running time, in particular the relationship between the running time and sample size, number of covariates, and number of latent classes.

We use the `Jointlcmm` function contained in the R package `lcmm` to fit JLCM. Since the objective function contains many local optima, careful initialization is required to ensure a meaningful maximum likelihood estimate [15]. We adopt the initialization techniques used in [15], which involves first fitting a JLCM model with

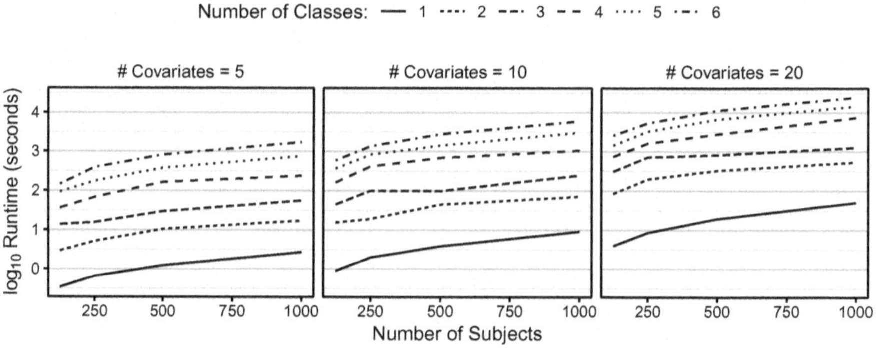

Fig. 3 JLCM running time (in seconds) on $\log_{10}$ scale versus number of subjects. The sub-panels correspond to three different numbers of covariates: 5, 10, 20. Each sub-panel contains six curves corresponding to number of latent classes $G = 1, 2, 3, 4, 5, 6$

only one class, and then using the one class result to initialize fitting JLCM with more classes. Thus, the initialization time is the same as fitting a JLCM with $G = 1$.

The data generation scheme in our simulation matches the JLCM model setup described in Sect. 2. In particular, we use one set of covariates to generate latent classes ($\mathbf{X}^{\mathrm{g}}$), and use another set of covariates for the remaining purposes ($\mathbf{X}^{\mathrm{f}} = \mathbf{X}^{\mathrm{r}} = \mathbf{X}^{\mathrm{s}}$). We consider scenarios in which there are 5, 10, and 20 covariates, and in which $\mathbf{X}^{\mathrm{g}}$ contains 3, 6, and 12 covariates, respectively. We also consider various numbers of subjects: $N = \{125, 250, 500, 1000\}$, and each subject has a random number of observations, with two observations per subject on average. For each simulated dataset, we fit JLCM with various numbers of latent classes: $G = 1, 2, 3, 4, 5, 6$. We use a proportional hazards model with class-specific Weibull baselines to simulate survival times, and use a linear mixed-effects model to simulate longitudinal outcomes.

We report the running time of JLCM under various simulation settings in Fig. 3. The simulations are performed on high performance computing nodes with 3.0 GHz CPU and 62 GB memory. The $\log_{10}$ running time increases linearly as a function of sample size, showing that the actual running time grows exponentially fast. Similarly, we observe that the running time grows exponentially fast as a function of number of covariates and number of assumed latent classes.

The study of running time shows that fitting JLCM can be computationally expensive. In practice, one would fit JLCM with various numbers of classes and choose the best number via a model selection criterion such as BIC. In our simulation, it takes JLCM about 3 h to fit all the values of $G \in \{1, 2, 3, 4, 5, 6\}$ on a sample of 1000 subjects with 10 covariates. For an even larger dataset, such as one based on electronic health records (which can number in the millions), JLCM can be prohibitively slow and impractical to use.

5 Limitation to Time-Invariant Covariates

In addition to its unscalable running time, the software implementation of JLCM has another limitation in that the modeling of both time-to-event and latent class membership are restricted to the use of time-invariant covariates. In this section, we discuss why it is helpful to relax both restrictions to allow the use of time-varying covariates.

Time-varying covariates are especially helpful in modeling time-to-event when treatment or important covariates change during the study; for instance, the patient receives a heart transplant [4], the longitudinal CD4 counts change during the study of AIDS [17], or the antibiotics exposure changes during an antibiotic resistance study [13]. Research shows that using time-varying covariates can uncover short-term associations between time-to-event and covariates [5, 10], and ignoring the time-varying nature of the covariates will lead to time-dependent bias [9, 13]. There exist several well-studied methods for dealing with time-varying covariates in survival analysis, and the most commonly used approach is the extended Cox model [3], which naturally extends the Cox model to use time-varying covariates. An alternative approach is landmarking [1], in which the sample of subjects is examined at an arbitrary future landmark time s, and the values of variables at time s are used as time-invariant covariates.

The other restriction of JLCM is that the latent class membership model only uses time-invariant covariates, and as a result the latent class membership of a subject is assumed to be fixed throughout the time. However, the stage of a disease of a patient is very likely to change during the course of clinical study; for instance, the disease would move from its early stages to its peak, and then move to its resolution. When the goal of joint modeling is to uncover meaningful clustering of the population that leads to definitions of disease stages, it is necessary to allow time-varying covariates in the latent class membership model, so that the model reflects the real-world situation.

6 Nonparametric Approach for Joint Modeling

JLCM is designed to give parametric descriptions of subjects' tendency of belonging to each latent class, and therefore, JLCM is a suitable model when the true latent class is indeed a random outcome with unknown probabilities for each class. However, fitting parametric models is in general computationally expensive.

In the situation where each observation deterministically belongs to a single latent class, in other words, where the population admits a deterministic (but potentially changing over time) partition, then a nonparametric approach would be more applicable. A tree-based approach [2] is the natural choice for this task, since it is very efficient to fit a tree, and the terminal nodes of a tree naturally represent a partition of the population. Observe that the deterministic partition is a special case of the probabilistic tendency model, where the probabilities of any latent class is either 0 or 1. Therefore, when the latent class membership is not entirely deterministic but the probabilities for each class are close to 0 or 1, a tree-based approach still serves as a good alternative to JLCM.

A tree-based approach for joint latent class modeling also addresses the time-invariant limitation of JLCM. In particular, with a carefully designed splitting criterion, one can easily use time-varying covariates to construct the tree. In addition, once a tree is constructed, it is up to the user to decide which type of survival models and which covariates to use within each terminal node.

In view of this, we have commenced investigation of a joint latent class tree (JLCT) model. The JLCT model, like JLCM, is based on the key assumption that conditioning on the latent class, time-to-event and longitudinal responses are independent. The JLCT model therefore looks for a tree-based partitioning such that within each estimated latent class defined by a terminal node, the time-to-event and longitudinal responses display a lack of association. Further details can be found in [18].

References

1. Anderson, J.R., Cain, K.C., Gelber, R.D.: Analysis of survival by tumor response. J. Clin. Oncol. **1**(11), 710–719 (1983)
2. Breiman, L., Friedman, J.H., Olshen, R.A., Stone, C.J.: Classification and Regression Trees. Wadsworth and Brooks, Monterey, CA (1984)
3. Cox, D.R.: Regression models and life-tables. J. R. Stat. Soc. Ser. B **34**(2), 187–220 (1972)
4. Crowley, J., Hu, M.: Covariance analysis of heart transplant survival data. J. Am. Stat. Assoc. **72**(357), 27–36 (1977)
5. Dekker, F.W., De Mutsert, R., Van Dijk, P.C., Zoccali, C., Jager, K.J.: Survival analysis: time-dependent effects and time-varying risk factors. Kidney Int. **74**(8), 994–997 (2008)
6. Dicker, R., Coronado, F., Koo, D., Parrish, R.G.: Principles of epidemiology in public health practice. US Department of Health and Human Services (2006)
7. Garre, F.G., Zwinderman, A.H., Geskus, R.B., Sijpkens, Y.W.: A joint latent class changepoint model to improve the prediction of time to graft failure. J. R. Stat. Soc. Ser. A **171**(1), 299–308 (2008)
8. Hajiro, T., Nishimura, K., Tsukino, M., Ikeda, A., Oga, T.: Stages of disease severity and factors that affect the health status of patients with chronic obstructive pulmonary disease. Respir. Med. **94**(9), 841–846 (2000)
9. Jongerden, I.P., Speelberg, B., Satizábal, C.L., Buiting, A.G., Leverstein-van Hall, M.A., Kesecioglu, J., Bonten, M.J.: The role of systemic antibiotics in acquiring respiratory tract colonization with gram-negative bacteria in intensive care patients: A nested cohort study. Critic. Care Med. **43**(4), 774–780 (2015)
10. Kovesdy, C.P., Anderson, J.E., Kalantar-Zadeh, K.: Paradoxical association between body mass index and mortality in men with CKD not yet on dialysis. Am. J. Kidney Dis. **49**(5), 581–591 (2007)
11. Laird, N.M., Ware, J.H.: Random-effects models for longitudinal data. Biometrics **38**(4), 963–974 (1982)
12. Lin, H., Turnbull, B.W., McCulloch, C.E., Slate, E.H.: Latent class models for joint analysis of longitudinal biomarker and event process data: Application to longitudinal prostate-specific antigen readings and prostate cancer. J. the Am. Stat. Assoc. **97**(457), 53–65 (2002)
13. Munoz-Price, L.S., Frencken, J.F., Tarima, S., Bonten, M.: Handling time-dependent variables: antibiotics and antibiotic resistance. Clin. Infect. Dis. **62**(12), 1558–1563 (2016)

14. Proust-Lima, C., Joly, P., Dartigues, J.F., Jacqmin-Gadda, H.: Joint modelling of multivariate longitudinal outcomes and a time-to-event: a nonlinear latent class approach. Comput. Stat. Data Anal. **53**(4), 1142–1154 (2009)
15. Proust-Lima, C., Philipps, V., Liquet, B.: Estimation of extended mixed models using latent classes and latent processes: The R package lcmm. J. Statist. Softw. Articles **78**(2), 1–56 (2017)
16. Rizopoulos, D.: Joint Models for Longitudinal and Time-to-Event Data: With Applications in R. CRC Press (2012)
17. Tsiatis, A., Degruttola, V., Wulfsohn, M.: Modeling the relationship of survival to longitudinal data measured with error. Applications to survival and CD4 counts in patients with AIDS. J. Am. Stat. Assoc. **90**(429), 27–37 (1995)
18. Zhang, N., Simonoff, J.S.: Joint latent class trees: a tree-based approach to joint modeling of time-to-event and longitudinal data (2018). https://arxiv.org/abs/1812.01774

To Rank or to Permute When Comparing an Ordinal Outcome Between Two Groups While Adjusting for a Covariate?

Georg Zimmermann

Abstract The classical parametric analysis of covariance (ANCOVA) is frequently used when comparing an ordinal outcome variable between two groups, while adjusting for a continuous covariate. However, the normality assumption might be crucial and assuming an underlying additive model might be questionable. Therefore, in the present manuscript, we consider the outcome as truly ordinal and dichotomize the covariate by a median split, in order to transform the testing problem to a nonparametric factorial setting. We propose using either a permutation-based Anderson–Darling type approach in conjunction with the nonparametric combination method or the pseudo-rank version of a nonparametric ANOVA-type test. The results of our extensive simulation study show that both methods maintain the type I error level well, but that the ANOVA-type approach is superior in terms of power for location-shift alternatives. We also discuss some further aspects, which should be taken into account when deciding for the one or the other method. The application of both approaches is illustrated by the analysis of real-life data from a randomized clinical trial with stroke patients.

Keywords Nonparametric covariate adjustment · Nonparametric combination method · NPC · Rank-based inference · Nonparametric analysis of covariance

---

Electronic supplementary material The online version of this chapter (https://doi.org/10.1007/978-3-030-57306-5_48) contains supplementary material, which is available to authorized users.

G. Zimmermann (✉)
Department of Mathematics, Paris Lodron University, Hellbrunner Str. 34, 5020 Salzburg, Austria

Team Biostatistics and Big Medical Data, IDA Lab Salzburg, Paracelsus Medical University, Strubergasse 16, 5020 Salzburg, Austria

Department of Neurology, Christian Doppler Medical Centre, Ignaz-Harrer-Strasse 79, 5020 Salzburg, Austria
e-mail: georg.zimmermann@pmu.ac.at

M. La Rocca et al. (eds.), *Nonparametric Statistics*, Springer Proceedings in Mathematics & Statistics 339, https://doi.org/10.1007/978-3-030-57306-5_48

1 Introduction

In many clinical studies, the main research interest is focused on the comparison of two groups with respect to a certain outcome variable of interest. Examples include the classical parallel-group design (e.g., verum vs. placebo) in randomized clinical trials, but also observational studies, which are aimed at studying the association between a particular factor (e.g., two different pathologies) and a health-related variable (e.g., a laboratory measurement or a quality of life score). Frequently, the two-sample t-test is used for analyzing data from such studies. If the comparison is adjusted for one or several covariates, the analysis of covariance (ANCOVA) is regarded as the method of choice. A comprehensive overview about the classical ANCOVA and some of its alternatives is given by [11]. However, most methods require that the outcome is continuous, or at least that the effects are additive. Therefore, applying these approaches in settings where the outcome is ordinal might be inappropriate. However, ordinal outcomes are quite frequently encountered in applied research: One example, among others, is the modified Rankin Scale (mRS), which is used to quantify the functional outcome of stroke patients by assigning a rating on a scale from 0 (no symptoms) to 6 (dead) [9, 21, 27]. Although some evidence suggests that converting the ordinal factor levels to scores and subsequently applying the standard parametric ANCOVA does not affect the performance of the test substantially [25], there are two crucial issues with this approach. Firstly, with respect to the normality assumption, a highly discrete variable might be particularly challenging to deal with. Secondly, careful thoughts concerning whether or not the resulting effect measure (e.g., the difference of the adjusted means) can be interpreted in a sensible way are required. Therefore, analyzing the outcome as a truly ordinal variable might be an attractive option. Again, however, classical approaches such as proportional odds models (for an introduction, see [1]) require assumptions which might be too restrictive.

Nonparametric rank-based methods may serve as a remedy in order to overcome the aforementioned difficulties. In particular, approaches where the so-called relative effect is used as the effect measure are also applicable to ordinal outcomes. One well-known example is the Wilcoxon-Mann-Whitney test [14, 28], which is the classical nonparametric counterpart of the two-sample t-test. Several extensions to general factorial designs have been proposed (e.g., [2, 4, 6, 8]). However, only a few approaches allow for adjustment for a continuous covariate [3, 26].

Alternatively, using permutation methods might be an appealing option, mainly due to their finite-sample properties (e.g., exactness) under relatively mild assumptions. For an overview of the underlying theory, we refer, for example, to [18]. Like with the rank-based methods that have been mentioned previously, there is a broad variety of permutation approaches, which cover various practically relevant settings (see, for example, [16, 19]). Again, however, the present setting, that is, the comparison of an the comparison of an ordinal outcome between two groups while adjusting for a continuous covariate, seems to be a somewhat difficult problem. The reason is that many permutation tests (e.g., the so-called synchronized permutation approach, see [10, 22]), implicitly assume an additive model, and hence, are only applicable if the outcome variable is metric. However, using the so-called using the so-called "nonparametric combination method" might be a promising alternative to a classical parametric approach: For the purpose of the present manuscript, the comparison of two samples with respect to an ordered categorical outcome, which was studied in [17], is of interest. Nevertheless, like with other approaches, continuous covariates cannot be directly accounted for. One straightforward solution might be to categorize the covariate, because an additional adjustment for a categorical covariate could be done by another application of the nonparametric combination method, then. Indeed, in applied research, a continuous covariate is often transformed into a binary variable by, for example, applying a median split. Alternatively, there might be some well-established cutoffs available, or subject-matter expertise could help to define appropriate categories. Therefore, it might be of interest especially for biostatisticians to have some empirical guidance at hand concerning which of the two approaches— permutation tests and the nonparametric combination method, or rank-based tests—should be used in practice.

The manuscript is organized as follows: In Sect. 2, we introduce the pseudo-rank version of the nonparametric ANOVA-type test for general factorial designs [4] and the Anderson–Darling type permutation test [17], as well as some fundamental pieces of the respective theory. Section 3 contains the results of an extensive simulation study, covering balanced and unbalanced settings, as well as different distributional assumptions. The application of the two methods under investigation is illustrated by the analysis of a real-life data example in Sect. 4. Finally, Sect. 5, contains some concluding remarks and a brief synopsis of the respective advantages and drawbacks. This hopefully helps applied researches to choose an appropriate statistical analysis approach. All tables and figures are provided in the Online Supplement.

2 The Nonparametric Combination Method and a Pseudo-rank-based Approach

Let $(X_{11}, Y_{11}), \ldots, (X_{1n_1}, Y_{1n_1}) \overset{iid}{\sim} F_1$ and $(X_{21}, Y_{21}), \ldots, (X_{2n_2}, Y_{2n_2}) \overset{iid}{\sim} F_2$ denote two independent bivariate samples. Thereby, the first and second components are the continuous covariate and the ordinal outcome, respectively. Let $\mathcal{R}_Y := \{C_1 \leq \ldots \leq C_K\}$ denote the support of the outcome Y, which consists of the ordered categories $C_1, \ldots, C_K$. For example, in medical research, but also in psychology and other fields, the outcome is frequently assessed by a rating on some scale (e.g., modified Rankin Scale, Glasgow Coma Scale, Visual Analog Scale, or Functional Gait Assessment, just to name a few). We have the impression that quite frequently, such a sort of outcome is analyzed by using classical methods for metric variables (e.g., t-test, ANOVA). This might indeed be appropriate in case that the interpretation as a metric variable is sensible from the respective subject matter point of view. Nevertheless, we would like to emphasize that this issue requires careful case-by-case considerations.

Moreover, especially in case of a highly discrete variable (i.e., the cardinality of the support of Y is quite "small"), assuming a normal distribution might not be justified.

Regarding the covariate, the continuous random variables $X_{11}, \ldots, X_{2n_2}$ are categorized by applying a measurable function $g : \mathcal{R}_X \to \{1, 2\}$, where $\mathcal{R}_X$ denotes the range of the covariate. The choice of g is either guided by subject-matter expertise or relies on statistical considerations (e.g., median split). Hence, in the sequel, we shall partition the outcomes $Y_{11}, \ldots, Y_{2n_2}$ according to the transformed covariate values $Z_{11} := g(X_{11}), \ldots, Z_{2n_2} := g(X_{2n_2})$. Doing so, and after some re-indexing, we get

$$Y_{111}, \ldots, Y_{11n_{11}} \overset{iid}{\sim} F_{1|Z=1}, Y_{121}, \ldots, Y_{12n_{12}} \overset{iid}{\sim} F_{1|Z=2},$$
$$Y_{211}, \ldots, Y_{21n_{21}} \overset{iid}{\sim} F_{2|Z=1}, Y_{221}, \ldots, Y_{22n_{22}} \overset{iid}{\sim} F_{2|Z=2}.$$

It should be noted that actually, n_{ij} is a random variable, $i, j \in \{1, 2\}$, since the covariate and its categorized version are random quantities. Nevertheless, for ease of presentation, we consider the cell sizes as fixed in the sequel, which means that everything has to be understood conditionally on a fixed set of covariate values $z_{11}, \ldots, z_{2n_2}$. This does not restrict the generality of the two approaches proposed in Sects. 2.1 and 2.2, because the formal procedures work completely analogously in case of random covariates. However, caution is needed concerning the simulation setup. Therefore, in all settings discussed in Sect. 3, the covariate will be considered as random again.

For the sake of notational simplicity, let $F_{ij} := F_{i|Z=j}$, $i, j \in \{1, 2\}$. Recall that our main aim is to compare the outcome Y between the two treatment groups (i.e., $i = 1$ and $i = 2$). In the following section, we propose two different approaches.

2.1 The Nonparametric Combination Method, Applied to an Anderson–Darling type Permutation Test

The basic idea underlying the nonparametric combination (NPC) method is quite simple: The hypothesis is split up into a finite set of partial hypotheses, and subsequently, a hypothesis test is constructed for each of these partial testing problems. In a second step, the resulting test statistics or p values are combined by using an appropriate combination function (e.g., Fisher's combination function). It follows immediately from the underlying theory that the basic properties of the separate tests (e.g., consistency, exactness) are carried over to the combined test, then. For an overview, we refer to [18, 19].

We consider the hypothesis $H_{0,NPC} : \{F_{11} = F_{21}\} \cap \{F_{12} = F_{22}\}$ vs. $H_{1,NPC} : \{F_{11} \neq F_{21}\} \cup \{F_{12} \neq F_{22}\}$ and construct a corresponding test by a suitable nonparametric combination of two partial permutation statistics T_1 and T_2. Let $n_{.j} = n_{1j} + n_{2j}$ denote the number of observations with transformed covariate value j,

$j \in \{1, 2\}$. For even total sample size N, dichotomizing the covariate by applying a median split yields $n_{.1} = n_{.2} = N/2$. For a permutation $s \in \mathcal{S}_{n_{.j}}$, we define the corresponding permutation of the pooled observations $\mathbf{Y}_j = \{Y_1, \ldots, Y_{n_{.j}}\}$ within covariate subgroup $Z = j$ by $\mathbf{Y}_j^* := \{Y_{s(1)}, \ldots, Y_{s(n_{.j})}\}$, $j \in \{1, 2\}$. Now, we use an Anderson–Darling type test statistic for each of the two partial tests, that is

$$T(\mathbf{Y}_j^*) := \sum_{k=1}^{K-1} (\hat{F}_{1j}^*(C_k) - \hat{F}_{2j}^*(C_k))^2 (\hat{F}_{.j}(C_k)(1 - \hat{F}_{.j}(C_k)))^{-1}, \; j \in \{1, 2\}. \quad (1)$$

Thereby, $\hat{F}_{ij}^*$ and $\hat{F}_{.j}$ denote the permutation version of the empirical CDF within treatment group i, given $Z = j$, and the marginal empirical CDF, given $Z = j$, respectively. This Anderson–Darling type permutation test has already been considered for the two-group comparison setting without adjustment for covariates by [17]. The main idea in the present setting is to just apply that test to the observations within each of the two covariate subgroups separately. For the subsequent nonparametric combination of $T(\mathbf{Y}_1^*)$ and $T(\mathbf{Y}_2^*)$, there are several choices available (see, for example, [18]). We would like to mention two of them: On the one hand, the direct combination can be used, that is,

$$T_{AD,dir}(\mathbf{Y}_1^*, \mathbf{Y}_2^*) := T(\mathbf{Y}_1^*) + T(\mathbf{Y}_2^*),$$

and subsequently, the permutation p value is calculated by

$$p_{AD,dir} = \frac{1}{n_p} \sum_{m=1}^{n_p} \mathbf{1}\{T_{AD,dir}(\mathbf{Y}_1^{*(m)}, \mathbf{Y}_2^{*(m)}) \geq T_{AD,dir}(\mathbf{Y}_1, \mathbf{Y}_2)\}, \quad (2)$$

where n_p denotes the number of Monte Carlo replications (e.g., $n_p = 2000$), and $\mathbf{Y}_j^{*(m)}$ denotes the m-th permuted dataset, $j \in \{1, 2\}$. Alternatively, one may calculate the respective p values first and combine them by using the Fisher combination function, then. Hence, if we let $p_j := \frac{1}{n_p} \sum_{m=1}^{n_p} \mathbf{1}\{T(\mathbf{Y}_j^{*(m)}) \geq T(\mathbf{Y}_j)\}$, $j \in \{1, 2\}$, the Fisher combination p value is obtained by

$$p_{AD,F} = 1 - H(-2log(p_1 p_2)), \quad (3)$$

where H denotes the CDF of a central Chi-square distribution with 4 degrees of freedom.

Observe that each summand in (1), might be regarded as a separate test statistic, which essentially compares the cumulative frequencies up to category C_k between the two treatment groups, conditionally on $Z = j, k \in \{1, 2, \ldots, K - 1\}, j \in \{1, 2\}$. Hence, the two Anderson–Darling type test statistics are again direct combinations of partial tests, thus representing another application of the NPC method.

2.2 A Nonparametric (Pseudo-)Rank-based Method for Factorial Designs

As an alternative to the permutation approach, the nonparametric rank-based ANOVA-type test proposed by [4], might be used. Analogously to the parametric linear model, the hypothesis corresponding to the main effect of the binary treatment factor is stated as $H_0 : F_{11} - F_{21} + F_{12} - F_{22} = 0$ vs. $H_1 : F_{11} - F_{21} + F_{12} - F_{22} \neq 0$. Let R_{ijl} denote the rank of Y_{ijl} (i.e., the outcome of subject l in treatment group i with dichotomized covariate value $Z = j$, for $l \in \{1, 2, \ldots, n_{ij}\}$, $i, j \in \{1, 2\}$) within all $N = \sum_{i,j} n_{ij}$ observations. Let $\bar{R}_{ij.} = n_{ij}^{-1} \sum_{l=1}^{n_{ij}} R_{ijl}$ and $S_{ij}^2 = (n_{ij} - 1)^{-1} \sum_{l=1}^{n_{ij}} (R_{ijl} - \bar{R}_{ij.})^2$ denote the empirical mean and variance of the ranks, $i, j \in \{1, 2\}$. We consider the test statistic

$$T_A(\mathbf{Y}) = \frac{(\bar{R}_{11.} - \bar{R}_{21.} + \bar{R}_{12.} - \bar{R}_{22.})^2}{S_0^2}, \tag{4}$$

where $S_0^2 := \sum_{i=1}^2 \sum_{j=1}^2 S_{ij}^2 / n_{ij}$. Under H_0, this test statistic has, asymptotically, a central Chi-square distribution with 1 degree of freedom. For small samples, however, the distribution of T_A can be approximated by a F-distribution with numerator degrees of freedom equal to 1 and denominator degrees of freedom

$$\hat{f}_0 = \frac{S_0^4}{\sum_{i,j} (n_{ij} - 1)^{-1} (S_{ij}^2 / n_{ij})^2}.$$

We would like to add some important remarks. Firstly, in order to allow for establishing a unified framework regardless of whether ties are present or not, the normalized CDF $F := (F^+ + F^-)/2$ should be used. Thereby, F^+ and F^- denote the right and left continuous versions of the CDF, respectively. Accordingly, the so-called mid-ranks are used in (4). For the sake of notational simplicity, however, we have not explicitly used the normalized CDF (mid-ranks) in the formal considerations above. Secondly, it has been noticed recently that using ranks might lead to paradoxical results [5]. Replacing the ranks by the so-called pseudo-ranks has been shown to serve as a remedy [6]. Operationally, one just uses pseudo-ranks instead of ranks when calculating $T_A(\mathbf{Y})$. This corresponds to replacing the weighted mean distribution function $W := N^{-1} \sum_{i=1}^2 \sum_{j=1}^2 n_{ij} F_{ij}$ by the unweighted mean CDF $U := 1/4 \sum_{i=1}^2 \sum_{j=1}^2 F_{ij}$, as proposed by [8]. For the reasons discussed in [5], we recommend using pseudo-ranks and denote the corresponding test statistic by $T_A^{\psi}(\mathbf{Y})$ in the sequel. Finally, since the numerator degrees of freedom of the distribution of T_A are equal to 1, one could also consider the linear (pseudo-)rank statistic $\sqrt{T_A^{\psi}(\mathbf{Y})}$, which has a large-sample standard normal distribution. For small samples, the same approximation idea as outlined above may be used (see [7] for details). In particular, using the linear rank statistic would allow for testing one-sided hypotheses. Likewise, it is straightforward to construct the one-sided counterparts of the Anderson–Darling type permutation tests that have been discussed in the previous section [17].

3 Simulations

The simulation scenarios are based on the data from the SIESTA (Sedation vs Intubation for Endovascular Stroke Treatment) trial, where conscious sedation and general anesthesia were compared with respect to early neurological improvement in patients undergoing stroke thrombectomy [23, 24]. We considered the functional outcome, which was assessed at the end of the study (i.e., 3 months after the intervention) by using the modified Rankin Scale (mRS). Since this was one of the secondary outcomes in the original study, no covariate-adjusted analysis had been conducted. Nevertheless, for the present purpose, we compared the mRS at 3 months between the two treatment groups, while adjusting for "door-to-arterial-puncture time" as a covariate. The respective group-specific empirical means and variances for both variables were extracted from Table 3 in [24], and subsequently averaged over the two treatment groups, yielding $\mu_{\tilde{Y}} = 3.6\,(\sigma^2_{\tilde{Y}} = 3.425)$ for the continuous variable $\tilde{Y}$, which was assumed to underlie the outcome, and $\mu_X = 70.6\,(\sigma^2_X = 627.25)$ for the covariate X (time), respectively. Observe that for the sake of simplicity, we assumed that the distribution of the covariate was the same in both groups. This assumption is met in (well-designed) randomized clinical trials (note that we refer to the equality of the distributions at the population level, so, empirical imbalance especially in small samples is not an issue). With these specifications, and assuming a correlation of 0.5, the realizations of $\tilde{Y}$ and X were simulated from a bivariate normal distribution. Note that for power simulations, the means of the outcome in the first and in the second group were set to $\mu_{\tilde{Y}}$ and $\mu_{\tilde{Y}} - \delta$, where $\delta \in \{0.5, 1.0, 1.5\}$, respectively. Secondly, independent and identically distributed error terms $\tilde{\xi}_{ij}, i \in \{1, 2\}, j \in \{1, 2, \ldots, n_i\}$, were drawn from one out of several different distributions (standard normal, standard lognormal, $exp(1)$, $t(3)$, Cauchy and Pareto(1)) and standardized by

$$\varepsilon_{ij} = \frac{\tilde{\xi}_{ij} - \mathrm{E}[\tilde{\xi}_{ij}]}{(\mathrm{Var}[\tilde{\xi}_{ij}])^{1/2}},$$

for $i \in \{1, 2\}, j \in \{1, 2, \ldots, n_i\}$ (of course, provided that the second moments were finite). Then, we calculated the sum of the realizations of the variable underlying the outcome and the errors, that is, $\tilde{Y}_{ij} + \varepsilon_{ij}$, and rounded the resulting values. Finally, in order to obtain outcomes Y_{ij} within the range of the mRS $(0 - 6)$, we set negative values to 0 and values ≥ 7 to 6, respectively. Since doing so yielded relatively large proportions of 0 and 6 values, we reduced the variance $\sigma^2_{\tilde{Y}}$ of the underlying continuous variable by 1 (it should be noted that this might resemble the real-life data more closely, because adding the error terms increases the variance by 1). So, summing up, we at first, generated samples from a bivariate normal distribution, subsequently added the error terms to the first coordinates (i.e., the outcomes) and manipulated the resulting values accordingly, in order to eventually obtain integer values between 0 and 6. Furthermore, due to construction, shift effects were considered for power simulations.

Our main aim was to examine the empirical type I error rates and power of the direct and the Fisher combination of the Anderson–Darling type permutation tests—the corresponding formulas for calculating the p values are given in (2) and (3)—as well as the performance of the pseudo-rank-based nonparametric ANOVA-type test $T_A^{\psi}(\mathbf{Y})$. The latter method is implemented in the `rankFD` package [13], in `R` [20]. The code that was used for the simulations and the real-life data analysis (Sect. 4), is available upon request. As a competitor for benchmarking the results, we also tested for a significant group effect by employing a probabilistic index model (PIM; see [26]). It should be noted that analogously to the aforementioned setup, the median split version of the covariate (i.e., door-to-arterial-puncture time) was included, in order to ensure comparability of the results. Furthermore, preliminary simulations revealed that including the covariate as a metric variable in the PIM might lead to considerable power loss (results not shown). For carrying out the PIM simulations, we used the `pim` package [15]. For all scenarios and tests, we considered three balanced and three unbalanced group size configurations, namely $\mathbf{n}_1 = (20, 20)$, $\mathbf{n}_2 = (40, 40)$, $\mathbf{n}_3 = (80, 80)$, $\mathbf{n}_4 = (20, 40)$, $\mathbf{n}_5 = (20, 60)$, and $\mathbf{n}_6 = (20, 80)$. For each combination of the simulation parameters, we conducted $n_{sim} = 10,000$ simulations. The number of permutations within each run was set to $n_p = 2,000$. The significance level was specified as $\alpha = 5\%$.

Both NPC- and rank-based tests maintained the target type I error level very well (Table S1). However, the PIM test tended to be slightly liberal, especially in small samples, as well as in case of severe group size imbalance. With respect to power, the ANOVA-type approach showed either a similar or a better performance compared to the two permutation-based tests, which were almost empirically equivalent. Depending on the scenario, the difference in power was up to about 13% points. The power of the PIM approach was lower compared to our proposed methods, especially for moderate to large effect sizes. The results were very similar across most error distributions, as obvious from Figs. S1 and S2, except for some power loss in case of errors from a Cauchy or a Pareto(1) distribution (Fig. S3). For the latter settings, the simulation study also revealed that there might be some computational problems when conducting the permutation-based tests for small balanced group sizes, and when using the ANOVA-type test with substantially unbalanced groups, due to degenerated empirical variances in one of the subgroups. However, this problem was only present in very few simulation runs (permutation-based tests: 13 runs for type I error simulations, 7 and 1 runs for $\delta = 0.5$ and $\delta = 1.0$, respectively; ATS: 1 run for each $\delta \in \{0, 0.5, 1.0, 1.5\}$). The PIM test was not affected by these problems, yet being inferior to the other approaches in terms of power again. However, for unbalanced Pareto(1) scenarios, the PIM approach outperformed the NPC-based tests. Apart from that, for unbalanced settings, interestingly, the empirical power values of all tests under consideration did not change substantially as the group allocation ratios were becoming more extreme, despite the fact the total sample sizes were thus increased (Fig. S2).

Moreover, in order to explore whether the aforementioned findings may depend on the particular choices of the test statistics and the alternatives, respectively, we conducted a number of further sensitivity analyses. Firstly, all simulations were repeated using the Fisher combination method, but with the Anderson–Darling type test being replaced by the Wilcoxon-Mann-Whitney (WMW) test [14, 28]. To this end, we used the corresponding function `rank.two.samples` in the `rankFD` package [12]. The underlying rationale for using the WMW test was to examine whether or not the aforementioned power discrepancies between the rank- and permutation-based approaches may be at least partially explained by the different structures of the respective test statistics and effect measures. Overall, the results were very similar to the Anderson–Darling-based combination tests, with small gains in power in some scenarios (Table S1, Figs. S1–S3). However, it should be mentioned that computational problems were present in up to 5–10

Secondly, in order to compare the methods under consideration for other alternatives than location shifts, we modified the data generation process in the following way: Both group means were set to $\mu_{\tilde{Y}} = 3.6$, but the variances of the outcome in group 1 and 2 were specified as $\sigma_1^2 = \sigma_{\tilde{Y}}^2 - 1 = 2.425$ and $\sigma_2 = d\sigma_1^2$, where $d \in \{4, 8, 12, 16, 20\}$. All other simulation parameters that have been described at the beginning of this section were left unchanged. For ease of presentation, only sample size scenarios $\mathbf{n}_1$ and $\mathbf{n}_5$ with normal and lognormal errors were considered. The results are summarized in Fig. S4. The Anderson–Darling combination tests clearly outperformed the PIM approach, which, in turn, was more powerful than the rank-based test. Hence, the latter method should not be used in settings where scale alternatives are of interest. Observe that the Fisher combination of WMW-tests would have been a suboptimal choice either, because like the ATS approach, the WMW test is based on the so-called relative effect and, therefore, lacks power for scale alternatives.

Finally, we conducted a small simulation study that was aimed at investigating the performance of our proposed approaches in comparison to a classical parametric proportional odds (PO) model (see, for example, [1]). Analogously to the notations in the previous sections, let Y, Z, and G denote the outcome (i.e., 3-months mRS), the covariate (i.e., door-to-arterial-puncture time), and a dummy variable representing the group indicator (i.e., $G = i - 1$ for $i = 1, 2$), respectively. We considered the PO model

$$logit(P(Y \leq C_k)) = \alpha_k + \beta_1 G + \beta_2 Z + \beta_3 (GZ),$$

where C_k denotes the k-th mRS category, so, $C_k = k - 1, k \in \{1, 2, \ldots, 6\}$. Observe that $P(Y \leq 6) = 1$, so, this probability does not have to be modeled. Moreover, (GZ) denotes the covariate representing the group-covariate interaction. Data were simulated from this model as follows:

1. Simulate covariates $Z_{ij} \stackrel{iid}{\sim} \mathcal{N}(\mu_X, \sigma_X^2)$, where μ_X and σ_X^2 denote the (group-averaged) empirical mean and variance of "door-to-arterial-puncture time", respectively (as defined at the beginning of Sect. 3).
2. Calculate $P(Y_{ij} \leq C_k) = logit^{-1}(\alpha_k + \beta_1 G_{ij} + \beta_2 Z_{ij} + \beta_3 (GZ)_{ij})$, where the parameters were specified as follows: $\beta_1 \in \{0, 2, 4\}$, $\beta_2 = -0.1$, $\beta_3 = 0$, and $\alpha_k = k + 1, k \in \{1, \ldots, 6\}$.
3. Finally, the realization of Y_{ij} was sampled from the corresponding probability distribution $\{P(Y_{ij} = C_1), \ldots, P(Y_{ij} = C_7)\}$.

It should be noted that the probability distribution in step 3 depends on the covariate, which nevertheless has not been stated explicitly for the sake of notational simplicity. The resulting empirical type I error and power rates of the PO-based test for $H_0 : \beta_1 = 0$ and the aforementioned competitors are reported in Table S2. Note that for the PIM and PO tests, the original covariate instead of its categorized counterpart was included in the model, because using the latter led to computational errors. Obviously, the permutation approaches, as well as the rank-based ANOVA-type test outperformed their competitors in terms of power, while maintaining the prespecified 5% level. It has to be mentioned, however, that computational problems due to degenerated variances were present in a considerable number of simulation runs (2–5%).

4 Real-Life Data Example

In order to illustrate the different approaches under consideration, we analyzed the data from the SIESTA trial that has been mentioned in Sect. 3. The outcome variable was the modified Rankin Scale (mRS) at 3 months post intervention, and the age at baseline was considered as the (continuous) covariate. Each patient had been randomly assigned to either conscious sedation or general anesthesia. Firstly, the direct combination method yielded the combined Anderson–Darling type statistic $T_{AD,dir} = 0.53126$, with the corresponding permutation p value $p_{AD,dir} = 0.556$. The Fisher combination p value was very similar ($p_{AD,Fi} = 0.570$). However, the pseudo-rank-based ANOVA-type approach yielded a somewhat smaller p value ($p_{ATS} = 0.374$, $T_A^{\psi} = 0.79539$, $\hat{f}_0 = 139.69$). Summing up, the results might point to some gain in power when using the ANOVA-type statistic. For the sake of completeness, we also conducted a test for the group indicator in a PIM model with the dichotomized covariate age, as well as the covariate-group interaction as additional explanatory variables. The resulting p value was 0.5513, which is also in line with the findings from our simulation study (see Sect. 3).

5 Discussion

In the present manuscript, we have considered two different nonparametric approaches for comparing an ordinal outcome variable between two groups while adjusting for a continuous covariate. By contrast to existing methods, we did not incorporate the covariate directly, but dichotomized it by applying a median split. In applied research, covariates are frequently categorized, where the choice of the categories is either guided by subject matter expertise or based on certain empirical quantiles. Hence, our proposed method can be regarded as a representative of what is frequently done in practice. The simulation results showed that type I error rates were neither inflated nor deflated. With respect to power, the pseudo-rank-based ANOVA-type statistic outperformed the permutation-based approaches for location-shift alternatives. However, since multiple aspects should be considered when deciding for or against a particular statistical method, we would like to briefly discuss further advantages and drawbacks now.

Firstly, in addition to the gain in power at least for location-shift alternatives, another argument in favor of the ANOVA-type approach is that it provides a corresponding effect measure, the so-called relative effect, which can be estimated from the data. In fact, the ANOVA-type statistic is based on these estimated relative effects [4]. By contrast, the permutation-based methods are designed for testing hypotheses rather than for the purpose of estimation. Of course, one could calculate the estimated relative effects in addition to the permutation p value, yet with the drawback of introducing some discrepancy between what is tested and what is considered as the effect measure.

Secondly, on the other hand, although the relative effects are straightforward to interpret, the ANOVA-type approach might be somewhat more difficult to apply in so far, as there are several options for specifying the hypotheses, the test statistics and the ranks that are used. Although the `rankFD` package is a very convenient implementation in `R`, it might be easier to understand what is going on exactly when using the permutation-based approaches. Apart from that, we have to acknowledge that our empirical results, like any simulation study, only provide evidence for some specific settings. Although the range of sample size scenarios and distributional assumptions is quite broad, we would like to emphasize that we only considered shift effects. But, we have demonstrated in Sect. 3, that, for example, considering scale alternatives might yield very different results in terms of power. Hence, we recommend conducting further simulations that appropriately reflect the particular setting of interest, including various types of alternatives, before actually applying the one or the other method. Moreover, especially in case of very small sample sizes, we conjecture that the permutation-based approaches might be superior to the pseudo-rank-based method, because the former is finitely exact. Apart from that, despite the somewhat suboptimal performance of the PIM-based tests in our simulations, it should be emphasized that the PIM model clearly warrants further examination in future research on analysis methods for ordinal outcomes, due to its attractiveness in terms of the broad range of potential applications. Likewise, it might be worthwhile to consider employing a proportional odds model at least in particular settings where the underlying assumptions are tenable, due to the straightforward interpretation of the results.

Finally, we would like to emphasize that the approaches under consideration can be easily applied to settings with multiple (categorical or categorized) covariates, too. Moreover, we would like to briefly sketch how the permutation methods that have been proposed in the present manuscript can be extended to the case of multivariate outcomes. For example, in the SIESTA trial (see Sect. 4), it would be of interest to compare the mRS, as well as the National Institute of Health Stroke Scale (NIHSS), between the two treatment groups. Note that standard parametric tests (e.g., Wilks' Lambda) rely, in particular, on the assumption of multivariate normality, which might be restrictive and is even more difficult to justify than univariate normality. Therefore, using the nonparametric combination (NPC) method could be an appealing alternative. However, as the number of subgroups and/or dimensions increases, the permutation-based method is getting more and more demanding with respect to computational resources. In addition to that, a thorough empirical examination of the properties of the resulting tests has to be done in future work, in order to ensure that this approach can be safely used in applied research.

References

1. Agresti, A.: Categorical Data Analysis, 3rd edn. Wiley, Hoboken, New Jersey (2013)
2. Akritas, M., Arnold, S., Brunner, E.: Nonparametric hypotheses and rank statistics for unbalanced factorial designs. J. Am. Stat. Assoc. **92**, 258–265 (1997)
3. Bathke, A., Brunner, E.: A nonparametric alternative to analysis of covariance. In: Akritas, M., Politis, D. (eds.) Recent Advantages and Trends in Nonparametric Statistics, pp. 109–120. Elsevier, Amsterdam (2003)
4. Brunner, E., Dette, H., Munk, A.: Box-type approximations in nonparametric factorial designs. J. Am. Stat. Assoc. **92**, 1494–1502 (1997)
5. Brunner, E., Konietschke, F., Bathke, A.C., Pauly, M.: Ranks and pseudo-ranks—paradoxical results of rank tests. arXiv preprint arXiv:1802.05650 (2018)
6. Brunner, E., Konietschke, F., Pauly, M., Puri, M.L.: Rank-based procedures in factorial designs: hypotheses about non-parametric treatment effects. J. R. Stat. Soc.: Ser. B (Statistical Methodology) **79**(5), 1463–1485 (2017)
7. Brunner, E., Munzel, U.: Nichtparametrische Datenanalyse: Unverbundene Stichproben. Springer, Berlin, Heidelberg (2013)
8. Brunner, E., Puri, M.L.: Nonparametric methods in factorial designs. Stat. Papers **42**, 1–52 (2001)
9. Farrell, B., Godwin, J., Richards, S., Warlow, C.: The United Kingdom transient ischaemic attack (UK-TIA) aspirin trial: final results. J. Neurol., Neurosurg. Psychiatry **54**, 1044–1054 (1991)
10. Hahn, S., Salmaso, L.: A comparison of different permutation approaches to testing effects in unbalanced two-level ANOVA designs. Stat. Papers **58**(1), 123–146 (2017)
11. Huitema, B.: The Analysis of Covariance and Alternatives: Statistical Methods for Experiments, Quasi-Experiments, and Single-Case Studies. Wiley, New York (2011)
12. Konietschke, F., Bathke, A., Harrar, S., Pauly, M.: Parametric and nonparametric bootstrap methods for general MANOVA. J. Multi. Anal. **140**, 291–301 (2015)

13. Konietschke, F., Friedrich, S., Brunner, E., Pauly, M.: rankFD: Rank-Based Tests for General Factorial Designs (2016). https://CRAN.R-project.org/package=rankFD. R package version 0.0.1
14. Mann, H., Whitney, D.: On a test of whether one of two random variables is stochastically larger then the other. Ann. Math. Stat. **18**, 50–60 (1947)
15. Meys, J., De Neve, J., Sabbe, N., Guimaraes de Castro Amorim, G.: pim: Fit Probabilistic Index Models (2017). https://CRAN.R-project.org/package=pim. R package version 2.0.1
16. Pesarin, F.: Multivariate Permutation Tests: With Application in Biostatistics. Wiley, Chichester (2001)
17. Pesarin, F., Salmaso, L.: Permutation tests for univariate and multivariate ordered categorical data. Austrian J. Stat. **35**(2–3), 315–324 (2006)
18. Pesarin, F., Salmaso, L.: The permutation testing approach: a review. Statistica **70**(4), 481–509 (2010)
19. Pesarin, F., Salmaso, L.: Permutation Tests for Complex Data. Applications and Software. Wiley Series in Probability and Statistics. Wiley, Chichester, Theory (2010)
20. R Development Core Team: R: A language and environment for statistical computing (2018). http://www.R-project.org. R Foundation for Statistical Computing, Vienna, Austria
21. Rankin, J.: Cerebral vascular accidents in patients over the age of 60. ii. prognosis. Scottish Med.J. **2**(5), 200–215 (1957)
22. Salmaso, L.: Synchronized permutation tests in 2^k factorial designs. Commun. Stat.: Theory Methods **32**(7), 1419–1437 (2003)
23. Schönenberger, S., Möhlenbruch, M., Pfaff, J., Mundiyanapurath, S., Kieser, M., Bendszus, M., Hacke, W., Bösel, J.: Sedation vs. intubation for endovascular stroke treatment (SIESTA)—a randomized monocentric trial. Int. J. Stroke **10**(6), 969–978 (2015)
24. Schönenberger, S., Uhlmann, L., Hacke, W., Schieber, S., Mundiyanapurath, S., Purrucker, J., Nagel, S., Klose, C., Pfaff, J., Bendszus, M., Ringleb, P., Kieser, M., Möhlenbruch, M., Bösel, J.: Effect of conscious sedation vs general anesthesia on early neurological improvement among patients with ischemic stroke undergoing endovascular thrombectomy: A randomized clinical trial. JAMA **316**(19), 1986–1996 (2016)
25. Sullivan, L., D'Agostino, R.: Robustness and power of analysis of covariance applied to ordinal scaled data as arising in randomized controlled trials. Stat. Med. **22**, 1317–1334 (2003)
26. Thas, O., Neve, J., Clement, L., Ottoy, J.: Probabilistic index models. J. R. Stat. Soc.: Ser. B (Statistical Methodology) **74**(4), 623–671 (2012)
27. Van Swieten, J., Koudstaal, P., Visser, M., Schouten, H., van Gijn, J.: Interobserver agreement for the assessment of handicap in stroke patients. Stroke **19**(5), 604–607 (1988)
28. Wilcoxon, F.: Individual comparisons by ranking methods. Biometrics **1**, 80–83 (1945)

Printed by Printforce, the Netherlands